COMMUNICATION FAILURE IN DIALOGUE AND DISCOURSE
Detection and Repair Processes

COMMUNICATION FAILURE IN DIALOGUE AND DISCOURSE

Detection and Repair Processes

Edited by

Ronan G. REILLY

Educational Research Centre
St. Patrick's College
Dublin, Ireland

1987

NORTH-HOLLAND
AMSTERDAM · NEW YORK · OXFORD · TOKYO

© ELSEVIER SCIENCE PUBLISHERS B.V., 1987

All rights reserved. No part of this publication may be reproduced, stored in a retrieval system, or transmitted, in any form or by any means, electronic, mechanical, photocopying, recording or otherwise, without the prior permission of the copyright owner.

ISBN: 0 444 70112 5

Published by:
ELSEVIER SCIENCE PUBLISHERS B.V.
P.O. Box 1991
1000 BZ Amsterdam
The Netherlands

Sole distributors for the U.S.A. and Canada:
ELSEVIER SCIENCE PUBLISHING COMPANY, INC.
52 Vanderbilt Avenue
New York, N.Y. 10017
U.S.A.

Library of Congress Cataloging-in-Publication Data

Communication failure in dialogue and discourse.

Bibliography: p.
Includes indexes.
1. Interactive computer systems.   2. Man-machine
systems.   I. Reilly, Ronan G., 1955-
QA76.9.I58C65   1987   006.3'5   86-19868
ISBN 0-444-70112-5 (U.S.)

PRINTED IN THE NETHERLANDS

PREFACE

As the use of computers becomes more widespread, the need for more flexible and more natural person-computer interaction becomes crucial. The most important medium of interpersonal communication is natural language. It has been a goal of researchers for many years to harness the ease and flexibility of natural language in computer-based applications. An essential consideration in this endeavour has been not only to provide a wide linguistic coverage, but also to provide robust understanding abilities. Interpersonal communication is error prone, yet we manage to overcome errors without too much difficulty by means of a variety of detection and repair techniques. It is an assumption underlying the structure and content of this volume that a study of interpersonal communication and miscommunication can form a useful basis for the development of robust computer-based natural-language understanding systems. Included, therefore, are not only contributions from the areas of artificial intelligence and cognitive science, but also from psycholinguistics, linguistics, and sociolinguistics. In order to provide a realistic backcloth to what are, in the main, theoretical papers, a contribution from the area of human factors is also included.

The papers presented here are from three somewhat overlapping sources. The core of the papers originate from a symposium held at St Patrick's College in Dublin, in July 1985 on the theme of discourse processing. The symposium was partially funded by the European Community's ESPRIT programme, the Irish National Board for Science and Technology, and the Reading Association of Ireland. Additional papers are based on work done as part of the European Community's ESPRIT programme (sub-area AIP, project number P527). The ESPRIT related papers are those of Ronan Reilly, Jerry Harper, Owen Egan, Giacomo Ferrari and Irina Prodanof, and Bruce Christie and Margaret Gardiner. Finally, contributions were solicited from other researchers in Europe and the United States.

Needless to say, the preparation of this volume owes much to many people. I would like to thank the contributors for replying promptly to my request for papers. In many cases, authors significantly revised their symposium contribution to fit in with the theme of the book. In chronological order, thanks are due to: Vincent Greaney, Robert Cochrane, and Mike Rogers, who helped get the symposium off the ground; Edward Lynch who was indispensable in helping with the task of proof reading; Hilary Walshe who typeset the book; Sheilanne O'Donoghue who prepared some of the figures; and Ruth McLoughlin who helped in the preparation of the indices. Special thanks are due to Thomas Kellaghan who provided a suitable working environment for the execution and completion of this project.

Ronan Reilly
Educational Research Centre
St Patrick's College, Dublin.

LIST OF CONTRIBUTORS

Anthony Anderson
MRC Applied Psychology Unit
15 Chaucer Road
Cambridge CB2 2EF
ENGLAND

Ann Cahill
Department of Psychology
University of Exeter
Exeter, EX4 4QC
ENGLAND

Sandra Carberry
Department of Computer Science
University of Delaware
Newark, DE 19715
USA

Arthur Cater
Department of Computer Science
University College
Belfield, Dublin 4
IRELAND

Bruce Christie
Department of Office Technology
  and Administration
City of London Polytechnic
London EC3A 7BU
ENGLAND

Owen Egan
Linguistics Institute of Ireland
31 Fitzwilliam Place, Dublin 2
IRELAND

Giacomo Ferrari
Department of Linguistics
University of Pisa
Via Santa Maria, 36
56100 Pisa
ITALY

Margaret Gardiner
Human Factors Technology Centre
ITT Europe ESC
Harlow, Essex CM20 2BN
ENGLAND

Simon C. Garrod
Department of Psychology
University of Glasgow
Adam Smith Building
Glasgow G12 8RT
SCOTLAND

Bradley Goodman
BBN Laboratories
Bolt Beranek and Newman Inc.
10 Moulton Street
Cambridge, MA 02238
USA

Jerry Harper
Educational Research Centre
St Patrick's College, Dublin 9
IRELAND

John Harris
Linguistics Institute of Ireland
31 Fitzwilliam Place, Dublin 2
IRELAND

Claire Humphreys-Jones
81a Friary Park
Ballabeg
Isle of Man
UNITED KINGDOM

Aravind Joshi
Department of Computer &
  Information Science
University of Pennsylvania
Philadelphia, PA 19104
USA

Kathleen McCoy
Department of Computer Science
University of Delaware
Newark, DE 19715
USA

Henry McLoughlin
Department of Computer Science
University College
Belfield, Dublin 4
IRELAND

Michael McTear
Department of Communication
University of Ulster at
  Jordanstown
Co. Antrim, BT37 0QB
NORTHERN IRELAND

Don Mitchell
Department of Psychology
University of Exeter
Exeter, EX4 4QC
ENGLAND

Christopher Owens
Department of Computer Science
Yale University
New Haven, CT 06520
USA

Martha Pollack
Artificial Intelligence Centre
SRI International
333 Ravenswood Avenue
Menlo Park, CA 94025
USA

Irina Prodanof
c/o Department of Linguistics
University of Pisa
Via Santa Maria, 36
56100 Pisa
ITALY

Ronan Reilly
Educational Research Centre
St Patrick's College, Dublin 9
IRELAND

Roger Schank
Department of Computer Science
Yale University
New Haven, CT 06520
USA

Noel Sharkey
Cognitive Science Centre
University of Essex
Colchester, Essex
ENGLAND

Amanda Sharkey
Cognitive Science Centre
University of Essex
Colchester, Essex
ENGLAND

Noel Sheehy
Department of Psychology
University of Leeds
Leeds LS2 9JT
ENGLAND

Brian Torode
Department of Sociology
Trinity College, Dublin 2
IRELAND

Bonnie L. Webber
Department of Computer &
  Information Science
University of Pennsylvania
Philadelphia, PA 19104
USA

Ralph Weischedel
BBN Laboratories
Bolt Beranek and Newman Inc.
Cambridge, MA 02238
USA

John Wilson
Department of Communication
  and Speech
Southern Illinois
University of Carbondale
Carbondale, IL 62901
USA

CONTENTS

Preface   v

List of contributors   vii

1   FRAMEWORKS AND MODELS

Types of communication failure in dialogue
Ronan Reilly   3

The structure of misunderstandings
Claire Humphreys–Jones   25

Communication failure: a development perspective
Michael F. McTear   35

Towards a framework for the formal treatment of dialogue
Jerry Harper   49

Conceptual primitives and their metaphorical relationships
Arthur W.S. Cater   59

2   LEXICAL AND SYNTACTIC FAILURE

Speech comprehension and lexical failure
John Harris   81

Meta-rules as a basis for processing ill-formed input
Ralph M. Weischedel and Norman K. Sondheimer   99

3   REFERENCE FAILURE

Repairing reference identification failures by relaxation
Bradley A. Goodman   123

Generating responses to property misconceptions using perspective
Kathleen F. McCoy   149

The dynamics of referential meaning in spontaneous conversation: some preliminary studies
Anthony Anderson and Simon C. Garrod   161

4 INFERENCE AND EXPECTATION FAILURE

The use of inferred knowledge in understanding pragmatically ill-formed queries
M. Sandra Carberry — 187

Understanding by explaining expectation failures
Roger C. Schank and Christopher C. Owens — 201

Some aspects of default reasoning in interactive discourse
Aravind K. Joshi, Bonnie L. Webber, and Ralph M. Weischedel — 213

Induction and decision making in the vicinity of dialogue failure
Owen Egan — 221

Personae: models of stereotypical behavior
Henry B. McLoughlin — 233

5 PLANS AND GOALS

Some requirements for a model of the plan inference process in conversation
Martha E. Pollack — 245

Plans and goals in story comprehension
Ann Cahill and Don C. Mitchell — 257

6 DIALOGUE IN CONTEXT

Discourse situation misunderstandings
Giacomo Ferrari and Irina Prodanof — 271

KAN: a knowledge access network
Noel E. Sharkey and Amanda J.C. Sharkey — 287

Communication failure at the person–machine interface: The human factors aspects
Margaret M. Gardiner and Bruce Christie — 309

Nonverbal behaviour in dialogue
Noel P. Sheehy — 325

7 A SOCIO-LINGUISTIC PERSPECTIVE

Negotiation and breakdown in speech event construction
John Wilson — 341

Computational socio-linguistics and communication failure: on the resolution of incompleteness in automatic discourse parsing
Brian Torode — 361

Name Index	391
Subject Index	399

1

FRAMEWORKS AND MODELS

TYPES OF COMMUNICATION FAILURE IN DIALOGUE[1]

Ronan Reilly

Educational Research Centre
St Patrick's College, Dublin 9

This chapter describes a taxonomy of miscommunication which is derived from a form of information-processing analysis of the communication process. It is intended that the system of classification be used as an aid in classifying current research in the dialogue and discourse area and as a tool in the construction of robust natural-language interfaces. It should provide a means of assessing the frequency of occurrence of various types of miscommunication in natural and person-machine dialogue. The relationship of the taxonomy to other research in the miscommunication area is also discussed and a number of examples of its application to communication failures in real dialogues from various sources are given.

1. Introduction

A useful approach to the study of miscommunication in dialogue is to view dialogue as a set of inter-related *processes*. This approach is typically that of psychology and cognitive science, though not exclusively so. From this perspective, the locus of miscommunication can be considered to be the processes involved in mapping one dialogue structure onto another. These processes are internal to the dialogue participants. They can be likened to the processing modules of an information-processing model (Estes, 1978), although we acknowledge that there are some drawbacks inherent in this approach (Reilly, 1985).

In what follows, we will describe the structures pertinent to the dialogue process, indicate the processes which map to and from these structures, and give examples of miscommunication which can be localised to one or other of the processes described.

2. Dialogue Structures

We will distinguish here between two main classes of data structure which we consider relevant to the dialogue process: structures derived directly from the external, shared world of the dialogue participants, and those derived internally, by transformations of internal representations. This is a fairly uninformative distinction, but it is helpful to clarify our terms in advance of a more detailed analysis.

2.1 Representations of the External World

There are two forms of internal representation of the external world with which we will be concerned. These are the representation of the acoustic and/or visual signal (S) which constitutes the primary information source, and the representation of the external physical and social context in which the communication takes place (XC). The latter can be thought of as a secondary information source. We consider both S and XC to be an unsegmented stream of visual and/or acoustic energy. In the case of XC, however, the information being conveyed in the energy stream is primarily non-verbal. We would include in XC such things as the visual and auditory surroundings of the dialogue participants as well as the physical actions performed by the participants.

We will use XC and S to indicate the external signals, and XC_a and S_a to indicate their internal representations for the receiving dialogue participant. We will refer to the receiving participant as A and to the initiating participant as B. XC and S are mapped onto XC_a and S_a by the processes p_1 and p_2 respectively. This can be represented as follows:

(1a) $p_1(S) \rightarrow S_a$

(2a) $p_2(XC) \rightarrow XC_a$

The simplifying assumption is made here that the processes p_n are the same for both dialogue paritcpants. This is in keeping with a general cognitive modeling approach which assumes that individual differences arise from a difference in data structures across individuals, rather than from differences in processes.

There are a number of possibilities for communication breakdown at this level of processing. A breakdown at p_1 would result in either a failure to map S onto S_a:

(1b) $p_1(S) \rightarrow \emptyset_a$

or in an incorrect mapping:

(1c) $p_1(S) \rightarrow S'_a$

Examples of (1b) would be where a section of an utterance is unanalysable or where the lettering in a text is illegible. Another possibility is that the receiving participant's attention strayed momentarily and the signal event was missed. In the context of a person-machine interface, a type (1b) failure might occur if the screen was overwritten by a system message, or if screen intensity, or some other physical characteristic of the display, interfered with legibility. Type (1b) miscommunications are potentially detectable at an early stage, although they have to be ultimately corrected by processes operating further up the processing hierarchy.

Type (1c) miscommunications occur when S is incorrectly mapped. What do we mean by an incorrect mapping? In order to define correctness we are obliged to refer to the intentions of the dialogue participants. A mapping such as (1c) indicates that the representation that the receiving participant (A) generates from the signal is not the representation that the initiating participant (B) intended. From now on, when we refer to an incorrect mapping or mis-mapping we mean that the representation generated does not match that intended by the initiating participant. As with type (1b), type (1c) miscommunications can occur when the visual or acoustic signal is distorted. Whereas in the case of type (1b) failures, the distorted elements of the signal are not in the phonetic or orthographic repertoire of the process, in type (1c) the distortions fall within the repertoire of the process, and the signal is mapped to a representation unintended by the initiating participant. Unlike type (1b) failures, recovery can usually only be effected by higher level processes.

The mapping external context onto its internal representation can also breakdown. Again, we have a similar set of possible outcomes to the signal mappings:

(2b) $p_2(XC) \rightarrow \emptyset_a$

(2c) $p_2(XC) \rightarrow XC'_a$

XC can be considered to be an amalgam of non-verbal acoustic and/or visual signals. A null mapping, as shown in (2b), may occur if their is a lapse in attention on the part of one participant, or if the non-verbal signals are not in the repetoire of the p_2 process. A mis-mapping can occur if those aspects of XC which participant B intends to be taken into account are not represented in XC_a. This may be stretching the use of the term "intend" somewhat. What we want to indicate here is that the representation XC_a does not correspond to XC_b.

Unlike S_a, XC_a can be thought of as the "meaning" of the contextual signals. S_a, on the other hand is merely an echoic or iconic representation of S, depending on modality. This is in keeping with the notion that the information in a visual scene or a non-verbal sound is apprehended directly, rather than through some intermediate decoding process.

Ringle and Bruce (1980) identify two main categories of miscommunication in ordinary conversation which have some relevance here: input failures and model failures. An input failure occurs when the participant is unable to obtain a complete, or at least a coherent, interpretation for an utterance; a model failure occurs when the listener is unable to assimilate inputs to a coherent belief model as intended by the speaker. Ringle and Bruce define input failures to include:

1. Perceptual failures: A word or phrase is not clearly perceived and no interpretation results, or a word or phrase is misperceived and an inconsistent interpretation results.

2. Lexical failures: A word or phrase is clearly perceived but the participant either fails to produce the correct semantic interpretation or is unable to produce any interpretation at all.

3. Syntactic failures: Individual words and phrases are correctly perceived and interpreted, but the participant's intended meaning is misconstrued.

All three of the above categories have corresponding categories in our taxonomy. The failure types illustrated in (1b) and (1c) can be classed as perceptual failures within the Ringle and Bruce taxonomy. However, Ringle and Bruce do not address the issue of external context perception (or misperception). Therefore, the mappings in (2a-c) do not correspond to any category in their taxonomy.

2.2 Internally Derived Representations

The other large class of dialogue structures are what we will refer to as internally derived representations. For the purposes of our classification system we will distinguish between five different types of internally derived representations: lexical (L), syntactic/semantic (S/S), internal context (IC), goal structure (GS), and discourse model (DM). In the case of L we will use capital letters to indicate sets of lexical items and 1 to indicate elements within the sets. Obviously, these internal representations vary in their complexity, as do their associated mapping processes.

2.2.1 Lexical Representations

The signal representation S_a is mapped onto a set of lexical representation(s):

(3a) $\quad p_3(S_a) \rightarrow L_a$

This can be a one-to-one or many-to-one mapping, in that a given signal may provide information about a number of separate lexical items. The process p_3 is responsible for segmenting the utterance into its component lexical items. It is assumed that the elements in L_a will consist not just of simple lexical entries, but will also contain information about each word's syntactic and semantic properties.

As with earlier examples, there are a number of mapping possibilities apart from (3a):

(3b) $\quad p_3(S_a) \rightarrow \emptyset_a$

(3c) $\quad p_3(S_a) \rightarrow L'_a$

(3d) $\quad p_3(S'_a) \rightarrow L_a$

(3e) $\quad p_3(S'_a) \rightarrow L'_a$

(3f) $\quad p_3(S'_a) \rightarrow \emptyset_a$

In (3b) we have a failure to generate any lexical entries from the signal. In (3c) an incorrect set of lexical entries is associated with the signal. A correct set of lexical items is associated with an originally incorrect signal in (3d). This might occur where a word has been slightly misspelled and the lexical retrieval process retrieved the originally intended word on the basis of its similarity. The mapping in (3e) illustrates an incorrect set of lexical items being associated with an already incorrect signal representation. The effects of this mapping are similar to those of (3c). Finally, (3f) illustrates a failure to map an incorrect signal onto a valid lexical item (or set of items). This mapping is equivalent to trying to find a lexical entry for a non-word or for a word not in the person's/system's vocabulary, and its effects are similar to (3b).

In the context of the person-machine interface, this is the stage of the dialogue process where the problems begin. Weischedel and Sondheimer (this volume) cite work by Thompson (1980) who, in an extensive study of ill-formed input in a database query environment, found that of 1615 inputs, an overall total of 446 (28%) contained various kinds of input errors: 161 (10%) of these were vocabulary problems and 61 (4%) were due to misspellings. Therefore, about 14% of inputs in Thompson's study involved mappings of type (3b) to (3f).

For a more detailed discussion of the issues involved in the analysis of communication failure at the lexical level, particularly in speech comprehension, we refer the reader to Harris (this volume).

2.2.2 Syntactic/Semantic Representations

The set of lexical items are mapped onto a meaning representation. We will refer to this as a syntactic/semantic representation. We do not find it useful to distinguish between the two components of sentential representation because both levels of representation interact in complex ways during dialogue and during communication breakdown. As has been pointed out in earlier studies (Cohen, 1983; Grosz, 1978; Reichman, 1981; Polanyi & Scha, 1983), certain clue words, many of them function words, are used to indicate a shift in focus in a dialogue. It is their syntactic role, rather than their meaning that is crucial to their proper understanding. Thus a purely semantic representation would lack a dimension critical to its integration into the discourse model. Therefore, both the syntactic properties of certain words in an utterance and the semantic representation of the utterance as a whole share equal status in controlling the flow of information in a dialogue.

The relevant mappings are as follows:

(4a) $\quad p_4(L_a) \rightarrow S/S_a$

(4b) $\quad p_4(L_a) \rightarrow \emptyset_a$

(4c) $\quad p_4(L_a) \rightarrow S/S'_a$

(4d) $\quad p_4(L'_a) \rightarrow S/S_a$

(4e) $\quad p_4(L'_a) \rightarrow S/S'_a$

(4f) $\quad p_4(L'_a) \rightarrow \emptyset_a$

A failure to derive any coherent representation is illustrated in (4b). A failure to derive a full and/or correct representation is depicted in (4c). Sentences such as "Colourless green ideas sleep furiously" or "Sleep green furiously ideas colourless" would result in a mapping of type (4b). Although the first sentence has a legal syntax, it is meaningless. A criterion for a mapping of type (4a), (4c), (4d) and (4e) is that it is possible to derive some meaning from a set of lexical items. In the case of (4c) and (4e), it is not the meaning intended by the initiating participant. In (4d) the intended meaning is reconstructed from *local* syntactic and semantic properties of the utterance. Context does not play a role in recovery at this stage in the model. Mappings of type (4e) and (4f) illustrate further attempts to map an incorrect set of lexical items onto a syntactic/semantic representation. Incorrect in this case can mean any set of lexical items which do not correspond to the set intended by the initiating participant. In general, this will mean that the set is either incomplete, or that it contains extraneous items. In (4e) the attempt succeeds in producing a coherent, but incorrect, representation. In (4f), the attempt fails.

In the study already mentioned above by Thompson (1980) only 1093 (68%) of the 1615 inputs were parsable. Of the overall error total of 446 (28%), 72 (4%) were punctuation errors and 62 (4%) were due to ungrammaticality. Furthermore, 211 (13%) inputs were fragmentary.

In another study (Eastman & McLean, 1981), an analysis was made of 693 English queries to a database. Grammatical violations, including subject-verb disagreement, tense errors, apostrophe problems, and possessive/plural errors occurred in 12.3% of the queries. Omitted words, extraneous words and phrases, telegraphic ellipsis, and incomplete sentences arose in 14% of the queries. Taking the above two studies together we can say that type (4b) mappings account for between 8% and 12% of inputs. It is interesting to note the consistency of the rate of occurrence of fragmentary input (13-14%). Based on this figure we can only hypothesise that mappings of type (4d-f) account for no *more* than 14% of input. Unfortunately, it is not possible to give a finer-grained breakdown than this.

Carbonell and Hayes (1983) distinguish between several types of what they refer to as sentential extragrammaticality, and what Ringle and Bruce (1980) would term "syntactic failure". We will try and relate their descriptions to the taxonomy being developed in this chapter. The following are the four main categories of extragrammaticality which Carbonell and Hayes describe (we provide examples in parentheses): (1) Missing words (copy new files my directory); (2) spurious words (copy if you would be so kind the new files to my directory please; delete I mean copy the new files to my directory); (3) words out of order (new files to my directory copy); and (4) constraints violation (copy the two new file to my directory; copy my directory to the new file).

Category 1 is a good example of the commonly observed phenomenon that when people communicate with a computer via a natural language interface they employ a form of "computerese", leaving out function words, articles, and so on. This is probably in the belief that a telegraphic form of input is easier for the computer to understand. Within category 1, Carbonell and Hayes also group ellipses and other fragmentary inputs. Although these are entirely legal phenomena, Carbonell and Hayes maintain that from a computational point of view they must be treated by the same mechanisms that treat sentential ill-formedness because they have the same appearance as ill-formed sentences. Within our framework, we do not mark out ellipses for special treatment, and they are not considered as L' structures unless they fail to accurately reflect the initiating participant's intentions. If the intentionally telegraphic input cannot be parsed then we have a case of a type (4b) mapping. If the input is unintentionally telegraphic or elliptic and also unparsable we have a mapping of type (4f).

Category 2 illustrates the opposite problem to that of missing words, the occurrence of spurious words. In some cases these may be valid phrases that cannot be dealt with by a computer-based parser, as in the first example. Limitations on computer-based parsers are not directly relevant to our undertaking here. Within our framework, if an utterance is unparsable, as in (4b) and (4f), we mean it is unparsable by any *human* parser. Therefore we do not distinguish between type (4b) and (4f) mappings which result from inadequate coverage by the parser and those which are humanly unparsable. Of more relevance to us here is the problem of spurious words which result from the user breaking off an utterance and starting a new one. In some cases an explicit signal is given that a phrase is being restarted, such as "I mean" in the second example of category 2. Strictly speaking, the set of lexical items is unparsable by a parser which is unable to deal with the metastatement embedded in the utterance. However, we will assume that our human-like parser has this ability. Therefore, the second example in cateogry 2 would be considered to be an L_a structure, and subject to any of the (4a-c) mappings.

Out of order constituents, as illustrated in category 3, are similar from a parsing point of view to sentences in which the function words, or the "syntactic glue" of the utterance is omitted. Much syntactic information is carried by word position (although this is true of English more than, say, Italian). We can assume, unless the initiating participant is behaving perversely, that most examples in this category are L'_a structures, and subject to the (4d-f) mappings.

Constraint violations cover a range of ill-formed input, from the violation of syntactic constraints (category 4, first example), to the violation of semantic constraints (second example). The latter type of constraint violation does not fall within the scope of the (4a-f) mappings, but is dealt with in the set described below. However, it can be assumed that a reasonable and competent participant will not intentionally violate syntactic constraints. Therefore, the relevant lexical structure is L'_a which is subject to the (4d-f) mappings.

2.2.3 Discourse Model

An additional form of internal representation is that of the discourse model. We consider this to be a representation which is cummulatively constructed as the dialogue proceeds. We envisage it to be a form of semantic net which is progressively modified by processes which map onto it long-term stored knowledge (internal context) as well as new linguistic input. When we refer to discourse model (DM) we mean that aspect of the overall discourse representation currently being attended to and augmented. We have in mind here the concept of focussing put forward by Sidner (1979), Grosz (1981), and in particular Grosz and Sidner (1985). Thus, in the mappings described below the term discourse model refers to that aspect of the overall discourse representation currently in focus. All mappings onto the DM are considered to be incremental, in the sense that the DM is in a particular state prior to the mapping and this in some way determines the state it will be in after the mapping. We represent an incremental mapping by a processing function which takes two arguments with the understanding that a correct output, DM_a, may not be identical to the current DM_a which is input.

The relevant mappings are:

(5a) $p_5(XC_a, DM_a) \rightarrow DM_a$

(5b) $p_5(XC_a, DM_a) \rightarrow \emptyset_a$

(5c) $p_5(XC_a, DM_a) \rightarrow DM'_a$

(5d) $p_5(XC'_a, DM_a) \rightarrow DM_a$

(5e) $p_5(XC'_a, DM_a) \rightarrow DM'_a$

(5f) $p_5(XC'_a, DM_a) \rightarrow \emptyset_a$

(5g) $p_5(XC_a, DM'_a) \rightarrow DM_a$

(5h) $p_5(XC_a, DM'_a) \rightarrow \emptyset_a$

(5i) $p_5(XC_a, DM'_a) \rightarrow DM'_a$

(5j) $p_5(XC'_a, DM'_a) \rightarrow DM_a$

(5k) $p_5(XC'_a, DM'_a) \rightarrow DM'_a$

(5l) $\quad p_5(XC'_a, DM'_a) \to \emptyset_a$

Mappings (5a-d) illustrate the correct use of external context given a correct DM, the failure to use external context, the inappropriate use of external context, and the correct use of incorrectly perceived external context. Mappings (5e-f) indicate failures to map incorrect external context into the DM either correctly (5e) or at all (5f). The mappings (5g-l) are identical to the (5a-f), but they assume an incorrect DM.

The work of Goodman (this volume) in analysing miscommunications in the construction of a toy water-pump is especially relevant to the types of failure described above. His work focuses in particular on failures of reference, where the description of an object in the external context is misunderstood. In terms of our framework, we are unable to correctly map the external context onto the DM. We can illustrate this in the following excerpt from Goodman (this volume):

S: 1. And now take the little red
              2. peg
[P takes PLUG]
              3. Yes.
              4. and place it in the hole at the
              5. green end,
[P starts to put PLUG into OUTLET2 of MAINTUBE]
              6. no
              7. the – in the green thing
[P puts plug into green part of PLUNGER]
P: 8. Okay.

Utterances 4 and 5 are successfully mapped onto the DM (see 6a below), but there is then a failure on P's part to map the external context successfully onto the DM, resulting in a (5c) mapping. We will return to this excerpt in our discussion of the goal structure.

In the following set of mappings we deal with the incorporation of the syntactic/semantic representation into the discourse model:

(6a) $\quad p_6(S/S_a, DM_a) \to DM_a$

(6b) $\quad p_6(S/S_a, DM_a) \to \emptyset_a$

(6c) $\quad p_6(S/S_a, DM_a) \to DM'_a$

(6d) $\quad p_6(S/S'_a, DM_a) \to DM_a$

(6e) $\quad p_6(S/S'_a, DM_a) \to DM'_a$

(6f) $\quad p_6(S/S'_a, DM_a) \to \emptyset_a$

(6g) $\quad p_6(S/S_a, DM'_a) \to DM_a$

(6h) $\quad p_6(S/S_a, DM'_a) \to \emptyset_a$

(6i) $\quad p_6(S/S_a, DM'_a) \to DM'_a$

(6j) $\quad p_6(S/S'_a, DM'_a) \to DM_a$

(6k) $\quad p_6(S/S'_a, DM'_a) \to DM'_a$

(6l) $\quad p_6(S/S'_a, DM'_a) \to \emptyset_a$

The mapping at (6a) illustrates the successful incorporation of the meaning of the utterance into a correct DM. At (6b) we have a failure to incorporate any aspects of the utterance into the DM. However, this does not mean that the utterance is meaningless with respect to the dialogue as a whole, only with respect to the segment of the dialogue currently in focus. A successful incorporation, but one which results in a mismatch between the participants' DMs is illustrated in (6c). The mapping at (6d) illustrates a successful incorporation of an incorrectly or partially parsed utterance. The (6e-f) mappings involve, respectively, the unsuccessful incorporation of incorrect syntactic/semantic representation and a complete failure to execute a mapping. The mappings (6a-f) involve transformations of an already correct DM. The remaining (6g-l) mappings are identical but for the fact that the base DM is incorrect.

Additionally, we propose a broad class of mappings which transform the DM by mapping long-term knowledge (i.e., internal context, IC) onto the DM. This is equivalent to the use of an inferring process which fills in the gaps in the discourse model on the basis of the participants' shared, but implicit, knowledge. The various mappings can be illustrated schematically as follows:

(7a) $p_7(IC_a, DM_a) \rightarrow DM_a$

(7b) $p_7(IC_a, DM_a) \rightarrow \emptyset_a$

(7c) $p_7(IC_a, DM_a) \rightarrow DM'_a$

(7d) $p_7(IC'_a, DM_a) \rightarrow DM_a$

(7e) $p_7(IC'_a, DM_a) \rightarrow DM'_a$

(7f) $p_7(IC'_a, DM_a) \rightarrow \emptyset_a$

(7g) $p_7(IC_a, DM'_a) \rightarrow DM_a$

(7h) $p_7(IC_a, DM'_a) \rightarrow \emptyset_a$

(7i) $p_7(IC_a, DM'_a) \rightarrow DM'_a$

(7j) $p_7(IC'_a, DM'_a) \rightarrow DM_a$

(7k) $p_7(IC'_a, DM'_a) \rightarrow DM'_a$

(7l) $p_7(IC'_a, DM'_a) \rightarrow \emptyset_a$

The mapping at (7a) illustrates a correct application of internal context to a correct DM which produces a match between the DM of the initiating participant (B) and that of the receiving participant (A). A failure to infer anything on the basis of IC is illustrated in (7b). A mismatch between the participants' DMs resulting from the application of shared IC is illustrated in (7c). Mappings (7d-e) illustrate situations in which the internal context of the participants is not shared. In (7d) this does not have a deleterious effect on the DM, however in (7e-f) it causes a disparity between participants' DMs. The mappings (7g-l) have the same pattern as the above, but they involve the use of an incorrect DM as a base.

2.2.4 Goal Structure

The dialogue structure which mediates between the participants' internal representations and their external manifestation is the goal structure (GS). We consider the goal structure to be a hierarchically organised structure of goals and sub-goals. In line with findings of Egan, Gardiner, and Reilly (1986) we acknowledge that the GS in any dialogue can be seen to consist of three main subcomponents: a substantive GS, a communicative GS, and a procedural GS. Substantive goals are, as their name suggests, concerned with the achievement of the overall aim of the dialogue. In other words, they are concerned with the "why" of the dialogue. Another way of looking at substantive goals is to consider them to be a situation-independent set of goals.

Procedural goals are concerned with how to go about achieving the substantive goals and are dependent on the situation. In other words, they are the "how" of the dialogue. Communicative goals are concerned with ensuring that efficient communication takes place. For the sake of simplicity we will give mappings for each of these goal types separately. However, certain of the mappings involving the GS will be more relevant to one goal type than to another. We will indicate this where appropriate. The relevant mappings are:

(8a) $\quad p_8(DM_a, GS_a) \rightarrow GS_a$

(8b) $\quad p_8(DM_a, GS_a) \rightarrow \emptyset_a$

(8c) $\quad p_8(DM_a, GS_a) \rightarrow GS'_a$

(8d) $\quad p_8(DM'_a, GS_a) \rightarrow GS_a$

(8e) $\quad p_8(DM'_a, GS_a) \rightarrow GS'_a$

(8f) $\quad p_8(DM'_a, GS_a) \rightarrow \emptyset_a$

(8g) $\quad p_8(DM_a, GS'_a) \rightarrow GS_a$

(8h) $\quad p_8(DM_a, GS'_a) \rightarrow \emptyset_a$

(8i) $\quad p_8(DM_a, GS'_a) \rightarrow GS'_a$

(8j) $\quad p_8(DM'_a, GS'_a) \rightarrow GS_a$

(8k) $\quad p_8(DM'_a, GS'_a) \rightarrow GS'_a$

(8l) $\quad p_8(DM'_a, GS'_a) \rightarrow \emptyset_a$

In the above set of mappings we have the DM being mapped onto the GS in the usual way. However, the meaning of GS and GS' requires a little clarification. We cannot say that GS' represents a mismatch between the initiating and receiving participant's goals. Strictly speaking the participants' goals must differ. For example, in a task-oriented dialogue, such as Goodman's example excerpted above, the substantive goal of the instructor is to *teach* the apprentice how to construct the water pump, and the substantive goal of the apprentice is to *learn*. Similarly, the communicative and substantive goals of the participants will differ as the dialogue proceeds. However, the ultimate substantive goal of the participants is successfully to construct the water pump. The ultimate communicative goal is to effect clear communication. Therefore at a certain level of abstraction we can say that the participants have identical goals. It is only at the level of sub-goals and of actions to achieve these sub-goals that there is a difference between their goal structures. Therefore, a mapping resulting in GS indicates that both participant's share the same ultimate goals, and a GS' mapping indicates that the participants' ultimate goals differ.

Both the internal and external contexts are also influential in structuring the GS. Information about the physical layout of the dialogue situation or about the non-verbal actions of the other participant can play an important role in shaping the goal structure of both participants. We can account for this by mapping the representation of external context (XC) onto the goal structure. As well as external context, information about similar situations encountered in the past can also play a role in determining the nature of the goal structure. This is accounted for in our taxonomy by mappings from internal context (IC) onto the goal structure. We consider the goal structure to be incrementally transformed, as with the DM, therefore the processes involved in the mappings take two arguments. Below, are the two sets of mappings dealing with both types of context. First, we give the XC mappings:

(9a) $\quad p_9(XC_a, GS_a) \rightarrow GS_a$

(9b) $p_9(XC_a, GS_a) \to \emptyset_a$

(9c) $p_9(XC_a, GS_a) \to GS'_a$

(9d) $p_9(XC'_a, GS_a) \to GS_a$

(9e) $p_9(XC'_a, GS_a) \to GS'_a$

(9f) $p_9(XC'_a, GS_a) \to \emptyset_a$

(9g) $p_9(XC_a, GS'_a) \to GS_a$

(9h) $p_9(XC_a, GS'_a) \to \emptyset_a$

(9i) $p_9(XC_a, GS'_a) \to GS'_a$

(9j) $p_9(XC'_a, GS'_a) \to GS_a$

(9k) $p_9(XC'_a, GS'_a) \to GS'_a$

(9l) $p_9(XC'_a, GS'_a) \to \emptyset_a$

The IC mappings are identical:

(10a) $p_{10}(IC_a, GS_a) \to GS_a$

(10b) $p_{10}(IC_a, GS_a) \to \emptyset_a$

(10c) $p_{10}(IC_a, GS_a) \to GS'_a$

(10d) $p_{10}(IC'_a, GS_a) \to GS_a$

(10e) $p_{10}(IC'_a, GS_a) \to GS'_a$

(10f) $p_{10}(IC'_a, GS_a) \to \emptyset_a$

(10g) $p_{10}(IC_a, GS'_a) \to GS_a$

(10h) $p_{10}(IC_a, GS'_a) \to \emptyset_a$

(10i) $p_{10}(IC_a, GS'_a) \to GS'_a$

(10j) $p_{10}(IC'_a, GS'_a) \to GS_a$

(10k) $p_{10}(IC'_a, GS'_a) \to GS'_a$

(10l) $p_{10}(IC'_a, GS'_a) \to \emptyset_a$

The above mappings follow the usual pattern, and do not require much further explanation. We will illustrate specific examples of some in a later section. There is one further set of processes which maps onto the GS, and these relate to the $\emptyset$ mappings. When a null mapping takes place it may or may not be detected. If it is detected the participant will perform some action to get round the problem. Since actions are mediated by the GS, we require a mapping from the null structure to the GS. The following are the relevant mappings:

(11a) $p_{11}(\emptyset_a, GS_a) \to GS_a$

(11b) $p_{11}(\emptyset_a, GS_a) \to \emptyset_a$

(11c) $p_{11}(\emptyset_a, GS_a) \to GS'_a$

(11d) $p_{11}(\emptyset_a, GS'_a) \to GS_a$

(11e) $p_{11}(\emptyset_a, GS'_a) \to \emptyset_a$

(11f) $p_{11}(\emptyset_a, GS'_a) \to GS'_a$

Note that there is no such structure as $\emptyset'$, therefore we have half the usual number of mappings for this type of process.

At the beginning of this section it was stated that GS is the bridge between the internal representations and the outside world. The GS is involved in a series of mappings onto the two external structures in our taxonomy S, and XC. The following are the mappings involved:

(12a) $p_{12}(GS_a) \to S$

(12b) $p_{12}(GS_a) \to XC$

(12c) $p_{12}(GS_a) \to \emptyset_a$

(12d) $p_{12}(GS'_a) \to S$

(12e) $p_{12}(GS'_a) \to XC$

(12f) $p_{12}(GS'_a) \to \emptyset_a$

Note that none of the mappings result in an S' or an XC'. This is because both S and XC are shared, external phenomena. The ' notation is used to indicate disparity in *internal* representations. Consequently, we have half the usual number of mappings.

Figure 1 provides an overview of the interrelationships between the different internal and external dialogue structures. The directions of the arrows indicates the direction of information flow.

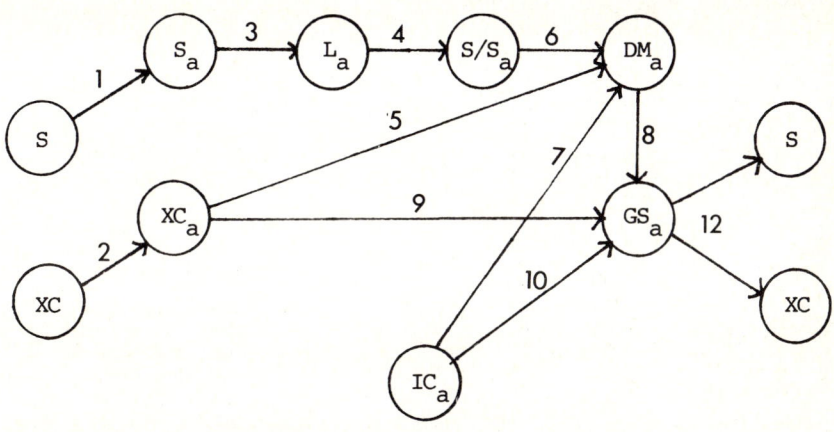

FIGURE 1: The flow of information between structures in the taxonomy, with associated mapping reference numbers.

We have now covered all the major structures and processes involved in the taxonomy. In order to determine whether our method of classification is to serve a useful purpose we must show that it can be applied to realistic dialogue, and that it can differentiate between various types of dialogue miscommunications. The next section will address these issues.

3. Some Sample Applications

The following dialogue is an excerpt from one of the large number of dialogues gathered as part of an ESPRIT project (ref. AIP P527) on communication failure in dialogue in which the Educational Research Centre is involved. The set of miscommunication examples which are presented here were collected as part of a study of the type of dialogue that takes place between a human database expert and a naive user of the database (Egan, Forrest, Gardiner, Reilly, & Sheehy, 1986). The interaction is in fact three way, involving not just the two human participants but also the database system itself. The task of the user was to interrogate the database for information. Two databases were used, a student records database, and a geographical database.

3.1 Example 1

In the following example the user (A) is accessing a simple two-dimensional database of population data with the help of an expert (B). The database being used here is Minitab (Minitab Inc., 1985). The rows in the database represent different countries, and the columns represent various measures of population (rural, urban, etc.). Minitab only permits the separate display of columns of the data matrix rather than rows, and this is the source of the first misunderstanding. The second one concerns the failure of the user to see that the data she has requested is already on display.

The following is the transcript:

1 A: And if I want to know one element of that matrix, how do I call that up?

2 B: You can ask it to TABLE that – that portion.

3 A: Can I ask it to TABLE Belgium?

4 B: No. No.

5 A: So I can ask it to TABLE population?

6 B: You can.

7 A: So can I ask it to TABLE rural population? Or is that unnecessary?

8 B: Population – If you ask it to – well that's the information you've got. You've got the total population in millions, the city population and the rural population.

Table 1 provides a process analysis of the above dialogue from both A's and B's perspective. The sequence of mappings in Table 1 is not exhaustive, the dialogue involves many more of them. We just show the ones that are involved explicitly in the breakdowns and in their repair. Table 1, and subsequent tables of the same type should be read as follows: when the utterance is ascribed to a participant, for example "2 B", the mappings under B are those that gave rise to the utterance, and those under A arise from its interpretation. If we take A's sequence first, we see the first breakdown occurring at utterance 2, when A comprehends B's response (hence DM_a rather than DM'_a) but because B's instructions are not specific enough the mapping of the DM onto the GS results in an incorrect GS. B detects the error at utterance 3, when the mapping from DM_b to GS_b fails. B detects this failure and attempts to correct it at utterance 4 where a the null mapping is mapped onto GS and the GS in turn gives rise to the correctional response at 4.

A's second misunderstanding arises from a failure to take proper account of the external

Utterance	User (A)	Expert (B)
1 A	$p_{12}(GS_a) \rightarrow S$	
2 B	$p_8(DM_a, GS_a) \rightarrow GS'_a$	$p_{12}(GS_b) \rightarrow S$
3 A	$p_{12}(GS'_a) \rightarrow S$	$p_8(DM, GS_b) \rightarrow \emptyset_b$
4 B	$p_8(DM, GS'_a) \rightarrow GS_a$	$p_8(\emptyset_b, GS_b) \rightarrow GS_b$
		$p_{12}(GS_b) \rightarrow S$
.		
.		
7 A	$p_9(XC'_a, GS_a) \rightarrow GS'_a$	$p_8(DM, GS_b) \rightarrow \emptyset_b$
8 B	$p_8(DM, GS'_a) \rightarrow GS_a$	$p_8(\emptyset_b, GS_b) \rightarrow GS_b$
		$p_{12}(GS_b) \rightarrow S$

TABLE 1: A process analysis of dialogue example 1.

context (utterance 7). This is represented by the mapping of XC_a onto GS_a resulting in a GS'_a. The incorrect GS is corrected at utterance 8 by a mapping from DM_a onto the faulty GS resulting in a corrected one. B detects and corrects this failure in exactly the same way as the first one.

3.2 Example 2

The following is another extract from the same session. One of the tasks that the user (A) has to perform is to calculate the urban population as a percentage of the rural population on the basis of data already in the database, and then store the result of the calculation in the database. In Minitab this involves storing the data in a column of the data matrix (column eight). However, things are made difficult by the fact that data are in different scales (thousands, millions, etc.) and have to be converted to a common scale before the ratio can be calculated. A "/" preceding a speaker's utterance indicates an interruption by that speaker.

1 A: Well it seems to have given me a column eight.

2 B: It does. It does. I'm surprised the values are like that. They shouldn't be like that. Should they? I guess the reason

3 A: /I haven't multiplied by — I haven't multiplied by a 200 yet.

4 B: I've just realised what the problem is. Column two is expressed in millions, and that's expressed in thousands.

5 A: So I'm going to have to multiply ... emm ... that's in millions, that's in thousands, so I've got to get them both in the same units.

Utterance	User (A)	Expert (B)
1 A	$p_{12}(GS_a) \rightarrow S$	
2 B	$p_8(DM_a, GS'_a) \rightarrow GS'_a$	$p_9(XC_b, GS_b) \rightarrow \emptyset_b$
		$p_{11}(\emptyset_b, GS_b) \rightarrow GS_b$
		$p_{12}(GS_b) \rightarrow S$
3 A	$p_{12}(GS'_a) \rightarrow S$	$p_6(S/S_b, DM_b) \rightarrow DM'_b$
4 B	$p_8(DM_a, GS'_a) \rightarrow GS_a$	$p_{10}(IC_b, GS_b) \rightarrow GS_b$
		$p_8(DM'_b, GS_b) \rightarrow GS_b$
		$p_{12}(GS_b) \rightarrow S$
5 A	$p_{12}(GS_a) \rightarrow S$	$p_6(S/S_b, DM'_b) \rightarrow DM_b$

TABLE 2: A process analysis of the dialogue in example 2.

6 B: Yes.

An interesting aspect of this dialogue is that the two participants talk past each other, both trying to solve the column eight problem in their own way. Can we represent this in our framework? Table 2 provides an analysis of the dialogue in example 2. Note that the list concentrates only on those processes implicated in the failure and repair of the misunderstandings.

At utterance 1 we have A failing to detect that there has been any problem in the calculation of column eight. However, B's utterance at 2 indicates that he has detected an error. This is illustrated in our taxonomy by a failure to map the external context (XC) onto the goal structure (GS), resulting in a null mapping. This null mapping is detected by B and gives rise, ultimately, to utterance 2. However, A's processing of utterance 2 causes a mis-match between A's and B's goal strucutre, as indicated by the p_8 mapping. A's goal structure motivates her to provide an incorrect reason for the problem (that she had not multiplied by 200). This utterance is effectively ignored by B, who goes on to provide his own reason. We illustrate B's failure to take account of A's utterance by the p_6 mapping. The result of this mapping is that B's DM is not the one intended by A (represented by DM'_a). We do not treat this as a null mapping, because that would imply that B had not understood A fully. We must assume that he did, but did not agree with what she said. B then proceeds to solve the problem himself. We illustrate this by means of an internal context (IC) mapping, which is our general way of indicating problem solving behaviour. Ultimately B provides a correct solution to the problem; the data being manipulated were in a different scale, and that this had been forgotten about in the calculations. A's agreement with B's solution expressed in utterance 5 causes the DM discrepancy to be corrected by a p_6 mapping.

3.3 Example 3

We now provide a final example of the application of the taxonomy to naturally occurring dialogue. The subject and the database are different this time. The database in question is a student records database containing information about numbers of students taking various subjects, and their grades on these subjects. In the example below the expert (B) misunderstands what information the user (A) is requesting. She asks one of those problematic quantification questions beloved of logicians and linguists: "How many girls studied each of these subjects?" She wanted a subject breakdown for girls, the expert thought she wanted the overall number of girls in the database.

1 B: Well what would you like to find out?

2 A: How many girls studied each of these subjects?

3 B: Well

4 A: /I want to ask how many girls studied art.

5 B: Well, you already – OK. Yeah. Well, the easiest way to do that is to count the number of girls doing any subject you like. So if you ask it to count art for instance, it will count up just the number of girls who took art. It will give you the total number of girls who took

6 A: /Yeah (hesitantly).

7 B: Right.

8 A: So what do I have to

9 B: /Oh sorry. I misunderstand what you were going to – Yeah, I see – I'm sorry. I misunderstood. I thought you wanted to find out the number of girls in the file completely.

10 A: No. How many girls studied each subject.

11 B: /OK. OK.

Utterance	User (A)	Expert (B)
2 A	$P_{12}(GS_a) \rightarrow S$	$P_6(S/S_b, DM_b) \rightarrow DM'_b$
4 A	$P_{12}(GS_a) \rightarrow S$	$P_6(S/S_b, DM'_b) \rightarrow DM'_b$
5 B	$P_6(S/S_a, DM_a) \rightarrow DM_a$	$P_8(DM'_b, GS_b) \rightarrow GS'_b$
	$P_8(DM_a, GS_a) \rightarrow \emptyset_a$	$P_{12}(GS'_b) \rightarrow S$
6 A	$P_{11}(\emptyset_a, GS_a) \rightarrow S$	$P_6(S/S_b, DM'_b) \rightarrow DM_b$
7 B	$P_6(S/S_a, DM_a) \rightarrow DM_a$	$P_8(DM_b, GS_b) \rightarrow GS'_b$
	$P_8(DM_a, GS_a) \rightarrow \emptyset_a$	$P_{12}(GS'_b) \rightarrow S$
8 A	$P_{11}(\emptyset_a, GS_a) \rightarrow S$	$P_6(S/S_b, DM'_b) \rightarrow DM'_b$
9 B	$P_6(S/S_b, DM_a) \rightarrow DM_a$	$P_8(DM_b, GS'_b) \rightarrow GS_b$
		$P_{12}(GS_b) \rightarrow S$

TABLE 3: A process analysis of the dialogue in example 3.

Example 3 is of interest for a number of reasons, apart from the fact that it is a classic quantifier error. You will notice from reading the dialogue that it is fairly clear where the misunderstanding has taken place (at utterance 2 A), but it is not clear which, if any, of A's utterances precipitated the repair of the misunderstanding (4, 6, or 7). The eventual repair could have been a delayed response to 4 A. It could also have been prompted by the unenthusiastic go-ahead signal provided by A in utterance 6. Our problem here in classifying this example illustrates one drawback to our approach. We assume that communication failure is a discrete phenomenon, whereas in the above example it could be that an *accumulation* of inconsistent requests (utterance 4) and weak go-ahead signals (utterance 6) could have allowed B to repair his faulty DM. In our analysis of the exchange above we assume that the turning point in the dialogue was utterance 6, and that utterance 4 was incorrectly integrated into DM_b. Although utterance 6 corrects DM'_b, this does not manifest itself in the dialogue until utterance 9. We indicate this by the p_8 mapping which still gives a GS'_b. This situation is not corrected until utterance 9, when the same mapping results in a GS_b. What we have done here is to use the utterance nearest the repair which provides some information to counter the misunderstanding. Utterance 8, cannot be seen to fulfill this criterion. Therefore utterance 6 was chosen, although the information provided by 6 may not, of itself, have been sufficient to initiate the repair.

3.4 Conclusions

In these preliminary tests of the taxonomy, it appears to have functioned well. We have been able to apply it to a number of different types of miscommunication in natural dialogue. A process analysis of the above three examples yielded three distinct patterns of mappings. We expect, therefore, that our system should be useful in providing a concise "fingerprint" of dialogue miscommunications. This is useful in the analysis of large corpora of dialogue, and their associated miscommunications. In developing a computer-based dialogue system it is important to be able to determine the relative frequency of occurrence of various types of miscommunications in order to be able to allocate resources to the design of the system. It is all too easy to waste time in the development of a dialogue system by concentrating on theoretically interesting aspects of the design which cater for infrequently occurring linguistic utterances and dialogue situations. The taxonomy described above should help to highlight which broad categories of miscommunication are the most frequent, and which ones are less frequent.

4. Relationship to Other Work

We will now attempt to show how the taxonomy fits in which other research in the area of dialogue, and in particular with the work described in other chapters in this volume.

4.1 Is and Can-Be Misconceptions

Webber and Mays (1983) suggest that there are two main classes of misconceptions which a user can have when interacting with a database: (1) misconceptions about what *is* the case, and (2) misconceptions about what *can be* the case. The first class concerns misconceptions about the existence of objects in the database, or the existence of a set of objects which can be selected by a particular description. The following example from Webber and Mays (1983, p. 650) illustrates the point:

    A: Which female employees work in the shoe department?

In requesting this information the user must believe that there are employees, that some of them are female, and that there is a shoe department. It is possible that any or all of these assumptions is erroneous. If, for example, A was unaware that there were no female employees working in the shoe department, then the above utterance would be motivated by either of the following mappings:

(10e) $p_{10}(IC'_a, GS_a) \rightarrow GS'_a$

(10k) $p_{10}(IC'_a, GS'_a) \rightarrow GS'_a$

Both of these p_{10} mappings represent a mismatch between A's background information (the internal context) and that of B. This in turn leads to a mismatch in goals, thus giving rise to an erroneous request for information. One could determine from preceding exchanges in the dialogue which of the two mappings might be responsible for the error. For example, A might already have provided evidence that s/he was working with a faulty goal structure, indicating a (10k) mapping.

Another type of *is* misconception occurs when a user believes that an object has an attribute which it does not, or that one thing depends on another when it does not. An example of the latter is the following:

> A: What are the profit margins as a percentage of sales for each installation?
>
> B: Margins don't depend on sales. They are calculated as the difference between unit product cost and list price (Webber & Mays, 1983, p. 650).

As well as misconceptions about what is the state of the world, users can also have misconceptions about what can potentially be the state of the world. Webber and Mays consider that there are at least two major categories within this class of misconception. The first concerns the relations that entities in a database can have to each other. For example in a database of student and faculty at a university the request:

> A: Which undergraduates teach courses?

might be a deviant question if the relation "teach" can only hold between the objects "faculty" and "course".

The second type of *can be* misconception involves constraints between events and states and their relationships over time. An example of this type of misconception is given in the following exchange (Webber & Mays, 1983, p. 651):

> A: Is John registered for CSE220?
>
> B: No. He can't be registered for it because he hasn't yet taken CSE121.

To deal with these type of misconceptions the system needs knowledge of the past history of the database, and knowledge of the relationship between past events and what can be true afterwards.

All of the above miscommunications can be analysed in the same way as the "shoe department" example. They all represent the mapping of faulty internal context onto the goal structure, thus producing a faulty goal structure. However, there is nothing to prevent an elaboration of the IC mappings to take account of the distinctions that Webber and Mays make between different kinds of misconceptions.

4.2 Reference Failure

As has already been mentioned, Goodman (this volume) has studied miscommunication in the instructional dialogues used by Grosz (1977) in which an expert tells an apprentice how to construct a toy water pump. He has noted that a frequent source of error in this type of dialogue is the description used by a speaker to identify an object in the world. The description can be imprecise, confused, ambiguous, or overly specific. Goodman proposes that the primary means for repairing faulty descriptions is by reducing the specificity of parts of the description, i.e., by "relaxing" the reference.

Goodman has observed that different linguistic structures are used to convey different elements of a description. Relative clauses are used to provide complex information, prepositional phrases are used to express simpler information, and adjectives are used to express simple perceptual features. He has found that participants in a dialogue tend to relax their descriptions (when a problem has arisen) in the following order: adjectives, then prepositional phrases, and finally relative clauses and predicate complements.

The perceptual features of an object are used in a description to provide a means for distinguishing one from another. Goodman has observed that the features most likely to be relaxed are those that require the least active consideration, for example the colour or shape of an object. Features that require more active consideration on the part of the listener (such as relative size, distance, or weight) are usually the last features to be relaxed. People tend to be causal with less active features and more careful with active ones. As a result, the main cause of reference failure is found to be the inaccurate use of less active features.

The main aim of goodman's work is to try to incorporate in a computer-based system the tolerance for inaccurate description which he has observed to be used in human dialogue. He considers one of the main techniques used is a relaxation of certain parameters of the description in a principled way.

In order to see how our taxonomy deals with the types of miscommunication identified by Goodman, we will examine again the fragment of dialogue taken from Goodman (this volume) and already used in section 2.

1 B S: 1. An now take the little red
2. peg

2 A [p takes PLUG]

3 B 3. Yes.
4. and place it in the hole at the
5. green end,

4 A [P starts to put PLUG into OUTLET2 of MAINTUBE]

5 B 6. no
7. the − in the green thing

6 A [P puts plug into green part of PLUNGER]

7 B P: 8. Okay.

In order to provide a process analysis we have added our own utterance numbers to the dialogue. Table 4 provides a process analysis of the excerpt. The dialogue proceeds correctly until utterance (4 A). We will refer to A's actions as utterances, although this is stretching the definition of the term somewhat. Note that we represent actions as mappings from the goal structure onto the external context. Conversely, we represent interpretations of actions as mappings from the external context onto the goal structure. The breakdown in the above dialogue occurs when B fails to correctly resolve the reference "the hole at the green end". This is represented by a p_5 mapping which maps the external context onto the discourse model. This is the mapping we use to represent the resolution of references to the shared external context of the dialogue. In our analysis above, we assume that the XC_a representation matches that "intended" by B. This may not be the case. Another possible mapping is:

$$p_5(XC'_a, DM_a) \rightarrow DM'_a.$$

where we assume that the external context is incorrectly perceived, and is not the perception shared by B. However, from what we can see of the dialogue, in particular the loosely phrased reference, it is likely that both participants were working from a similar representation of the external context. The incorrect use of OUTLET2 more than likely stemmed from a choice taken from among a number of possibilities, rather than from a failure to perceive the other possibilities. The misunderstanding is detected by B in utterance (4 A), and this is illustrated as a failure by

Utterance	User (A)	Expert (B)
1 B	$p_8(DM_a, GS_a) \rightarrow GS_a$	$p_{12}(GS_b) \rightarrow S$
2 A	$p_{12}(GS_a) \rightarrow XC$	$p_9(XC_b, GS_b) \rightarrow GS_b$
3 B	$p_5(XC_a, DM_a) \rightarrow DM'_a$	$p_{12}(GS_b) \rightarrow S$
	$p_8(DM'_a, GS_a) \rightarrow GS'_a$	
4 A	$p_{12}(GS'_a) \rightarrow XC$	$p_9(XC_b, GS_b) \rightarrow \emptyset$
5 B	$p_5(XC_a, DM'_a) \rightarrow DM_a$	$p_{11}(\emptyset_b, GS_b) \rightarrow GS_b$
	$p_8(DM_a, GS'_a) \rightarrow GS_a$	$p_{12}(GS_b) \rightarrow S$
6 A	$p_{12}(GS_a) \rightarrow XC$	$p_9(XC, GS_b) \rightarrow GS_b$
7 B		$p_{12}(GS_b) \rightarrow S$

TABLE 4: A process analysis of the Goodman dialogue.

B to map A's actions onto B's goal structure. B's utterance at (5 B) then corrects A's faulty discourse model and goal structure.

4.3 Object Perspective

McCoy (1985) deals with a problem related to that of object description, namely misconceptions concerning the properties of objects. When a user interacts with a database or an expert system, he or she may attribute a property or property value to an object that the object does not have. She gives the following example of how a current database system and a human might respond to the following request containing a property misconception:

> A: Give me the HULL_NO of all DESTROYERS whose MAST_HEIGHT is above 190.

In the hypothetical database DESTROYERS cannot have that value for MAST_HEIGHT. Upon encountering such a query, a conventional system might give the response:

> B: There are no DESTROYERS in the database with a MAST_HEIGHT above 190. Would you like to try again.

However, as McCoy points out, this is not the way most humans would respond. A human respondent would attempt to get at the root of the misconception manifested in the utterance. In the database there is an object similar to DESTROYER that has the value of MAST_HEIGHT given. Therefore, the user's misconception may have resulted from a confusion of the two objects. Hence, a reasonable response would be:

> A: All DESTROYERS in the database have a MAST_HEIGHT between 85 and 90. Were you thinking of an AIRCRAFT_CARRIER?

The strategy adopted here is to implicitly deny the property/value given, to give the corresponding correct information, and to suggest an alternative query which might satisfy the user's goal.

The above exchange highlighted a confusion of objects by the user; it is also possible to confuse the attributes of the objects. Take, for example, the following exchange:

    A: Give me the HULL_NO of all DESTROYERS whose MAST_HEIGHT is above 3500.

    B: All DESTROYERS in the database have a MAST_HEIGHT between 85 and 90. Were you thinking of DISPLACEMENT.

This last response is similar to the one given above, except that the second query contains an attribute rather than an object confusion.

In order to develop a system capable of dealing with these types of misconception in a human-like manner it is necessary to devise a similarity metric for objects. However, McCoy points out that it is not sufficient to develop a metric, the metric has to be sensitive to context. Objects considered similar under one set of circumstances, might be dissimilar under another. McCoy proposes the notion of *object perspective* to provide a contextually sensitive measure of similarity. An object is represented in a hierarchical inheritance structure. In some contexts the superordinate (e.g., mammal is a superordinate property of human) of the object may be salient, in others it may not. When an object is viewed from a particular perspective, the perspective acts as a filter on the properties that the object may inherit from its superordinate. A number of perspectives are defined a priori for each domain that give salience values (e.g., high, medium, or low) for the attributes of each object in the domain. Properties of objects given a high salience value will be propagated through the inheritance network, those given a low rating will be suppressed. The key aspect of McCoy's proposal is that the perspective filter is orthogonal to the generalisation hierarchy of object attributes.

We can represent the miscommunications associated with property misconceptions in a number of ways. However, they all hinge on the mapping of a correct internal context (IC) onto the correct or incorrect goal structure (GS) of the participant making the error. There are just two different ways in which this might happen:

(10c) $p_{10}(IC_a, GS_a) \rightarrow GS'_a$

(10i) $p_{10}(IC_a, GS'_a) \rightarrow GS'_a$

It is not possible to determine from McCoy's data which of these two possible mappings underly each of her examples. However, it should be possible roughly to determine the precise mappings on the basis of earlier exchanges in the dialogue. For example, there may be utterances prior to the breakdown which indicated that A had already used a faulty goal structure in the generation of previous utterance. This would narrow the choice of mappings down to (10i).

It might be asked why we assume that misunderstandings based on property misconceptions are the result of incorrect mappings of a correct IC onto the GS, whereas in the case of the object misconceptions described in Section 4.1 we assume that the IC is incorrect. This is because the type of misunderstanding that McCoy refers to are primarily *unintentional* confusions of attributes of similar objects. This is supported by the fact that all of B's responses to A's property misconceptions in the above examples explicitly state that the erroneous utterances could not have been intended by A.

4.3 Pragmatic Overshoot

Carberry (this volume) describes an approach to repairing a class of *can be* miscommunication which she refers to as *pragmatic overshoot*. This term describes the phenomenon in which the speaker's utterance may be syntactically and semantically well-formed yet may violate the pragmatic rules of the listener's world model. She illustrates this with the example query:

    A: Which apartments are for sale?

In the real estate world model, single apartments are never for sale but are only ever rented. A characteristic of this type of misconception is that no matter how much more world knowledge is added to the system, the erroneously presumed relationship will still fail to exist. The main aim of her work is to develop a system which can offer a limited response to a pragmatically ill-formed query making use of the current context and the ill-formed utterance. The aim of the response would be to help satisfy the user's perceived goal. Her main contention is that by using the inferred user's plan, it is possible to choose a substitute for the proposition that gave rise to the pragmatic overshoot.

Within our taxonomy, the locus of the plan is the goal structure (GS). The plan is implicit in the organisational hierarchy of the GS. We avoid the potential for infinite regress associated with the conception of reflexive modelling (e.g., A's model of B's goals, or A's model of B's model of A's goals) by considering just two types of GS: GS_a and GS'_a. A faulty GS arises when one participant's GS is not the one intended by the other participant. As a convenient shorthand we refer to this as a mismatch between the participants' goal structures. However, as discussed earlier, even when there is no mismatch, we cannot say that the goal structures of the participants are identical. Carberry's approach focuses on the issue of repair of a set of incorrect mappings which encompasses those described in Section 4.1 and 4.2. The relevant mappings are:

(10c) $\quad p_{10}(IC_a, GS_a) \to GS'_a$

(10e) $\quad p_{10}(IC'_a, GS_a) \to GS'_a$

(10i) $\quad p_{10}(IC_a, GS'_a) \to GS'_a$

(10k) $\quad p_{10}(IC'_a, GS'_a) \to GS'_a$

The particular repairing mapping with which Carberry is concerned is:

(8g) $\quad p_8(DM_a, GS'_a) \to GS_a$

This involves the verbal correction of A's goal structure. B's corrective utterance is first mapped onto A's DM, and the DM in turn is then mapped onto the faulty GS. If B takes corrective action by some non-verbal means then the mapping would be:

(9g) $\quad p_9(XC_a, GS'_a) \to GS_a$.

5. Conclusions

We have described in this chapter a system of classification of dialogue processes which focuses on those points in the process where communication is liable to breakdown. The taxonomy we propose views dialogue as a set of processes which map one dialogue structure onto another. In most cases this involves mapping from one internal representation onto another. We do not make any strong claims about the order in which the various processes are executed. We are concerned primarily with the range of processes implicated in the occurrence of a particular miscommunication. We have shown that our taxonomy can distinguish between a wide variety of miscommunication distinctions made by other dialogue researchers.

Our approach does not purport to be the basis for automatically detecting dialogue miscommunications. In the dialogue examples given above it was not always possible to unambiguously determine which mappings were the locus of the breakdown. There was a certain amount of subjectivity and inference involved in some categorisations. However, given a more complete dialogue, including a videotape recording and possibly participant debriefing, it should be possible to considerably limit this problem.

The main role we see for our scheme is in the statistical analysis of corpora of dialogues prior to the development of a robust natural language interface. Here the taxonomy promises to be a useful guide in resource allocation by indicating, in the vast array of communication failures, which types are most frequent, and therefore, in matters of research and interface design, most important.

REFERENCES

Carberry, S. (in press). The use of inferred knowledge in understanding pragmatically ill-formed queries. In R. Reilly (Ed.), *Communication failure in dialogue and discourse*. Amsterdam: North-Holland.

Carbonell, J.G., & Hayes, P.J. (1983). Recovery strategies for parsing extragrammatical language. *American Journal of Computational Linguistics, 9,* 123-146.

Cohen, P.R. (1983). *A computational model for the analysis of arguments.* Technical Report CSRG-151, Computer Systems Research Group, University of Toronto.

Eastman, C.M., & McLean, D.S. (1981). On the need for parsing ill-formed input. *Americal Journal of Computational Linguistics, 7,* 257.

Egan, O., Forrest, M., Gardiner, M., Reilly, R., & Sheehy, N. (1986). *CFID project deliverable 3: Dialogue studies – pilot phase* (contract no. ESPRIT AIP P527). Commission of the European Communities, Brussels.

Egan, O., Gardiner, M., & Reilly, R. (1986). Methodological issues. In Egan, O., Forrest, M., Gardiner, M., Reilly, R., & Sheehy, N. (1986). *CFID project deliverable 3: Dialogue studies – pilot phase* (contract no. ESPRIT AIP P527). Commission of the European Communities, Brussels.

Estes, W.K. (1978). The information-processing approach to cognition: a confluence of metaphors and methods. In W.K. Estes (Ed.), *Handbook of learning and cognitive processes* (volume 5). Hillsdale, NJ: Erlbaum.

Goodman, B. (in press). Repairing reference identification failures by relaxation. In R. Reilly (Ed.), *Communication failure in dialogue and discourse*. Amsterdam: North-Holland.

Grosz, B. (1977). *The representation and use of focus in dialogue understanding.* Unpublished PhD thesis, University of California, Berkeley.

Grosz, B.J. (1978). Discourse analysis. In D. Walker (Ed.), *Understanding spoken language*. New York: Elsevier North-Holland.

Grosz, B.J. (1981). Focusing and description in natural language dialogues. In A. Joshi, & B. Webber (Eds.), *Elements of discourse understanding*. Cambridge, England: Cambridge University Press.

Grosz, B.J., & Sidner, C.L. (1985). *The structures of discourse structure.* Technical Report CSLI-85-39, Center for the Study of Language and Information, Stanford, CA.

Harris, J. (in press). Speech comprehension and lexical failure. In R. Reilly (Ed.), Communication failure in dialogue and discourse. Amsterdam: North-Holland.

McCoy, K.F. (in press). Generating response to property misconceptions using perspective. In R. Reilly (Ed.), *Communication failure in dialogue and discourse*. Amsterdam: North-Holland.

Minitab Inc. (1985). *Minitab reference manual: Minitab release 5.1.* State College, PA: Minitab Inc.

Polanyi, L.,& Scha, R.J.H. (1986). Syntactic and semantic aspects of discourse structure. In L. Polanyi (Ed.), *The structure of discourse*. Norwood, NJ: Ablex.

Reichman, R. (1984). Extended person-machine interface. *Artificial Intelligence, 22,* 157-218.

Reilly, R. (1985). Control processing vs. information processing in models of reading. *Journal of Research in Reading, 8,* 3-19.

Ringle, M.H., & Bruce, B.C. (1980). Conversation failure. In W. G. Lehnert, & M.H. Ringle (Eds.), *Strategies for natural language processing*. Hillsdale, NJ: Erlbaum.

Sidner, C.L. (1979). *Toward a computational theory of definite anaphora comprehension in English discourse.* Technical Report 537, MIT Artificial Intelligence Laboratory, Cambridge, MA.

Thompson, B.H. (1980). Linguistics analysis of natural language communication with computers. In *Proceedings of the Eighth International Conference on Computational Linguistics,* Tokyo.

Webber, B.L.,& Mays, E. (1983). Varieties of user misconceptions: Detection and correction. In *Proceedings of IJCAI '83,* Karlsruhe, Germany.

Weischedel, R.M.,& Sondheimer, N.K. (in press). Meta-rules as a basis for processing ill-formed input. In R. Reilly (Ed.), *Communication failure in dialogue and discourse*. Amsterdam: North-Holland.

# THE STRUCTURE OF MISUNDERSTANDINGS[1]

Claire Humphreys-Jones

Department of English Language,
University of Newcastle-Upon-Tyne

A misunderstanding is one of a number of potential or actual problems in communication and any attempt to detail it requires a theoretical framework which provides adequate terms of reference. We have to know how a misunderstanding is structured and how best to describe that structure before we can fully explore the complexity and consequences of misunderstandings. In this paper, the phenomenon of misunderstanding is discussed and an analytic system is presented which is capable of detailing the various components of misunderstanding, each of which is described and illustrated. This system relates both to textual features such as utterances or actions and to the psychological features such as the participants' intentions and understandings. It is argued that such an amalgamation of text, intention and understanding cannot be avoided in the analysis of a phenomenon which essentially comprises what is meant, what is said and what is understood.

1. Introduction

The process of communication minimally involves one participant who endeavours to express something which another participant endeavours to understand. The most frequently used channel is language and, in general, participants manage to communicate successfully. This success is sometimes the result of interactive work between them which establishes what the speaker has intended and also what the hearer has understood.[2] The fact that successful communication can never be guaranteed is due to three related problems: (i) the speaker knows what he wants to communicate and is often unable to appreciate that his hearer does not share this knowledge, (ii) the speaker communicates from an individual perspective and his use of language is influenced by this perspective, and (iii) speakers and hearers rarely use language in strict accordance with the rules and conventions of that language; this situation is further worsened by the fact that natural language is rife with ambiguity and imprecise deixis and usually draws heavily on contextual and situational information.

These problems are more pertinent to person-person dialogue but they are not avoided in person-machine dialogue because, although the machine may use language in precise accordance with the grammar in which it has been programmed, the person communicating with it will not be so precise. If the system requires rigorous linguistic precision it is hardly operating in

---

[1] This paper outlines the structure of misunderstandings and largely derives from Humphreys-Jones (forthcoming (a)), in which the phenomenon of misunderstandings and their structure are discussed in greater detail. I am grateful to Graham McGregor, Lesley Milroy and Brian Torode for their enlightened and enlightening criticism.

[2] 'Speaker' and 'hearer' are the terms adopted in this paper to represent the participant who endeavours to express something and the participant who endeavours to understand this something, respectively. The terms are used whether the dialogue in which they are participating is spoken or written.

natural language but rather in a compromise between formal and natural languages. It is certainly the case that the machine cannot appreciate that its hearer may not share its knowledge and may view the dialogue from a different perspective. Perspectives and appreciation are human attributes which a machine must be able to handle in order to ensure coherent dialogue. It is therefore an indication of how successful a machine language system is if it can tolerate and successfully deal with communication problems.

There are three fundamental outcomes to attempts by a speaker to communicate successfully to a hearer: correct understanding, lack of understanding and misunderstanding. If the hearer correctly understands what the speaker intends to express in an utterance their dialogue proceeds coherently. If the hearer does not understand what the speaker is endeavouring to express he is aware of his own lack of understanding and can either conceal it or seek clarification.[3] If the hearer misunderstands what the speaker is endeavouring to express in an utterance he has an understanding of that utterance which unbeknown to him is incorrect. In a misunderstanding the hearer believes he has a correct understanding which in fact he does not have, while the speaker believes that the hearer's understanding corresponds with what he intended his utterance to express which in fact it does not.

A misunderstanding can thus pose a serious threat to successful communication because it results in the speaker and the hearer attributing different meanings to an utterance while each believes that his attributed meaning is the only one which holds for that utterance on that occasion. It is quite possible for a dialogue to proceed without either participant ever becoming aware that they are or have been 'at cross purposes'; as Colby (1975) observes: "It is perhaps not fully recognized by students of language how often people misunderstand one another in conversation and yet their dialogues proceed as if understanding and being understood is taking place" (p. 48). On other occasions a misunderstanding may result in confusion or hostility if it remains undetected (Gumperz & Tannen, 1979; Milroy, 1984).

A misunderstanding is a complex phenomenon which influences the dialogue in which it occurs. In attempting to determine the structure of a misunderstanding one is confronted with a complex amalgam of utterances and the understandings which participants have of these utterances. An analysis of misunderstanding must therefore attempt to accommodate both the tangible linguistic presence of utterances and the intangible psychological concept of understandings. Such an analysis is far from straightforward because each participant has his own understanding of an utterance and furthermore these separate understandings may change as the dialogue proceeds.

The analysis presented in this paper aims to be a "process analysis" (Widdowson, 1979) which endeavours to account for the complexity of misunderstandings, a complexity which derives from the dynamism of dialogue between active and independently functioning participants, and which incorporates both textual and psychological features. The comprehensive discourse analysis of Labov and Fanshel (1977), applied to problematic communication by Grimshaw (1982), combines these textual and psychological features, but because it depends upon expansion of the dialogue it is not an economic analytic system and the analyst's interpretation of the dialogue is open to debate.[4] The analytic method discussed in this paper seeks to avoid these problems by focussing on references in the dialogue to the utterance which has

---

[3] E.g., Winograd's (1972) man-machine dialogue:
"(Person): how many things are on top of green cubes?
(Computer): I'M NOT SURE WHAT YOU MEAN BY "ON TOP OF" IN THE PHRASE "ON TOP OF GREEN CUBES" DO YOU MEAN:
1 – DIRECTLY ON THE SURFACE
2 – ANYWHERE ON TOP OF?" (p. 12).

[4] See e.g., Corsaro (1981) on Labov and Fanshel in particular and McGregor (1985) on utterance interpretation and the role of the analyst.

been misunderstood, to what was intended of that utterance, to how that utterance was understood and to how the misunderstanding was made evident in the dialogue. In addition to such references, the moments at which each participant becomes aware that a misunderstanding has occurred are identified. Prior to these moments, each participant believes his understanding of the utterance is the only one and subsequent to them, each participant has become aware that one understanding only is correct and that any others are incorrect.

2. Analysis of Misunderstandings

In this paper a misunderstanding is a phenomenon in dialogue which occurs when a hearer (H) fails to understand correctly the proposition (p) which speaker (S) expresses in an utterance (x). In order to limit the phenomenon to observable rather than intuitive instances, it is further required that H subsequently produces an utterance which is based on, or derives from, his misunderstanding of p. Without this requirement there would be no evidence that a misunderstanding had occurred, unless H admitted to it after the dialogue had ended. It would be possible for H to misunderstand an utterance and never produce an utterance based on this misunderstanding, such as a misunderstanding of the final utterance in a dialogue or of an utterance which is followed by a change of topic, thereby making reference to the misunderstood topic redundant or unacceptable. Any such misunderstanding would not affect the dialogue although it would affect H's ongoing and retrospective interpretations of the dialogue. Because the dialogue itself is not influenced in any way by such a misunderstanding and because there is no evidence in the dialogue that such a misunderstanding has occurred, examples of this type are not considered misunderstandings in this paper.

An utterance by H which questions his understanding of p is not an utterance based on or deriving from a misunderstanding. If H is aware of difficulty and queries what was intended or queries his understanding, he is forestalling a potential misunderstanding before it affects subsequent utterances.

The term 'proposition' is adopted to represent the "something" which S intends to express. It has been used in various senses by linguists and logicians (see Lyons, 1977) but in this paper a proposition is whatever S wishes to communicate; it can be declarative, interrogative, imperative, factual, descriptive, social or expressive. If, for example, S wishes to comment that he is bored or wishes to ask for a drink or wishes to demand silence, each of these wishes are propositions which must be expressed in utterances if they are to be communicated to H.[5] S has the task of devising an appropriate utterance which is in accordance with the appropriate syntactic, semantic and pragmatic conventions of his language and with phonemic and prosodic or orthographic conventions depending on whether he is speaking or writing.

When H hears an utterance it is not necessarily the exact x produced by S. The utterance, x^r, which H receives, that is, believes to have been produced, is determined by the sounds he hears and the way in which he processes these sounds. Similarly, the proposition, p^r, which H receives, that is, believes to have been expressed, is not necessarily the p which S intended to communicate. Given the fact that S and H are individuals with idiosyncratic auditory, vocal and cognitive systems, it is unlikely that exact replication between x and x^r and between p and p^r ever occurs. Close approximation therefore counts as equivalence and if x^r is equivalent to x and p^r equivalent to p correct hearing, processing and understanding obtain and consequently successful communication is achieved.

The states of equivalence between x and x^r and between p and p^r constitute the possible outcomes of communication. Thus, S expresses p in x; for H: x^r is either (a) = x, (b) = $\emptyset$, or (c) $\neq$ x and p^r is either (a) = p, (b) = $\emptyset$, or (c) $\neq$ p. The various possible combinations of these states of equivalence are detailed in Humphreys-Jones (forthcoming (a), (b)). If x^r is equivalent

---

[5] Propositions may, of course, be expressed in other ways, such as by gestures. This paper focusses solely on utterances as the means by which propositions are expressed.

to x, the utterance has been received correctly, that is, heard and processed correctly. If x^r is equivalent to $\emptyset$, no utterance has been received. If x^r is not equivalent to x, the utterance which H receives is not the same as the utterance which S produced, as in a mishearing. If p^r is equivalent to p, the proposition has been received correctly, that is, understood correctly. If p^r is equivalent to $\emptyset$, there is no understanding of the proposition. If p^r is not equivalent to p, the proposition which H receives is not the same as the proposition which S intended to express. The state of equivalence between x and x^r may influence but does not determine the state of equivalence between p and p^r; in other words, it is possible to understand correctly an utterance which has not been heard correctly and it is possible to misunderstand an utterance which has been correctly heard.

x, x^r, p and p^r provide reference points in this analysis of the structure of misunderstandings. A misunderstanding *originates* in S's utterance x, is *manifested* in another utterance, based on H's p^r, and develops through a variable number of other utterances, the contents of which depend on whether or not participants *realize*, that is, detect whether or not the misunderstanding has occurred. The structure of a misunderstanding extends from the utterance which is misunderstood, termed the *origin*, through to the *close*, which is when participants *realize* the misunderstanding has occurred or, in the absence of such *realization*, change the topic so that the dialogue is no longer affected by the misunderstanding.

A misunderstanding consists of primary and secondary components. The primary components, those which are essential to a misunderstanding, are as follows:

(i) the utterance by S, x, which is misunderstood by H, termed the *origin*;
(ii) the utterance by H which is based on or derives from his misunderstanding of the *origin*, termed the *manifestation*; and
(iii) the awareness which S, H and any other participant 0^1, 0^2, 0^n have of whether or not the misunderstanding has occurred, termed their *state of realization*. The secondary components, those which are optional to a misunderstanding, are paralinguistic and extralinguistic features and utterances other than the *manifestation* which occur between the *origin* and the *close*. The secondary components constitute the various ways in which participants develop, resolve, or fail to resolve a misunderstanding and are termed *devices*.

The components of a misunderstanding can be seen in the following example, in which S and H are discussing H's research:[6]

```
(1.1)                          H:  I've done the A.I. and the psychology part
  .2)                           :  Now I'm doing the linguistics side of it
(2)       ORIGIN               S:  Which is the hardest?
(3)       MANIF.               H:  No it's not actually ... no it's
                                                              + S real.
(4.1)     mus. sig.            S:  / laugh /
  .2)     p exp.                :  I meant a question
  .3)     x rpt. & x emph.      :  WHICH is the hardest?
                                                              + H real.
(5.1)     mus. ack.            H:  / laugh /
  .2)     mus. ack.             :  There you are ... a misunderstanding
(6)       mus. ack.            S:  Mmm
(7)          z                 H:  The A.I. was the hardest
```

In (2), the *origin*, S intends to ask H whether he found the A.I., the psychology or the linguistics the most difficult part of his research. S adopts conventional ellipsis in using the interrogative pronoun "Which". H, however, understands "Which" as a relative pronoun and believes that S

---

[6] This example of an actual misunderstanding was collected by the "diary method" and is one of a corpus of 100 data presented and analysed in Humphrey-Jones (forthcoming, (a)).

is commenting that the linguistics was the most difficult part. This incorrect p^r is made evident by (3), the *manifestation*, in which H responds to the *origin* by denying that the linguistics was the most difficult.

The *origin* and the *manifestation* are textual primary components. After (3) the third primary component is isolated, the *state of realization*. Initially in a misunderstanding that is '− real.' for all participants, neither S, H nor O *realize* that the misunderstanding has occurred until at least after the *manifestation*.[7] The inappropriateness of the *manifestation* might *effect* '+ S real.', as it does in this example, or *realization* might be *effected* as a result of other utterances, as it is for H in this example.

Following S's *realization* in the above example a number of *devices* are used. In (4.1) S signals the misunderstanding by laughing at the *manifestation*, in (4.2) S explains that she meant the proposition to be interrogative and in (4.3) she repeats the *origin* and emphasizes the part of it which was misunderstood. These *devices effect* '+ H real.' and evidence for H's *state of realization* is provided by (5.1) and (5.2), in which H acknowledges the misunderstanding. S also acknowledges the misunderstanding in (6) and finally, the misunderstanding having *closed*, H correctly answers the question posed in the *origin*.

The *states of realization* which may obtain are subjects to a binary classification, which is modified for occasions when it cannot be certain whether or not participants have realized the misunderstanding has occurred, that is, when there is no textual evidence and no subsequent confirmation from participants. The possible *states of realization* are therefore as follows: '+ S real.', '+ H real.', '+ 0 real.', '− S real.', '− H real.' and '− 0 real.' when the *states* are known; '? + S real.', '? + H real.', '? + 0 real.', '? −S real.', '? − H real.' and '? − 0 real.' when the *states* are assumed but cannot be confirmed; '? S real.', '? H real.' and '? 0 real.' when the *states* cannot be assumed. In any one misunderstanding it is possible for a combination such as the following to obtain: '+ S real. ? − H real. ? 0 real.'. If these *states* obtained at the *close* of a misunderstanding it is highly likely that at least two out of the three participants have different understandings of the *origin* utterance and probably of the subsequent dialogue insofar as it relates to and is affected by the *origin*.

Evidence for the *states of realization* which obtain is provided by retrospective commentaries from participants and/or the textual content of the dialogue, *devices* which constitute the secondary component of a misunderstanding, that is, utterances, paralinguistic and extralinguistic features. *Devices* are divided into four groups, as follows:

(i) *devices* which relate to the production and reception of the *origin* and to the intention behind and understanding of the proposition expressed therein, that is, to x, what is uttered; to x^r, what utterance is received (heard and decoded); to p, what S intends to express; and to p^r, what proposition is received (understood) by H. x can be completed, repeated, emphasized, amplified (that is, elaborated or augmented), explained, queried and refuted. x^r, p and p^r can be explained, queried and refuted.
(ii) *devices* which relate to the misunderstanding and to the awareness of an error or problem in the dialogue. These *devices* signal and acknowledge either misunderstandings or errors in H's p^r.
(iii) *devices* which relate to the *manifestation* and which involve other utterances. The manifestation can be completed, repeated, emphasized, amplified, explained, queried or refuted. Utterances other than those relating to x and to the *manifestation* are termed 'y' utterances if they precede *realization* and 'z' if they follow *realization;* this distinction is made because y utterances influence the development of the misunderstanding whereas z utterances do not.
(iv) *devices* which are extralinguistic, such as a positional move or other action which influences the development of the misunderstanding.

---

[7] The one exception to this is when H intentionally misunderstands an utterance for joke purposes; in intentional misunderstandings, H chooses to produce a *manifestation* and '+H real.' precedes that *manifestation*.

Participants choose to use some *devices* and use others unconsciously. If a participant *realizes* a misunderstanding has occurred, he can choose to draw the other participant's attention to it by explaining what was meant or what was understood, and so on. If a participant fails to *realize* he might, for example, use a y utterance which develops the topic of either the *origin* or the *manifestation*. This participant has not made a conscious effort to resolve the misunderstanding but his y utterance may well add sufficient information to enable the other participant to *realize* that the misunderstanding has occurred.

Devices can be used separately, as in S's 'misunderstanding acknowledgement' in (6) of the example, in combination, as in S's 'x repeated and x emphasized' in (4.3) or in sequence, as in H's 'misunderstanding acknowledgement, misunderstanding acknowledgement' in (5.1, 5.2). It is also possible that a *device* is used partially or incompletely. A *device* may contain glossing, which is a specification of the *device*, as in 'I meant ...' or 'I thought you meant ...'; these would be 'p explained' with glossing and 'p^r explained' with glossing, respectively.

In order to illustrate each *device* an actual misunderstanding is given below and then each *device* is given in the form it might have taken, had it been used. S and H are discussing a friend who is selling her car:[8]

```
(1)      ORIGIN        S: I wonder what she's selling it for
(2)      MANIF.        H: Going up in the world getting rid
                          of her bashed mini
                                  _____        + S real.
(3.1)  p^r ref. prtl.  S: No that's not what I mean
  .2)  p exp.           : I mean how much is she selling it for
                                  _____        + H real.
  .3)  z                : I'll have to remember that to tell it
                          to Clair for her misunderstandings
```

(i) *Devices* which relate to x, x^r, p and p^r:

1. x completed: 'It for' (if the *origin* had been interrupted, such as 'I wonder what she's selling').
2. x repeated: 'I wonder what she's selling it for'.
3. x emphasized: 'I wonder WHAT she's selling it for'.
4. x amplified: 'I wonder how much she's selling it for'.
5. x explained: 'I said I wonder what she's selling it for'.
6. x queried: 'Did you say I wonder what she's selling it for?'
7. x refuted: 'You didn't say I wonder what she's selling it for'.
8. x^r explained: 'I thought you said I wonder why she's selling it'.
9. x^r queried: 'Did you think I said I wonder why she's selling it?'.
10. x^r refuted: 'I didn't say I wonder why she's selling it'.
11. p explained: 'I meant how much is she selling it for'.
12. p queried: 'Did you mean you wonder how much she's selling it for?'.
13. p refuted: 'You can't mean you wonder how much she's selling it for'.
14. p^r explained: 'I thought you meant why is she selling it'.
15. p^r queried: 'Did you think I meant why is she selling it?'.
16. p^r refuted: 'I didn't mean why is she selling it' (cf. the actual use of the *device* in the example; the refution is partial because p^r is not cited).

(ii) *Devices* which relate to the misunderstanding or to an error in p^r:

17. p^r error signal: 'What?' or 'Pardon?'.

---

[8] This example of an actual misunderstanding was collected by the "diary method" and is one of a corpus of 100 data presented and analysed in Humphreys-Jones (forthcoming, (a)).

18. p^r error acknowledgement: 'Sorry' (assuming the p^r error acknowledgement stems, for example, from an objection by the other participant to the snide nature of the *manifestation's* comment on the social pretensions of the seller).
19. misunderstanding signal: 'We're at cross purposes'.
20. misunderstanding acknowledgement: 'Oh, I get you now'.

(iii) *Devices* which relate to the *manifestation* and which express new propositions:

21. y: 'I might be able to afford it'.
22. manifestation completed: 'Of her bashed mini' (if the *manifestation* had been interrupted after 'Going up in the world getting rid').
23. manifestation repeated: 'Going up in the world getting rid of her bashed mini'.
24. manifestation emphasized: 'GOING UP in the world GETTING RID of her bashed mini'.
25. manifestation amplified: 'Improving her status by getting rid of her battered old mini'.
26. manifestation explained: 'I said going up in the world getting rid of her bashed mini'.
27. manifestation queried: 'Did you say going up in the world getting rid of her bashed mini?'.
28. manifestation refuted: 'You didn't say going up in the world getting rid of her bashed mini'.
29. z: 'You don't want to buy that old thing' (after the misunderstanding has been resolved; cf. the actual use of the *device* in the example, which refers to the misunderstanding as an event in the dialogue).

(iv) *Devices* which involve extralinguistic activity:

30. action: looking at an advertisement for the car to see the price.

The structure of misunderstanding is thus a combination of primary and secondary components: ORIGIN + MANIFESTATION + STATES OF REALIZATION (+ *devices*). The *origin* is always the first component and the others then follow in a variable order. A misunderstanding has a *basic* structure if it consists of the minimal number of components required to ensure that both S and H *realize* the misunderstanding: ORIGIN → MANIFESTATION → + S REAL. → S device → H REAL. → (H device). Such a misunderstanding is dealt with in the most expedient way possible. S *realizes* the misunderstanding because of the inappropriateness of the *manifestation* and he uses one *device* to *effect* '+ H real.'; H then may use a *device* which, for example, acknowledges the misunderstanding. This *basic* structure can be *expanded* if either participant uses more than one *device* in the appropriate place in the sequence and can be *interactively expanded* if each participant uses one or more *devices* interactively in the places in the sequence where only one participant uses a *device* in the *basic* structure.

The structure of a misunderstanding is *diversified* if it fails to exhibit this *basic* structure. Such *diversification* may come about for a number of reasons, such as an utterance being produced between the *origin* and the *manifestation,* or *realization* not being *effected* in the order '+ S real.' before '+ H real.', or *realization* not being *effected* at all for one or more participants or the additional contributions of 0 altering the *basic* structure.

3. Discussion

The structure of a misunderstanding is not fixed and absolute but rather is wholly dependent on its participants and the ways in which they interact. The course of a misunderstanding is determined by individual participants: whether they find each other's utterances inappropriate responses to their own utterances, whether they *realize* misunderstandings have occurred, whether they choose to *effect realization,* which *devices* they choose to use, and so on. This variability poses problems for anyone seeking to analyse the structure of a misunderstanding and for anyone seeking to formulate ways of avoiding, detecting and resolving misunderstandings.

The structural features of the analytic method described above were developed by examining a corpus of data of actual misunderstandings and attempting to find the optimal way of describing this data. It rapidly became apparent that a purely textual analysis would not suffice because of the need to account for the fact that up to a certain point in the dialogue one belief, the misunderstanding, was held about the *origin* and after another point, *realization,* a different belief was held. The moment at which *realization* was *effected* crucially influenced subsequent utterances. To add to the complexity, the point at which S's belief changed, when '− S real.' became '+ S real.', tended to be at a different time to the point at which H's beliefs changed, when '− H real.' became '+ H real.' so that S and H did not appreciate at the same time that they had held different beliefs about the *origin* and the subsequent dialogue.

To dismiss the *states of realization* as being of no consequence to the textual structure of misunderstandings would have meant that one had no means of marking the extent of a misunderstanding, except in those instances when the misunderstanding had been acknowledged. Such overt acknowledgement is not always present and therefore in a number of data there would have been no means of determining the *close* of the misunderstanding. The incorporation into the analysis of the *states of realization* and the assessment of when these change for each participant enable the extent of a misunderstanding to be detailed for each participant, that is, from the *origin* to '+ real.' for each participant.

The *states of realization* are not the only facet of understanding to be included in the analysis. By taking p^r as H's understanding of what S intends to express, p, it is possible to analyse the various references in the dialogue to what is understood and what intended. By classifying the role of utterances in relation to the point at which the misunderstanding *originates* in terms both of the intended meaning, p, and the utterance produced, x, and of the reception of these, p^r and x^r, it is possible to analyse the process by which successful communication is achieved in the face of a problem in the communication attempt.

It is thus impossible to avoid an amalgamation of textual and psychological components in the analysis of communication failure. The crux of communication failure is that an attempt to express a proposition has not been successful and in seeking to analyse what has happened one must attend to what was intended and what was understood of the utterances produced in the communication attempt.

Having developed an analytic framework, it becomes possible to see how to manipulate misunderstandings. Knowing the cues to look for, one can *realize* when a misunderstanding has occurred and can take steps to resolve it. One can also be alert to the possibility of a misunderstanding's occurrence and can thus avert it. In man-machine dialogue it is less likely that the machine itself would misunderstand an utterance; when, for example, the cause of a misunderstanding is attributing an incorrect referent, the machine would inquire as to which referent was intended rather than assuming one referent only. However, a machine ought to be able to *realize* when the participant with whom it is communicating his misunderstood an utterance. Without this facility, communication between man and machine could have unfortunate consequences.

References

Colby, K.M. (1975). *Artificial Paranoia: A Computer Simulation for Paranoid Processes.* New York: Pergamon Press.
Corsaro, W.A. (1981). Communicative processes in studies of social organization: Sociological approaches to discourse analysis. *Test, 1,* 5-63.
Grimshaw, A.D. (1982). Comprehensive discourse analysis: An instance of professional peer interaction. *Language in Society, 11,* 15-47.
Gumperz, J.J., & Tannen, D. (1979). Individual and Social Differences in Language Use. In J. Fillmore, D. Kempler, & S-Y. Wang, *Individual Differences in Language Ability and Language Behaviour.* New York: Academic Press.
Humphreys-Jones, C.E. (forthcoming) (a). *A Study of the Types and Structure of Misunderstandings.* Thesis to be submitted for Ph.D. at the University of Newcastle-upon-Tyne.
Humphreys-Jones, C.E. (forthcoming)(b). Listen, Hear: Criteria for Successful Communication" In G. McGregor, & R. White (Eds.), *Arts of Listening.*
Labov, W., & Fanshel, D. (1977). *Therapeutic Discourse: Psychotherapy as Conversation.* New York: Academic Press.

Lyons, J. (1977). *Semantics* (2 vols.). Cambridge: Cambridge University Press.
McGregor, G. (1985). Utterance Interpretation and the Role of the Analyst. *Language and Speech, 28,* 1-28.
Milroy, L. (1984). Comprehension and Context: Successful Communication and Communicative Breakdown. In P. Trudgill, (Ed.), *Applied Sociolinguistics.* London: Academic Press.
Widdowson, H. (1979). Rules and Procedures in Discourse Analysis. In T. Myers, (Ed.), *The Development of Conversation and Discourse.* Edinburgh: Edinburgh University Press.
Winograd, T. (1972). *Understanding Natural Language.* Edinburgh: Edinburgh University Press.

COMMUNICATION FAILURE: A DEVELOPMENT PERSPECTIVE

Michael F. McTear

University of Ulster

Communication failure in natural language dialogue systems is studied for two main reasons — in order to eliminate communication problems and in order to better understand processes of communication. The second of these issues is the focus of the present paper. It is argued that empirical studies of communication failure can highlight the complex processes of communication which normally pass unnoticed. Furthermore, studies of children's communication can show what is involved in the acquisition of communicative skills. Two types of developmental studies are examined — experimental tasks which look at children's referential communicative abilities, and sociolinguistic studies of naturally occurring conversations. The former involve skills such as role-taking, description and comprehension monitoring, while the latter indicate that the ability to deal with communication problems emerges early in development. One of the implications of these studies for natural language dialogue systems is that greater account should be taken of the interactional nature of dialogue.

1. Introduction

Computer programs in general, and programs intended to model human performance in particular, suffer from an almost intolerable delicacy. If their users depart from the behaviour expected of them in the minutest detail, or if apparently insignificant adjustments are made in their structure, their performance does not usually change commensurately. Instead, they turn to simulating gross aphasia or death. (Bobrow, Kaplan, Kay, Norman, Thompson & Winograd, 1977, p. 172).

Human beings are able to talk about a wide range of topics, to anticipate potential misunderstandings and to repair miscommunication with apparent ease. Computer programs which use natural language as a medium of communication, on the other hand, as the above extract indicates, are still relatively deficient in this respect. The aim of producing more natural and more human-like human-computer interfaces can be motivated either by the quest for a better understanding of the human capacity by means of the *computational metaphor,* or by the commercially more expedient need to produce more robust and more user-friendly systems. These two approaches can be referred to respectively as *cognitive simulation* and *computational achievement* (Schank & Riesbeck, 1981). Whatever the motivation, however, it is now generally agreed in Artificial Intelligence circles that machine intelligence cannot be achieved in isolation from a theory of human intelligence.

The study of communication failure can contribute to this enterprise by providing greater insights into the processes of communication. The normally smooth transaction of communication between humans conceals its very complexities. It is only by looking at this process when it goes wrong that we can appreciate how it works normally. As Gumperz & Tannen (1979) point out, "by studying what has gone wrong when communication breaks down, we seek to understand a process that goes unnoticed when it is successful" (p. 308). In the present paper communication failure will be examined from a developmental perspective in the belief that the nature of a phenomenon can often be discerned more clearly through the study of its acquisition. This principle has proved beneficial in many aspects of the study of human behaviour, for example, in studies of language, cognition, or social interaction. In particular, the analysis of communication failure in children will illuminate what is involved in cases of unperceived and potential breakdown. Such cases are, if anything, more serious than the gross, but more obvious failures described in the above extract, largely because they can pass unnoticed, but with potentially dire consequences.

Communication failure can be defined broadly or narrowly. In the broad sense, it can refer to any breakdown in communication between people. Spouses can complain that they cannot communicate with their partner, who shows no interest in their concerns and activities. In the work situation there can be a breakdown in communication between management and workforce. If this is severe enough, it can lead to industrial action. At an international level breakdown in communication between nations can lead to the cessation of diplomatic relations, or even to war. If these situations have anything in common, it might be that they describe cases where communicative partners fail to accept each other's goals, especially where these are in conflict. Here, however, a narrower view of communication failure will be taken in which the focus will be on the processes whereby information is negotiated between a speaker and a hearer. If the hearer fails to understand what the speaker is trying to communicate, for whatever reason, then communication failure will have taken place.

Communication failure, in this narrower sense, can be divided into two main types – input failures and model failures (Ringle & Bruce, 1982; see also Reilly, this volume). Input failures occur when the listener is unable to obtain a complete or coherent interpretation for an utterance. They are local in origin, deriving usually from problems associated with a single word or phrase and involving perceptual, lexical or syntactic issues. Thus a mispronounced or misspelt word, an incorrect choice of a word, or a problematic syntactic structure can result in the listener being unable to construct the speaker's meaning. Model failures on the other hand occur when the listener is unable to assimilate the message to a coherent belief model. This may be due to an inability to assign the message to an appropriate conceptual framework because of deficiencies in the listener's background knowledge or from failures on the part of the listener to make necessary inferences.

Note that the emphasis so far has been on the role of the listener in the communication process. Failure occurs when the listener is unable to arrive at an appropriate representation of the speaker's meaning and comprehension is viewed as a process of construction in which the listener attempts to build a meaning on the basis of what the speaker has said as well as utilizing more general world knowledge. This perspective is, of course, particularly important in natural language understanding, where the aim is to program machines to "understand" a range of inputs in natural languages such as English and to eliminate deficiencies in the machine's comprehension abilities in respect of both input and model failures. However, what this perspective ignores is the fact that communication is a two-way process in which both speaker and listener are jointly involved in the process of constructing meanings. Where failure occurs, it might often be appropriate to assign blame to the speaker rather than the listener. For example, the input utterance may be unclearly articulated, an inappropriate lexical choice may have been made or the speaker may have failed to take account of the listener's state of knowledge by, for example, inadequately identifying a referent of the talk. Moreover, as will become clear presently, in communication between humans these problems are usually resolved interactively by means of collaboration between the speaker and the listener. An adequate model of human communication will need to take account of this two-way process, particularly if it is to achieve a capacity for conversational repair. This issue will be addressed further in the final section. In the meantime, the processes of human communication will be illustrated with research findings from empirical studies of communicative breakdown and repair in young children.

2. Children's Communicative Development

The study of children's communicative development has been carried out within two largely separate traditions. In the first the emphasis has been on the cognitive abilities associated with communication and with the development of skills such as description, comparison and questioning. Here communication has usually been investigated using artificial tasks in laboratory settings. In this tradition, which will be referred to as *referential communication,* results have tended to focus on the child's deficiencies in communication. The second tradition fails more within sociolinguistics. Here children have been studied in naturalistic settings with the emphasis on linguistic and social-contextual aspects of communication such as turn-taking and requesting behaviours. These studies have tended to emphasize the proficiency of children in the management of everyday talk. Although providing different perspectives, both traditions are in fact complementary rather than mutually exclusive. In what follows, a brief overview will be given of those aspects of both traditions relevant to the topic of communication failure. This will be followed by a discussion of some implications for the study of communication failure in artificial intelligence. For fuller accounts of work in referential communication the reader is referred to chapters in Dickson (1981). McTear (1985a) presents a detailed overview of work in the sociolinguistic tradition.

2.1 Referential communication

The study of the development of referential communication in children can be traced back to early work by Piaget (1959, first published in French in 1926). On the basis of his observational and experimental studies Piaget concluded that the communication of young children was *egocentric.* The notion of childhood egocentrism has influenced much of current work on referential communication. One aspect, which has been largely refuted in more recent naturalistic studies, was based on measurements of the extent to which children responded to one another in their spontaneous communications. For example, it was estimated that more than half of the utterances of three-year-olds were not socially adapted in the sense of being directed towards or responding to a conversational partner. This figure decreased with age, but even a quarter of the speech of seven-year-olds was egocentric in this respect.

The aspect of egocentrism which has been central in experimental studies of referential communication concerns the child's ability to take the listener's perspective into account in the transmission of information. This is often known as *role-taking.* Many studies have suggested that children are deficient in role-taking skills. An example of a task demanding these skills involves the experimenter demonstrating a game non-verbally to a child. The child is then required to explain the game verbally to a sighted and unsighted listener. The extent to which the child modifies the message as a function of the listener is taken as an indication of role-taking ability. For example, a child who accompanied explanations with nonverbal signals or with deictic expressions when talking to the unsighted listener would be judged deficient in role-taking ability. In recent work role-taking has been treated as a separate component of communication. Role-taking is, however, relevant to an account of communication failure and is moreover related to the concept of user modelling current in artificial intelligence. Furthermore, as Flavell (1974) has suggested, there are different levels of role-taking. These can be described in terms of Existence, Need, Inference and Application, as follows:

(1) Existence — the child becomes aware that others have perspectives, feelings and abilities different from his own;
(2) Need — the child appreciates the necessity to analyse the other's perspectives;
(3) Inference — the child is able to make appropriate inferences about the other's perspectives;
(4) Application — the child is able to utilise these inferences in communication.

The importance of this perspective is that it allows for the occurrence of problems in role-taking even where the existence of other perspectives is understood. Thus the difficulty may reside not in inferring other perspectives but in assessing them accurately. As Shantz (1981) argues, it is necessary to distinguish between assumed and inferred identity of perspectives. A speaker who assumes identity is performing egocentrically, whereas a speaker who communicates perhaps in the same way may be doing so on the basis of inferred identity, which may or may not be accurate. Communication failure is likely to arise in both cases, but the implications for repair are radically different.

As mentioned earlier, researchers have felt the need to distinguish role-taking from general communicative performance. Originally it was felt that children's poor performance in communication tasks could be attributed entirely to their lack of role-taking ability and that communicative ability would improve as the child became less egocentric. However, as studies of children in more naturalistic settings have shown, children display greater adaptability to the listener than is suggested by experimental studies. For example, in a classic study of four-year-olds, Shatz and Gelman (1973) found that the children adapted their messages according to whether they were addressing adults, peers or two-year-olds. But while children may be capable of sensitivity to the listener's needs, they may still experience difficulty in communication tasks involving skills of description, comparison and discrimination. Furthermore, studies which have assessed role-taking ability and communication skills separately have found little correlation between the two. There is also the problem that an emphasis on role-taking obscures many of the complex skills involved in accurate communication, some of which will be examined in greater detail in the remainder of this section.

In a task assessing communicative performance, children are usually required to describe objects (or referents) in such a way that they can be distinguished from non-referents. In a typical experiment two children are separated by a screen, and the first child, the speaker, has to describe certain objects and shapes from an array in front of him so that the second child, the listener, is able to pick out these objects from an identical array in front of himself. Inability to provide sufficient and accurate information would usually result in communication failure, as the listener would be unable to select the appropriate referent. For example, in giving instructions to select from a series of toy farm animals, the instruction *take the cow* would be sufficient only if there were only one cow in the listener's array. Additionally, the speaker has to determine which attributes of a referent are discriminating and which are redundant or unnecessary. If the array contained two brown cows, one large and one small, reference to size alone would discriminate, while reference to colour would be redundant. Generally, results have indicated that children have difficulty in providing such discriminating messages and in coping with unnecessary and redundant information.

Two aspects of referential communication have become prominent in recent research — the child's ability to cope with the specific features of a communication task (task analysis), and the child's ability to appraise the effectiveness of a message. Typical tasks include giving directions to a location and explaining the rules of a game. Studies which focus on comparison abilities involve discriminating among objects which differ in one or more critical features. One example is the triangle task in which children had to discriminate pairs of multidimensional triangles (Whitehurst & Sonnenschein, 1978). The differentiating attribute could be colour, size or pattern, and in some conditions irrelevant as well as relevant conditions were included. So, for example, in the simplest condition, the triangles would differ on one attribute only (e.g., red versus black triangle), so that the child had to choose only between the referent and a non-referent. In more complex conditions more than one nonreferent was involved. For example, there might be two red triangles, one large and one small, and a small black triangle. If the target were the large triangle, then the colour dimension would be irrelevant to its identification. Whitehurst and Sonnenschein found a developmental trend through the following stages:

(1) incomplete identification — the child fails to mention the critical attribute;
(2) redundant information — the child mentions essential as well as nonessential information;
(3) contrastive — the child mentions only the critical attribute.

What is interesting from this study for a general theory of communication is that the intermediate stage is probably closer to naturally occurring conversation. In this stage children distinguish referents from nonreferents but take the path of "least effort" by including redundant information. Naturally occurring conversation is similarly noted for its redundancy.

Children's ability to assess the adequacy of messages has usually involved tasks in which the child has to assign blame for perceived communication failure (Robinson, 1981). In one such study the child and the experimenter, separated by a screen and each having a set of six identical cards, had to describe a card so that the listener could identify it from the set. The cards depicted a man with a black pointed hat, a man with a top hat, a man with a red flower, a man with a blue flower, a man with a red flag up high and a man with a red flag down. Occasionally the experimenter would provide an inadequate message such as *a man with a flower* and ask the child to explain the communication failure by asking whose fault it was and what should have been said. In each case the reason for the failure derived from the fact that the message did not uniquely identify the object to which the speaker was referring. In other words, the blame for the failure was on the speaker's rather than the listener's side. What Robinson found, however, was that children aged five years tended to blame the listener rather than the speaker for the failure, while by seven years speaker blamers were more common and by eleven years all children located the problem with the speaker. In other studies Robinson found that listener blamers could be trained to assign blame more accurately as a result of explicit feedback on messages which were too general. For example, in one case the adult replied:

> well, there are four like that, I don't know whether it's got long sleeves or short sleeves, and I don't know whether it's got stripes or squares. (pp. 180-1)

Children given such feedback changed from listener to speaker blaming and, in addition, the quality of their own spontaneous messages improved.

Robinson's work on message adequacy addresses the issue of whether children understand the requirements of effective communication. Robinson argues that evidence for this understanding cannot be assumed from the child's performance in spontaneous conversation but must depend on assessment of the child's ability to actually explain communicative failure. This metacommunicative ability forms an integral part of what Markman (1981) has described as *comprehension monitoring*, i.e., the ability to realise when we do not understand. In various studies, Markman found that young children judged instructions to be complete when they contained glaring omissions, and essays to be comprehensible when they contained blatant contradictions. Markman argues that children may feel they understand a text if they can understand its component sentences. However, the comprehension of a text involves the ability to impose a structure on the text in order to see its overall coherence. The following is an example of a text which children did not judge to be inconsistent:

> Fish must have light in order to see. There is absolutely no light at the bottom of the ocean. It is absolutely dark down there. When it is that dark the fish cannot see anything. They cannot even see colours. Some fish that live at the bottom of the sea can see the colour of their food. (p. 62)

As Markman argues, comprehension of text requires the ability to recognize such inconsistencies and to discern relationships between the constituent parts of a text. This involves the formulation of expectations, predictions, and inferences during the processing of the text as well as the evaluation of these in the light of subsequent information. Coping with violation and inconsistencies is a means of recognizing comprehension failure and is a prerequisite for the ability to distinguish between what has been understood and what requires further clarification.

It might be helpful at this point to outline what is involved for speaker and listener in communicative tasks. In the ideal case the speaker encodes the message appropriately, the listener comprehends and communication is successful. Temporary failure occurs either when the speaker encodes the message inappropriately and the listener has to request clarification, or when the listener, for some reason, has a problem interpreting the message and similarly requests clarification. Communication will be successful if the speaker can adjust the message accordingly so that the listener can comprehend it. The dynamics of this process are illustrated in Figure 1.

However, as we have seen, further problems can arise at several points in this process. The listener may not recognize that he has failed to understand and if this is not picked up by the speaker then failure will result, although this may remain unnoticed by both participants. The listener may notice he has failed to understand but not know how to request clarification. Even if clarification is requested, the speaker may not be able to adjust the message accordingly. And finally, even if all this is achieved, the whole process might have to begin again if the listener is

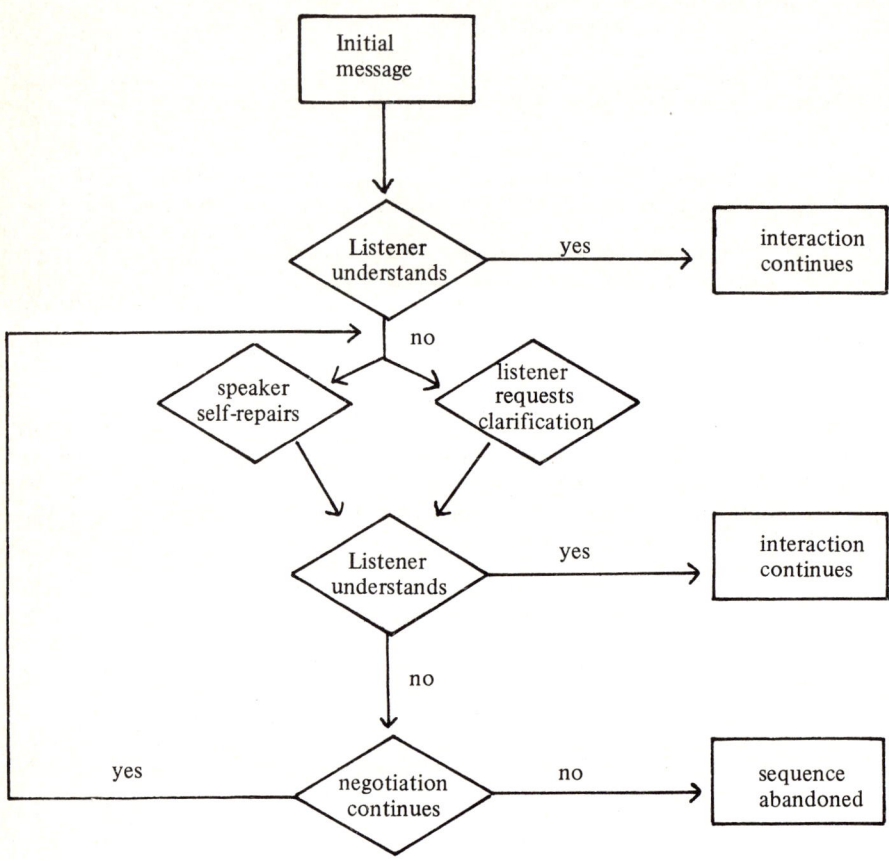

Figure 1. The Negotiation of Communication Failure.

still unable to understand. These processes will now be examined further in the light of sociolinguistic studies of children's communication.

2.2 Sociolinguistic studies

While studies of children's communicative abilities in the experimental paradigm have focussed on their deficiencies, studies in the sociolinguistic tradition have pointed to their remarkable communicative skills, which can be traced back through the prelinguistic period almost to birth. Much of this work is not, however, concerned with communication in terms of the transmission of messages but with aspects such as attention-getting, turn-taking, and the initiation and maintenance of conversational exchanges. Broadly speaking, the emphasis has been on interactional rather than transactional aspects of communication (Brown & Yule, 1983). However, transactional concerns such as the content of discourse cannot be ultimately divorced from interactional concerns and in what follows, emphasis will be mainly on those aspects of sociolinguistic studies which are relevant to the definition of communication failure proposed earlier.

Infants would appear to be biologically predisposed for communication from the moment of birth. Within hours of birth, they display the kind of motor activity, called *interactional synchrony*, which involves subtle movements in synchronization with the rhythm of another's speech. As a result of this, Condon and Sander (1974) have claimed that "the neonate participates immediately and deeply in communication and is not at birth a social isolate." Furthermore, close analysis of infant-caregiver interaction in the prelinguistic period has shown that infants often take the lead, initiating communication with gestures, looks, and vocalisations (Trevarthen, 1979).

It is necessary to sound two cautionary notes at this point. One concerns the extent to which intentionality can be ascribed to these early behaviours of the child. It is necessary, for example, to distinguish between reflex motor actions, such as early signs of smiling within the first few weeks, and behaviours which are intended to communicate and elicit a response. On the basis of detailed analyses of video-taped interactions, it has been claimed that gestural performatives such as reaching and indicating, and precursors of speech acts such as requests and statements, do not develop until around ten months (Bates, Camaioni & Volterra, 1979).

A second point concerns the assymetrical nature of early infant-caregiver interaction. Such interactions have the appearance of mature conversations — there is turn-taking with minimal overlap, and the infant and caregiven appear to produce conversation-like exchanges in which one participant initiates and the other responds. Much of this is due, however, to the caregiver who treats the infant's early behaviours as if they were communicative and effictively makes sense of the child's behaviours. In the following example of mother-child interaction with a three-month-old, the mother responds to the child's smiles, burps, and vocalizations as if they were proper contributions to the conversation and thus builds a conversational structure around these behaviours:

(1) Ann: (smiles)
     Mother: oh what a nice smile
             yes, isn't that nice?
             there
             there's a nice smile
     Ann: (burps)
     Mother: what a nice wind as well
             yes, that's better, isn't it?
             yes
     Ann: (vocalizes)
     Mother: there's a nice noise
                        (from Snow, 1977, p. 12).

Similarly, as the following example illustrates, an adult will help the child define a topic by means of asking questions which progressively provide more content and thus reduce the cognitive and interactional demands on the child, thus compensating for the child's conversational deficiencies and ensuring that the exchange has the appearance of being well formed:

(2) Child: daddy
     Father: what?
     Child: daddy
     Father: what've you done?

>   broken a toy?
>   (1.2)
>   eh?
>   (0.8)
>   little scamp
> Child: (laughs)

(from MacLure, 1981, p. 120).

Evidence such as this indicates the potential problems inherent in making assumptions about a child's abilities on the basis of behavioural evidence, particularly in the case of conversational interaction, where the child's partner can interpret and make sense of behaviours which may not have been endowed with meaning and intention in the first place. A similar point can, of course, also be made about adult communication, as conversational participants are generally predisposed to impose structure and meaning on blatantly irrational behaviour (see Garfinkel, 1967, for some striking examples). The response of many humans to computer programs such as ELIZA is a similar phenomenon (Weizenbaum, 1966).

Bearing these points in mind, we can now look at some cases of naturally occurring communicative breakdown involving children. There are two main sources of evidence, based on adult-child and child-child interaction respectively. In studies of adult-child interaction it has been shown that adults display considerable negotiation and accomodation in an effort to resolve communication failure, particularly in the early stages. Golinkoff (1983) found that children aged eight to twelve months were less likely to repeat their nonverbal signals when faced by indications of failure such as clarification requests, while slightly older children were prepared to alter the form of their original signal as a means of repair. However, the burden for the repair was mainly on the adult at this stage. Gallaher (1981), looking at slightly older children aged between 1;11 and 3;0, found that the children were generally able to respond to clarification requests, especially when these were in the form of requests for confirmation. Interestingly, it has been found that children often treat forms such as *what*? and *pardon*? not as requests to repeat but as opportunities to monitor for a source of trouble, the evidence for this conclusion coming from the fact that the children often reformulate their utterances rather than simply repeat them. The following is a typical example:

> (3)   Child: is Derek a nice boy?
>   Mother: pardon?
>   Child: is Derek a bad boy?

(from Langford, 1981, pp. 165-6).

Indeed, as Stokes (1977) points out, clarification requests may serve an educative function for children by motivating them towards a greater awareness of the form and use of language. In the following example, the child seems to be testing alternative rules for linguistic forms:

> (4)   Child: oh, she ate me
>   somebody else wants to be ates
>   Adult: what?
>   Child: eaten

Golinkoff (1983) proposes a developmental hierarchy for the function of episodes in which communication failure involving children is negotiated. At the earliest stage, the main goal of the adult is to discern the child's meaning, usually in order to respond to the child's goals. Subsequently such episodes have the function of maintaining communicative interaction. Requests for confirmation appear to serve this function primarily. Finally, at a later stage, the adult seems to use these episodes to emphasize the form of the child's utterance and thus to direct the child towards more conventionalized linguistic forms.

As far as children's production of clarification requests is concerned, the earliest forms are non-specific requests for repetition, realized by *hm*? and *what*? These emerge around two years, although some caution is necessary concerning their functional interpretation as children often use these forms indiscriminately with the apparent goal of simply forcing the adult to produce further talk. Perhaps because of this mothers rarely seem to provide exact repetitions for their young children but use the opportunity to elaborate their meaning using analogy or demonstration

(5) Child: dolly
Mother: the dollies ... they're called puppets
Child: hm?
Mother: they're puppets
Child: hm?
Mother: see, you put your hand in ...
Child: hm?
Mother: remember, we've got the owl at home?
Child: hm?
Mother: hello Jane
Child: (laughs)

(from Johnson, 1979, p. 7)

Children are often studied in peer interaction so that the problems which arise in assymetrical discourse with adults can be avoided. In peer interaction children cannot rely on the superior interpretive and structuring skills of an adult but have to depend more crucially on their own resources. In such cases communication failure is more likely and children may be forced to develop ways of dealing with it. In fact, it may be the case that exposure to peer interaction promotes the development of a greater awareness of communication breakdown and repair than is afforded by the more supportive interaction with an adult.

One source of evidence that young children are competent in this respect comes from work by Garvey (1977), who studied forty-eight children aged 2;10 to 5;7 in dyadic peer interaction. Garvey found that in general the children were able to respond appropriately to requests for clarification and were able to request clarification when necessary. Similar findings are reported in a detailed analysis of the form and structure of clarification request sequences in the interactions of two preschool girls (McTear, 1985a).

The distinction between input and model failure is not made explicit in sociolinguistic studies of communication failure, but, on the basis of cited examples, concern seems to be mainly with input problems. In most cases failure results from the child's immature linguistic system, whether as speaker or as listener. The other most usual source of trouble is the child's failure to adequately specify a referent of the talk, as in the following example, where child A refers to a lorry which child B is unable to identify. Child A is able to repair the failure with a more adequate description:

(6) A: I see shells on that lorry
B: what lorry?
A: that one that's blue

A further source of evidence comes from children's corrections of their partner's speech. In the next example, child B corrects her partner's lexical choice:

(7) A: there's some plasticine
B: that's not plasticine
it's Lego

Problems arising from references to objects and persons not present in the immediate situation are probably closest to the notion of model failure, as here the listener is required to search memory in order to identify these referents and if this search is unsuccessful, then failure occurs. The following set of examples shows development in one child from an initial inability to identify a referent adequately to a clear attempt to take account of the listener's knowledge. In the first example, the mother is unable to work out what the child means, partly because of the child's undeveloped phonological system as well as her inability to adequately specify the referent of her talk. In this case success is achieved as a result of the mother's interpretive work:

(8)   (Child aged 2;6)

    Child: daimen
    Mother: what's a daimen?
        (this exchange is repeated several times)
    Child: up steps
                bus in house
    Mother: oh Stephen
    Child: daimen

Here the child is forced to further specify the referent by recalling associated information about the child Stephen. Once the mother identifies the referent, the child brings the exchange to a satisfactory conclusion.

In the next example, the same child, now aged 4;6, refers to her playschool teacher and several clarification requests are required from her conversational partner until the referent is adequately identified:

(9)   Child A: well these are going to be for Emily instead of you then
        Child B: who are they going to be for?
        Child A: at the playschool
        Child B: Emily?
        Child A: yes
        Child B: who's she?
        Child A: the teacher at my playschool

A few months later, this child was able to take account of her partner's possible lack of background knowledge by adequately identifying the referent of her talk in advance:

(10)   (Child aged 4;9)

    Child: and my wee friend Andrea doesn't let me play with her toys

These examples have been cited because they raise several interesting points. The ideal reference would probably be as in example (10), where the speaker takes account of what the listener might not know. However, in naturally occurring conversation such listener assessments can be unsuccessful and referents can either be over- or under-specified. The former is just as problematic, as listeners are intolerant of hearing what they know already (see Grice, 1975, for a discussion of this issue). In the latter case the situation can be resolved with a simple request for clarification by the listener. For very young children much more supportive action is required on the part of an adult, as example (8) illustrates. Nevertheless, the main point is that in naturally occurring conversation failure is usually not irremediable, as communication success is achieved through the collaborative and co-operative work of both conversational participants. This basic insight into the interactional nature of conversation has important implications for computational modelling of conversation, as will become clear in the next section.

To summarize: it can be seen from sociolinguistic studies of children's communication that young children can handle communication failure by requesting clarification and by appropriately responding to such requests. For the most part failure occurs as a result of input problems. Indeed, it is not possible to control for model failure in the study of naturally occurring conversation, as such failures can often pass unnoticed. With both input and model problems it appears to be the case that listeners are prepared to let errors pass provided that they do not unduly obstruct communication. They may also postpone interpretation in the expectation that meanings will become clear as the talk proceeds (Cicourel, 1973). Indeed, overt correction is a dispreferred conversational activity and listeners will more usually formulate corrections as requests for confirmation or allow the speaker the opportunity to self-correct (Schegloff, Jefferson & Sacks, 1977). Much more stringent monitoring is required when the accuracy of the content of the message is at a premium, as in communication between humans and computers, an issue to which we now turn.

3. Implications for Artificial Intelligence

Natural language understanding is one of the major concerns of artificial intelligence. Getting a computer to "understand" the user's unrestricted input in natural language is seen as similar to the problems faced by the listener in natural communication. Research so far has treated the machine and the listener in isolation by focussing on the problems which arise in the comprehension of language. Studies of referential communication, which control for the separate contributions of speaker and listener to the success or failure of communication, are potentially helpful here. In a communication task the listener can only respond by selecting the appropriate referent or by carrying out the desired instructions, if the speaker's message is adequate. One important aspect of this involves comprehension monitoring, that is, the ability to realise when the input has not been understood and to know what remedial steps to take. The usual solution at present is to recognize the potential deficiencies of speakers and for the listener to somehow control what the speaker can say. As far as machine understanding is concerned, in the more extreme cases the speaker has to communicate through the machine's own language. Where natural language is permitted, input is constrained by limiting the discourse topic to one known by the machine. In this way various expectations can be brought to bear as an aid to the interpretation of the input, as in the script-based parsing associated with Schank and colleagues (Schank & Riesbeck, 1981). Another approach is user modelling, in which the characteristics of the potential users of a system can be represented as a means of solving, for example, problems which might arise out of differences between the user's understanding about information in a data base and how this information is actually stored. Current implementations are, however, static and applicable only to a general class of typical users, or are based on strong contextual expectations arising from a limited domain of discourse (Boguraev, 1985).

Future developments might see a more interactional approach to dialogue. Here the ultimate challenge would be unrestricted communication between human and machine or indeed between machine and machine. Communication at this level would require a more dynamic approach in which dialogue would be seen as a collaborative activity involving continual negotiation and interpretation between speaker and listener. Here greater redundancy is permitted than in experimental studies of referential communication, thus aiding the process of communicative repair. As far as breakdown is concerned, the system would have to be able to recognize what type of failure had arisen and to know how to deal with it. Indeed, if systems are to become more human-like, it will be necessary to adopt the perspective of *graceful interaction* in which communication failure is handled by much more sophisticated and less obtrusive techniques than are currently implementable (Hayes & Reddy, 1983). It will be necessary to examine the properties of naturally occurring conversation in order to understand these processes more fully. For example, a program will need to be able to assess, on the basis of the state of play in a dialogue and a dynamic construction of a model of the user, whether potentially minor conversational failures can be tolerated in the expectation that they are only of local importance or that they will be subsequently resolved (see McTear, 1985b, for further discussion of these issues). In other words, some sort of failure tolerance threshold will be necessary in order to achieve a satisfactory trade-off between a minimization of misunderstanding and a maximization of information transfer (Ringle & Bruce, 1982). Sociolinguistic studies of children's communication can illuminate the developing processes of naturally occurring conversation as an international rather than an individual achievement.

Communication failure is a vital area for further research as it is at the heart of our understanding of the whole process of communication. Often a system which normally works successfully can only be understood fully when it breaks down. In order to develop machines which can understand human language, we need first to understand humans. This is an interdisciplinary exercise. As Ringle and Bruce (1982, p. 219) have put it:

In the long run, the effort to understand the nature of human conversation and the aim of building a human-machine dialogue system will be seen not as merely complementary, but as inescapably interdependent.

The developmental perspectives presented here should further highlight these important issues.

References

Bates, E., Camaioni, L., & Volterra, V. (1979). The acquisition of performatives prior to speech. In E. Ochs & B. Schieffelin (Eds.), *Developmental Pragmatics*. New York: Academic Press.

Bobrow, D.G., Kaplan, R.M., Kay, M., Norman, D.A., Thompson, H., & Winograd, T. (1977). GUS: A frame-driven dialog system. *Artificial Intelligence, 8*, 155-173.

Boguraev, B. (1985). User modelling in cooperative natural language front ends. In G.N. Gilbert & C. Heath (Eds.), *Social action and artificial intelligence*. London: Gower.

Brown, G., & Yule, G. (1983). *Discourse analysis*. Cambridge: Cambridge University Press.

Cicourel, A.V. (1973). *Cognitive sociology*. Harmondsworth: Penguin.

Condon, W.S., & Sander, L.W. (1974). Neonate movement is synchronized with adult speech: Interactional participation and language acquisition. *Science, 183*, 99-101.

Dickson, W.P. (Ed.). (1981). *Children's oral communication skills*. New York: Academic Press.

Flavell, J.H. (1974). The development of inferences about others. In T. Mischel (Ed.), *Understanding other persons*. Totowa, NJ: Rowan and Littlefield.

Gallaher, T.M. (1981). Contingent query sequences within adult-child discourse. *Journal of Child Language, 8*, 51-62.

Garfinkel, H. (1967). *Studies in ethnomethodology*. New Jersey: Prentice Hall.

Garvey, C. (1977). The contingent query: A dependent act in conversation. In M. Lewis & L. Rosenblum (Eds.), *The origins of behaviour. Volume V: Interaction, conversation, and the development of language*. New York: Wiley.

Golinkoff, R.M. (1983). The preverbal negotiation of failed messages: Insights into the transition period. In R.M. Golinkoff (Ed.), *The transition from prelinguistic to linguistic communication*. New York: Academic Press.

Grice, H.P. (1975). Logic and conversation. In P. Cole & J.L. Morgan (Eds.), *Syntax and semantics. Volume 3: Speech acts*. New York: Academic Press.

Gumperz, J.J., & Tannen, D. (1979). Individual and social differences in language use. In C. Fillmore, D. Kempler, & W.S. Wang (Eds.), *Individual differences in language ability and language behaviour*. New York: Academic Press.

Hayes, P., & Reddy, D. (1983). Steps toward graceful interaction in spoken and written man-machine communication. *International Journal of Man-Machine Studies, 19*, 213-284.

Johnson, C. (1979). *Contingent queries: The first chapter*. Paper presented at Language and Social Psychology Conference, Bristol, July.

Langford, D. (1981). The clarification request sequence in conversation between mothers and their children. In P. French & M. MacLure (Eds.), *Adult-child conversation*. London: Croom Helm.

MacLure, M. (1981). *Making sense of children's talk*. Unpublished PhD dissertation, University of York.

Markman, E.M. (1981). Comprehension monitoring. In W.P. Dickson (Ed.), *Children's oral communication skills*. New York: Academic Press.

McTear, M.F. (1985a). *Children's conversation*. Oxford: Blackwell.

McTear, M.F. (1985b). Breakdown and repair in naturally occurring conversation and human-computer dialogue. In G.N. Gilbert & C. Heath (Eds.), *Social action and artificial intelligence*. London: Gower.

Piaget, J. (1959). *The language and thought of the child*. London: Routledge and Kegan Paul.

Ringle, M., & Bruce, G. (1982). Conversation failure. In W. Lehnert & M. Ringle (Eds.), *Strategies for natural language processing*. Hillsdale, NJ: Lawrence Erlbaum Associates.

Robinson, E.J. (1981). The child's understanding of inadequate messages and communication failure: A problem of ignorance or egocentrism? In W.P. Dickson (Ed.), *Children's oral communication skills*. New York: Academic Press.

Schank, R., & Riesbeck, C. (1981). *Inside computer understanding*. Hillsdale, NJ: Lawrence Erlbaum Associates.

Schegloff, E.A., Jefferson, G., & Sacks, H. (1977). The preference for self-correction in the organization of repair in conversation. *Language, 53*, 361-382.

Shatz, M., & Gelman, R. (1973). The development of communication skills: Modifications in the speech of young children as a function of listener. *Monographs of the Society for Research in Child Development, 38*, 5.

Shantz, C.U. (1981). The role of role-taking in children's referential communication. In W.P. Dickson (Ed.), *Children's oral communication skills*. New York: Academic Press.

Snow, C. (1977). The development of conversation between mothers and babies. *Journal of Child Language, 4*, 1-22.

Stokes, W.T. (1977). *Motivation and language development: The struggle towards communication*. Paper presented at Biennial meeting of the Society for Research in Child Development, New Orleans, March.

Trevarthen, C. (1979). Communication and co-operation in early infancy: A description of primary intersubjectivity. In M. Bullowa (Ed.), *Before speech: The beginning of interpersonal communication*. Cambridge: Cambridge University Press.

Weizenbaum, J. (1966). ELIZA – a computer program for the study of natural language communication between man and machine. *Communications of the ACM, 9*(1), 36-44.

Whitehurst, G., & Sonnenschein, S. (1978). The development of communication: Attribute variation leads to contrast failure. *Journal of Experimental Child Psychology, 25*, 453-490.

TOWARDS A FRAMEWORK FOR
THE FORMAL TREATMENT OF DIALOGUE

Jerry Harper

Educational Research Centre
St Patrick's College, Dublin

What follows is a prolegomenon to a formal theory of dialogue. The theoretical framework is developed with reference to both the structure of dialogue and the role of speaker presuppositions in miscommunication circumstances. The prerequisites for a formal theory are developed in terms of the notion of discourse situation and situation semantics.

1. Introduction

Devising a formalism to describe dialogue is a notoriously difficult task. Dialogues, in general, are unlike logical systems in that the relation between any utterance and its successor is not uniquely determined by either semantic or contextual parameters. For instance, a variety of perfectly acceptable responses is available to the following question:

    Q. Have you seen the kettle?
    A1. No I haven't.
    A2. Yes.
    A3. Yes, it's on the shelf.
    A4. Use the saucepan.
    A5. Why?

The line of responses could be extended indefinitely. The issue at hand, however, is whether any gradations can be imposed on such a set such that responses would be recognised as being more or less appropriate, more or less relevant. In particular, is there a criterion of "response-informativeness" which is formally representable? There is one assumption which, despite its initial plausibility, is not helpful: a question has a unique answer. If dialogues are broadly construed as co-operative question-answer exchanges, assuming that each question has a unique answer will not validate the natural understanding possessed by speakers of the use and extent of conversation. As is evidenced by the example above, a variety of acceptable answers is possible to any question, in general. However, there are questions to which only correctly unique answers are permissible. Consider the following:

    Q. What does 2+2 equal?
    A1. 4.
    A2. Nothing at all.
    A3. 7.

Quite obviously A2 and A3 are not appropriate answers to the posed question unlike the cases cited earlier, the former because it rejects the question and the latter because it is false. For the purpose of this exposition however, we will consider the set of questions with uniquely appropriate answers to be a minor subset of the set of all possible questions.

It is proposed to give an outline of a possible formal theory of dialogue. Accordingly, a discussion of some general characteristics of dialogue is required, followed by a list of elementary dialogue rules.

2. Some General Characteristics of Dialogue

A dialogue in the broadest sense is a directed series of moves, it has a linear quality, performed by two participants. Dialogues are unlike *conversations* in that the latter allow for a greater number of participants. Dialogues are, therefore, subsets of a set of possible conversations. The moves performed are linguistic and most commonly encompass the acts of uttering assertions, making retractions, asking questions, etc. The aim of a formal theory of dialogue is to specify a representation of such moves into a formal language as distinct from specifying a quasi-formal translation of the contents of the moves, i.e., the utterances themselves – though obviously there persists a certain degree of mutual dependence between the two tasks. By 'quasi-formal' we mean a methodological combination of certain natural linguistic phenomena and formal constructs. For instance, in a formal theory a linguistic production is represented via a pure formal translation scheme. Such schema function as sets of rewriting rules totally defined for the word alphabet under consideration, where the word alphabet defines the lexicon of the language. A quasi-formal theory, on the other hand, possesses a set of rewriting rules which is not totally defined for the word alphabet. Some translations will be supplemented by portions of the source language, as is the case with Situation Semantics.

In order to develop a theory of dialogue, the basic characteristics of a dialogue must be outlined since these will function as primitives in the theory. Borrowing from Carlson (1983), we can identify six basic features possessed by any dialogue which may be considered as the basic parameters defining a dialogue exchange:

1. The *author* of a move.
2. The *addressee* or audience of a move.
3. The *sentence* (utterance) of the move.
4. The *dialogue rules* which justify the move.
5. The *premises* (presuppositions) of the move.
6. The *dialogue(s)* the move is in.

The above six features, or dialogue constructs, can be grouped into three different sets (the reason for doing this will be evident when we reach a formal definition of dialogue) as follows:

(a) a set of two participants or agents who conduct the dialogue (1 and 2 above);

(b) a set of rules which govern the moves made in the dialogue, i.e. regulate the exchange between participants. The rules, however, can be flexibly applied in the sense that particular acts do not canonically demand the observance of rules (4 and 6 above); and

(c) a set of utterances or propositions which make up the informative core of the dialogue. The set of such utterances represents the varied commitments participants have made during the course of the dialogue (3 and 5 above).

From the perspective of communication failure, Carlson's condition five above is the most interesting. For instance, where a speaker has certain presuppositions concerning the knowledge base of his audience he will commit himself to certain dialogues. As is witnessed in daily conversation, where such presuppositions are false inappropriate moves result. Ferrari (this volume) gives an example wherein a tourist approaches a policeman in a railway station asking for time-table information. The tourist assumes that the policeman is a railway employee and initiates the dialogue move in accordance with that assumption. More important, perhaps, in the example cited, is the addressee's ability to recognise the invalidity of the premises of the move and thereby correct the tourist's presuppositions.

Taking a more formal approach, we will tentatively define a dialogue to be composed of the following sets:

(a) A two-element set of individuals, the dialogue participants.
(b) A set of expressions used in the dialogue.

(c) An ordered triple composed of a participant, the propositional attitudes of the participant as they govern the move made by an expression, and the expression itself.

Thus, formally: D is a dialogue iff,

There is a language L, a two-element set I of participants such that $I = \langle i_1, i_2 \rangle$; a set A of propositional attitudes such that $A = \langle a_1, ... a_n \rangle$ for $1 \leq j \leq n$; a set P of expressions such that $P = \langle p_1, ..., p_n \rangle$ for $1 \leq f \leq n$; and a natural number n such that $D = (d_1, ..., d_n)$ and for each i, $1 \leq i \leq n$, $d_i = (G_i, H_i)$, where G_i and H_i are triples of the form $\langle I_{1/2}, a_j, p_f \rangle$.

Note that each d_i states the dialogue exchange between participants in terms of the participants, their propositional attitudes and the expressions used. Thus, the definition embraces both Carlson's primitives and their set-theoretical grouping.

The above definition, however, lacks a certain precision in terms of a description of the language L. As it stands, it appears that a formal theory constructed on the above would operate in tandem with natural language expressions. To overcome any possible difficulties on this point another set should be added to the above, viz. a set of translation schemata. The function of such schemata would be the representation of the natural language in formal, or at least quasi-formal, terms. One suggestion is that a set of triples be expanded into a set of quadruples wherein a set K is contained giving a formal representation of the expression used. Choosing the correct formalism in such cases is a non-trivial task, though in the light of modern research some Montague-like system is, perhaps, most useful. Thus each d_i is of the form $\langle\langle I_1, a_j, k(p_f), p_f \rangle, \langle I_2, a_{j'}, k(p_{f'}), p_{f'} \rangle\rangle$ — one could of course remove p altogether once the formal translation has been accomplished.

Our next task is to give a list of uncontentious dialogue rules which may be considered as the basic primitive rules governing any dialogue (Carlson, 1983):

(D. say) Any participant may put forward any sentence in any order.

(D. ask) Any participant may ask a question of the other.

(D. question) Any participant may question any presupposition of the other participant's utterance.

(D. answer) When a participant has put forward a question an addressee may answer it.

(D. infer) When a participant has put forward a statement, he may infer another from it.

(D. explain) When a participant has made a statement, he may put forward an explanation for it.

(D. argue) When a participant has put forward a statement, he may argue for it using another statement (or set thereof) as evidence for it.

(D. reply) When a participant has put forward an answer to a question any participant may put forward another answer in reply.

The application of the above rules is governed by the following conditions:

(C. cons) Do not admit a sentence and its denial. In other words be consistent.

(C. self) Do not admit any self-contradictory sentences.

(C. consq) Do not accept an answer to a question you deny. Accept a consequence only if you accept the premises.

No attempt is being made to define a discourse grammar on the basis of the above rules. Consequently, the rules are not intended to demonstrate that in a particular context a sentence is either appropriate or relevant. Perhaps, the bane of formal theories of dialogue is precisely the assumption that 'relevance' is formalizable. No move in a dialogue logically determines its successor. A common attempt to force the issue and impose a 'logical' structure on exchange pairs is found in the following schema criticised by Carlson (1983),

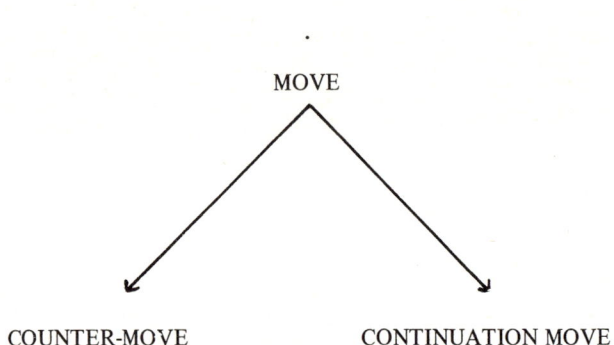

Carlson argues that such structural descriptions of dialogue are too restrictive in that they fail to account for subtle qualifications introduced by speakers. The use of "but" to qualify antecedent premises is a moot example. Consider the following:

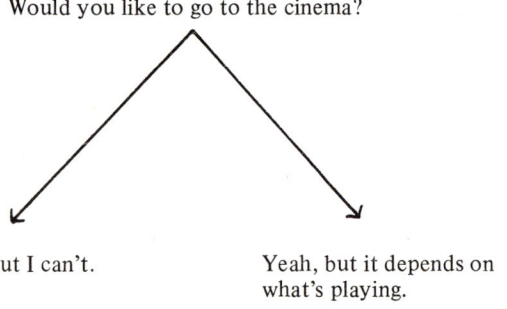

The counter-move does not contain an explicit denial of any of the question's presuppositions, likewise the continuation move is not fully endorsing the preceding question.

Lehnert (1984) gives further reasons why no fine grained logical calculus of dialogue can be produced by virtue of the multiplicity of combinations of exchange-pairs of moves possible. This was aptly demonstrated at the beginning of the chapter where a question move was succeeded by a variety of other moves ranging from rejection of the question's presupposition to a fully appropriate response. As a consequence of the above arguments, this chapter chooses not to attempt a formal theory of dialogue in terms of a formal language, but to opt for a quasi-formal approach based on two constructs:

(a) the notion of "discourse situation" as that which contains the dialogue exchange plus the accompanying contextual, pragmatic and linguistic presuppositions of the participants in the dialogue; and

(b) The notion of "situation semantics" as expounded in Barwise and Perry (1983).

3. Discourse Situations

The phrase "discourse situation" refers to the social context within which the dialogue is situated. Our concern is with the total contextual framework of a dialogue. The motivating principle is that situations contain information which either facilitates or militates against communicative efficiency. For instance, where a situation serves to disabuse a speaker of possible false presuppositions concerning the role of the addressee it is facilitating successful communication. The most difficult task in elucidating the notion of a discourse situation is that of giving a precise account of the role of context, particularly its disambiguating functions. Rather than approach this issue directly, we will give an account of the three fundamental properties inherent in every discourse situation.

1. Contextual presuppositions:
   These are presuppositions which both the speaker and addressee share to a greater or lesser degree concerning the physical locality of the discourse. Such presuppositions are of a spatio-temporal nature, e.g., where they are, when they met, etc. We may refer to such presuppositions as those which cover the participants' basic, or natural, world knowledge.

2. Pragmatic presuppositions:
   Such presuppositions are those regarding non-natural world knowledge, such as knowledge of social conventions, e.g., one does not spit in a policeman's face when he questions one. Also included among these presuppositions is knowledge concerning institutional roles and conventional linguistic practices. This type of knowledge provides, for example, the discrimination between what one tells one's spouse and says to one's lover.

3. Linguistic presuppositions:
   Finally, this set of presuppositions encapsulates participants' knowledge of their language, both its syntactic and semantic properties. This enables speakers of the language to decide whether a sentence is well formed, or whether it is meaningless.

Of the three types of communication failure possible in a discourse situation (syntactic, semantic and pragmatic), it is pragmatic failure that occurs most commonly in a linguistically competent community. Before detailing the nature of the failure involved in these cases, it is as well to mark the type of failure which can occur where dialogue participants do not share the same set of contextual presuppositions. This phenomenon, though rare in ordinary dialogue exchanges, is often commented upon in psychiatric reports. The therapist is confronted with a patient who is severely psychologically dislocated and who may deny the validity of a whole series of contextual presuppositions shared by the community at large. Consequently, dialogues concerning the common objects of experience may frequently breakdown. Where such communication failure occurs, repairing it is a far from simple matter. Further comment on the issues and techniques involved is outside the scope of our competence, however.

At a more mundane level the other types of failure are more readily corrected by the average competent speaker. Syntactic failure arises where a participant produces an ungrammatical sentence which the addressee cannot parse in terms of any legitimate grammatical structure. In computational terms the problem amounts to that of making sense out of ill-formed input. Detecting syntactic ill-formedness is a relatively straightforward task for most speakers in the linguistic community. However, correcting such anomalies relies on an appeal both to matters of grammar and matters of meaning. Quite crucial in many such cases is the addressee's capacity to infer the speaker's intended meaning, and only then correcting the speaker's ill-formed productions. An obvious example of this process at work occurs when one attempts a series of dialogue exchanges with a tourist whose knowledge of one's own language is slight.

Semantically induced communication failure occurs where a participant fails to recognise the meaning of the speaker's utterance. This can be brought about in many ways. Firstly, there is the case where the audible sounds produced by the speaker do not accord with recognised phonetic patterns in the linguistic community. Consequently, since the addressee is unable to recognise any legitimate phonetic pattern in the utterance he must consider the utterance to have no identifiable *public* meaning. Accordingly, he must also regard the production as ungrammatical. The speaker is thus unable to have himself understood, as his production is strictly meaningless. There are, however, exceptions to this rule. In particular the speaker may be disabled in some way, but the addressee will infer the speaker's intended meaning on the basis of his pragmatic knowledge. Secondly, a speaker may use legitimate lexical items but grammatically arrange such items so as to render the meaning of the resulting production obscure to the point of inscrutability. Classic examples are the following:

Quadruplicity drank procrastination.

Colourless green ideas sleep furiously.

Once again certain grammatical and, primarily, semantic rules have been violated. Finally, there is the case where the speaker produces a perfectly well-formed meaningful expression, but the addressee fails to infer the intended meaning. Strictly speaking this type of failure is due more to one participant's lack of familiarity with conventions of language usage, rather than to an explicitly semantic factor. As such it should be considered as a type of pragmatic failure, but it is convenient to treat it here. An example of this type of failure occurs frequently where innuendo is used in dialogue, or where a question is seemingly posed when in fact a statement is being made. The addressee processes the literal meaning of the sentence but fails to recognise the intended meaning of the speaker. The intended meaning is commonly referred to as the *illocutionary force* of the utterance. Usually in dialogue, speakers rely upon intonation to deliver their intended meaning. Take the following example: suppose the government has decided to raise taxes and this is a very unpopular measure with the participants in our dialogue. Thus

Wasn't it very considerate of the government to raise taxes?

should not be understood literally. The speaker is in fact expressing dissatisfaction with the government's policies. Where the addressee fails to recognize the intended meaning, communication failure is likely to occur. In the cases cited above, repairing the breakdown is only possible where the addressee interrogates the speaker as to precisely what he means. This is particularly the case where the addressee is unaware of the meaning of a certain word or phrase used by the speaker. The problematic phrase must be explained to the addressee by the speaker, as the former is unaware of its lexical content.

Syntactically and semantically induced failures can generally be brought under the heading of linguistic presupposition failure: other types of failure are of a more pragmatic nature. Thus, in the case cited earlier of the tourist and the policeman, while both share the same set of contextual presuppositions concerning the nature of physical reality, both do not share the same set of pragmatic presuppositions. However, we assume they share the same set of linguistic presuppositions. Invoking our earlier definition of a dialogue we may say that the set of propositional attitudes of the speaker concerning the discourse situation are not shared by the addressee.

Consider the following dialogue between the two participants:

(1) I_1 When is the next train leaving?
(2) I_2 I don't know.
(3) I_1^1 Couldn't you find out?
(4) I_2^2 No, I'm not a railway guard. I'm a policeman.
(5) I_1^2 Oh, I'm sorry. Can you direct me to one?

Invoking our earlier definition of dialogue we have I as the two element set of participants, A as the set of propositional attitudes of the participants, P the set of their utterances, and K is the formal translation of the members of P (hypothetical in all the following cases):

$I_1\ ; a_1$ = believes I_2 is a railway guard; $k_1\ ; p_1$ = (1).

$I_2\ ; a_2$ = believes I_1 is a tourist; $k_2\ ; p_2$ = (2).

$I_1\ ; a_3$ = believes I_2 is an unhelpful guard; $k_3\ ; p_3$ = (3).

$I_2\ ; a_4$ = believes I_1 is unaware of his status; $k_4\ ; p_4$ = (4).

$I_1\ ; a_5$ = believes he has false presupposition concerning $I_2\ ; k_5\ ; p_5$ = (5).

The above example is a clear account of how communication failure, based on false pragmatic presuppositions, is both detected, in this case by the addressee, and repaired, also by the addressee. Note also that once the failure has been repaired other dialogue moves are permissible for the continuation of the dialogue. But if I_2 had decided to be uncooperative after (3) then the dialogue would have collapsed with no further exchanges possible. It is to be noted that in many cases of communication failure the addressee determines the best method of repair by attempting to infer the speaker's plan, or by focusing on what the addressee supposes are the speaker's dialogue intentions (Grosz, 1981).

In the next section we examine certain quasi-formal tools which may enable us to get a better purchase on dialogue structures.

4. Situation Semantics

The original motivation behind situation semantics was the provision of a theory of meaning based on certain naturalistic assumptions about the principles of language usage and comprehension. Meaning, it is argued, is a product of man's interaction with his environment, consequently it is not to be found "in the head" but in the world. Therefore, particularly close attention must be paid to actual situations where linguistic communication occurs if meaning is to be elucidated. However, many of its key concepts are not fully developed. But it is not our intention here to assess the merits, or demerits, of situation semantics as a theory of linguistic meaning. Rather, we are interested in borrowing certain tools and constructs from the theory which will assist our task of fleshing out of the structure of discourse situations.

Consider a discourse situation DU, a location l, a speaker a, an addressee b, and an utterance u. In situation semantics a addressing b is represented as follows:

DU: = at $\dot{l}$; speaking, $\dot{a}$; yes
   addressing, $\dot{a}, \dot{b}$; yes
   saying, $\dot{a}, \dot{u}$; yes

When dots appear above characters the characters are said to represent "indeterminates". These are, perhaps, best considered intuitively as variables, i.e., 'slots' yet to be filled by determinate objects.

The DU is considered to possess the following four linguistic roles:

speaker = $<\dot{a}, DU>$
addressee = $<\dot{b}, DU>$
discourse-location = $<\dot{l}, DU>$
expression = $<\dot{u}, DU>$

Roles are uniformities, or similarities, which persist across classes of events (situations) and indeterminates.

Situation semantics recognizes the following objects in its universe of discourse: individuals, properties, relations and locations. The basic principle of the theory is that a situation is composed from a course of events and a location. Meaning is held to be a relation obtaining between utterances and described situations, i.e., between utterances and events. At its simplest, where u = utterance and e = event, the representation of "John is tall", said by me, is as follows:

u [John is tall] e
iff
there is a location l and an individual a, such that
in u: at l; speaks, a; yes
in e: at l; says, a, u; yes
        John, b; yes
        is-tall, b; yes

A course of events which has no indeterminates is said to be an "event type", written as E. Each course of events is a triple composed of individual indeterminates, relational indeterminates and location indeterminates. Where all indeterminates receive a determinate value in a course of events the latter is an event type provided there is some function which "anchors" the indeterminates. The process of "anchoring" indeterminates in situation semantics is analogous to the process by which an assignment function assigns a value to a variable in standard model-theoretic semantics.

One use of the above is in dealing with pronoun ambiguity. Consider "John moved his car":

in E: = at l; moves, a, b; yes
        owns, a, b; yes
        car, b; yes

But if "his" refers to other than that which is owned by John,

in E: = at l; moves, a, b; yes
        owns, a, b; no
        owns, c, b; yes
        car, b; yes

The above examples are given not as arguments for the expressive power of situation semantics, but to demonstrate the format of the analyses it entertains.

Consider "John moved his car" represented in standard predicate logic form:

$$\exists x \ \exists y \ [(\text{John}(x) \wedge \text{car}(y) \wedge \text{owns}(x,y)) \rightarrow \text{moves}(x,y)]$$

or

$$\exists x \ \exists z \ \exists y \ [(\text{John}(x) \wedge \text{car}(y) \wedge \text{owns}(z,y)) \rightarrow \text{moves}(x,y)]$$

In either case there is a certain counter-intuitive quality about the 'if-then' representations. Situation semantics, though less formal in its representation of the sentence, has a more intuitive appeal as the epistemic "blocks" which compose the sentence are clearly listed. However, this is not a particularly effective argument for situation semantics for it is precisely at the level of fine-grained analysis that the semantic complexity of a sentence is unravelled. Thus, it is our opinion that whereas a system such as Montague Grammar (Dowty, Wall, & Peters, 1981) is capable of giving the necessary precise analysis of each sentence in a dialogue, situation semantics can but give a fairly low-level analysis of the same sentences. Nevertheless, we hope to show in what follows that situation semantics falls quite naturally to the task of representing dialogue as opposed to sentence structures, a basic representation of which would be:

at $\dot{l}$; $\dot{a}$ and $\dot{b}$ share a common language
    $\dot{a}$ has/needs information concerning $\dot{x}$
    $\dot{a}$ asserts/queries $\dot{x}$
    $\dot{b}$ hears $\dot{a}$ assert/query $\dot{x}$
    $\dot{b}$ receives/gives the information that $\dot{x}$

Consider the sample dialogue mentioned earlier between the tourist and the policeman:

in E: = at l; speaks, a, b; yes   "When is the next train leaving?"
; questions, a, b; yes
; utters, a, $u1$; yes

in E: = at l; speaks, b, a; yes   "I don't know".
; replies b, a; yes
; utters, b, $u2$; yes

Though we have specified the nature and content of the dialogue, we have not interjected any account of the propositional attitudes of the participants as our original definition of dialogue requires. The following elaborates this point:

in E: = at l; speaks, a, b; yes   "When is the next train leaving?"
; questions, a, b; yes
; believes-is-railway-guard, a, b; yes
; is-railway-guard, b; no
; utters, a, $u1$; yes

in E: = at l; speaks, b, a; yes   "I don't know".
; replies, b, a; yes
; believes-is-tourist, b, a; yes
; is-tourist, a; yes
; utters, b, $u2$; yes

in E: = at l; speaks, a, b; yes   "Couldn't you find out?"
; questions, a, b; yes
; believes-is-indolent, a, b; yes
; is-indolent, b; no
; ...

in E: = at l; ...   "No, I'm not a railway guard.
; ...   I'm a policeman."
; believes-unaware-of-addressee's-status, b, a; yes
; is-unaware-of-addressee's-status, a; yes
; ...

in E: = at l; ...   "Oh, I'm sorry. Can you direct
; ...   me to one?"
; believes-not-a-railway-guard, a, b; yes
; believes-is-a-policeman, a, b; yes
; ...

The virtue of this approach lies not alone in its representation of dialogue structure but in the ready identification of speaker presuppositions. Situation semantics so used can explicate the role of extra-linguistic pointers which, together with semantic information, allow participants to both detect and repair dialogue failure.

5. Conclusion

In conclusion three main points have been made in this chapter. Firstly, dialogue theory cannot approach the sophistication and rigour of formal calculi. The best that can be achieved is a quasi-formal theory.

Secondly, the notion of relevance or appropriateness is, unlike entailment, a quality of efficient communication which formal theories can have only the most tentative purchase upon. The problem of whether a particular utterance is appropriate is best settled by appeal to the total discourse situation and the goals of the participants.

Finally, we have argued that the quasi-formal system of situation semantics sketched here is adequate for describing dialogue moves and the presuppositions of the participants. In paritcular, the theory describes communication failure and the techniques used for repairing the dialogue miscommunication. At this point there seems to be no objection in principle to representing the utterances in some appropriate formalism (though illocutionary force has a problematical status in this regard due to the difficulties involved in quantifying intonational effect), such as Montague Grammar, and thereby bolstering the formal nature of the theory. The ultimate test of the tractability of the combination of the context handling facilities (planning mechanisms) of situation semantics and the formal translation scheme of Montague Grammar, lies in the successful computational implementation of the two approaches in a natural language processing system which allows both the detection and repair of communication failure in dialogue.

References

Barwise, J., & Perry, J. (1983). *Situations and attitudes.* Cambridge, Mass: MIT Press.
Barwise, J., & Perry, J. (1984). *Shifting situations and shaken attitudes.* Report no. CSLI-84-13: Stanford University.
Carlson, L. (1983). *Dialogue games.* Dordrecht: D. Reidel.
Dowty, D., Wall, R., & Peters, S. (1981). *Introduction to Montague semantics.* Dordrecht: D. Reidel.
Ferrari, G., & Prodonoff, I. (1986). Discourse situation misunderstandings. This volume.
Grosz, B. J. (1981). Focusing and description in natural language dialogues. In A. Joshi, B. Webber, & I. Sag (Eds.), *Elements of discourse understanding.* Cambridge: CUP.
Lehnert, W. (1984). Problems in question answering. In L. Vaina, & J. Hintikka (Eds.), *Cognitive constraints on communication.* Dordrecht: D. Reidel.

CONCEPTUAL PRIMITIVES AND THEIR
METAPHORICAL RELATIONSHIPS

A.W.S. Cater

Department of Computer Science
University College Dublin

Metaphor is one of the trickier phenomena that a natural language understanding program must handle. This chapter first surveys some previous work on metaphor comprehension, from the perspectives of psychology, linguistics, and artificial intelligence. The primitive actions of Conceptual Dependency are described, and their implicit relationships are examined. From them are abstracted three very general primitive actions. Specific meaning is conveyed by using these general actions in conjunction with particular states. The merit of these general actions is that sense extension can be partially accounted for by mutation of conceptual representations. This would allow great simplification of the lexicon for a program, because common words would no longer need to be given numerous senses. Finally, the paper describes in outline how, and why, mutations of the conceptual representations might be carried out.

1. Introduction

Natural use of natural language is rich in metaphor. This point has been forcefully demonstrated by Lakoff and Johnson (1980), who list many examples of metaphors serving to structure the way we conceive of the world. Their examples are what may be called "large metaphors", in the sense that a basic analogy underlies a whole family of related metaphorical statements which are in common currency. While these metaphors are in a sense productive, and hence not "dead metaphors", they are widely known and accepted. Small metaphors (Waltz, 1982), in contrast, are typically novel to both the speaker and the hearer.

Numerous proposals from linguistics and psychology have been made concerning human processing of metaphor (Rumelhart, 1979; Paivio, 1979; Lyons, 1977; Miller, 1979; Levinson, 1983), and these will be discussed in the next section. Several proposals have also been advanced (Carbonell, 1982; Hobbs, 1977, 1979; Indurkhya, 1985; Wilks, 1977b) concerning the comprehension of metaphorical utterances by natural-language processing programs. These will be discussed in Section 3. In Section 4, I propose the thesis that the phenomenon of "sense extension" in "small metaphors" can be partially accounted for in terms of semantic or conceptual primitives. The "primitive acts" of Schank (1975) will be reduced to only three, yielding a predominantly state-based representation language for use in natural language processing systems. Finally, shortcomings of this approach will be examined, and suggestions made for further work.

2. Metaphor Comprehension

Figurative language, and indeed language in general, has been a topic of interest in many disciplines over many years: Philosophy, Linguistics, Psychology, and recently Artificial Intelligence, have all contributed to its study. Over the last few decades, the aura of disrepute which shadowed figurative language has lifted (Ortony, 1979; Sacks, 1979), to the point where the view has been expressed that literal and figurative uses of language are not essentially different Rumelhart, 1979; Paivio, 1979; Miller, 1979). Striking examples of how pervasive is the use of metaphor are given by Lakoff and Johnson (1980).

Several arguments have been advanced to explain why metaphor should be used at all. Most would agree that there are several reasons. One, that in many domains there are useful concepts for which there is no literal term. Abstract domains in particular often have a property of inexpressibility. An example is the functioning of the human mind: we have the concept of attention, but what we do with attention is 'pay', 'lose', and 'switch' it. We have no literal terms for these concepts, so we must make do with terms applied metaphorically. A second reason is commonly agreed to be that metaphorical utterances correspond to analogies, and that the ease with which they are produced and comprehended is merely a reflection of the centrality of analogy in human cognitive processes. Third, metaphors are economical. Even where literal speech would suffice, a short metaphorical remark can convey ideas that would take several sentences to convey by literal language alone. A fourth interesting hypothesis, which may not have such wide support, is that a metaphor prompts the hearer to deepen his inferences (Beaugrande, 1980). For a novel metaphor, the hearer is forced to make more inferences in order to establish continuity: and with a well-chosen metaphor, the speaker can guide the inference processes of the hearer along just the lines he desires.

Metaphor is but one of the figures of language. For many purposes, however, it is convenient to conflate the classic categories of figurative speech. For instance, metonymy and synechdoche are conflated by Lakoff and Johnson (1980); metaphor and metonymy are conflated by Lyons (1977); and Fraser (1979) points out that to distinguish metaphor from irony requires, in the general case, full access to the conversational context.

In one view, comprehension of figurative language can be conceived as proceeding in stages: recognition, classification, reconstruction, and interpretation. The first stage involves recognising that an utterance is figurative, that is, is not intended literally. It is variously suggested that semantic anomaly or pragmatic infelicity will serve for this purpose. Classification of the utterance as one or other of the figures is then required, because the appropriate strategy for reconstruction of intended meaning differs. Reconstruction then occurs, in which the literal meaning is transformed to a new, presumably felicitous, meaning; and finally this transformed meaning is comprehended as though it were the content of a literal utterance.

The contrasting view, that figurative and literal uses of language are not fundamentally different, has also been proposed (Rumelhart, 1979; Miller, 1979). I shall first discuss the serial-process view, both because of its naive appeal and because it readily provides names for aspects of comprehension of metaphors. The second "homogeneous" view will be treated later.

2.1 The Serial-process View

The basic assumption underlying the serial-process view of metaphor comprehension is that figurative language in general is deviant: that is, that meanings inhere in literal utterances, but that figurative language cannot have direct meaning. Therefore, if a figurative utterance is to be credited with meaning, it must first be recast in literal terms. To achieve this, the figure first must be recognised as such, and then manipulated in such a way as to render it literally acceptable.

2.1.1 Recognition of Figurality

Two broad classes of mechanism, semantic anomaly and pragmatic infelicity, have been suggested for recognition that an utterance is intended figuratively. According to the semantic anomaly theory, the violation of a selectional restriction results in a nonsensical utterance. This may arise in two ways: a predicate may be applied to a semantically inappropriate term, or two terms of different semantic character may be equated. For example:

(1) Inflation robbed me of my savings.
(2) John is a rock.

When an utterance is semantically anomalous, it cannot be interpreted literally and so must be figurative. Unfortunately, utterances intended figuratively do not always exhibit semantic anomaly: they can make perfect sense as they stand, and yet be intended to mean something different. An example given by Levinson (1983, p. 151) is:

(3) Your defence is an impregnable castle.

which, in the context of a chess game, may be meant either literally or figuratively, or even both simultaneously.

Pragmatic considerations can account for recognition of some further cases of figurative intent. In particular, noticing a violation of the "Maxim of Quality" or "Maxim of Relevance" strongly suggests that a figurative meaning should be inferred. That is, if an utterance is clearly false, or if it wantonly introduces a new topic, then it was not meant literally. Levinson (1983, p. 153) gives the following example of irrelevance:

(4) Q: What kind of mood did you find the boss in?
    A: The lion roared.

However, the distinction between an irrelevant utterance and a genuine attempt to change topic is difficult to formalise. Even if this were done, the problem of recognising (3) as potentially figurative remains unresolved, since it can obey both maxims. Hence, there appears to be no set of criteria which can reliably indicate whether or not an utterance is intended figuratively. This is not unduly surprising, since examples like (3) may or may not be meant to be taken figuratively.

2.1.2 Classification of Figurative Utterances

If and when figurative intent is recognised, the figurative meaning must be extracted from the literal meaning. Because the relationship between these two depends on the trope used, classification of the utterance is required. For example, in irony the meanings are typically opposite; in metonymy one entity must be replaced by some related entity. The clues which led to recognition can often assist also in classification, but extra clues are available. Irony, for example, is usually delivered with a characteristic intonation. The reliability of these clues has not, to my knowledge, been convincingly demonstrated. Nevertheless, in what follows the success of classification will be assumed.

2.1.3 Reconstruction of the Meaning of a Metaphor

Reconstruction of the meaning of a metaphor has been widely discussed. There are many schools of thought, of which two are especially prominent: these are the "comparison theory" and the "interaction theory". Levinson (1983, p. 148) characterises them as follows:

The comparison theory
Metaphors are similes with suppressed or deleted predications of similarity.

The interaction theory
Metaphors are special uses of linguistic expressions where one 'metaphorical' expression (or *focus*) is embedded in another 'literal' expression (or *frame*), such that the meaning of the focus interacts with and *changes* the meaning of the *frame,* and vice versa.
(Emphasis in original).

Notable among the "comparison theory" school is the work of Miller (1979), which proposes rules for rewriting various classes of metaphor as analogies involving (usually) four terms. For this purpose, metaphors are classed as one of: predicative metaphor, nominal metaphor, sentential metaphor. Examples of each class have been given already. Miller calls (1) a predicative metaphor, (2) a nominal metaphor, and (4) a sentential metaphor. For each class, Miller gives rules for reconstructing an explicit analogy. He maintains that the reader of a text is obliged to interpret the text as true for its author; and that this drives the reader to give the text an interpretation which could be true; and that an explicit restatement in terms of similarity fulfils this need.

Theories of the interaction type are most easily stated in terms of manipulations of semantic features. Cohen (1979) contrasts two methods of handling metaphor, which he calls "cancellation" and "multiplication". The multiplication method involves providing multiple lexical entries for each word, such that different entries for the same lexeme would bear different combinations of features. As a general method, this approach would clearly result in unwelcome growth of the lexicon, with the concomitant need to select between all senses on all occasions of use of a word. The "cancellation" method avoids this growth of the lexicon: features attached to words or phrases may be cancelled by other parts of an utterance. Cohen implicates the "topic-comment" distinction to characterise the direction in which this cancellation operates: in metaphorical utterances features of the topic may eliminate features from the comment. Thus in (2), features of "John", the topic, cancel features from the comment, "is a rock". In literal usages, cancellation may also operate, but in the reverse direction. Similarly, features of a head noun may cancel features of a modifier in metaphors, but in literal use the direction of cancellation is again reversed.

Reconstruction of intended meaning by the cancellation method requires that some subset of semantic features be selected for cancellation. Cohen (1979) suggests that features be ranked by semantic importance, where importance is inversely correlated with prior probability of occurrence. Thus, a general or probable feature like +ANIMATE is not important, and may be cancelled in even a literal sentence; while improbable and hence highly informative features like +METALLIC may be cancelled, but not with such ease. Cohen seems then to suggest that an utterance will appear metaphorical if important features must be cancelled. However, Cohen explicitly denies that the importance of semantic features needs to be uniform throughout the lexicon (Cohen, 1979, p. 75).

An alternative approach to reconstruction, still within the interaction class of theories, is hinted at by Paivio (1979). Paivio states that long-term "semantic" memory is implicated in comprehension, and that the comprehension of metaphor in particular is crucially dependent on the organisational structure of semantic memory: this structure determines which attributes of the constituent parts of a metaphor mediate the metaphorical relationship between them. Paivio further suggests that episodic memory is involved: the metaphorical utterances, the linguistic context, and extralinguistic factors, may all provide information useful for guiding the search through semantic memory.

2.1.4 Interpretation of a Metaphor

In the serial-process model, the processes of recognition, classification, and reconstruction ultimately provide a "literalised" utterance which can be interpreted in just the same way as a "normal" literal utterance. This is not to say that interpretation is then straightforward, only that the process of interpretation needs no new special machinery for metaphorical language.

2.2 The Homogeneous Model

The serial-process model sketched above is founded on the assumption that metaphor, and figurative language in general, is somehow deviant. That assumption is rejected in the homogeneous model: figurative language is accepted as normal. Therefore, any account of semantics or pragmatics which requires additional processes specifically to handle metaphor is severely deficient.

The arguments for the normalcy of metaphorical usage are supported by the demonstration (Lakoff & Johnson, 1980) that much of everyday language is metaphorical. For if, as Lakoff and Johnson suggest, the way we conceive of abstract concepts is shaped after our understanding of relatively concrete ones, it is counter-intuitive that linguistic use of conceptual metaphors should require additional processing.

Further, if literal and metaphorical usages are comprehended by substantially different mechanisms, it should be easy to distinguish instances of each class. Yet, as Levinson (1983, p. 150) asks, where in the following sequence is the first metaphorical sentence?

> John came hurriedly down the stairs.
> John ran down the stairs.
> John rushed down the stairs.
> John hustled down the stairs.
> John shot down the stairs.
> John whistled down the stairs.

Third, recasting metaphors in literal form does not seem to capture the same meaning as the original conveyed. Quite aside from questions of rhetorical force, literal restatements are lame in that they do not readily provoke the same inferences: metaphors seem to provide both new vocabulary within a domain, and to suggest appropriate inferences by analogy with another domain. To do these things with literal language is long-winded by contrast. Real-world knowledge intuitively plays a strong role in the comprehension of metaphor, but is out of place in a theory of semantics.

The homogeneous model stresses the interpretation of utterances in context. For Rumelhart (1979), for example, comprehension of an utterance consists in finding a "schema" that matches the utterance; Rumelhart maintains that this process is the same for metaphorical utterances as for literal ones. Indeed, he claims that language is seldom, if ever, used literally. To support this startling claim, he cites the sentence:

(5) The policeman raised his hand and stopped the car.

This does not seem to involve metaphor, and looks perfectly literal. Yet Rumelhart asserts that its proper comprehension involves knowledge of the way policemen signal to the drivers of cars: that is explaining the utterance in terms of an already stored schema. This comprehension is so clearly inferential in nature, and so far beyond the meanings of individual words, that Rumelhart concludes that neither compositional semantics, nor the notion of literalness itself, is of any help. For Rumelhart, the difference between "literal" and "metaphorical" language is only a difference in quality of match between utterances and their explanatory schemas.

Some of these points are made also by Miller (1979). While Miller holds to a version of the comparison theory in which metaphors are explicitly recast as similes (as mentioned in section 2.1 above), he nevertheless suggests that the processing of utterances may be homogeneous. For Miller, a reader assumes an obligation to treat a text as representing truth for its author: which may require the application of some rewriting rules if the text is to be kept coherent with the reader's own knowledge. Some of these rewriting rules are useful for turning blatant falsehood into possible truth, for example the transformation of the semantic representation for "John is a rock" into that for "John is like a rock". For Miller, recognition of metaphor is post-hoc: the need to apply certain such rules indicates that a statement was only metaphorically true.

As has been argued (Levinson, 1983; Searle, 1979), to rewrite metaphors as explicit similes or analogies merely shifts the burden of comprehension. Sometimes, indeed, the corresponding similes or analogies do not seem fair paraphrases of the original. This is particularly so when metaphor is being used to fill a gap in the lexicon, which is one of the widely acknowledged functions of metaphor (Miller, 1979; Paivio, 1979; Lyons, 1977; Beaugrande, 1980). In such cases, the metaphor still lingers because some word is being used in an extended sense. The extended usage carries along with it a cloud of associated information from the source domain, which Levinson (1983) has nicely termed a "connotational penumbra". In a straightforward simile paraphrase, such usage is maintained. An attempt to rephrase literally requires us to make explicit all the relevant associated information, which makes such a translation verbose.

3. Metaphor Comprehension by Computer

Several attempts have been made to describe how Natural Language Processing (NLP) systems might handle metaphor. Some of these have also been implemented as computer programs, usually in a program capable of handling both literal and metaphorical usages. Such programs must cope with many diverse problems: they must use syntactic and semantic knowledge; they must formalise and apply pragmatic principles; and they must model the psychological process of analogical inference. No single program can yet perform satisfactorily in all those tasks, but several of the subproblems have been extensively considered.

Metaphors seem to range from "small" metaphors, in which a word may be identified as being used in an extended sense, to "large" metaphors, where both vocabulary and inferences from one domain are used to describe and reason about a second domain. Computational models exist for this entire range, and these models will be described in order of increasing size of the metaphors they try to process.

At the smallest level, where individual words are used in extended senses, the relevant works are those of Wilks (1977b) and Hobbs (1977, 1979). Wilks proposed that the *preference semantics* model, rather than merely tolerating the use of words in contexts which violated preference restrictions, should seek to impose on these words a new reading derived from real-world knowledge. To this end he suggested the use of *"pseudo-texts"* associated with entries in a thesaurus. The thesaurus would have a hierarchical structure, with the topmost general entries being semantic primitives, and with the most specific entries being *formulas* of primitives corresponding to senses of individual words. The pseudo-text associated with an entry would give common-sense encyclopaedic knowledge about it. The semantic primitives were themselves mostly common English words (Wilks, 1977a), but would have very little detail in their pseudo-texts. Very specific words could, in principle, have very complex and detailed pseudo-texts.

The proposal was that when the system had no alternative but to construct a semantic substructure which violated the preference restrictions, it should be replaced with a substructure found in a pseudo-text. For example, given the sentence

(6) My car drinks petrol

the preference of 'drink' for an animate agent is unavoidably violated. Pseudo-texts associated with the constituents of the offending fragment should be scanned for fragments which almost match the offending one. In this example, part of the pseudo-text for 'car' would contain a fragment saying, in semantic-primitivese, "car consume petrol". This would be the best-matching fragment, and would be substituted for the problematic fragment in the program's interpretation of a real text.

The major problem with Wilks's proposal is that the two sentences convey somewhat different meanings. The "consume" version is straight and factual, but the "drink" version implies that consumption is excessive. Wilks pleads the case that this is idiomatic, and so beyond a general inference procedure (Wilks, 1977b). However, the work reported by Indurkhya (1985), discussed later, may prove to provide an analogical inference procedure capable of capturing this idiomatic implication.

Hobbs (1977, 1979) presents a method for handling metaphor "in the small", yet which is based on the grandest metaphorical view of all. Everything in Hobbs's model is based on a spatial or visual metaphor.

Hobbs uses predicate calculus expressions as the semantic representation delivered by a parser. The predicates in these expressions figure in axioms used in two ways, as *inference rules* and as *rewriting rules*. In their role as rewrite rules, the axioms show how expressions with domain-specific predicates may be re-expressed in terms of general visual/spatial predicates. For example, to relate the specific concept of variable to more general predicates, the following axiom is used:

(i) $(\forall x,y)$ $variable(y) \rightarrow [equal(y,x) \rightarrow at(y,x)]$

that is, if Y is a variable, then when Y is equal to the value X we may say that Y is "at" the value X.

The parser's lexicon contains information showing how words, and specifically verbs, are to be represented in the predicate-calculus expressions. The standard approach to polysemy, exemplified by common dictionaries as well as by most NLP systems, is to list the various senses of each word. For some common words, the list is long. For such words, Hobbs's lexicon does not detail the various senses, but rather gives a general visual/spatial predicate. For example, the definition for 'go' would be:

(ii) Build: go (a,b,c)   ; A goes from B to C
     Require: at (a,b)   ; A is at B, but ...
     Infer: become (at(a,b), at(a,c)) ; ... then A is at C.

The process of *predicate interpretation* involves discovering in just what sense a general predicate is being used. So, for a sample sentence

(7) V goes from 1 to 100

the desired interpretation is that V is a variable, initially it is equal to the value 1, but later comes to equal the value 100.

In constructing this interpretation, the program notes that the predicate 'go' has a precondition that, for this example, *at (V,1)* should be true. It attempts to find some chain of inferences which allow this to be established, and finds rule (i): V is a variable, so if *equal (V,1)* is true, *at (V,1)* could be asserted. So "at (V,1)" can be replaced by "equal (V,1)". In the *Infer* part of the definition of "go", the same reinterpretation of *at* as *equal* is done, yielding *become (equal (V,1), equal (V,100))*.

Hobbs's approach is similar to that of Wilks, (1977b) in that in both cases domain knowledge is used to rewrite parts of a representation. Hobbs appears to concentrate on rewriting of predicates, and this approach may not prove readily extensible to cases of metonymy. I suspect that examples like the following may be misunderstood:

(8) (Waiter to Manager) The hamburger went off without paying.

Other problem areas would be phrasal verbs, like some uses of "take", where the presence of a particular preposition can greately constrain the possible meanings; and texts where several ambiguous words are mutually constrained. This is an area where Wilks's work is particularly good.

The major practical advantage of the method proposed by Hobbs is the simplification of lexical entries for highly-polysemous words. Rather than listing all possible senses and methods for choosing between them, which is probably impossible anyway, Hobbs represents such words in terms of a small number of highly general senses. In the ideal case, probably also unattainable, such a word would have only one entry in the lexicon. The problem of selecting a sense congruent with the domain is tackled *in the domain representations*, rather than in the lexicon.

Indurkhya (1985) gives a computational model of analogical reasoning, and shows how it may be exploited for interpreting metaphorical utterances. The emphasis is on large-scale metaphors, of the kind discussed by Lakoff and Johnson (1980), yet the approach adapts well to small novel metaphors.

The two central features of Indurkhya's model are a formalism for representing domains, and methods for creating *transfer maps* between those domains. A *domain* comprises a set of tokens and a set of axioms, referred to as structural constraints. Indurkhya claims that metaphors function by a selective juxtaposition of the content of two domains, named the *source* domain and the *target* domain. He proposes that this be modelled by the use of a transfer map. A transfer map has two components. The first is a pairing of tokens in the source domain to tokens in the target domain; the second is a subset of the structural constraints of the source domain. The basic idea is that this subset should have all source-domain tokens replaced by target-domain tokens, and should then be added to the target-domain's own structural constraints.

One novelty in Indurkhya's notion of maps is that new tokens, not already existing in the target domain, can be created for use in such a map. Such a new token should be related to existing target-domain tokens by one or more of the transferred structural constraints. This achieves three purposes. First, it explains the extension of senses that so often accompanies a metaphor, because a word is given an interpretation in a foreign domain. Second, it conforms with the intuition that metaphors are often used because there is no literal term for what is meant. Finally, the multiplicity of constraints that can sometimes be mapped across corresponds well with the verbosity of literal restatements, when they are possible, compared with the conciseness of the original metaphor.

As an example, consider the use of family-relationship terms in talking about directed acyclic graphs: we freely talk of parent nodes, daughter nodes and sister nodes. Indurkhya shows how, for the two ideal domains "family" and "directed acyclic graphs", maps can be created which import tokens from "family". Some, like parent, will map onto existing tokens in the target domain, while others like "sister" serve to create new tokens. The constraints which are mapped across give meaning, in the target domain, to these newly created tokens.

Indurkhya describes a method, *Constrained Semantic Transference,* in which consistency must be maintained between the structural constraints of the target domain and the constraints being imported by mapping. However, this method can be applied only in special cases: real-world domains are not in general describable by consistent sets of axioms which are finite under closure. The method is modified, to a form called *Approximate Semantic Transference,* which is computationally feasible. In this modified version, the sets of axioms may be infinite under closure, and addition of further constraints is permitted provided that an added constraint is not, within some complexity limit, provably inconsistent with any existing axiom.

Indurkhya thus provides a method for augmenting knowledge about a domain by transferring knowledge from other domains. He proceeds to discuss how this method might be applied by a higher-level process, for example a text understanding program. One of the problems is that identification of the source and target domains must be performed, and Indurkhya provides sketches of algorithms for partitioning tokens and associated constraints into two disjoint sets, thereby creating ad-hoc domains. Because his method does depend critically on the identification of domains, and because it is unlikely that the existing domains are ideally suited to each new analogy (or metaphor interpretation problem), the provision of such algorithms is clearly desirable. It does however raise a question about the status of the pre-existing domains, and of the transferred structure: if knowledge is stored by domain, and analogical reasoning involves transferring structure to new ad-hoc domains, where in long-term memory should the new structure be kept? To what *existing* domain should structure be added? Without a clear answer to that question, the chunking of knowledge into domains is suspect. So too is a method of reasoning which depends on it.

Carbonell (1982) describes a method for handling known large-scale metaphors. This method seeks to *recognise* metaphors, rather than to *reconstruct* the meanings of metaphors. As Carbonell notes, there are approximately fifty large mataphors catalogued by Lakoff and Johnson (1980). The task of recognising an utterance as an instance of one of these large metaphors should be much easier than the task of reconstructing the meaning of a novel metaphor; and, because these fifty metaphors are both small in number and very common in use, a method for interpreting them can be both fairly cheap and very useful.

Carbonell proposes four components involved in understanding utterances which relate to these common large metaphors. (1) A *recognition network* which identifies which metaphor, if any, is implicit in the utterance; (2) a *basic mapping* of semantic features of the utterance onto features of the intended meaning; (3) an *implicit-intention component* encoding the reasons for choosing a particular metaphor; and (4) a *transfer mapping* which specifies how other features of the utterance are to be used in the transformed meaning.

The difference between "basic mapping" and "transfer mapping" can be illustrated with an example:

(9) Prices are soaring.

The recognition network would identify this as an instance of the "more is up" metaphor. The basic mapping states that "upward movement of a thing" is to be construed as an increase of some directly quantifiable aspect of that thing. The transfer mapping specifies how other aspects should be construed: for this example, reading "soaring" as "rising high and fast", the transfer map dictates that the altitude descriptor "high" should be mapped to "large" quantity, and that the rate descriptor "fast" should be carried over unmodified.

The "implicit intention component" is used to enrich the construal of the utterance with information about the speaker's hidden meaning. For Carbonell, metaphors convey more than literal restatements precisely because inferences can be drawn from the selection of a particular metaphor. These inferences are provided by the implicit-intention component.

These four components together provide a means of handling common large metaphors. One of Carbonell's main points is that metaphors often do not simply extend the sense of some word, but rather permeate an entire text. In such a case, Carbonell proposes that a text understander should note the use of a known metaphor, and try to use the associated information in processing subsequent utterances. Among other benefits, this allows for certain cases of sense extension: with information about the "conduit metaphor" (Reddy, 1979), a phrase like "censorship barriers" can be understood with only the central physical-object sense of "barrier" listed in the lexicon. However, Carbonell's method for dealing with sense-extension would appear adequate only when the extension is a manifestation of one of the precoded large metaphors.

Carbonell offers some thoughts on the processing of novel metaphors, although as stated above his primary concern is with large well-known metaphors. To account for some properties of metaphors, and particularly "nominal metaphors", he proposes an *invariance hierarchy* which governs the mapping process. In using a common metaphor, for example *"Inflation is War"*, only some concepts from the source domain (War) are carried across. Carbonell gives ten categories in the invariance hierarchy. At one end of this scale are goals, almost always transferred invariant: thus the metaphor gives rise to the goal of "defeating" inflation. At the other end, identity of objects is only rarely preserved: the fight against inflation uses neither tanks nor submarines.

4. Relationships between Primitives

The models described in the preceding section have one factor in common: all of them seek to handle metaphor by using specific domain knowledge. Certainly such knowledge does have a crucial role to play: the current trend in all branches of Artificial Intelligence is toward widespread application of specific knowledge at all points of every process. Nevertheless, the phenomenon of sense extension seems amenable, at least in part, to a general "weak" method. There are two reasons for electing to use such a weak method.

The first reason concerns the role of a parser in a natural language understanding system. To understand an utterance is, among other things, to relate that utterance to previously stored world knowledge. This knowledge is not stored as text, but is stored in a highly processed form which facilitates inferential and matching operations. In order to relate a text or text fragment to that knowledge, the parser "translates" an utterance into a similar processed form. To achieve this, it uses other knowledge (of grammar, semantics, etc.); but this knowledge is not of the same kind as the world knowledge to which the text is assimilated. But to use world knowledge in the parsing process presupposes that the analysis is sufficiently complete; thus there are early stages of analysis where this knowledge cannot apply.

The second reason is based on an intuition about the lexical knowledge that humans have. Common dictionaries provide definitions of many words and phrases, and these definitions are typically organised around several distinct senses of individual words. On looking at the definitions given by any dictionary, it is easy to imagine contexts in which the words could be used with the senses listed for them; yet there are two ways in which the organisation seems unnatural. First, trying to list for oneself all the distinct senses of a word (playing lexicographer, as it were) is very difficult: whether or not two usages in fact reflect different senses is often an arbitrary decision. Second, taking a piece of text, and selecting which dictionary-given sense is the correct interpretation of a certain word in a certain context, is similarly difficult. The skeptical reader is urged to try this as an exercise.

This is not to criticise dictionaries or real lexicographers, but rather to challenge the assumption that human organisation of lexical knowledge mirrors the organisation of a dictionary. That assumption is carried further when writers of NLP programs provide "dictionaries" with multiple discrete senses for each word, as in previous work by this author (Cater, 1981) and others. A project currently underway (Alshawi, Boguraev and Briscoe, 1985) which aims to provide an interface for NLP systems to the *Longman Dictionary of Contemporary English* takes to an extreme the assumption that such a system will find it both feasible and useful to discriminate between the many senses of numerous common words.

The "weak method" proposed here is to fashion a set of conceptual primitives to be used for representing meanings. These primitives are inspired by those of Schank (1975), but consist primarily of state-primitives. Only three act-primitives are proposed, replacing the eleven adopted by Schank. The claim is that by spreading the burden of inferentially adequate representation between states and a very small number of acts, a significant portion of the phenomenon of sense-extension can be accounted for without recourse to encyclopaedic world knowledge.

This is argued over the following few subsections. First, I present Schank's "conceptual acts", which are the cornerstone of the theory of Conceptual Dependency. Next, the similarities between several of these acts are examined, together with Schank's reasons for keeping them separate. Third, a much smaller set of acts is proposed, and it is shown how the desirable aspects of Schank's primitives are reproduced.

4.1 The Acts of Conceptual Dependency

Conceptual Dependency, henceforth "CD", is a theory of meaning representation which has proved useful in numerous NLP programs. It has three principal facets:

(1) A set of primitives.
(2) Rules of "conceptual syntax", which dictate how the primitives may join to form larger structures called *conceptualizations*.
(3) Semantic restrictions on the fillers of "argument slots".

To state the semantic restrictions, it is necessary to postulate "semantic features" as part of the representation of concrete nouns. While CD does not explicitly restrict these features, it is always tacitly assumed that the set of features will also be small: the set (HUMAN, ANIMATE, PHYSICAL-OBJECT, LOCATION, BODY-PART, SENSE-ORGAN, MENTAL-LOCATION) is typical of the feature sets used in practice.

The primitives of CD are of two kinds, *acts* and *states*. The acts are intended to represent activities (or operations) which can be performed, while the states are used to describe properties of entities. Taken together with a representation of concrete nouns and a few classes of causation, these primitives are thought adequate to represent a large fraction of the subject matter of everyday texts.

Writings on CD invariably emphasise the primitive acts, with relatively little detail on the states. With each act is associated a "case frame", where different acts may need a different collection of cases. In building a representation of the meaning of an utterance, the presence of an act requires (via rules of conceptual syntax) that fillers be found for each conceptual case; and it requires (via rules of semantic restriction) that these fillers conform to certain requirements. If, as is often the case, a text fragment does not specify appropriate fillers, inference is required to fill in the gaps. This inference process is strongly guided by the semantic restrictions on possible fillers of the act's cases.

The acts have been listed and described in many papers. The list and description below is loosely based on that in Schank, (1975, pp. 41-44). Every act has an "Agent" case, to be filled by a person, thing, or agency responsible for performing the act. The "directive case" used by Schank usually consists of two components, corresponding to "source" and "destination"; for ease of the subsequent discussion, these are specified separately as "TO" and "FROM" in the list below.

PROPEL: means "apply a force to".
   AGENT: Animate, or Inanimate (eg., a car), or Agency (eg., gravity).
   OBJECT: Physical object.
   TO: Location.

MOVE: means "move a body part".
   AGENT: Animate.
   OBJECT: Body part of agent.
   Direction: Path followed by body part.

INGEST: means "take something to the inside of an animate object".
   AGENT: Animate.
   OBJECT: Physical object.
   TO: Body opening.
   FROM: Location.

EXPEL: means "force something to the outside of an animate object".
   AGENT: Animate.
   OBJECT: Physical object.
   TO: Location.
   FROM: Body opening.

GRASP: means "to physically grasp an object".
   AGENT: Animate.
   OBJECT: Physical object.
   TO: Body part doing the grasping.

PTRANS: means "to change the location of something".
   AGENT: Animate, or Inanimate (eg., a car), or Agency (eg., gravity).
   OBJECT: Physical object.
   TO: Location.
   FROM: Location.
   Instrument: an act-based conceptualization, typically using PROPEL.

ATRANS: means "to change some abstract relationship with respect to an object".
   AGENT: Animate.
   OBJECT: Composite of (i) Physical object, and
                              (ii) Abstract relationship (eg., possession).
   TO: Animate.
   FROM: Animate.

SPEAK: means "to produce a sound".
   AGENT: Animate, or Inanimate (eg., a car).
   OBJECT: Type of sound.

TO: Direction in which sound is projected.
FROM: Physical object, often a body part.

ATTEND: means "to direct a sense organ towards a stimulus".
  AGENT: Animate.
  OBJECT: Sense organ.
  TO: Location (of stimulus).

MTRANS: means "to transfer information".
  AGENT: Animate.
  OBJECT: Conceptualization.
  Instrument: an act-based conceptualization, typically using SPEAK or ATTEND.
  FROM: Mental location.
  TO: Mental location.
  > For this purpose, human memory is divided into three "mental locations": long-term memory, intermediate memory, and "Conceptual Processor".

MBUILD: means "to create or combine thoughts".
  AGENT: Animate.
  TO: "Conceptual Processor" of agent.
  OBJECT: Composite of (i) the thoughts, if any, being combined;
                        (ii) the new thought.
  > The destination is always the agent's "Conceptual Processor": a new thought is always thought of.

Schank (1975) justifies this particular set of action primitives on two grounds. First, the members of this set are claimed to provide adequate coverage for a large fraction of the actions that people talk about: but Schank does not claim that the set is complete in this sense. Second, the presence of a particular act in a conceptualization licenses a unique set of inferences. These inferences can be said to give "meaning" to the primitive, distinct from the meanings of all other act primitives. These claims are discussed by Wilks (1977a), who documents a different system of primitives. For present purposes, however, Schank's primitive acts will be accepted to be largely adequate as stated.

4.2 Relationships between Acts.

The meaning of an utterance is represented by a conceptualization, which is a structure containing one or more of the primitives, together with fillers for the required conceptual cases. The primitive act[1] plays two major roles in a conceptualization. First, like the verb of a typical English sentence, it is central to the meaning; second, again like an English verb, it imposes an organisation upon the surrounding constituents: it requires fillers for certain conceptual cases, and further constrains their semantic character.

When the primitives are actually used to represent meanings, it is common for two case slots to have the same filler, or to have fillers which are closely related and which derive from the same surface constituent of an utterance. For each act, several subclasses may be discerned depending on which pair of cases (if any) are filled with the same item.

For example, the act PTRANS (Physical transfer) is used in a small number of ways, as listed below. Strictly, its *TO* and *FROM* slots may only be filled by concepts of type LOCATION; but since a location is usually specified by naming an object at that location, the distinction between objects and their locations is more conveniently ignored.

---

[1] As has been noted, CD also uses numerous state primitives, used to represent descriptive or relational information. These state primitives play just the same roles in a conceptualization as do the primitive acts.

Subclasses of PTRANS

Mnemonic phrases	Condition	Examples of corresponding English
PTRANS/SELF:	Agent = Object	Walk; Go to the pub;
PTRANS/EMIT:	Agent = From	Throw a ball; Send a letter;
PTRANS/ABSORB:	Agent = To	Get your coat; Pull a chair;
PTRANS/IMPEL:	Agent ≠ any	Siphon petrol; put away your toys;
PTRANS/WAVER:	To = From	Swing a bucket; Hop.

This classification is less than ideal in two respects. First, the classes "IMPEL" and "WAVER" should properly be analysed in terms of either (i) a sequence of PTRANS acts, or (ii) a fuller specification of trajectory than is given simply by endpoints. For example, "Put away your toys" requires that each toy be taken from its location and then be placed elsewhere: that is, first an "ABSORB" and then an "EMIT" must be performed.

The second defect is that, for both EMIT and ABSORB, the location may be related to the agent in one of two ways: the location may be inside the agent, or simply close to the agent. The classification can be expanded by distinguishing EMIT/IN from EMIT/BY, and ABSORB/IN from ABSORB/BY.

The merit of the classification is that it allows other of CD's acts to be compared against PTRANS. Do other acts exhibit the same pattern, or similar patterns, or quite different patterns? Clearly, the acts INGEST and EXPEL are related to PTRANS, because all three involve physical movement. In fact, EXPEL can be seen as simply a PTRANS, where the *From* slot is a body aperture of the agent; and similarly for INGEST's *To* slot. So EXPEL is just PTRANS/EMIT/IN, and INGEST is PTRANS/ABSORB/IN.

More adventurous comparisons can be made between PTRANS and MTRANS, and between PTRANS and ATRANS. For MTRANS, the SELF subclass cannot exist, because the *Object* is necessarily an idea or conceptualization, while the *Agent* must be animate. (In practice, we might like the possibility of certain other objects as agent, for example televisions or computers: but even so, there can be no overlap with conceptualizations.) Further, the *To* and *From* slots must be filled by mental locations. These mental locations naturally bear a close relationship with individuals, which I propose to exploit in a similar fashion to the relationship of physical locations with physical objects. Thus, the subclasses of MTRANS are:

Subclasses of MTRANS

Mnemonic	Condition	Example of corresponding phrases
MTRANS/EMIT:	Agent = From	Tell a story; Write a letter;
MTRANS/ABSORT:	Agent = To	Hear a rumour; Read a book;
MTRANS/IMPEL:	Agent ≠ any	Copy; Translate; Play a record; TV show;
MTRANS/WAVER:	To = From	Recall; learn.

In a similar fashion to the distinction of /IN and /BY for PTRANS, there is a need to distinguish which of the mental locations is meant. Also, the "IMPEL" and "WAVER" classes are suspect. For example, in MTRANS/IMPEL, there are two problems. First, if a person translates or copies a text, they do so by a sequence of "absorbing" and then "emitting" the information. While it may seem that the operation of a television is not properly represented in this sequential fashion, it is clear also that it does not transmit ideas at all: it transmits only signals, which are interpreted as ideas by a viewer. The act SPEAK is used to represent production of sound, but is inadequate for this problem on two counts: it is restricted to sound only, the agent must be animate, and the inferential basis for SPEAK holds the production of sound to be under autonomous control. None of these holds in the case of a television set. But there is no CD act which can represent the transfer of general signals (or transfer of energy), so this MTRANS/

WAVER representation will be adopted here on a temporary basis.

The relationship between PTRANS and ATRANS is fairly straightforward, largely because there is no problem of relating *Agent* with *To* or *From*: for ATRANS, all three slots may be, and indeed usually are, filled by Animate entities. The only problem is posed by the composite object required by ATRANS: the *Object* slot is to be filled by a physical object and an abstract relationship. To facilitate comparison with PTRANS, these two components must be separated: the physical-object part will be called *Object,* and the relation will be called *Rel.* Then we may list:

Subclasses of ATRANS

Mnemonic	Condition	Examples of corresponding phrases
ATRANS/SELF:	Agent = Object	Elect premier; Join club;
ATRANS/EMIT:	Agent = From	Give object; Delegate authority;
ATRANS/ABSORB:	Agent = To	Take object; Seize power;
ATRANS/IMPEL:	Agent ≠ any	Award damages; Amputate limb;
ATRANS/WAVER:	To = From	Lend book; Recover property.

The examples in the above list demand certain abstract relationships, such as possession, political power, membership of a set, and part-of-whole. While most work with CD has emphasised mainly possession, ownership, and immediate control over objects, there is no reason why other binary relationships may not also be used. But whatever the relationship, the WAVER subclass should yet again be rewritten as a sequence of other subclasses.

The ATRANS act is more general than either PTRANS or MTRANS, in the sense that it may be applied to many binary relationships. In fact, if we ignore the semantic constraints of the various case slots, PTRANS may be regarded as an ATRANS where the relationship is location; and similarly, MTRANS may be regarded as an ATRANS using the relationship of "mental location". In section 4.3, I propose to do just that: ATRANS, MTRANS, PTRANS, INGEST, and EXPEL will all be reduced to a single primitive, TRANSFER, defined as a generalised ATRANS:

TRANSFER: means "to change some relationship of an object"
   Agent: Animate, or physical object, or agency.
   State: The name of CD state (binary relationship).
   Object: Object of type appropriate for the state.
   From: Previous value of the state for the object.
   To: New value of the state for the object.

Schank claims that each primitive act leads to a distinct set of inferences, and that this is why the acts themselves are distinct (Schank, 1975, p. 68). For an example, consider the inferences from:

   PTRANS: Agent X; Object Y; From Z; To W; At-Time T
   MTRANS: Agent X; Object Y; From Z; To W; At-Time T

For the PTRANS example, the following inferences (among others) would be drawn:

   Y Physically-Located-At Z Until-Time T
   Y Physically-Located-At W Since-Time T

For the MTRANS example, some inferences would be:

   Y Mentally-Located-At Z Around-Time T
   Y Mentally-Located-At W Since-Time T

The difference between the set of inferences is in two parts. First, the relationship is different

(Physically-Located-At vs. Mentally-Located-At); second, for PTRANS the object vanishes from its original location, whereas for MTRANS it persists.

I shall claim, in section 4.3, that similarly different sets of inferences can be drawn by having the inference procedure consult both the act, TRANSFER, and the relationship, be it "Mentally-Located-At", "Physically-Located-At", or something else.

The remainder of this section sketches, in much less detail than was given earlier, how the other acts of CD might be related to one another. The other acts are MBUILD, MOVE, PROPEL, GRASP, ATTEND, and SPEAK. In keeping with the very general spirit of TRANSFER, two other general primitives will be proposed: TRANSFORM and CONTROL.

I view MBUILD as an example of TRANSFORM; a new thought is formed from nothing or from several old thoughts. However, while MBUILD is the only one of the existing CD acts which does reflect transformation, this merely indicates that those acts do not form a complete set. To represent many common activities in CD, one is forced to forgo the use of act primitives, and instead specify "Person performs unspecified act, with such-and-such result".

The new act TRANSFORM is envisaged as follows:

TRANSFORM: means "to change the relationship of a whole to its parts".
AGENT: Animate
STATE: A relationship between the parts and the whole.
PARTS: Collection of objects, of type appropriate for STATE.
WHOLE: A single object, of type appropriate for STATE.
CHANGE: A boolean, indicating whether the STATE relating PARTS to WHOLE is made to become TRUE or FALSE.

The table below shows how several examples could be represented in terms of this general TRANSFORM act. The names given to the states should be clear, with the exception of MPART and PSTATE: MPART is an abbreviation for Mentally-Part-Of, and would lead to inferences about a Mentally-Located-At relation (MLOC); while PSTATE is an abbreviation for Physical-State, leading to inferences about Physically-Located-At.

Representations using the act TRANSFORM.

State	Agent	Parts	Whole	Change	Example
MPART	Me	(Premise1, Premise2)	Conclusion	True	Decide
PART	Me	(Egg, Milk, Herbs)	Omelette	True	Make, Cook
PSTATE	Me	(Smoke, Ash)	Cigarette	False	Burn
FAMILY	Judge	(Fred, Mary)	Couple	False	Divorce

The remaining acts of CD are seldom used except as instrumental acts. Thus ATTEND and SPEAK are normally instruments for MTRANS; and MOVE, PROPEL and GRASP are normally instruments for PTRANS or sometimes ATRANS. Nevertheless, there are many English verbs which demand representation in terms of these acts: for example, Listen, Sing, Sit, Press, and Hold. Also, there are many expressions with apparently closely related meanings, but for which the only representation possible in CD is the DO-CAUSE-STATECHANGE trick: for example, Aiming a camera is similar to "ATTEND", Dominating a crowd is similar to "GRASP".

The proposal here is that these acts should be replaced by one, CONTROL. Essentially, what these acts have in common is that some object is being made to obey some constraint. This constraint is naturally represented in CD as a state: some states, like LOC, relate an object to some entity; while others relate an object to a point on a notional scale (Schank, 1975). So the act CONTROL must have cases for the object being constrained, the dimension of constraint (that is, the name of a CD state), and the value (entity or scale point, depending on the state) which the object is constrained to have for that state. To summarise, and give examples of this:

CONTROL: means "to constrain something to have some property".
   Agent: Animate.
   Object: Physical object.
   State: The name of some CD state.
   Value: A value appropriate for the State.

Representations using CONTROL .

Agent	Object	State	Value	Example phrases
Person	Body	SHAPE	"SITTING"	Sit; Lie; Cross the legs;
Person	Thing	PRESSURE	Direction	Press; Tug; Fall;
Person	Thing	LOCATION	Bodypart	Hold, Cup head in hands;
Person	Eyes	DIRECTION	Thing	Listen; Point a telescope;
Person	Voice	MODULATION	Laughter	Laugh; Sneeze; Whisper;
Teacher	Pupils	BEHAVIOUR	Quiet	Dominate; Control a crowd.

In this section, the three proposed acts have been introduced and partly motivated. The only valid motivation for a set of primitives, however, comes from their computational effectiveness: if a set of primitives allows a program to manipulate its knowledge easily for its intended tasks, then those primitives are paying their way. In the next section, I hope to describe the utility of these primitives for one task of metaphor comprehension, namely, interpreting known words when used in unfamiliar extended senses.

4.3 Acts with Better Properties

There is presently no program which uses the three primitive acts described above. It is therefore quite likely that errors have been made in the descriptions. Nevertheless I believe that a system of representation based on them will, on a broad scale, prove adequate. The acts are intended to be effective for two tasks: to be at least as expressive as CD, and to guide the inference process responsible for interpreting extended usages of known words.

Each act of CD is merely a specialisation of one of the primitives presented here. For example, PTRANS is TRANSFER with state LOC (or Physically-Located-At). The semantic constraints on CD's case slot fillers can be motivated in the new system by reference to the state. Thus, rather than specifying that PTRANS needs a location as its *From* slot, we say that the *From* slot of TRANSFER must be filled by something acceptable as the *Value* slot of the given state: in the case where LOC is the state, its *Value,* and hence both the *To* and *From* slots of TRANSFER, must be filled by a location.

Each act of CD provokes a unique set of inferences, and this is the reason given by Schank (1975) for separating the acts. It has however been found useful in previous work (Cater, 1981, 1983) to make the inference process more selective: some inferences are valid only when a case slot filler satisfies certain conditions. The inferences drawn from a representation depend not only on the presence of a particular act, but also on the outcomes of several idiosyncratic tests. It will be a simple matter to extend this approach, so that some of the inferences from TRANSFER, for example, are drawn only when the state is "Physically-Located-At", and others drawn only when the state is "Mentally-Located-At", and so on.

The new primitives can mimic CD in expressiveness, in inferential effect, and in motivated semantic constraints. The purpose of introducing them is not to rival CD, but to improve upon it. The new acts should support the interpretation of metaphorical uses of known words by providing explicit dimensions along which a known sense can be altered. This is expected to have two major benefits.

On occasion, speakers knowingly and deliberately use words in a novel sense. In such a case, neither human hearer nor machine can be expected to have pre-stored a definition of the word appropriate to this new use. Yet people normally, and often unconsciously, are able to interpret the word in approximately its intended sense. By using a set of primitives, such as those outlined in this paper, it is hoped to guide a machine to a good guess at this intended sense. An act-based conceptualization will now always make explicit mention of a state: if the conceptualization is ill-formed, for example by violation of semantic constraints, there are several ways that it can be deformed, or mutated, to produce new conceptualizations.

Three classes of deformation of a conceptualization can be distinguished. First, a troublesome role filler can be replaced with a filler of the expected semantic category. Presumably the replacement filler would be closely related to the old one. A trivial example would be substitution of a location for a physical object: the obvious relationship between location and physical object is that the object is at the location. A description of a location in these terms could then be placed into the conceptualization.

Example: Fred went to his desk.
Assuming that the dictionary lists "went" as TRANSFER of the relation "Physically-Located-At", it is necessary to construct a location from the filler "his desk".

The second class of deformation, replacing a state name with another, is more drastic. This process would be guided by two factors: the semantic character of the other role fillers would serve to identify the permissible replacement states; and an inherent similarity ranking between the states, if such can be found, would allow a choice between those.

Example: Send my regards to Mary.
Assuming the "Send" is also TRANSFER of "Physically-Located-At", "regards" is not a physical object, nor is one readily associated with it. "Mentally-Located-At" should be substituted.

The final class of deformation is replacing one act with another. This does appear to be sometimes necessary, but would require considerable computational effort. It is hoped that it is to be used only as a last resort.

Example: Drugs sent John crazy.
Here, the apparent syntactic object is itself a state description. The proper interpretation replaces TRANSFER with CONTROL, and uses the state described by "crazy" to fill the other slots.

It may be objected that the above examples do not extend the senses of words, but that the lexicon will list senses appropriate for each usage. Even if this is so, there is advantage in using general primitives which can be deformed. All the separate senses of words need not be listed. We need only provide core senses for words and handle the peripheral senses of words through the mechanism of mutation. By doing so, we have no need either to provide many senses for common words, nor to take great care in defining them so that the "right" sense can be chosen on each occasion. The second benefit of using the general primitives then is that they permit a great simplification of the lexicon.

5. Conjectures and Conclusions

Conceptual Dependency, or "CD", has been used with success in many NLP systems, far too numerous to list here. This paper proposes to generalise the eleven primitive acts to only three, TRANSFER, TRANSFORM, and CONTROL. Using these generalised acts requires explicit mention in act-based representations of some state. The motive for this drastic step is to facilitate the proper interpretation of phrases in which some word is used metaphorically, that is, in an extended sense. A second benefit is that the lexical entries for individual words can be made much smaller and much simpler, because only core senses of words need be listed in the lexicon. However, the proposal has not yet been tested in a NLP program, and so must be viewed as speculative.

Metaphorical language can be crudely classified as "large" or "small". Large metaphors correspond to analogies between entire domains, where for example "Argument" is discussed using terminology from the domain of "War". Interpreting these analogies clearly requires recourse to general world knowledge, and the primitives described here would be neither better nor worse than many other representation schemes for that task. Small metaphors in contrast are local, involving only the extension of the sense of a word to apply in a new situation. It is this problem, of interpreting a word in a nonliteral fashion, that the present proposal addresses.

Small metaphors may be found in many forms. Sowa (1984) cites the example "key employee", saying that to interpret this phrase will require either a large dictionary or a metaphor-handling ability. More generally, a highly productive class of metaphors has been identified by Levi (1978). Levi's concern is to account for the syntactic and semantic properties of complex nominals, which includes what are commonly called compound nouns. Her thesis is that a small set of "Recoverably Deletable Predicates" can account for these properties; but that one of these predicates, BE, is quite frequently intended non-literally.

If even the interpretation of a noun phrase can require interpretation of metaphors, it does not seem reasonable that metaphor processing should have to wait until the completion of syntactic and semantic processing. The last decade has been marked by two trends in natural language understanding programs, toward deterministic parsing and toward integrated processing. Combining these leads to the conclusion that syntactic and semantic knowledge, and world knowledge used to draw inferences and to make plausibility judgements during parsing, conspire to make possible the interpretation of an utterance on the fly. This implies that the process of interpreting a word in context, even when it is used metaphorically, should not be delayed. The primitives described in this paper, and the deformation procedure associated with them, should allow extended senses to be correctly interpreted as soon as the local context can be gathered.

This paper has addressed only the metaphorical relationships between acts. One consequence of the proposed new set of acts is that the representation of meaning will rely more heavily on states, of which there are many. It is left for future work to assess how, if at all, the states are related to one another. One suggestion is that there can be found an inherent similarity metric between the states: such a measure would provide one of the heuristics needed to guide the mutation procedure described above for act-based conceptualizations, and presumably would be useful for a similar procedure operating on state-based representations.

Standard Conceptual Dependency places great emphasis on the primitive acts, while giving little attention to the states (Schank, 1975). Yet the classifications in *Roget's Thesaurus* (1962) are overwhelmingly of a "state" character, in Schank's sense. Cercone and Schubert (1975) explicitly propose a state-based representation system. Hobbs (1977) describes another state-based system, where the primitives have a visual or spatial character, and domain-specific terms are set in a metaphorical correspondence with those primitives.

The proposal here is that some actions, of a very general nature, should be retained in the representation language, but that these should be designed to facilitate metaphor interpretation. Three action primitives are described, which are general but can carry specific meaning when associated with a state in a conceptual representation. Procedures for interpreting extended senses are outlined.

Further work is clearly needed, on clarifying the relationships between state primitives so that the approach can be made more general, and on implementing a system which tests the practical adequacy of the proposal.

References

Alshawi, H., Boguraev, B.K., & Briscoe, E J. (1985). Towards a dictionary support environment for real-time parsing. In *Proceedings of the Second European Meeting of the Association for Computational Linguistics*. Menlo Park, CA: Association for Computational Linguistics.

Beaugrande, R. de. (1980). *Text, discourse and process: Toward a multidisciplinary science of texts*. Norwood,

NJ: Ablex.

Carbonell, J.G. (1982). Metaphor: An inescapable phenomenon in natural-language comprehension. In W.G. Lehnert and M.H. Ringle (Eds.), *Strategies for natural language processing*. Hillsdale, NJ: Lawrence Erlbaum.

Cater, A.W.S. (1981). *Analysis and inference for English.* (Technical Report No. 19) Cambridge: University of Cambridge, Computer Laboratory.

Cater, A.W.S. (1983). A unified approach to ad-hoc inferencing. In K.P. Jones (Ed.), *Intelligent information retrieval: Informatics 7.* London: Aslib.

Cercone, N., & Schubert, L. (1975). Toward a state based conceptual representation. In *Proceedings of the Fourth International Joint Conference on Artificial Intelligence*. Menlo Park, CA: SRI International.

Cohen, L.J. (1979). The semantics of metaphor. In A. Ortony (Ed.), *Metaphor and thought*. Cambridge: Cambridge University Press.

Fraser, B. (1979). The interpretation of novel metaphors. In A. Ortony (Ed.), *Metaphor and thought*. Cambridge: Cambridge University Press.

Hobbs, J.R. (1977). Coherence and interpretation in English texts. In *Proceedings of the Fifth International Joint Conference on Artificial Intelligence.* Menlo Park, CA: SRI International.

Hobbs, J.R. (1979). *Metaphor, metaphor schemata, and selective inferencing.* (Technical Note 204). Menlo Park, CA: SRI International.

Indurkhya, B. (1985). *A computational theory of metaphor comprehension and analogical reasoning.* (Technical Report No. 84-31). Amherst: University of Massachusetts, Department of Computer and Information Science.

Lakoff, G. & Johnson, M. (1980). *Metaphors we live by.* Chicago: University of Chicago Press.

Levi, J.N. (1978). *The syntax and semantics of compound nominals.* New York: Academic Press.

Levinson, S.C. (1983). *Pragmatics.* Cambridge: Cambridge University Press.

Lyons, J. (1977). *Semantics* (Vol. 2). Cambridge: Cambridge University Press.

Miller, G.A. (1979). Images and models, similes and metaphors. In A. Ortony (Ed.), *Metaphor and thought*. Cambridge: Cambridge University Press.

Ortony, A. (Ed.). (1979). *Metaphor and thought.* Cambridge: Cambridge University Press.

Paivio, A. (1979). Psychological processes in the comprehension of metaphor. In A. Ortony (Ed.), *Metaphor and thought*. Cambridge: Cambridge University Press.

Reddy, M.J. (1979). The conduit metaphor – a case of frame conflict in our language about language. In A. Ortony (Ed.), *Metaphor and thought*. Cambridge: Cambridge University Press.

Roget's Thesaurus. (1962). London: Longman.

Rumelhart, D.E. (1979). Some problems with the notion of literal meanings. In A. Ortony (Ed.), *Metaphor and thought*. Cambridge: Cambridge University Press.

Sacks, S. (Ed.). (1979). *On metaphor.* Chicago: University of Chicago Press.

Schank, R.C. (1975). *Conceptual information processing.* Amsterdam: Elsevier.

Searle, J.R. (1979). Metaphor. In A. Ortony (Ed.), *Metaphor and thought*. Cambridge: Cambridge University Press.

Sowa, J.F. (1984). *Conceptual structures: Information processing in mind and machine.* Reading, MA: Addison-Wesley.

Waltz, D.L. (1982). The state of the art in natural-language understanding. In W.G. Lehnert and M.H. Ringle (Eds.), *Strategies for natural language processing.* Hillsdale, NJ: Lawrence Erlbaum.

Wilks, Y.A. (1977a). *Good and bad arguments about semantic primitives* (Technical Report No. 42). Edinburgh: University of Edinburgh, Department of Artificial Intelligence.

Wilks, Y.A. (1977b). Knowledge structures and language boundaries. In *Proceedings of the Fifth International Joint Conference on Artificial Intelligence.* Menlo Park, CA: SRI International.

2

**LEXICAL AND
SYNTACTIC FAILURE**

SPEECH COMPREHENSION AND LEXICAL FAILURE

John Harris

Institiúid Teangeolaíochta Éireann

Existing research on communication failure has paid almost no attention to speech comprehension and speech production. The purpose of the present paper is to show that the processes by which communication failures arise and are repaired are intimately bound up with the structure and operation of the speech comprehension system itself. In particular, we consider what recent research on spoken language understanding can tell us about the likely processes of lexical failure and failure repair. The first half of the paper describes the main characteristics of the speech comprehension system which have been documented by research over the last fifteen years. Results of some unpublished research on these issues are also reported. The second half of the paper describes the likely course of lexical failure and its repair in such a speech comprehension system. The paper concludes with a reconsideration of existing definitions of lexical failure and a discussion of the relationship between the structure of the speech comprehension system and the temporal coordination of conversation.

1. Introduction

Much of the recent interest in conversation failure has been prompted by the desire to improve man-machine dialogue systems. The success of this psycholinguistic research, not to mention the usefulness of its findings in the context of man-machine dialogue, may crucially depend on not defining the phenomena to be explained in too narrow a fashion. Thus, it may be just as significant to discover why serious conversation breakdown is relatively uncommon, as it is to document the kinds of failures which do occur and the manner in which they are repaired. It is fairly obvious, for example, that in many ordinary conversations the potential for lexical failure, due to wide differences in the lexical competence of the participants, does not have serious consequences for the smooth and coherent progress of communication. It is only necessary to point to conversations between parents and children or to conversations between native and non-native speakers of a language to confirm this. Or we can consider the false starts and slips of the tongue which seem to occur regularly in conversations and try to explain why the result is not more often a catastrophic breakdown in communication. We can account for the relative infrequency of communication failure in a more general way, of course, by pointing to the apparent indeterminacy of conversation. But to recognise this is not in any way to imply a lessening of the need to search for underlying precision, order and regularity in conversation, or to prejudge the issue of the degree of indeterminancy actually present.

At the moment, the formal study of conversation failure is still quite new so that relatively little theory-building and data collection has been possible (Ringle & Bruce, 1982). In this kind of situation, it is often useful to consider to what extent existing research in closely related areas may be able to contribute theoretical insights or actual data relevant to the new issue of interest. For example, there is now a substantial body of research on speech errors – slips of the tongue and slips of the ear – occurring spontaneously in conversation (Fromkin, 1973, 1980). This data has usually been studied with a view to evaluating theories of speech production and speech perception. But, they can equally well be considered as data on input failure or communication failure in dialogue. Their value is enhanced by the fact that in some cases the conversational contexts in which the errors or failures occurred have been carefully recorded (Celce-Murcia, 1980). Other areas of research which are directly relevant to conversation failure include the work on 'foreigner talk' and investigations of the differences between 'native-speaker/native-speaker' and 'native-speaker/non-native speaker' discourse (Day, Chenoweth, Chun & Luppescu, 1984; Long, 1983; Tarone, 1980).

In the present paper, however, the body of research whose relevance to conversation failure will be considered is that of speech comprehension. As I hope to show, recent work on this topic has particularly important implications for an understanding of lexical failure. In the earlier part of the paper I will review research related to word recognition in continuous speech, on the assumption that an adequate account of lexical failure must proceed from a consideration of the normal process of spoken word recognition. This review will include an account of some research of my own which deals with the manner in which informational dependencies in discourse affect the basic process of spoken word recognition and speech comprehension. I will then go on to consider in detail some of the implications of this research for an understanding of lexical failure in dialogue. In particular I will try to explain why lexical failure is not a more troublesome problem in conversation. Finally, I will examine the relationship between the temporal structure of speech comprehension and the temporal coordination of conversation.

2. The Definition of Lexical Failure

First, however, a number of questions relating to the definition of lexical failure must be considered. Lexical failure is the name given to one of a number of different kinds of input failure by Ringle and Bruce (1982). An input failure occurs when the listener in a conversation is unable to obtain a "complete, or at least a coherent interpretation for an utterance". Three kinds of input failure are distinguished. Perceptual (acoustic) failures are defined as instances where a word or phrase is not clearly perceived or is misperceived and either no interpretation or an erroneous one results. Lexical failures are those where a word or phrase is clearly perceived but the listener is either unable to produce any interpretation at all or fails to produce the correct one. Syntactic failures consist of cases where individual words and phrases are correctly perceived and interpreted but the speaker's intended meaning is misconstrued by the listener.

There are at least two problems with these definitions. First, it is in practice extremely difficult to separate these different kinds of failure. This is shown by the example which Ringle and Bruce themselves give as an illustration of syntactic failure i.e., they instance a situation where a listener does not hear a negative and thus perceives the speaker to assert the opposite of what the speaker intended to assert. But even in Ringle and Bruce's own terms this is primarily a case of perceptual (acoustic) failure and only secondarily a syntactic failure. More fundamentally, the definitions of the different types of failure seem to implicitly assume a speech processing system which is strongly serial in nature, an assumption which is no longer justified by the available research evidence. Speech processing is now known to be highly interactive in character, with all available knowledge sources being employed simultaneously to develop an interpretation of the utterance on-line (Marslen-Wilson & Tyler, 1980). An examination of the implications of this research for an understanding of potential and actual lexical failure is, of course, one of our main concerns here.

Second, a number of critical words and phrases which appear to be taken as equivalent in the Ringle and Bruce definitions, such as "interpretation", "correct semantic interpretation" and "the speaker's intended meaning" are not in fact equivalent. For example, the speaker may choose the incorrect word to convey his particular intended meaning, yet speak that incorrect word quite clearly. The listener may then either correctly identify the word actually spoken, or may fail to identify it due to his own linguistic limitations. Finally, the listener may or may not, irrespective of which word was actually spoken and recognised, correctly identify the word (meaning) *intended* by the speaker. The Ringle and Bruce definitions of input failures, in short, do not adequately distinguish between deficiencies in linguistic knowledge, errors in speech processing and failures to identify the intended meaning of the speaker. Here again an examination of recent work on the structure of the speech processing system in general, and research on the process of spoken word recognition in particular, can help to clarify the issues involved in distinguishing different kinds of input failure.

3. Speech Comprehension

Until the mid 1970's, the study of spoken language understanding was dominated by a 'serial-autonomous' view of the processing system (Fodor, Bever & Garrett, 1974; Forster, 1979; Marslen-Wilson, 1976). A basic assumption of this model was that the flow of information between the different components or levels in the system was in one direction only – from the bottom to the top. This also implied that the processor at each level was autonomous, that it had access only to its own knowledge source and to the output from the processor at the next lowest level. Thus, the lexical processor only had available its own stored lexical knowledge which was used to analyse the output of the acoustic phonetic processor. In this kind of strictly serial system, high level syntactic and semantic analyses could not intervene in the internal decision mechanisms of the lower level lexical processor. Consequently, the well known context effects in word recognition could only be interpreted as occurring after the completion of the normal initial word recognition process. More specifically, the argument had to be that semantic or syntactic constraints on word recognition only come into operation at some later stage and involve selecting among a number of word candidates which have already been output by the lexical processor. Thus, there would have to be a real time lapse between the detection of an ambiguity by a processor at one level and the resolution of that ambiguity by a higher level processor.

Much of the evidence on which the serial autonomous model is based has been challenged in recent years, however, mainly on the grounds that it involved the use of post-sentence measures of processing (Harris, 1978; Marslen-Wilson, 1976; Tyler, 1981). In addition, numerous experiments carried out over the last decade conflict with the serial-autonomous view and support instead an 'on-line interactive' view (Harris, 1978; Marslen-Wilson, 1975, 1980, 1984, 1985; Marslen-Wilson & Tyler, 1980; Tyler & Marslen-Wilson, 1977). This alternative approach assumes a flexibly structured processing system in which analyses corresponding to one knowledge source can be made available to affect the operations of any other knowledge source. The acoustic-phonetic analysis of the sensory input is, however, excluded from the kind of communication and interaction which is assumed elsewhere in the system. But all other analyses – lexical, structural and interpretative – are assumed to communicate and interact freely. This means that, in principle, an utterance can be fully interpreted as it is heard. No time lapse need be assumed between the progress of analyses at different linguistic levels. For example, syntactic and semantic analyses of an accumulating string of words do not have to be delayed until the clause boundary – as the most common version of the serial-autonomous model would suggest. Instead, in the on-line interactive view, the structural and interpretative implications of a word choice are immediately determined on-line. Another crucial implication is that during the analysis of an utterance, even the basic process of word recognition can be influenced by the interpretative context in which the word occurs. Experimental results in recent years not only provide evidence for this general possibility that lexical, structural and interpretative processes interact, but also that these interactions occur right from the beginning of an utterance (Harris, 1978; Marslen-Wilson, 1975; Marslen-Wilson & Tyler, 1980; Tyler & Marslen-Wilson, 1977).

Having described the on-line interactive view in these general terms, I will now review some of the main studies which support it and which further clarify the nature of the word recognition process. One of the early experimental tasks providing evidence for the on-line interactive view is speech shadowing in which the subject is required to repeat back as quickly as possible speech heard over headphones (Marslen-Wilson, 1973, 1975, 1985; Marslen-Wilson & Welsh, 1978). The repetition delay or shadowing latency is the time between the onset of a word in the taped material and the onset of the same word in the subject's repetition. The early experiments (Marslen-Wilson, 1973) showed that some subjects, the 'close' shadowers, could reliably and accurately repeat back or shadow connected prose at a mean delay of only about 250 milliseconds, a little more than the duration of a syllable. By examining the quality of the subject's repetition, and by studying the variables which affected it, it was possible to discover what kind of analysis of the input was conducted by the subjects within a quarter of a second of hearing the material.

One approach, for example, was to examine the spontaneous errors made by close shadowers — e.g., instances where they changed a word in the original version to a new word. It was found that virtually all of these spontaneous errors were entirely consistent with the preceding context, including the word immediately before the error. This shows that even at a repetition delay of only quarter of a second, the subjects must have processed the material not just at a lexical level but also at structural and interpretative levels. Otherwise the errors would not have been constrained by the structural and interpretative properties of the speech input. Other analyses confirmed that the close shadowers were not different to more distinct shadowers as regards their perceptual processing of the material, but only in the strategy they adopted towards the production of an output. Thus, we may accept the performance of the close shadowers as revealing the nature of speech comprehension (Marslen-Wilson, 1985).

Another way of studying the on-line analysis of speech at structural and interpretative levels is to experimentally manipulate the spoken material to be shadowed and to study how the subject's performance changes as a result of these manipulations. For example, by replacing words in a spoken text in a principled way, it is possible to create a version which is syntactically equivalent to the original but semantically uninterpretable. This semantically uninterpretable version can then be scrambled to produce a further version which incorporates neither syntactic nor semantic constraints, but consists simply of unconnected strings of words. Marslen-Wilson (1985) shows that close shadowers have a longer repetition latency and produce more errors with the semantically uninterpretable prose whan with the normal prose, while their performance deteriorates still further with the material lacking both semantic and syntactic constraints. These studies, then, confirm that speech is analysed at structural and interpretative levels as it is heard. They also provide general evidence that the results of processing at one level guide and constrain processing at other levels.

Another task used to study the nature of on-line speech processing is 'word-monitoring'. This task, which is used as a measure of word recognition, involves the subject listening to a sentence for a prespecified target word. As soon as he encounters and recognises the target word, he presses a key which allows the time lapse between the acoustic onset of the target and his own recognition response to be measured. In general, recognition latencies for words in context average a quarter of a second to one-third of a second. Marslen-Wilson and Tyler (1980) used word monitoring to determine what kind of higher level structural and interpretative analyses were available to facilitate word recognition as speech was heard. To explain the results, it will be useful to present a sample of the test sentences used in their study:

Normal Prose
The church was broken into last night. Some thieves stole most of the *lead* off the roof.

Semantically Uninterpretable Prose
The power was located into great water. No buns puzzle some in the *lead* off the text.

Syntactically and Semantically Uninterpretable Prose
Into was power water the great located. Some the no puzzle buns in *lead* text the off.

(Marslen-Wilson & Tyler, 1980, p. 8)

The target words, italicised in the samples above, always occurred in the second (test) sentence. Each target word, it can be seen, was heard in three different prose contexts (1) normal prose (2) semantically uninterpretable prose and (3) semantically and syntactically uninterpretable prose. Note that in the case of (2) and (3) not only are semantic (or both syntactic and semantic) constraints absent from the text sentence, but the lead-in sentence does not provide a sensible context for the test sentence.

Results showed that word recognition was fastest in the normal prose condition. The advantage was present right from the beginning of the test sentence. This early advantage of normal prose over semantically uninterpretable prose at the very beginning of the sentence, of course, could only be due to the presence of a context sentence. This was confirmed by removing all the context sentences and rerunning the experiment. Deprived of the prior context sentence, word recognition latency at the beginning of the test sentence was the same for all three prose types. The results also showed that where the prose material had either a syntactic interpretation, or both a syntactic and a semantic interpretation, word recognition became faster as more of the test sentence was heard. This serial-position effect was most marked for normal prose where both syntactic and semantic constraints were available to facilitate word recognition.

In summary, these experiments suggest that the interpretation of a sentence begins right from the first word. They also show that the development of this interpretation involves not only the extraction of the syntactic and semantic implications of the initial words of the current sentence but also the immediate mapping of the unfolding fragment of that sentence onto the interpretative context provided by any prior discourse. Finally, they demonstrate that the larger interpretative domain in which a sentence occurs can be brought to bear on the recognition of even the first word of that sentence.

4. Lexical Processing, Structural-Interpretative Processing and Discourse

Given this research, which places the unfolding interpretation of the utterance directly in a discourse as well as a sentence context, it becomes essential to investigate the general nature of lexical, structural and interpretative operations in this larger linguistic-informational domain, and to determine what kind of informational dependencies affect these operations. And, in doing so, it is critical to use experimental tasks that are appropriate to the types of analysis process being studied. In fact a major claim of the unpublished research to be described in this section is that certain linguistic and informational variables can have quite different effects on lexical and structural-interpretative processing, but that these differences only become apparent when the appropriate experimental tasks are used. A complete account of the research is presented in Harris (1978).

As we have seen, research within the on-line interactive approach has tended to examine the properties of structural and interpretative processes only via their effects on word recognition. The typical tasks here, speech shadowing and, in particular, word monitoring, tap the relative speed of word recognition under different conditions. This means that the properties of structural and interpretative processes are only measurable in so far as they affect — and, in particular, facilitate — the basic process of word recognition. This places limitations on the kind of information about higher level processes that can be detected.

The limitations arise because the sensitivity of a word recognition task to higher level processes necessarily terminates as soon as the word recognition response is made. But being able to recognise a word need not signal the completion of all the structural and interpretative operations prompted by the input. As a word is being recognised, it may be assumed that this makes available a cluster of syntactic and semantic information which has to be integrated and interpreted with respect to the syntactic and interpretative properties of the current representation. Clearly, these higher level analyses can vary greatly in complexity and time course. But these variations may fail to be fully reflected in word recognition and even shadowing tasks since the response here can be made as soon as sufficient information (from whatever source) has accumulated to uniquely identify the word, independent of the completion of the further analyses associated with the word. To avoid these limitations of word-recognition-based tasks, the research now described uses a speeded sentence continuation task in which, we assume, the response cannot be made until the structural and interpretative ramifications of a particular input have been fully computed.

The basic question we are asking, then, is how the lexical processing of an utterance on the one hand, and the structural and interpretative processing of an utterance on the other, are affected by the informational dependencies which exist between the utterance and its larger discourse context. To provide data relevant to this question we must, therefore, not only use experimental tasks which distinguish lexical and structural-interpretative processing, but we must also experimentally manipulate different kinds of utterance-discourse dependencies. Before I try to describe the predictions in any more detail, it will be useful to first introduce a sample set of experimental sentences which represent the informational dependencies which are to be studied. I will then proceed to a description of the experimental tasks which are designed to distinguish lexical from structural-interpretative processing.

4.1 Informational Dependencies

The informational dependencies which were experimentally varied in the present research relate to the left-to-right (or temporal) order of information within an utterance, the presence or absence of a larger discourse context beyond the sentence and the 'given-new' relationships between utterance and discourse.

Context

Normal
(1) The boss has been displeased since employing the plumbers.

Irrelevant
(2) Valley in needed the to land badly rain this irrigate is.

Syntactic forms of target sentence

Pseudo-cleft
(3) What annoys him is their wasting time.

Passive
(4) He is annoyed by their wasting time.

Conjunction
(5) They waste time and that annoys him.

Reverse Pseudo-cleft
(6) Their wasting time is what annoys him.

It can be seen that there are two types of context material and four types of target sentence. All 72 sets of sentences used in the experiments were very similar to this set, except that the people and events described were changed. One of the two context types is a normal relevant sentence. The other is an irrelevant sentence in which the word order has been scrambled. The latter represents a "no-context" situation and simply provides a standard of comparison against which the effect of adding normal relevant context can be judged.

Looking at the target sentences now, it can be seen that each one contains both "old" and "new" information in relation to the normal context sentence. For example, the first clause of Sentence 3, the words "what annoys him", restates old information which occurs in the normal context sentence in the form "the boss has been displeased". In two types of target sentence, that is (3) and (4), the old information occurs early in the sentence while in the other two types, that is (5) and (6), it occurs late. It should be mentioned in passing that the form of these context and target sentences was determined by a number of other factors which are not central to the present discussion but which are discussed fully elsewhere (Harris, 1978).

The main prediction is that in the normal context condition, the complexity of structural-interpretative processing has to increase at some point as new information clauses are processed. In the case of lexical processing, in contrast, the addition of a normal prior context sentence should only have a facilitating effect (or else have no effect), whether information being encountered is old or new. The argument for this revolves around what I believe is a crucial difference between the nature of lexical processing on the one hand and structural interpretative processing on the other. The point is that structural-interpretative processing is essentially *relational* in character, and so it cannot proceed without making constant reference to, and identifying the changing interconnections within, the developing partial representation of both context and target sentences. In contrast, though it is facilitated by the constraints deriving from processing of earlier occurring material (Marslen-Wilson & Tyler, 1980), lexical processing does not have to wait on the *completion* of higher level analyses of the speech stream. A word may well have been recognised even while other levels of analyses are still in progress.

According to this line of reasoning, then, the addition of a normal prior context sentence can only have the effect of facilitating lexical processing or leaving it unchanged. But the addition of the same normal context sentence must at some point increase the complexity of on-line structural-interpretative processing, because of the increased number of components in the existing representation to which incoming new information must be related. If incoming information at a particular moment is old, that is, if it has already been encountered in some form in the context sentence, structural-interpretative processing may instead be facilitated at that point in the target sentence. Thus, we predict that, given that the target sentence contains some new information, context will, at least sometimes, have opposite effects on immediate structural-interpretative and lexical processing.

4.2 Experimental Tasks

To test these predictions, we needed separate measures of immediate (a) structural-interpretative and (b) lexical processing. Word monitoring latency provided a straightforward measure of lexical processing. None of the existing on-line tasks, however, can be considered reasonably direct measures of structural-interpretative processing and, consequently, none of them would be likely to be sensitive to the kind of context effects predicted.

Previous studies using on-line linguistic tasks, such as word monitoring and speech shadowing, have indeed tried to describe the progress of structural-interpretative processing. But they have done this in an indirect fashion, by examining and making inferences from, changes in the number of constraints on lexical processing. In the present case, however, we wanted to measure changes in the complexity of structural-interpretative processing itself — not just changes in the state of on-line representation being constructed or consequential changes in the number of constraints on lexical processing. The measure of structural-interpretative processing we developed was a modified, short-latency version of the familiar sentence-fragment completion task. In this on-line version of the task, subjects hear a fragment of a sentence, with twin signal lights being instantly activated at the acoustic offset of the last word of the fragment. Their task is to generate a one-word appropriate continuation as quickly as possible and, then, without time pressure, to produce a completion for the sentence which incorporates that one-word continuation. Latency to initiate continuations is, on average, between 600 and 700 milliseconds for the sentences in these experiments.

Space does not permit me to go into too much detail about the various arguments for considering sentence continuation latency as a measure of structural-interpretative processing. It may be noted at least, however, that in this task subjects have to generate the appropriate continuation word themselves, so that they cannot begin to prepare and initiate a response, until material up to the last word heard has been fully analysed at structural and interpretative levels. In contrast, responses on the various lexically based monitoring and shadowing tasks can be made without necessarily waiting for the completion of analyses at these levels. Consequently, response latencies on these latter tasks may not directly or fully reflect fluctuations in the complexity of structural-interpretative processing. We assume, incidentally, that an increase in continuation latency reflects an increase in the complexity of structural-interpretative processing at that point. And a decrease in sentence continuation latency is assumed to reflect a reduction in the complexity of structural-interpretative processing.

Lastly, we come to the 'target position' factor, that is, the various positions in the target sentence at which processing activity is measured. There are six levels of target position, corresponding to the first six words of the target sentence. Thus, target position 3, for example, refers to a situation where the subject hears the first three words of the target sentence before receiving a visual signal to provide a continuation. At the corresponding position 3 in word monitoring the prespecified target word is the third word of the sentence.

Exactly the same linguistic materials were run under word monitoring and sentence continuation conditions in separate experiments. In both experiments subjects were presented with only one of the eight possible context-sentence/target-sentence combinations from each of the 72 sets in order to avoid the problem of their becoming familiar with the materials. Subjects also heard practice and filler sentences. A different group of 24 male and female University of Chicago students took part in each experiment, making a total of 48 subjects in all.

4.3 Results

4.3.1 Sentence-Discourse Dependencies

All statistical analyses reported here take account of error due to sampling of subjects and sentences. Looking at the last column in Table 1 it will be seen that while the presence of a normal prior discourse context reduces word monitoring latency overall, it increases sentence continuation latency. This fact was also reflected in a significant *context x task* interaction in an ANOVA where *task* referred to word monitoring or sentence continuation. Looking at Figures 1-4 now, it can be seen that at some positions sentence continuation latency is actually reduced by the addition of a normal prior context sentence. In other words the overall direction of the context effect for sentence continuation is reversed at some target positions. All these reversals occur in what I have called old information clauses. This result is reflected in a number of ways in the various statistical analyses I conducted. For example, in an analysis of variance of the sentence continuation latencies alone, the interaction *context x position x target syntactic type* was significant. I interpreted this interaction as being due to the fact that old information clauses, and hence reversals, occurred early in two target types and later in the other two.

Experimental Task	Context	Target position						All positions
		1	2	3	4	5	6	
Word Monitoring	Normal	359	381	321	288	255	336	323
	Irrelevant	416	438	340	298	273	341	351
Sentence Continuation	Normal	732	692	590	632	745	556	658
	Irrelevant	701	655	570	605	704	555	632

Table 1. Mean Latency in Milliseconds according to Task, Context, and Target Position.

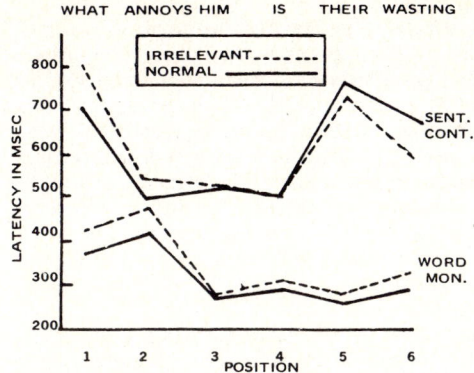

Figure 1. Mean latency at each position for pseudocleft.

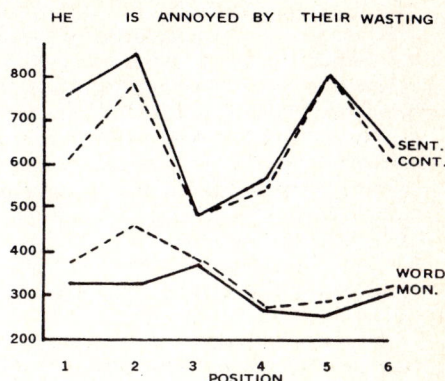

Figure 2. Mean latency at each position for passive.

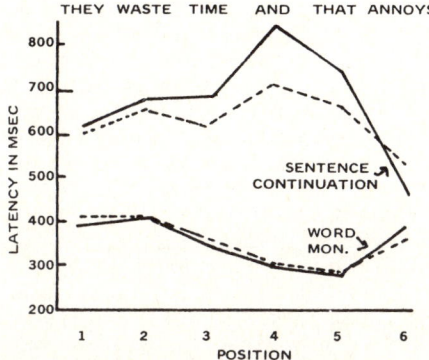

Figure 3. Mean latency at each position for 'conjunction'.

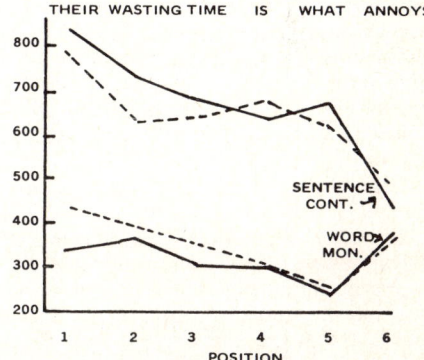

Figure 4. Mean latency at each position for reverse pseudocleft.

To check this interpretation I conducted a planned comparison of mean sentence continuation latencies for all old information clauses and for all new information clauses. When I tested the difference between these means under the irrelevant context condition the result was not significant, since, of course, the old/new distinction could not be relevant when prior context is irrelevant. Not surprisingly, however, a similar test under the normal context condition showed that old information clauses have significantly shorter mean sentence continuation latencies than have new information clauses. In other words, the reoccurrence of old information in the target sentences leads to a reduction in the complexity of structural-interpretative processing while that old information is unfolding.

Normal discourse context, then, affects both the immediate structural-interpretative and lexical processing of subsequent sentences. At a given position in a sentence, discourse context either facilitates lexical processing or leaves it unchanged. But that very same discourse context may either significantly increase or significantly decrease the complexity of structural-interpretative processing at that same position, depending on whether the discourse context renders incoming information old or new.

4.3.2 Sentence Internal Dependencies

So far we have been looking at informational dependencies between the target sentence and the discourse context. I would now like to turn briefly to one or two aspects of order of information within the target sentence itself and its effects on structural-interpretative and lexical processing. One incidental thing to note first is that the pattern of latencies across target positions for the two tasks supports the claim that word monitoring and sentence continuation measure different aspects of processing. Examining the curves in Figures 1-4, and mentally averaging across contexts at each target position, it can be seen that the pattern of change in mean latency across target positions is quite different for the two tasks. This difference is shown to be significant by a *task x position* and *task x position x target syntactic type* interaction in an ANOVA of the data from these two tasks. This adds further weight to the claim that the experimental tasks distinguish two aspects of on-line processing.

There are two aspects to within-target sentence order of information which I would like to mention. The first concerns the role of the clause boundary versus order of information, and the second concerns the effect of the interpretation of the sentence becoming more established as more of the sentence is heard.

Reference to Figures 1-4 shows that, contrary to strict serial-autonomous models of spoken language processing, the clause boundary, position 3, is not associated with any consistent pattern of change in the complexity of processing as reflected in the mean sentence continuation latencies. In particular, mean latencies for the four syntactic forms do not coincide at *position 3* as they would be expected to if the clause boundary as such marked a consistent stage in the progress of structural processing (Fodor, Bever & Garrett, 1974). In fact, significant differences in mean latency between target syntactic forms at the clause boundary can better be accounted for in terms of a specific feature of the order of information contrasts. Note that at position 3 the means are very similar for each member of those pairs of syntactic forms whose first clause contains basically the same information. The first clause of the "passive" and "pseudocleft" form one such same-information pair, and the first clause of the "conjunction" and "reverse pseudocleft" form another same-information pair. When comparisons are made between these pairs, however, first clause information is quite different. Planned comparisons between mean sentence continuation latencies at position 3 showed that when target syntactic forms contained basically the same information up to the clause boundary, mean latency for these syntactic forms did not differ at the clause boundary. When first clause information was dissimilar, mean latencies for the target syntactic forms *were* different at the clause boundary. Order of information, then, rather than the clause boundary as such, seems to be the crucial factor in explaining variation across word positions in the complexity of structural-interpretative processing.

The second main issue concerns the effect on structural and lexical processing of how much of the sentence has already been heard. Two types of analyses were conducted. First, a linear component was subtracted from the overall target position curves. It was found that latency for both word monitoring and sentence continuation was a decreasing function of position, the regression coefficient being significant for each task. This is evidence that there is a tendency for the complexity of structural-interpretative processing to decrease, and for lexical processing to be increasingly facilitated, as sentences unfold.

In case this tendency might be confounded with word length or frequency effects, another more carefully controlled test was made. This involved comparing mean latencies at early positions and late positions, selecting from the target syntactic forms in such a way that the target words monitored, or the words which occurred at the end of sentence fragments, were identical. Planned comparisons showed that sentence continuation latencies and word monitoring latencies were significantly shorter at later compared to earlier positions in target sentences. This supports the on-line interactive view that expectations generated on-line facilitate later structural-interpretative as well as lexical processing. This is the first time, however, that the effect has been directly demonstrated in the case of structural-interpretative processing, although as was pointed out earlier, the effect has been previously demonstrated in the case of lexical processing by Marslen-Wilson and Tyler (1980).

In conclusion, then, these experiments seem to confirm the importance of both sentence-internal informational structure and sentence-discourse informational dependencies in determining the progress of immediate lexical and structural-interpretative processing as speech is understood. To this extent, they add to the other evidence already presented for an on-line interactive model of speech comprehension.

5. Temporal and Contextual Factors in Word Recognition

We turn now to a number of experiments which provide a more detailed description of how words in continuous speech are recognised. These experiments have two main themes (a) the balance between reliance on the sensory input and reliance on discourse context during word recognition and (b) the role of context in determining the speed and earliness of word recognition. A number of experiments show that listeners rely less on the actual sensory input as various kinds of constraints, lexical, structural or interpretative, increase. For example, Marslen-Wilson and Welsh (1978) asked subjects to shadow spoken material in which certain critical trisyllabic words had been mispronounced. The predictability of each of these words from the preceding context had been independently established. Results showed that the more predictable the word, the less likely the shadowers were to perceive a mispronunciation and, correspondingly, the more likely they were to fluently restore the word to its correct form in repeating it. It was also clear that, irrespective of contextual constraints, minor mispronunciations of words (one feature changes) in continuous speech tended not to be noticed and the words were repeated in their correct form. Major mispronunciations (three-feature changes), however, were likely to be missed only when they were located later in the word and when that word was highly predictable. The results suggest, then, that the acoustic phonetic input corresponding to the beginnings of words is attended to relatively carefully irrespective of the strength of the contextual constraints. They also show that the acoustic signal is attended to less carefully later in the word, particularly in highly-constrained contexts. An earlier shadowing experiment (Marslen-Wilson, 1975) produced broadly similar findings to these.

The study by Marslen-Wilson and Tyler (1980) mentioned earlier also provides evidence on the balance between the listener's dependence on the sensory input and the strength of the contextual constraints which are available. It will be recalled that the Marslen-Wilson and Tyler experiment involved monitoring target words in three different kinds of prose context: normal prose, semantically uninterpretable prose, and semantically and syntactically uninterpretable prose. The same target words were monitored in each of the three kinds of prose. The result of interest to the present discussion concerns the strength of the relationship between the duration of the target words and word monitoring latency. Regression analyses showed significant linear effects of word duration in all three prose contexts, but these effects were much stronger in semantically uninterpretable prose than in normal prose. The effects were stronger still in the case of prose which was both semantically and syntactically uninterpretable, that is, where the sensory input itself was all that the subject had available as a basis for recognising the word. This suggests that as contextual constraints decrease, dependence on the sensory input increases. And as dependence on the sensory input increases, then sensory or word level variables (such as word duration in this case) tend to be more closely related to word recognition time.

These and other experiments have also been used to show that spoken words in context are recognised before sufficient sensory information has become available to actually allow a decision about their identity (Grosjean, 1980; Marslen-Wilson, 1984; Marslen-Wilson & Tyler, 1980; Tyler & Wessels, 1983). This has led to the development of a model of spoken word recognition (Marslen-Wilson & Tyler, 1980) which proposes that the utterance context plays a direct role by narrowing down the choice of words which are still consistent with the amount of sensory input available at the moment of identification. Grosjean (1980), for example, used the gating paradigm to study the amount of sensory information necessary to identify words. The gating paradigm involves presenting listeners with successively larger fragments of a word and asking them each time to try to identify the word. Starting from word onset the duration of the acoustic fragments added at each step was 30 milliseconds. Grosjean found that when these words were presented in normal sentence contexts, the average duration of the sensory input required for their successful identification was only 199 milliseconds. Since the average total duration of the words was 510 milliseconds, this means that listeners were able to recognise words in normal context on the basis of only half the total acoustic signal available. This of course is broadly consistent with the evidence described earlier showing that it was only at the beginnings of words that the sensory input was attended to with sufficient care to produce a high detection rate for mispronunciations. Incidentally, Marslen-Wilson (1980, 1984), basing his estimate on word monitoring and shadowing data, arrives at the same average figure as Grosjean (200 milliseconds) for the recognition of words in continuous speech.

As Marslen-Wilson points out, 200 milliseconds is equivalent to the duration of a consonant and a following vowel in words taking that initial form. This would suggest that word recognition in continuous speech occurs before sufficient acoustic-phonetic information has accumulated to actually specify the word uniquely. Marslen-Wilson and Tyler (1980) and Marslen-Wilson (1980, 1984), confirm that this is so by reference to data on initial phonemic sequences in English words. They show in the case of their word monitoring experiment that if a listener has available only the first two phonemes in a word, there will be a median of about 26 words that show this initial sequence. In fact, even after three phonemes there would still be a median of six words consistent with the sequence. Thus, if listeners typically have identified words in continuous speech within 200 milliseconds of the acoustic onset, they must be doing so on the basis of sensory information which is actually insufficient. Clearly, then, contextual information must be providing the additional constraints on the identity of word necessary to permit such early identification.

It was in order to account for results such as these that Marslen-Wilson and Tyler (1980) proposed a new "cohort"-based interactive theory of spoken word recognition. This theory argues that as soon as the initial sound sequence of a word is heard, a group of recognition elements become active. The number of elements so activated equals the number of words in the language that begin with that sound sequence. This relatively large group of word candidates is called the word-initial cohort. The remaining task of the word recognition system then is to reduce this word-initial cohort to one, i.e., the target word. If the word happens to be heard in isolation, this reduction of the cohort will be achieved by the detection of mismatches between the internal acoustic-phonetic specifications of the recognition elements and the results of the on-going analysis of the unfolding acoustic-phonetic signal. As the Marslen-Wilson and Tyler results mentioned above show, isolated words can be uniquely identified in this way even when only part of the acoustic-phonetic signal corresponding to them has become available.

The recognition elements are also capable of interacting with contextual constraints when words are heard in continuous speech. In that case, recognition elements whose specifications do not match those of the current discourse context can be excluded from the initial cohort. This accounts for the fact that words in context can be identified before sufficient acoustic-phonetic information to justify recognition has become available. Thus, it appears that an optimum balance is maintained between acoustic-phonetic and contextual information sources. No more of the acoustic-phonetic information need be analysed than is required to achieve identification given the contextual constraints available. Finally, it may be noted that the word-initial cohort is generated from the acoustic-phonetic input only, thereby setting important limits on the powerful contribution of context. Context alone, therefore, cannot override the evidence provided by the actual speech signal in facilitating word recognition.

6. Speech Comprehension and Lexical Failure

We will now consider some of the implications of this research on speech comprehension for a number of issues related to lexical failure and the organisation of conversation. The general possibility we will explore is that the relative smoothness and coherence of conversation, and its relative freedom from failure, may in large part be explainable in terms of the structure and operation of the speech comprehension system. We will begin by identifying some of the features of the speech comprehension system which ensure that the considerable potential for perceptual and lexical failure which exists in most conversations rarely results in actual communication failure. We will then go on to show how the temporal structure of speech comprehension is critical to the precise coordination of the various elements in conversation and to the smooth and efficient exchange of information. First, we consider the relationship between speech comprehension and lexical failure.

We can begin by listing just a few examples of the kinds of factors and situations in which, other things being equal, a relatively high level of perceptual and lexical failure might be expected: conversations between native speakers who have widely different accents or dialects; conversations conducted under noisy conditions; conversations in which one or more of the participants is a non-native speaker; conversations between experts and non-experts involving the use of technical language; conversations between children and adults; conversations between native speakers whose lexicons are only partly shared due to such factors as educational background. Such conversations are quite commonplace and do not seem to be markedly more vulnerable to serious communicative failure than other kinds of conversations. Part of the reason for this, as has been noted by many authors, may be that the structure of discourse itself changes in various ways as potential failures develop. There are a number of features of the speech comprehension system itself, however, which appear to be particularly well adapted to the circumvention of potential failure and to the minimization of its communicative consequences. These very same features of the speech comprehension system also allow the precise and early identification of those failures which do occur so that the need for repair can be promptly signalled and the appropriate procedures executed.

The research already reviewed identifies three related features of the speech comprehension system which seem to have particularly important implications for conversation failure: (1) the immediate and direct involvement of all available contextual information sources in the facilitation of spoken word recognition; (2) the flexible balance between reliance on context and reliance on the sensory input depending on local conditions; and (3) the earliness of word recognition and the speed with which the structural and interpretative implications of a word choice are extracted. We have already referred, for example, to research showing that when contextual constraints are strong, minor mispronunciations anywhere in a word, and major mispronunciations later in a word, are often not detected by the listener. And these experimental results were obtained under rather ideal conditions where both speaker (on tape) and listener were native speakers with very similar linguistic systems. Clearly, in conversations involving either non-native speakers or speakers with different dialects or accents, or in conversations conducted under noisy conditions, this feature of the processing system is likely to become more significant and to be a major factor in reducing what would otherwise be a considerable source of failure. While these kinds of "restorations" of the input are, strictly speaking, perceptual errors (failures of acoustic phonetic processing), they must also be counted as instances of the avoidance of potential communication failure since they lead to the successful recovery of the speaker's intended meaning. In any case, this kind of beneficial "tidying-up" of the input seems to be part of the normal process of speech comprehension.

Equally important from the point of view of potential conversation failure is the research showing the earliness of word recognition and the speed with which new words are incorporated into ongoing structural and interpretative analyses. The significance of this is the corresponding earliness with which the occurrence of perceptual or lexical failure during ordinary first pa processing becomes apparent to the listener. Consider just two examples of lexical failu Suppose, first that the correct word is clearly spoken but that the listener either recovers an incorrect meaning stored for this word in his mental lexicon or else fails to recover any meaning at all. Given the research reviewed earlier, we know that in the former case all the candidate meanings generated in the word-initial cohort are likely to have been declared mismatches long before the acoustic offset of the word. Or this early decision about the word might be that even the best of the candidate-meanings in the initial cohort provides an implausible or anomalous development of discourse. Needless to say, if the listener has no meaning at all stored for the target word in his mental lexicon, the occurrence of lexical failure is likely to become obvious even earlier. The important point is that in each case the listener has been alerted with optimal speed to the fact that a problem exists and that corrective measures are needed.

As a second example, consider a situation where the speaker unknowingly uses the wrong word to convey his intended meaning. Assume also that the word actually spoken is accurately identified by the listener and its conventional meaning (though not the speaker's intended meaning) recovered. Again, the listener will realise well in advance of the acoustic offset of the signal corresponding to that word that the word he is hearing does not provide a sensible extension of discourse. In fact, given that the system normally arrives at a final word choice by detecting mismatches between the initial cohort candidates and the available discourse constraints, this kind of lexical failure might become apparent marginally before the normal recognition point for that word when heard in context.

But if lexical failure can be identified so early, it implies that remedial action can also be initiated while the sensory input is still accumulating. The real time lapse between the initial decision point in a word and the acoustic offset of that word provides an opportunity for corrective action to be undertaken before processing of the next word has to begin. There are numerous possible remedial measures which could be implemented, the choice depending to some extent on the listener's perception of the origins of the failure.

7. The On-Line Repair of Lexical Failure

One possibility would be for the listener to reconsider whether the initial acoustic-phonetic analysis of the word could conceivably be consistent with a different, perhaps larger, word-initial cohort. This response is probably most likely to be favoured under noisy listening conditions, or where the speaker's accent or dialect is unfamiliar to the listener. Nevertheless, it is quite possible that this kind of revaluation is occasionally necessary even with familiar accents. Note, for example, the potential for error in acoustic-phonetic processing which is suggested by the fact that minor mispronunciations of words are often not even noticed by listeners during first pass processing (Marslen-Wilson & Welsh, 1978). In effect, the listener in this type of attempted repair examines the possibility that the original signal, though no longer directly available, may have been consistent with a greater range of acoustic-phonetic analyses. A new, larger word initial cohort, consistent with this greater range of possible acoustic-phonetic analyses, might then be evaluated against the available context to see if a more plausible word candidate could be identified.

Another possibility is that the listener may conclude that the meaning for the word which he has stored in his mental lexicon is, in fact, incorrect. This response is most likely in situations where the listener has a high level of confidence both in the accuracy of his own acoustic-phonetic processing and in the adequacy of the speaker's command of the language, but where the listener is also relatively uncertain of his own mastery of the language. In these circumstances, the listener will probably be obliged to resort to more radical remedies such as guessing the word on the basis of context, waiting for subsequent discourse to resolve the issue, or signalling lack of comprehension to the speaker. The success of some of these more extreme on-line corrective measures depends once again on the nature of the normal speech comprehension system. For example, the immediate availability of a comprehensive set of contextual constraints already being brought to bear on the basic word recognition process should serve to maximise both the speed and accuracy of any guess which has to be made about the correct meaning of the word. The on-line sentence continuation data examined above (Harris, 1978) suggests that the maximum amount of time to generate such a guess on the basis of context would be less than about half a second. An even shorter time to generate a guess is suggested by the fact that the spontaneous restoration of words at a shadowing latency of about quarter of a second almost always yields new words which are contextually appropriate in every way (Marslen-Wilson, 1975, 1985). It should be emphasized, incidentally, that the kind of guessing under discussion just now is guessing undertaken as a remedial measure. The available evidence suggests that normal first pass processing does not involve the 'top-down' generation of word candidates (Marslen-Wilson, 1984).

Another response to lexical failure would be to assume that it was due to the speaker using the wrong word to convey the intended meaning. The listener will most readily adopt this assumption, presumably, where the speaker's command of the language is obviously incomplete and where the listener himself has a high level of confidence both in his own knowledge of the language and in the accuracy of his acoustic-phonetic processing. Among the remedies he would be likely to consider under these conditions again would be guessing on the basis of existing discourse constraints or signalling failure to the speaker.

One other situation which is worth considering briefly is where the listener, for whatever reason, has some degree of doubt about the word-meaning initially recovered. His judgement is that the best candidate-meaning arrived at on-line provides an implausible development of discourse but one which, nevertheless, is not clearly anomalous or incorrect. In that case, he can resort to one or more of the remedies already mentioned, e.g., tag the recovered meaning as questionable and wait for subsequent discourse to resolve the issue, or, signal failure to the speaker.

It must be admitted that this account of lexical failure and its repair is both speculative and incomplete. For example, only the more extreme forms of lexical failure have been listed. Very often the meaning of a word may only be vaguely known to the listener or he may only be able to set rather broad limits on its possible meanings. Alternatively, a word which is inappropriately used by a speaker may nevertheless provide the listener with a clue to the intended meaning because of some acoustic similarity to another word which would indeed be consistent with the available discourse context. The kind of corrective action taken by listeners in such situations may be different to, perhaps less radical than, that taken in more clear-cut cases of failure. We have not discussed either the possibility that a number of different types of corrective action (e.g., 'signal failure', 'guess on the basis of context') might be initiated simultaneously. Nor have we tried to indicate whether any remedial processes — such as the re-evaluation of the acoustic-phonetic analysis in order to generate a new word-initial cohort — might be automatically triggered by initial lexical failure. And, of course, we have not mentioned at all what is potentially the most serious kind of failure — where the wrong word meaning is initially recovered and is mistakenly accepted by the listener because it provides a satisfactory development of discourse. Ringle and Bruce (1982, p. 209) provide a telling example of the communicative consequences of this kind of failure.

The problem is that the time course of lexical failure and of its attempted remediation during speech comprehension has not been systematically studied. In addition, no detailed account is available of how the speech comprehension system copes with non-native speech or with unfamiliar accents and dialects (Harris, 1980). Nor are there any on-line studies of how speech comprehension processes in native and non-native speakers might differ. In the absence of such research, the present attempt to assess some of the implications of existing speech comprehension research can be justified on two grounds: (a) the need to indicate the importance of a neglected area of research — the basic processes of lexical failure and its repair; and (b) the value of even a loose definition of the range of problems and processes which might be involved in lexical failure and repair.

Quite apart from these considerations, however, there are some reasons for thinking that extrapolating from research on ordinary speech comprehension, as we have done here, may not lead us too far astray. It seems reasonable to assume, for example, that the structure of the speech comprehension system has developed in response to the range of conditions with which it must typically contend, e.g., noise (including extraneous speech), unfamiliar accents and dialects, speech in languages that are only partly known, the speech of children and adults, speech incorporating specialized vocabularies, jargon or slang. We might also expect the distinctive or unique features of the speech comprehension system to contain the key to how the system itself adapts to these different conditions. The existing on-line research which we discussed earlier has documented the essential characteristics of the speech comprehension of competent native speakers under good listening conditions. The argument here, then, is that the characteristics of the system identified by that research — its interactive nature, the earliness of word recognition, the on-line employment of all available knowledge sources at the earliest possible moment in the interpretation of the signal, and so on — that these are also the keys to the system's adaptability in the face of the more unfavourable conditions where failure is likely to threaten. In fact, the more or less constant possibility of encountering conditions which provoke lexical failure could plausibly be advanced as a major factor in determining the structure of the speech comprehension system. And we have already noted that the speech comprehension system seems to be particularly well geared to the repair of lexical failure.

This is a convenient point at which to return briefly to the question of defining input failure which we raised early in the paper. In view of the highly interactive nature of speech comprehension, and the probability that many of the failures which do occur are rapidly repaired on-line, it seems clear that the Ringle and Bruce distinctions between perceptual, lexical and syntactic input failure are overly simple if not actually misleading. If we are to adequately account for input failure, we surely need to develop a descriptive system which does justice not just to those failures which do get through the net, as it were, but also to the large variety of invisible failures which are repaired on-line. In fact, unless we can first explain the complex and subtle mechanisms by which these on-line repairs are effected, it is difficult to see how we can be successful in accounting for those failures which persist. The point is that failures to recover the speaker's intended meaning, or the short-circuiting of such failures on-line, are really products of the speech comprehension system as a whole. Thus, attempts to locate the origin of different kinds of failure within particular linguistic or other knowledge sources can only be useful if the definitional system being employed takes full account of the true processing structure of speech comprehension.

8. The Temporal Coordination of Conversation and Speech Comprehension

So far we have been discussing the relationship between the structure of the speech comprehension system and the process of lexical failure and repair. At a more general level, however, it is clear that the whole temporal coordination of conversation, and the tracking of the various dynamic elements which contribute to it, are also intimately bound up with the structure and operation of the speech comprehension system. Critical to this aspect of the management of conversation is the close temporal relationship which exists between, (a) the production of the speaker's output, (b) the interpretation of that output by the listener, and (c) the rapid, on-line adjustment and updating of both the speaker's and listener's model of discourse. We know from the speech comprehension data that even as the speaker produces the acoustic signal corresponding to the end of a word, the listener will often have already identified the word and will have assessed its implications in terms of the model of discourse constructed up to that point. A consequence of this, as we pointed out above, is that any developing communication failures can be identified by the listener and signalled to the speaker with great speed. In fact, visual signals to indicate failure can be initiated by the listener while the speaker is still delivering the very word which caused the failure or which brought it to light. The close temporal association between speech production and comprehension also means that visual signals can pinpoint the critical word for the speaker with great accuracy. The usefulness of such signals is further enhanced, of course, by the fact that they have as a referential background speaker and listener models of discourse which have been precisely tracked and confirmed up to that point.

Needless to say, this kind of rapid, accurate and non-disruptive listener-speaker feedback during dialogue is only possible with an on-line interactive speech comprehension system. In a serial-autonomous system – which would necessarily involve a real time lapse between the initiation of processing at different linguistic levels – such fine temporal meshing of the various elements contributing to conversation could not be achieved. There is little doubt either that considerably more elaborate and vulnerable methods of tracking input failure and model failure would have to be implemented. The visual signalling of failure, for example, would have to be more complex, and would almost certainly be much more prone to error, under a serial autonomous system than under an on-line interactive system.

At the most general level, then, all this points to the fact that the process of conversation failure and repair can only be fully understood in terms of the basic operations of the speech production and speech comprehension systems. It also highlights the importance of studying the precise, temporal coordination of conversation and speech processing events. Finally, it draws attention to the need for research to be sensitive to the fundamental differences between auditorily-based and visually-based dialogue and processing systems – for example, the fact that in auditorily-based systems processing of a word can begin while the signal is still accumulating (Marslen-Wilson, 1984). Significantly, some of the most important developments in speech comprehension research over the last twenty years occurred when researchers first switched from using post-sentence measures of processing (which could not capture the dynamic, temporal aspects of speech) to using on-line immediate measures. There is every reason to expect that in the case of communication failure and repair, also, really worthwhile progress will have to wait until researchers are prepared to turn their attention more directly to the fine temporal sequence of events which occur as conversation proceeds.

References

Celce-Murcia, M. (1980). On Meringer's corpus of "slips of the ear". In V.A. Fromkin (Ed.), *Errors in linguistic performance: Slips of the tongue, ear, pen, and hand*. London: Academic Press.
Day, R.R., Chenoweth, N.A., Chun, A.E., & Luppescu, S. (1984). Corrective feedback in native-nonnative discourse. *Language Learning, 34*(2), 19-43.
Fodor, J.A., Bever, T.G., & Garrett, M.F., (1974). *The psychology of language*. New York: McGraw-Hill.
Forster, K. (1979). Levels of processing and the structure of the language processor. In W.E. Cooper and E.C.T. Walker (Eds.), *Sentence processing: Psycholinguistic studies presented to Merrill Garrett*. Hillsdale, NJ: Erlbaum.
Fromkin, V.A. (Ed.) (1973). *Speech errors as linguistic evidence*. The Hague: Mouton.
Fromkin, V.A. (Ed.) (1980). *Errors in linguistic performance: Slips of the tongue, ear, pen, and hand*. London: Academic Press.
Grosjean, F. (1980). Spoken word recognition processes and the gating paradigm. *Perception and Psychophysics, 28*, 267-283.
Harris, J.W. (1978). *Levels of speech processing and order of information*. Unpublished PhD dissertation, University of Chicago, Chicago.
Harris, J.W. (1980). Why the second language speaker comprehends more slowly. *Teangeolas, 11*, 18-26.
Long, M.H. (1983). Native speaker/non-native speaker conversation and the negotiation of comprehensible input. *Applied Linguistics, 4*(2), 126-141.
Marslen-Wilson, W.D. (1973). Linguistic structure and speech shadowing at very short latencies. *Nature, 244*, 522-523.
Marslen-Wilson, W.D. (1975). Sentence perception as an interactive parallel process. *Science, 189*, 226-228.
Marslen-Wilson, W.D. (1976). Linguistic descriptions and psychological assumptions in the study of sentence perception. In R.J. Wales and E.C.T. Walker (Eds.), *New approaches to language mechanisms*. Amsterdam: North-Holland.
Marslen-Wilson, W.D. (1980). Speech understanding as a psychological process. In J.C. Simon (Ed.), *Spoken language generation and understanding*. Dordrecht, Holland: D. Reidel.
Marslen-Wilson, W.D. (1984). Function and process in spoken word recognition. In H. Bouma and D. Bouwhuis (Eds.), *Attention and performance X: Control of language processes*. Hillsdale, NJ: LEA.
Marslen-Wilson, W.D. (1985). Speech shadowing and speech comprehension. *Speech Communication, 4*, 55-73.
Marslen-Wilson, W.D., & Welsh, A. (1978). Processing interactions and lexical access during word recognition in continuous speech. *Cognitive Psychology, 10*, 20-63.
Marslen-Wilson, W.D., & Tyler, L.K. (1980). The temporal structure of spoken language understanding. *Cognition, 8*, 1-71.
Ringle, M.H., & Bruce, B.C. (1982). Conversation failure. In W.G. Lehnert and M.H. Ringle (Eds.), *Strategies for natural language processing*. Hillsdale, NJ: LEA.
Tarone, E. (1980). Communication strategies, foreigner talk, and repair in interlanguage. *Language Learning, 30*, 417-431.
Tyler, L.K. (1981). Serial and interactive theories of sentence processing. *Theoretical Linguistics, 7*, 29-65.
Tyler, L.K., & Marslen-Wilson, W.D. (1977). The on-line effects of semantic context on syntactic processing. *Journal of Verbal Learning and Verbal Behaviour, 16*, 683-692.
Tyler, L.K., & Wessels, J. (1983). Quantifying contextual contributions to word recognition processes. *Perception and Psychophysics, 34*(5), 409-420.

Communication Failure in Dialogue and Discourse
R.G. Reilly (Editor)
© Elsevier Science Publishers B.V. (North-Holland), 1987

META-RULES AS A BASIS FOR PROCESSING ILL–FORMED INPUT[1]

Ralph M. Weischedel[2]

BBN Laboratories
10 Moulton Street
Cambridge, MA 02238

Norman K. Sondheimer

USC/Information Sciences Institute
4676 Admiralty Way
Marina del Rey, CA 90292

If natural language processing systems are ever to achieve natural, cooperative behavior, they must be able to process input that is ill-formed lexically, syntactically, semantically, or pragmatically. Systems must be able to partially understand, or at least give specific, appropriate error messages, when input does not correspond to their model of language and of context. We propose meta-rules and a control structure under which they are invoked as a framework for processing ill-formed input. The left-hand side of a meta-rule diagnoses a problem as a violated rule of normal processing. The right-hand side relaxes the violated rule and states how processing may be resumed, if at all. Examples discussed in the paper include violated grammatical tests, omitted articles, homonyms, spelling/typographical errors, unknown words, violated selection restrictions, personification, and metonymy. An implementation of a meta-rule processor within the framework of an augmented transition network parser is also described.

1. Introduction

Natural language understanding systems have improved markedly in recent years, and natural language interfaces have even begun to enter the commercial marketplace, for example, the INTELLECT system of Artificial Intelligence Corporation (Harris 1978). These systems promise to make major improvements in the ease of use of data base management and other computer systems. However, they have only begun to consider the problems of truly natural input. The emphasis has been, and continues to be, on the understanding of well-formed inputs. True natural language input is often ill-formed *in the absolute sense* of being filled with misspellings, mistypings, mispunctuations, tense and number errors, word order problems, run-on sentences, extraneous forms, meaningless sentences, and impossible requests. In addition, natural input is ill-formed *in the relative sense* of containing requests that are beyond the limits of either the computer system or the natural language interface.

Evidence indicates that absolutely ill-formed input regularly occurs in a data base query environment. For instance, in an extensive study (Thompson 1980) including 1615 inputs, only

---

[1] This material is based upon work supported in part by the National Science Foundation under Grant Nos. IST-8009673 and IST-8311400 and in part by the Defense Advanced Research Projects Agency under Contract No. MDA 903-81-C-0335, ARPA Order No. 2223. Views and conclusions contained in this paper are the authors' and should not be interpreted as representing the official opinion or policy of DARPA, the U.S. Government, or any person or agency connected with them.
[2] Currently visiting at the Department of Computer &Information Science, University of Pennsylvania, Philadelphia, PA 19104.

Reprinted with permission of the Association of Computational Linguistics.

1093 were parsable, and an overall total of 446 contained various kinds of errors: 161 with vocabulary problems, 72 with punctuation errors, 62 with ungrammaticality, and 61 with spelling errors. Furthermore, 211 inputs were fragmentary, including 91 parsed terse replies and 67 parsed terse questions.

In another experiment (Eastman and McLean 1981), 693 English queries to a data base system were analyzed. Co-occurrence violations, including subject-verb disagreement, tense errors, apostrophe problems, and possessive/plural errors occurred in 12.3% of the queries. Omitted words, extraneous words/phrases, telegraphic ellipsis, and incomplete sentences arose in 14% of the queries.

Our conjecture is that wherever natural language interfaces are employed, ill-formed input will occur. Any natural language interface, when faced with ill-formed input, must either intelligently guess at a user's intent, request direct clarification, or at the very least accurately identify the ill-formedness. As Wilks (1976) states, "Understanding requires, at the very least, ... some attempt to interpret, rather than merely reject, what seem to be ill-formed utterances."

Though experience has shown that users can adapt to the limitations of the system's well-formed anticipated input (Harris 1977b, Hendrix et al. 1978), we feel that relying on such user adaptation ignores one of the most powerful motivations for English input: enabling infrequent users to access data without an intermediary and without extensive practice. Even the person who frequently uses such a system will be exasperated if it cannot explain why it misunderstands an input.

In some circumstances, ill-formed input may be less frequent. For instance, Fineman (1983) reports that in an experiment where users were constrained to *discrete speech,* ill-formedness occurred in as little as 2% of the input. Another unusual environment can be created by informing the users that the system cannot really understand natural language, thus biasing language use.

In addition to natural language interfaces, processing ill-formed input is critical to correcting language use. Prototype systems have already been investigated in the language-learning environment (Weischedel et al. 1978) and in the office automation environment of document preparation (Miller et al. 1981). Even in published text unintentional ungrammaticalities occur.

Most natural language understanding systems deal with a few types of ill-formedness. Out of our own work, and that of others, we have produced a framework for processing ill-formed input. This approach treats ill-formedness as *rule-based.* First, natural language interfaces should process all input as presumably well-formed until the *rules* of normal processing are violated. At that point, error handling procedures based on *meta-rules* relating ill-formed input to well-formed structures through the modification of the violated normal rules should be employed. These meta-rules correspond to types of errors.[3]

The rest of the paper argues for this rule-based approach. Section 2 characterizes both the types of ill-formed input, and the types of possible approaches to them, including our proposal. Section 3 gives examples of meta-rules for processing ill-formed input. Section 4 describes how some heuristics developed by others fit within our paradigm. An implementation is sketched in Section 5. Section 6 discusses limitations of the proposal. Sections 7 and 8 present directions for future work and conclusions.

2. Approaches to Ill-formedness

This section introduces the problems of interpreting ill-formed input. First, we discuss the types of ill-formed input briefly. Then we consider the range of approaches that have been tried for allowing for such input.

---

[3] The purpose of these meta-rules is therefore quite distinct from that of Gawron et al. (1982).

Ill-formedness phenomena can be divided into two sets. The first defines what we call *absolute ill-formedness*. An utterance is absolutely ill-formed if the typical listener considers it ill-formed. The definition unfortunately appeals to subjective evaluations; these are known to differ widely (Ross 1979). But it seems to include the majority of typical cases and exclude the majority of types of good English sentences.

The second set defines *relative ill-formedness*. This is ill-formedness with respect to the normal processing rules of the formal computing system including the natural language interface and the underlying application system. The set of ill-formed inputs for an interface can be defined as the union of these two sets for that interface.

The set of ill-formed input captured by these definitions can also be seen through the four typical phases of interpretation in natural language interfaces: lexical, syntactic, semantic, and pragmatic processing. In lexical processing, absolute ill-formedness can come from misspelling, mistyping, and mispronunciation; relative ill-formedness can arise from unknown words. In syntactic processing, absolute ill-formedness is seen in faulty subject-verb agreement, word order errors, omitted words, run-on sentences, etc; relative ill-formedness is seen in grammatical combinations of words that exceed the interface's grammar.

Semantic processing can be defined as the interpretation of the input in isolation. Knowledge of the task domain can be applied, but the context of input with respect to previous interactions and the state of the underlying computing system are only considered in pragmatic processing.

Absolute ill-formedness in semantics includes omitting needed information and violating of selectional restrictions. Absolute ill-formedness in pragmatics includes breaking the rules of conversation, as when answering a question with a question, having presuppositions of the speaker fail, and failing to make clear an anaphoric reference. Relative ill-formedness in both cases includes "overshoot", requesting capabilities or information not covered by the system in its current state, and parenthetical expressions incomprehensible to the system.

2.1 Four alternative approaches to ill-formedness

There are at least five approaches one can take to ill-formedness. This section outlines the four alternatives to the approach we have formulated; our approach is covered in Section 2.2. In describing the five approaches, we use the following informal notation. SYSTEM[s] refers to a system designed to process a set of sentences. WELL–FORMED is a set of well-formed utterances; ILL–FORMED is a set of ill-formed utterances. Naturally, an approach that covers the broadest range of linguistic behaviour should be preferred.

One alternative is to treat the processing of ill-formed and well-formed inputs identically, by ignoring constraints. That is, one designs SYSTEM[WELL-FORMED U ILL-FORMED]. Schank et al. (1980) and Waltz (1978) have taken this approach toward grammatical constraints. CASPAR (Hayes and Carbonell 1981) exhibits this approach for grammatical constraints as well. Since there is much redundancy in language, the practice of not using certain constraints will often work. However, this will fail on many utterances, since it ignores rules that not only constrain search but also eliminate unintended interpretations. One can see this by considering subject-verb agreement, a grammatical constraint that people sometimes violate and that is often left out of natural language systems. Though other constraints, such as semantic (selection) restrictions between a verb and its subject, often indicate the intended interpretation, it is easy to think of examples where subject-verb agreement is crucial to understanding. Comparing examples (1) and (2) below, subject-verb agreement is crucial to determining whether the company or assets were purchased.

(1) List the assets of the company that was purchased by XYZ Corp.
(2) List the assets of the company that were purchased by XYZ Corp.

A second approach is to build systems for well-formed input and for ill-formed input together; that is, one designs SYSTEM[ILL-FORMED] merged with SYSTEM[WELL-FORMED]. Unlike the first approach, well-formedness constraints are employed on well-formed input. LUNAR (Woods et al. 1972), an early English interface to a question-answering system, and SOPHIE (Burton and Brown 1977), an intelligent tutoring system with an English interface, both used this approach. The problem with this approach is that it does not reflect the fact that constraints indicate preferences in interpretation. For instance, though example (3) below has two legitimate syntactic interpretations, the one that violates our model of the world is rejected, causing us to reject the "garden path" interpretation.

(3) I saw the Statue of Liberty flying to New York.

As another example, the two pronouns in "He shot him" are normally considered to refer to different people; the alternative that the speaker meant "He shot himself" does not arise unless there are strong expectations ahead of time that that is the correct proposition.

A third approach is to build two systems, but to use SYSTEM[ILL-FORMED] only if SYSTEM[WELL-FORMED] finds no interpretation. A commercially available English interface to data bases (Harris 1977) has taken this approach. The EPISTLE project (Jensen and Heidorn 1983, Miller et al. 1981) employs this alternative for grammatical violations. DYPAR (Carbonell et al. 1983) has taken this approach in an interface to an expert system. Kaplan (1978) developed a strategy to give more useful responses when a data base query yields a negative response, for example, when no entity satisfies the desired conditions. Chang (1978) created a heuristic for inferring missing joins in incomplete queries to relational data bases. The defect in this model is that there is no means of relating strategies for processing ill-formedness explicitly to the strategies for processing well-formedness. We argue here that one can explicitly relate the two classes of strategies.

A fourth approach is to build only one system, SYSTEM[WELL-FORMED], and to employ a metric to measure how far a postulated interpretation is from satisfying all well-formedness constraints. Charniak (1983) has advocated this for grammatical processing; Wilks (1975) has made this the basis of semantic processing during the interpretation phase. Of course, the notion of weighing alternatives and using metrics has been used for phenomena other than ill-formedness, such as parsing (Robinson 1982) and speech understanding (Walker 1978, Woods et al. 1976). Clearly, ranking alternative interpretations is necessary. However, if one relies *solely* on a metric and SYSTEM[WELL-FORMED], then an account of the fact that the ill-formedness often has specific implications is still needed. In example (4), the selection restriction that "like" requires animate agents is violated; a reasonable inference is that the speaker somewhat personifies the computer in question.

(4) My home computer doesn't like to run BASIC.

Nor does a metric reflect the fact that there are clear patterns of error, such as those that have been reported in linguistic studies (Fromkin 1973) and in application studies (Thompson 1980, Eastman and McLean 1981). Table 1 summarizes these four approaches.

2.2 Our approach

Based on previous work, both our own and that of others, we propose a framework employing meta-rules to relate the processing of ill-formed input to well-formedness rules. This framework may be stated as follows:

1. Process the input using SYSTEM[WELL-FORMED].
2. If no interpretation is found by SYSTEM[WELL-FORMED], apply a meta-rule to the well-formedness rules, based on a ranking of the alternatives, in order to
   (a) diagnose the problem, that is, the rule that is violated and how it is violated,
   (b) relax the rule,
   (c) add a "device note" to the interpretation recording the violation,
   (d) resume processing via the well-formedness rules, if possible.
3. Repeat step 2 as necessary.

Approach 1
    Characterization: Do not encode well-formedness constraints.
                Flaw: Well-formedness rules convey meanings by constraining interpretations.

Approach 2
    Characterization: Design systems for well-formed input and ill-formed input together.
                Flaw: This gives no preference of well-formed interpretations over ill-formed ones.

Approach 3
    Characterization: Search for well-formed interpretations prior to considering any ill-formed ones.
                Flaw: This does not explicitly relate handling ill-formedness to processing well-formedness.

Approach 4
    Characterization: Use a metric to rank ill-formedness interpretations, and select the one that comes closest to satisfying all constraints.
                Flaw: This does not state what the deviation is so that one may draw inferences from the ill-formedness, nor does it capture actual patterns of error.

Table 1. Four Rejected Approaches.

Each meta-rule should correspond to a pattern of ill-formedness and should account for utterances corresponding to only that pattern. SYSTEM[ILL-FORMED] is therefore implicit in the meta-rules.

This framework has advantages lacking in one or more of the other approaches. Well-formedness constraints, whether syntactic, semantic, or pragmatic, are employed to eliminate unintended interpretations. Well-formed interpretations are always preferred. Ill-formedness processing is explicitly related to the well-formedness rules. Only the constraint that seems to be violated is relaxed; all other well-formedness constraints are still effective. Furthermore, the deviance notes record the aspect that deviates from well-formedness, thus allowing pragmatic inferences by later processing.

In the next two sections, we propose a handful of primitives for syntactic and semantic problems and also propose a formalism for writing meta-rules. As supporting evidence, we state meta-rules for a number of problems and describe approaches for several others. These phenomena include the following:

failed grammatical tests,
word confusions,
spelling errors,
unknown words,
restarted sentences,
resumptive pronouns and noun phrases,
contextual ellipses,
selection restriction violation,
metonymy,
personification, and
presupposition failure.

3. Examples of Meta-rules

To further argue for meta-rules as a uniform framework for processing ill-formed input, we describe a wide variety of meta-rules in this section and the next. We adopt the following notation for meta-rules in this paper:

$$C1\ C2\ ...\ Cn\ \text{-->}\ A1\ A2\ ...\ Am$$

The left-hand side (LHS) diagnoses what the problem might be: the right-hand side (RHS) states how to relax the failed constraint. The C_i are conditions on the computational state of the system; all must be true if the meta-rule is to apply. The A_i are actions stating how to rewrite the violated constraint and resume processing; all will be executed if the rule applies. Many of the actions we suggest here can be viewed as rewriting a rule of the normative system, for example, a grammar rule or case frame. However, some are more appropriately viewed as changing the computational state when blockage occurs; an example is correcting the spelling of a word. In Section 5 we will argue that it is best to implement all of the actions as modifications of a blocked alternative.

Naturally, in rewriting a rule, pattern-matching and substitution are fundamental. We adopt the following definition of patterns. A pattern is a LISP s-expression. Atoms preceded by a question mark are variables. Expressions preceded by a dollar sign are evaluated using the LISP rules; their values are treated as patterns.[4] If a period appears before a pattern variable that is the last item in a list, that pattern variable matches the tail of a list. All other items in patterns are literals. The scope of a pattern variable is the whole meta-rule. The first time a variable is bound in a meta-rule, it retains the binding throughout the rule.

Potentially, there may be many places where relaxation can occur. If a meta-rule applies to more than one configuration, it will be applied to each in turn, creating a list of possibilities for processing after recovery is complete. Consequently, the meta-rules will refer to only one failed configuration at a time.

3.1 Meta-rules related to syntax

First, let us consider meta-rules dealing with the grammar. Many of our examples here are reformulations of our earlier work (Weischedel et al. 1978, Kwasny and Sondheimer, 1981, Weischedel and Black 1980) within the uniform framework of meta-rules. All meta-rules pertaining to syntax should have a first condition which is (SYNTAX-FAILED?); this is true if the parser is blocked. Since all rules in this section would contain that predicate, we will not include it in the examples.

Many syntactic formalisms have a similar framework for expressing rules: these include context-free grammars, augmented phrase structure grammars (Heidorn 1975), Programmar (Winograd 1972), the linguistic string parser (Sager 1981), Lifer (Hendrix et al. 1978), and augmented transition networks (ATNs) (Woods 1970). In fact, all of these can be viewed as formally equivalent to ATNs, and we will describe our techniques in that framework.

Figure 1 gives several predicates that should be useful in the LHS of meta-rules. The LHS of the meta-rule is matched against the environment in which an ATN arc failed. The environment is called a configuration and includes the current ATN state, the arc, all ATN registers, and the remainder of the input string.

An important action for the RHS is NEW-CONFIGURATION, which defines a new parser configuration, thus replacing the failed configuration that the meta-rule matched. It may take

---

[4] The expression $expr could be implemented as (*EVAL* expr). The pattern variable ?atom could be implemented as (*VAR* atom).

(IN-STATE? state)	Did the configuration block in state?
(CAT? category)	Is the current word in category?
(WRD? list)	Is the current word a member of list?
(NEXTCAT? category)	Is the word after the current one in category?
(NEXTWRD? list)	Is the word after the current one in list?
(FAILED-TEST? pattern)	Is the pattern a predicate expression in the test of the arc and did both the expression and the test evaluate to false?
(FAILED-ARC? pattern)	Does the failed arc match pattern?
(HOLDLIST-NOT-EMPTY?)	Is the hold list empty?
(CONFUSION-WORD? x)	Is x a word frequently confused with another? If so, the related word is returned.

Figure 1. Useful Conditions for Syntactic Meta-rules.

(EMPTY-HOLD)	defines the hold list to be empty.
(FAILED-CONSTRAINT pattern)	adds the instantiation of the pattern to a list of violated constraints stored in the configuration. The position of the parser within the input string will automatically be recorded as well.
(SUBSTITUTE-IN-ARC pattern 1 pattern 2)	changes the arc in the failed configuration by replacing all expressions matching pattern 1 by pattern 2.
(REPLACE-* x)	makes x the current word in the blocked configuration.

Figure 2. Useful Actions for Syntactic Meta-rules.

State	Arcs	(Figure 3.)

S/
    (PUSH NP/T ...
      (SETR SURFACE-SUBJECT *)
      (* We think this is the subject)
      (TO S/NP)
    (VIR NP T (SETR SURFACE-SUBJECT *)
      (* In relative clauses, this identifies a noun phrase from the hold list as surface subject)
      (TO S/NP))

S/NP
    (CAT VERB (SUBJECT-VERB-AGREE? (GETR SURFACE-SUBJECT) *)
      (* The predicate SUBJECT-VERB-AGREE? is true iff the number and person of the subject and verb are compatible)
      ... (SETR VERB *)
      (TO S/V))

S/V
    (JUMP S/POP (INTRANSITIVE-VERB? (GETR VERB)))
    (CAT ADV T ... (TO S/V))
    (PUSH NP T (SETR OBJECT *) ... (TO S/POP))
    (VIR NP T (SETR OBJECT *)
      (* Identifies a noun phrase from the hold list as the direct object in relative clauses)
      (TO S/POP))

S/POP
    (POP (BUILDQ ...)
      (AND (TRANSITIVE-VERB? (GETR VERB))
        (NOT (REQUIRES-INDIRECT-OBJ? (GETR VERB)))))

NP/
    (CAT PRO T ...
      (SETR PRO *)
      (TO NP/POP))
    (CAT DET T ...
      (SETR DET *)
      (SETR NUMBER (GETNUMBER DET))
      (TO NP/DET))

NP/DET
    (CAT ADJ T ...
      (* collecting adjectives before head noun)
      (TO NP/DET))
    (CAT N T ...
      (SETR N *)
      (TO NP/N))

NP/N
    (CAT N T ...
      (ADDR COMPOUND (GETR N)) (SETR N *)
      (* a possible nominal compound)
      (TO NP/N)
    (JUMP NP/POP
      (DET&NOUN-AGREE? (GETR NUMBER) (GETR N))
      (* The predicate DET & NOUN-AGREE? is true iff the determiner used is incompatible with the head noun) ...)
    (PUSH RELCL/T
      (SENDR TRACE (TRACE))
      (* This sends a trace of a noun phrase to a relative clause)
      (SETR RELATIVE-CLAUSE *) (TO NP/POP))

NP/POP   (POP (BUILDQ ...) T)

```
RELCL/   (CAT RELPRO T (HOLD (GETR TRACE)) ...
         (TO S/))
         (PUSH NP T (SETR SURFACE-SUBJECT *)
         (HOLD (GETR TRACE)
         (* This allows for relative clauses where the subject is present)
         (TO S/NP))
         (JUMP S/NP (SETR SURFACE-SUBJECT (GETR TRACE))
         (* Allows for elided subjects in reduced relative clauses))
```

Figure 3. A Simple ATN Graph.

any number of arguments which set parts of the configuration. For example, SETR will define the new value of an ATN register. A list of useful arguments to NEW-CONFIGURATION is given in Figure 2. Failed constraints fill the role of the deviance notes of Kwasny and Sondheimer (1981). All parts of the failed configuration that are not explicitly changed in NEW-CONFIGURATION remain the same. Our implementation assumes that there is only one NEW-CONFIGURATION per meta-rule, though one could generalize this so that executing NEW-CONFIGURATION n times in a meta-rule gives n new configurations to replace the failed one. If a new configuration is generated, the parse can be resumed.

Figure 3 gives a trivial ATN which will be used for the sample meta-rules. The start state is S/. A list beginning with an asterisk in the actions of an arc is a comment.

3.1.1 Simple grammatical tests

Our earlier work showed how to relax tests that appear on ATN arcs. In one study (Eastman and McLean 1981), subject-verb disagreement occurred in 2.3% of the English queries. Meta-rule (i) relaxes that agreement test. The new configurations here are the result of replacing the agreement test in each failed arc by the predicate T. Since the new configuration is generated, parsing is resumed using it. Though the substitution was trivial in this case, SUBSTITUTE-IN-ARC is a general pattern-matching and substitution facility. As an example, consider "A curious problem showing unusual conditions appear ..." A top-down, left-to-right parse using a grammar such as the one in Figure 3 would block at the word "appear". One of the blocked configurations would correspond to the agreement test failure in the arc leaving state S/NP; meta-rule (i) would apply, allowing the sentence to be parsed.

3.1.2 Omitted articles

Another frequently occurring problem is omitting required articles from count nouns. In the study by Eastman and McLean (1981) this occurred in 3.3% of all queries. In the grammar of Figure 3, blockage would occur at NP/N because of the test DET&NOUN-AGREE? on an example such as "Print price of P27 over the last five years". Rule (ii) relaxes the test. When the parser starts on the new configuration, the modified test will be checked, verifying that no determiner is present. If none is, the message from FAILED-CONSTRAINT is available for error recovery.

The meta-rule approach allows for more sophisticated actions. Suppose that a linguistic study of utterances with missing determiners showed that a default assumption of definite reference is a good heuristic. In this case, one could simply add the action (SETR DET 'the) to the actions in NEW-CONFIGURATION.

One could argue that, in a data base environment, the grammar should simply treat omitted determiners as a normative construction. Even though determiners are frequently omitted in data base contexts, preferring well-formed interpretations can eliminate some ambiguities in complex noun phrases such as "a machine running programs". The determiner constraint suggests that "running programs" modifies the head noun "machine" rather than "machine" and "running" both modifying "programs".

3.1.3 Confusion words

A number of word pairs are frequently confused, such as homonyms and "good" for "well". Meta-rule (iii) allows for such errors, since REPLACE-* will modify the current word in the blocked configuration. MR-SETQ binds the value of its second argument to the pattern variable appearing as its first argument. Hence, "You performed good" would block at S/V, and the meta-rule would substitute "well" for "good".

3.1.4 Resumptive pronouns

Another kind of ill-formedness is resumptive pronouns and resumptive noun phrases. These occur in relative clauses where the entity referred to by the relative pronoun is improperly repeated in the relative clause as a pronoun or noun phrase. An example is "John's friend Mary married the man that she planned to marry him", since there is no syntactic slot in the relative clause for the relative pronoun "that" to fill. A typical ATN strategy for interpreting relative clauses is to put a place holder or trace on a "hold list"; the ATN processor prevents POPping from a level if the hold list is non-empty. The test prevents accepting clauses where traces are not used. Meta-rule (iv) provides for resumptive pronouns and resumptive noun phrases. One can imagine more complicated tests, since there are specific conditions (Kroch 1981) under which resumptive pronouns and resumptive noun phrases are more likely.

(i)   (FAILED-TEST? (SUBJECT-VERB-AGREE? ?X ?Y))
      --> (NEW-CONFIGURATION
           (FAILED-CONSTRAINT (SUBJECT-VERB-AGREE? ?X ?Y))
           (SUBSTITUTE-IN-ARC (SUBJECT-VERB-AGREE? ?X ?Y) T))

(ii) (FAILED-TEST? (DET&NOUN-AGREE? ?X ?Y))
      --> (NEW-CONFIGURATION
           (SUBSTITUTE-IN-ARC (DET&NOUN-AGREE? ?X ?Y) (NULL ?X))
           (FAILED-CONSTRAINT (DETERMINER&NOUN ?Y -- MISSING DETERMINER)))

(iii) (MR-SETQ ?X (CONFUSION-WRD *))
      ---> (NEW-CONFIGURATION
            (REPLACE-* ?X)
            (FAILED-CONSTRAINT (?X SUBSTITUTED FOR *)))

(iv) (FAILED-ARC? (POP ?VALUE . ?Z))
      (IN-STATE? S/POP)
      (HOLDLIST-NOT-EMPTY?)
      ---> (NEW-CONFIGURATION
            (FAILED-CONSTRAINT (Resumptive clause $?VALUE))
            (EMPTY-HOLD))

(v)   (IN-STATE? S/POP)
      --> (PRINT-RESPONSE-PATTERN
           (I DO NOT UNDERSTAND YOUR USE OF THE VERB $(GETR VERB) /. WOULD YOU
            LIKE EXAMPLES OF WHAT I UNDERSTAND?))
          (SELECTQ (READ)
           ((YES Y) (PRINT-EXAMPLES (GETR VERB))) NIL)

3.1.5 Error messages

Of course, the parser may not be able to recover at all due to either absolute or relative ill-formedness. Weischedel and Black (1980) presented a technique for associating error messages with states where the parser blocked. The only way to block in S/POP is if the verb complement expected for the main verb is not present. Meta-rule (v) could handle this simple case. Notice that this is a different class of meta-rule, for it does not resume computation. Naturally, such rules should be tried only after no other meta-rules are available. One could define different classes of meta-rules by appropriate declarations; alternatively, this class can be recognized easily, since none of the actions resume processing.

This is not the only alternative in the face of failure to parse even with relaxation; Jensen and Heidorn (1983) present heuristics for what to pass to the semantic interpreter in this case, given bottom-up parsing.

3.2 Meta-rules related to semantics

In addition to these syntactic examples, semantic problems can also be addressed within the formalism. If some semantic tests are included in the parser, say a certain arc test contains calls on the semantic component, specific semantic tests can be relaxed by the general mechanism we described for relaxing tests on ATN arcs.

Instead, suppose that semantic constraints are encoded in a separate component. Semantic constraints may be expressed in several formalisms, such as semantic nets (Bobrow and Webber 1980a,b; Sondheimer et al. 1984), first-order logic, and production rules (for example, PROLOG, Warren et al. 1977). It is generally agreed that all are formally equivalent to first-order logic. For the purposes of this paper, we assume that the selection restrictions are encoded in first-order logic.

One of the most common designs for a semantic interpreter is based on selection restrictions and case frames (Bruce 1975). At least five kinds of constraints may be violated:

(1)   what may fill a given case,
(2)   which cases are required for a complete constituent,
(3)   which may have multiple fillers without conjunction,
(4)   which are allowed for a given case frame, and
(5)   what order cases may appear in.

Figure 4 lists tests useful for diagnosing failures in such a semantic interpreter. Assume that any predicate on the semantic class of a constituent is encoded simply in LISP notation, for example, (HUMAN x) is true iff x is of class human. All meta-rules in this section can be assumed to include an initial test (SEMANTICS-FAILED?.)

For convenience, we have used the same names for some of the actions as in the syntactic cases (for example, FAILED-CONSTRAINT, NEW-CONFIGURATION, etc.). When implemented in a particular system, different names may be used, since the concept of configuration, blockage, etc., is usually different for the types of processing (for example, lexical, syntactic, semantic, and pragmatic). Figure 5 lists several actions useful in semantic meta-rules.

3.2.1 Universal relaxation of semantic class

Meta-rule (vi) is a very general rule. Assuming that semantic class tests are organized in a hierarchy, it states that the failed test is to be replaced by its parent in the hierarchy, yielding the next most general test.

An example of the use of meta-rule (vi) is "My car drinks gasoline". The restriction on the AGENT case could be the predicate ANIMATE. A fragment of a semantic hierarchy appears in Figure 6. In that, ANIMATE has a parent predicate of CONCRETE that would include cars.

(FAILED-SEMANTIC-TEST? pattern)	Is the pattern a predicate expression (on a constituent) and is the predicate false?
(MANDATORY-CASE-MISSING?)	Is a required case absent?
(TOO-MANY-FILLERS?)	In trying to fill case, does it already have the maximum number assigned?
(SEMANTIC-CLASS? class)	Is class a semantic class predicate, for example, human?
(MATRIX-TYPE? class)	Is the matrix to which we are trying to assign the current constituent in class?
(VIEWABLE? x y z)	Can entity x be viewed as a y in the context z? (This is a question for the pragmatic component.)
(CASE? case)	Is the constituent supposed to fill role case?

Figure 4. Predicates for Semantic Meta-rules.

(FAILED-CONSTRAINT pattern)	adds the instantiation of the pattern to a list of violated constraints stored in the configuration.
(SUBSTITUTE-IN-CASE pattern 1 pattern 2)	replaces pattern 1 by pattern 2 everywhere in the constraint.
(SUBSTITUTE-FOR-CASE case)	tries assigning the constituent as case.

Figure 5. Actions for Semantic Meta-rules.

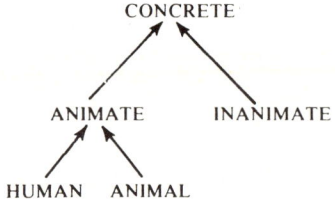

Figure 6. A Fragment of a Semantic Hierarchy.

The failure of the initial sentence and the subsequent processing using the meta-rule would accept a sentence with the special deviance note identifying the semantic oddity.

3.2.2 Personification

In a way similar to our arguments against approach 4 in Section 2.1, we feel that general meta-rules such as (vi) will prove less valuable than specific rules. A particular test that can be relaxed is the requirement for a human; for instance, the verbs of saying and those of propositional attitude, such as "believe" and "think", normally have a restriction that their agent be human. Nevertheless, such a constraint is regularly violated through personification of pets, higher animals, machines, etc.

Since personification is infrequent compared to the norm of descriptions designating humans, a case constraint of "human" can trim the search space. Since personification conveys particular inferences (Lakoff and Johnson 1980), a relaxation rule that records the detected personification can trigger appropriate inference processes. Figures of speech certainly are not absolutely ill-formed; we argue here that it is useful to treat them as relatively ill-formed.

Meta-rule (vii) is one simple relaxation for personifying animals. More specific ones may prove preferable, if classes of personification are taxonomized.

3.2.3 Metonymy

There are at least seven classes of metonymy (Lakoff and Johnson 1980), including a part for the whole, the producer for the product, the object for its user, the controller for the controlled entity, the institution for the people responsible, the place for the institution, and the place for the event. This analysis suggests two kinds of strategies. A particular class of descriptions may occur in exactly the same linguistic environments as their class of metonymous descriptions. For instance, institutions and people appear interchangeable as the logical subject of the verbs of saying and of propositional attitude. That can be encoded directly in the case frames of those verbs.

However, many types of metonymy are conditioned on a highly specialized relationship. For instance, places can be used metonymously for events only if the speaker believes an event is identifiably associated with the location. For instance, compare the following examples:

Pearl Harbor caused us to enter the war.
*Fifth and Lombard caused us to reconsider graduated income taxes.

Therefore, a meta-rule such as (viii) seems appropriate to prefer the normal, but accept metonymous descriptions of events by places. In meta-rule (viii), we have assumed that there is a variable FILLER of the semantic interpreter that holds the constituent to be assigned. Also, in the call to VIEWABLE?, CURRENT indicates that the pragmatic component should use its current context.

3.2.4 Phrase ordering

Failure in selection restrictions can indicate other semantic errors. These include ordering problems, for example, "John killed with a gun Mary", and unexpected prepositions, for example, "John killed Mary by a gun". The LHS of the appropriate meta-rules would begin with identification of selectional restriction failures but would also include other tests. The RHS would change the assumed case. A rule for the first example is (ix). Here the assumption is that the ordering problem will be first noted when "Mary" is tried as a time modifier. Using SUBSTITUTE-FOR-CASE postulates the constituent "Mary" to fill the object case and attempts to do so.

3.3 Generality of the approach

Though we have experience in implementing our framework for ATN parsers only, we believe the framework to be applicable over a broad range of parsers. It assumes only that a "configura-

(vi) (FAILED-SEMANTIC-TEST? (?Y ?Z)
    (SEMANTIC-CLASS? ?Y)
    --> (NEW-CONFIGURATION
        (FAILED-CONSTRAINT (?Y TOO RESTRICTIVE -- USING PARENT $(PARENT-OF ?Y)))
        (SUBSTITUTE-IN-CASE (?Y ?Z) ($(PARENT-OF ?Y) ?Z)))

(vii) (FAILED-SEMANTIC-TEST? (HUMAN ?Z))
    --> (NEW-CONFIGURATION
        (FAILED-CONSTRAINT
            ((HUMAN ?Z) -- ASSUMING PERSONIFICATION OF ANIMAL))
        (SUBSTITUTE-IN-CASE (HUMAN ?Z)))

(viii) (FAILED-SEMANTIC-TEST? (EVENT ?X)) (LOCATION FILLER) (VIEWABLE? FILLER 'EVENT
    'CURRENT)
    --> (NEW-CONFIGURATION
        (SUBSTITUTE-IN-CASE (EVENT ?X) T) (FAILED-CONFIGURATION
        (METONYMY PLACE-FOR-EVENT FILLER)))

(ix) (FAILED-SEMANTIC-TEST? (TIME ?Z)) (MATRIX-TYPE? 'CLAUSE) (CASE? 'TEMPORAL)
    --> (NEW-CONFIGURATION (FAILED-CONSTRAINT (ORDERING PROBLEM -- OBJECT CASE
    ASSUMED)))

tion" or "alternative" representing a blocked, partial interpretation can be stored, modified, and restarted. No assumption regarding the direction of processing (for example, left to right), the nature of search (for example, top-down vs. bottom-up), nor the class of problem (for example, lexical, syntactic, or semantic) is made. For instance, the design of an implementation for semantic meta-rules as in Section 3.2 is complete. The underlying semantic component is based on searching case frames breadth-first with both top-down and bottom-up characteristics. Except for the one meta-rule regarding incorrect phrase ordering in Section 3.2.4, the semantic meta-rules themselves are independent of whether proposing a phrase for a given case in a frame is based on syntactic considerations or other criteria (for example, Schank et al. 1980). Naturally, the primitive conditions and actions of a given set of rules will depend on a particular formalism. In the next section, we relate our framework to a variety of parsers and problems.

4. Additional Supporting Evidence

Many natural language interfaces have some heuristics for processing one or more classes of ill-formed input. Since an exhaustive analysis would be impossible here, we will review only a handful of techniques that have inspired us to develop the meta-rule framework. We describe each technique by showing how it would be phrased as a meta-rule within our paradigm.

The LADDER system (Hendrix et al. 1978) implements three major techniques for processing ill-formed input. All fit within the framework we suggest. One deals with recovery from lexical processing. In this system, the developer of a question-answering system prepares only a dictionary of well-formed words. If a sentence contains a word that is not in the dictionary, the parser will fail. The system localizes the area of failure to the ATN state associated with the partial interpretation that has proceeded rightmost in the input and that is shallowest (in terms of incompleted ATN PUSH arcs). Candidates for the correct spelling are limited to the words that would permit the parser to proceed and that are close to the spelling that appears. An equivalent meta-rule would check in the LHS that the parser failed. The RHS would compute a list of words expected next for each type of arc leaving that state, for example, the category members and literal words expected next. The next action would apply the Interlisp spelling correction algorithm to postulate a known word that was expected next. This word would replace the unrecognized one in the input and parsing would resume. A similar heuristic is running in our current implementation, with the addition that, if the unrecognized word appears to have an inflected ending, spelling correction is performed on the possible root.

A second technique in LADDER deals with understanding contextual ellipsis, if no parse for the input is found. This heuristic interprets "the fastest submarine" as "To what country does the fastest submarine belong", if it occurs after a query such as "To what country does each merchant ship in the North Atlantic belong". In Weischedel and Sondheimer (1982), we extended that heuristic to allow for turn-taking in dialogues and to allow expansions as well as substitutions, such as the elliptical form "Last month" following "Did you go to Chicago?"

A third technique in LADDER is the printing of error messages, in the same sense that meta-rule (v) above prints a message when all attempts have failed. We could phrase this heuristic as a meta-rule whose LHS would check that the parser has blocked. This meta-rule would be ordered strictly after the ones for spelling correction and contextual ellipsis. A state would be postulated as the locale of the problem by the same heuristic as for spelling correction. The RHS would print for each arc that leaves that state the category, constituent, or word that was expected by that arc.

Hayes and Mouradian (1980) emphasize recovery techniques for blocking during left-corner parsing (Aho and Ullman 1972). Their strategies are invoked only if the parser blocks. Two of them can be reformulated as meta-rules as follows. One meta-rule would check in its LHS that the parser was blocked and that a special parser variable (call it BLOCKED-PARSE) was empty. The RHS would save the blocked configuration in BLOCKED-PARSE, and start parsing as if the current word were the first word of the input. This would enable the system to ignore initial strings that could not be understood. A useful example of this is restarted inputs, such as "Copy all print all headers of messages". A second meta-rule is related. The LHS would check whether the parser was blocked and BLOCKED-PARSE had a configuration in it. Furthermore, the LHS would check to see that another parser variable (call it DONE-ONCE) was NIL. If so, the RHS would set DONE-ONCE to T. The RHS would then swap the current configuration with BLOCKED-PARSE and would try resuming the parse form and current word with that configuration. This heuristic is designed to ignore incomprehensible material in the middle of an input. For instance, it would enable skipping the parenthetical material in "List all messages, assuming there are any, from Brown".

In the area of pragmatics, solutions that could fit within our paradigm have been suggested for two classes of problems. One problem is the failure of presuppositions of an input. In the environment of an intelligent tutor for computer-assisted language instruction, a technique suggested in Weischedel et al. (1978) could be formulated as a meta-rule as follows. The LHS would check whether processing was blocked due to a presupposition being false. Since that system would have a more complete knowledge of language than a beginning student of a foreign language, the system could treat the input as absolutely ill-formed. A sophisticated RHS could paraphrase the false presupposition for the student and indicate which word or syntactic construction was used inappropriately. Thus, the tutor could point out mistakes such as "Das Fraeulein ist Student", indicating that the student should look up the meaning of "Student" (which applies only to males).

Kaplan (1978) suggests an alternative heuristic for false extensional presuppositions in a data base environment. One can reformulate it as a meta-rule whose LHS would check that the query had requested a set as a response and that the set was empty. The RHS would compute queries corresponding to subsets that the original query presupposed would have a non-empty extension. The RHS would paraphrase the most general such query with an empty response set, reporting to the user that the system knew of no such entities.

5. Implementation

We implemented a grammatical meta-rule processor first for an ATN interpreter and more recently for an ATN compiler (Burton and Brown 1977). Our experiments have used RUS (Bobrow 1978), a broad coverage grammar of English with calls to a semantic component to block anomalous interpretations proposed by the grammar.

Design and implementation of a meta-rule processor for violation of semantic constraints is currently underway in two different semantic interpreters. In one, case constraints are expressed as sets of logical formulas; in the other, KL-ONE is used to encode case frames (Sondheimer et al. 1984).

Four design issues are considered in the following sections.

5.1 Applying meta-rules

The set of meta-rules dealing with the grammar or semantic system could be viewed formally as a function f from a component's rules S to a new component's rules S'.

$$f(S) = S'$$

S' is the transitive closure of applying every meta-rule pertaining to the system rules in every possible way. (Since it is the transitive closure, S is contained in S'). There are three alternatives. One is to compute S' and use it, rather than S, as the basis for processing, assuming that the transitive closure S' is a finite closure. The second is to apply meta-rules only as needed, thus making S' a virtual system. The third alternative is a combination of applying some meta-rules as needed and applying others in advance.

The first alternative is superficially similar to approach 2 of Section 2.1, where ill-formedness processing is embedded in the normative system; however, S' will maintain the preference for normal interpretations over ill-formed ones. We have rejected this alternative because of the combinatorial growth of rules needed for S'. For instance, one can write meta-rules for handling relaxation of word categories and relaxation of predicates on ATN arcs. Since both can occur throughout the grammar, they should not be expanded ahead of time. A similar argument is used to justify treating conjunction processing as a separate process rather than building it directly into the grammar (Woods 1973). Since the classes of ill-formedness can occur in combination, the number of relaxed rules in S' can be very large. Furthermore, since utterances where many, many combinations of errors occur should be rare, computing the transitive closure seems uncalled for.

The second alternative, generating a relaxed rule each time it is needed, is the one we implemented first in the context of an ATN interpreter. This alternative provides a kind of virtual system and avoids the increased memory necessary to hold S'.

The third alternative, applying some rules ahead of time and using others only as needed, offers the greatest flexibility and a variety of alternatives. We have implemented a version in which the underlying parser is the output of an ATN compiler. When the meta-rule processor applies a meta-rule at a given arc, the relaxed version of the arc is compiled and saved.[5] If the meta-rule is to be tried by the meta-rule processor at that arc again, the form of the relaxed arc need not be re-computed; it can simply be executed.

This third alternative also offers the potential of adapting the system to the idiosyncrasies of an individual's language and also the potential of extending its own model of language. Obviously, this is an area for future research.

Alternatives two and three assume only that the processor applying well-formedness rules is able to store a "configuration" in a queue or agenda. No assumption about the type of processing (for example, bottom-up or top-down), nor the class of violated rule (for example,

---

[5] The current implementation is limited somewhat; it saves the relaxed arc only if the RHS of the meta-rule modifies only the arc itself. Our misspelling meta-rule, for example, does not modify the arc at all, but rather the input string.

lexical, syntactic, semantic, or pragmatic) is necessary.

5.2 What to store

When a configuration blocks because of the well-formedness rules, should the blocked configuration be stored or the results of applying each relevant meta-rule? Both the implementations in the ATN environment save only the blocked configuration, namely, a blocked arc at the end of a path. The number of blocked configurations can be large. At present, there is insufficient evidence to determine whether a well-tuned set of meta-rules will yield a substantially larger (or smaller) number of relaxed configurations compared to the set of blocked configurations.

Some types of problems, for example, subject-verb agreement, may be so common, and some types of relaxation, for example, an unrecognized word, may be so diagnostically clear that the corresponding meta-rule should be applied immediately. In the case of subject-verb agreement, hand-compiling the meta-rule into the grammar may be appropriate (that is, writing an arc whose test is that subject-verb agreement failed and whose action places the new configuration on a queue that is tried only after all normal configurations have failed).

5.3 Localizing the problem

When processing ill-formed inputs, some means of ranking alternatives is appropriate, since the system must determine what is intended in the face of violated constraints and possible error. Also, the number of relaxed configurations may be large, even with a set of well-tuned meta-rules designed to open the search space minimally.[6] The ideal solution is that the ranking of alternatives should be based on syntactic, semantic, and pragmatic evidence, in addition to the diagnosis and recovery strategy.

The current implementation uses only some of those bases and employs a rather simple ranking. Since both grammatical constraints and selection restrictions are employed while parsing with RUS, both syntactic and semantic evidence is used. Blocked configurations are ordered on the amount of input processed; there is also a partial order on the meta-rules.

One of our students, Amir Razi, is designing an experiment to collect data on the performance of the system. The current system can be run in one of two modes: saving all blocked configurations or using only ones that proceeded rightmost in the input. One aspect of the experiment is to determine the frequency with which the interpretations covering the most input in a left-to-right parse block at the true source of the problem. Some preliminary evidence (Weischedel and Black 1980) indicates that this heuristic frequently does indicate where the problem is, if the normative system is nearly deterministic, for example, because the grammar is a fairly constrained subset of English or because semantic criteria filter out parses that have no meaning in the application domain.

Our long-term goal is accurate determination of both the problem in an ill-formed utterance and what was intended. The current implementation represents the first step toward that by employing both syntactic and semantic evidence. We are investigating the use of pragmatic evidence for that purpose as well. In addition, we wish to explore techniques for examining both the left and right contexts of a blocked interpretation, for instance, by employing bottom-up processing.

5.4 A meta-rule index

Hand-compilation of meta-rules as mentioned in Section 5.1 is just one way to pinpoint the configurations to which a meta-rule applies; another is providing an index from blocked configurations to the meta-rules that could apply. We have implemented a preprocessor that

---

[6] However, it is not clear whether the combinatories alone for typical inputs will be a problem, given the rapid increase in processor power/cost and the prospect of multi-processing.

builds an index from an ATN arc to the meta-rules that can apply to it. When loading the ATN grammar, our preprocessor localizes the syntactic meta-rules having IN-STATE?, FAILED-TEST?, and FAILED-ARC? in their LHS to the few arcs to which they could possibly apply. Clearly, if IN-STATE? is an LHS, that meta-rule can apply to only the handful of arcs leaving one state. Since FAILED-ARC? and FAILED-TEST? require the arc to match a given pattern, meta-rules using these tests can be identified with the arcs satisfying those patterns.[7] Such preprocessing provides an index into the possible rules that apply to a blocked configuration, since the state and the arc will be part of the configuration. Furthermore, the pattern-matching operations in the LHS need not be repeated at run-time, since the preprocessor stores for each are an altered form of the meta-rule (without the calls to the pattern matcher) and the bindings that pattern matching created.

Some meta-rules will not have any tests that localize their applicability; an example is the one for confusion words, which can appear almost anywhere. These are stored separately, and must be checked for any arc to which relaxation is to be tried.

6. Limitations

There are a number of points of caution. It should be clear that relaxation does not necessarily guarantee understanding. After all, relaxing any arc to (TST X T ...) will accept any word syntactically; yet that is no guarantee that the word will be understood. Relaxing constraints introduces additional potential for confusion.

What one classifies as "absolutely ill-formed" is clearly open to dispute, as Ross (1979) points out. Therefore, the system may classify something as ill-formed, ranking it behind other interpretations, even though the user views it as well-formed. We suspect that categorizing almost any particular constraint as normative could be the basis for argument. The criteria for deciding whether a constraint should be included in the normative system should include at least the following:

a) whether a native speaker would edit inputs that violate it,
b) whether violating the constraint can yield useful inferences,
c) whether examples exist in which the constraint carries meaning,
d) whether the constraint, if classified as normative, trims the search space, and
e) whether a processing strategy for the constraint can be stated more easily as a modification of normative processing, as in the case of conjunction (Woods 1973) or the case of contextual ellipsis in the data base environment (Weischedel and Sondheimer, 1982).

Thus far we have considered only constraints that are associated with a single point in the processing, such as relaxing a single case frame or relaxing a single ATN arc. Obviously, this need not be the case if, for instance, word or phrase order is permuted. At present, we have no general way of dealing with such problems.

7. Future Work

The problems of processing ill-formed input require several substantial research efforts. One is collecting additional corpora to determine patterns of errors and their frequency of occurrence. This is particularly important for two reasons. First, the more detail uncovered on patterns of error, the tighter the meta-rules for relaxing constraints. In our experience, the effort to make relaxation procedures as constrained and accurate as warranted by the patterns of occurrence is highly worthwhile, not only in trimming the search space, but also in eliminating senseless interpretations. Second, the patterns of ill-formedness will depend on the user community and the modality of input. For instance, non-native speakers of a language make different errors than native speakers. Typed input has a predominance of typographical/spelling errors; spoken input may have more restarted utterances.

---

[7] Of course, this preprocessing assumes that no patterns in the LHS contain a form $expr.

As a correlate to the need for more corpora of ill-formed natural language, there is an obvious need to define highly specific heuristics (as meta-rules) to diagnose and recover from each type of ill-formedness. Some of the heuristics should involve clarification dialogue, another area for research.

There are many possible responses given a diagnosed problem. Consider a simple problem: violation of selection restrictions. In German, the verb "fressen" presupposes that the one eating is an animal. To an input such as "Dieser Mann frisst oft",[8] several recovery strategies could apply:

a) The selection restriction could be ignored.
b) The selection restriction could be generalized for future use.
c) The system could conclude that an error has occurred, as in the aforementioned language learning environment.
d) The system could engage in clarification dialogue to determine whether the user intended to use that word.
e) The system could assume the user believes that the man referred to eats like an animal.

The conditions for selecting a strategy need to be studied. An explicit model of the user is needed for deciding the intent of the user and for appropriate recovery from ill-formedness.

Learning the idiosyncrasies of particular users and automatic extension of the system (based on detecting relatively ill-formed input) is very challenging. Some initial steps in this direction have been taken in Carbonell (1979), but there is much to be done. A significant aspect of the learning problem in this environment is the substantial uncertainty about whether the system has the intended interpretation, and the effect on both the functional and time performance of the system as the abnormal is viewed as more normal (and the search space correspondingly grows).

For syntactic ill-formedness, pure bottom-up parsing is intuitively very appealing, since one has descriptions of what is present both to the left and the right of the problem(s). The EPISTLE project (Jensen and Heidorn 1983) is employing bottom-up parsing. The advantage of employing top-down strategies, including left-corner parsing strategies, is the strong expectations available when a configuration blocks. Consequently, many relaxation strategies and systems in the literature (for example, Hendrix, et al. 1978; Kwasny and Sondheimer 1981; Weischedel and Sondheimer 1983) have been proposed and implemented in that framework. Use of bottom-up strategies offers interesting new classes of relaxation, such as rearranging constituents for ordering problems. It is not obvious how the combinatorics of bottom-up strategies will compare to those of top-down strategies. However, developing relaxation techniques for bottom-up processing and extensive empirical studies comparing them to top-down are certainly needed.

One of the most critical problems is control. The need to relax the very rules that constrain the search for an interpretation is like opening Pandora's box. This affects not only the time required to understand an ill-formed input, but also ambiguity through the additional alternatives the system is prepared to accept. There are several aspects to controlling this search. First, the well-formedness constraints should reflect strictly what is normative. Second, the relaxation rules should be made as tight as warranted by patterns of ill-formedness in language use. Third, a partial order on the relaxations should be established. Fourth, not only syntactic constraints and selection restrictions should be used (as in our system) but also pragmatic information to suggest the most promising alternatives. We have begun research on how to use pragmatic knowledge in an information-seeking environment for this purpose; see Carberry (1983, 1984) and Ramshaw and Weischedel (1984). In the environment of messages reporting events, Granger et al. (1983) reports on using expectations based on stereotypical events for this purpose. Extensive empirical studies regarding effective control of the search space are needed.

The acid test for a framework, relaxation heuristics, and control strategies is not relaxing

---

[8] "This man eats often."

simple tests like subject-verb agreement or diagnosing obvious problems like a word not in the dictionary. Rather the acid test is a wide spectrum of problems, including examples like misspellings/typographical errors that result in a known word, because in this type of example, all of the local evidence can indicate that the incorrect word is perfectly correct. Trawick (1983) has initiated work on such misspelling problems.

8. Conclusions

Ill-formed input cannot be ignored by natural language processing systems. This paper has suggested a uniform framework for processing ill-formed input in the hope of providing a basis for standardizing work on ill-formedness.

Our framework has several advantages:

a) Well-formed interpretations are always preferred.
b) Ill-formed processing is explicitly related to the well-formedness rules.
c) Only the constraint that seems to be violated is relaxed; all other well-formedness constraints are still effective for eliminating senseless interpretations and trimming search.
d) Deviance notes record the aspect that deviates from well-formedness, thus allowing pragmatic inferences by later processing.
e) Though our approach is uniform, it permits encoding as much specific knowledge into the diagnosis and recovery procedure as one desires for highly specialized cases.
f) Though this paper has drawn most of its examples from ATN grammars and from case frame processing, as argued in Section 3.3., the framework is not dependent on a particular model of language processing.
g) The framework should be applicable to lexical, syntactic, semantic, and pragmatic constraints.

Acknowledgements

The authors appreciate the helpful suggestions of Sudhir Advani, Robert J. Bobrow, Madeleine Bates, Sandra Carberry, Sheila Coyazo, Giorgio Ingargiola, Stan C. Kwasny, William Mann, William Mark, Geoff Pullum, Lance Ramshaw, Amir Razi, Richard L. Wexelblat, and William A. Woods in discussions about the research.

Much of the coding for the implementation described was done by Amir Razi.

References

Aho, A.V., & Ullmann, J. (1972). *The theory of parsing, translation, and compiling,* (Vol. 1). Englewood Cliffs, NJ: Prentice Hall.
Bates, M. (1976). Syntax in automatic speech understanding. *American Journal of Computational Linguistics.* Microfiche 45.
Bobrow, R.J. (1978). The RUS System. In B.L. Webber & R. Bobrow, (Eds.), *Research in natural language.* (Tech. Rep. No. 3378). Cambridge, MA: Bolt Beraneck and Newman Inc.
Bobrow, R.J., & Webber, B. (1980a). PSI-KLONE: Parsing and semantic interpretation in the BBN natural language understanding system. *Proceedings of the 1980 Conference of the Canadian Society for Computational Studies of Intelligence.* CSCS1/SCEIO.
Bobrow, R.J., & Webber, B. (1980b). Knowledge representation for syntactic/semantic processing. *Proceedings of the National Conference on Artificial Intelligence.* AAA1.
Bruce, B. (1975). Case systems for natural language. *Artificial Intelligence, 6,* 327-360.
Burton, R.R., & Brown, J.S. (1977). *Semantic grammar: A technique for constructing natural language interfaces to instructional systems.* (Tech. Rep. No. 3587). Cambridge, MA: Bolt Beraneck and Newman Inc.
Carberry, S. (1983). Tracking user goals in an information-seeking environment. *Proceedings of the National Conference on Artificial Intelligence,* (pp. 59-63). Washington, DC.
Carberry, S. (1984). Understanding Pragmatically Ill-formed Input. *Proceedings of COLING 84,* (pp. 200-206). Sandford, CA.
Carbonell, J.G. (1979). Towards a self-extending parser. *Proceedings of the 17th Annual Meeting of the Association for Computational Linguistics,* (pp. 3-8). La Jolla, CA.

Carbonell, J.G., Boggs, W.M., Mauldin, M.L., & Anick, P.G. (1983). The XCALIBUR Project: A natural language interface to expert systems. *Proceedings of the Eighth International Joint Conference on Artificial Intelligence,* (pp. 653-656). Karlsruhe, West Germany.

Chang, C.L. (1978). *Finding missing joins for incomplete queries in relational data bases.* (Research Report RJ2145). San Jose, CA: IBM Research Laboratory.

Charniak, E. (1983). A parser with something for everyone. In M. King, (Ed.), *Parsing Natural Languages.* New York: Academic Press.

Eastman, C.M., & McLean, D.S. (1981). On the need for parsing ill-formed input. *American Journal of Computational Linguistics, 7,* 257.

Finemann, L. (1983). Questioning the need for parsing ill-formed inputs. *American Journal of Computational Linguistics, 9,* 22.

Fromkin, V.A. (Ed.). (1973). *Speech errors as linguistic evidence.* Janua Linguarum, Series maior 77. The Hague: Mouton.

Gawron, J.M., King, J., Lamping, J., Leobner, E., Paulson, E.A., Pullman, G.K., Sag, I.A., & Wasow, T. (1982). The GPSG linguistic system. *Proceedings of the 20th Annual Meeting of the Association for Computational Linguistics,* (pp. 74-81). Cambridge, MA.

Granger, R.H., Staros, C.J., Taylor, G.B., & Yoshii, R. (1983). Scruffy text understanding: Design and implementation of the NOMAD system. *Proceedings of the Conference on Applied Natural Language Processing,* (pp. 104-106). Santa Monica, CA.

Harris, L.R. (1977a). *ROBOT: A high performance natural language interface for data base query.* (Tech. Rep. No. TR 77-1). Hanover, NH: Department of Mathematics, Dartmouth College.

Harris, L.R. (1977b). User oriented data base query with the ROBOT natural language query system. *International Journal for Man-Machine Studies, 9,* 697-713.

Harris, L.R. (1978). The ROBOT system: Natural language processing applied to data base query. *Proceedings 1978 Annual Conference Association for Computing Machinery,* 1965-1972. Washington, DC.

Hayes, P.J., & Carbonell, J.G. (1981). Multi-strategy construction-specific parsing for flexible data base query and update. *Proceedings of the Seventh International Joint Conference on Artificial Intelligence,* (pp. 432-439). Vancouver, BC, Canada.

Hayes, P., & Mouradian, G. (1980). Flexible parsing. *Proceedings of the 18th Annual Meeting of the Association for Computational Linguistics and Parasession on Topics on Interactive Discourse,* (pp. 97-103). Philadelphia, PA.

Heidorn, G.E. (1975). Augmented phrase structure grammars. *Proceedings of the Workshop: Theoretical Issues in Natural Language Processing,* (pp. 1-5). Cambridge, MA.

Hendrix, G.G., Sacerdoti, E.E., Sagalowicz, D., & Slocum, J. (1978). Developing a natural language interface to complex data. *ACM Transactions on Database Systems, 3,* 105-147.

Jensen, K.E., & Heidorn, G.E. (1983). The fitted parse: 100% parsing capability in a syntactic grammar of English. *Proceedings of the Conference on Applied Natural Language Processing,* (pp. 93-98). Santa Monica, CA.

Kaplan, S.J. (1978). Indirect responses to loaded questions. *Theoretical Issues in Natural Language Processing 2.* University of Illinois at Urbana-Champaign.

Kroch, A.A. (1981). On the role of resumptive pronouns in amnestying island constraint violations. *The Proceedings of the 17th Regional Meeting of the Chicago Linguistic Society.*

Kwasny, S.C., & Sondheimer, N.K. (1981). Relaxation techniques for parsing ill-formed input. *American Journal of Computational Linguistics, 7,* 99-108.

Lakoff, G., & Johnson, M. (1980). The metaphorical structure of the human conceptual system. *Cognitive Science, 4,* 195-208.

Linde, C., & Labov, W. (1975). Spatial network as a site for the study of language and thought. *Language, 51,* 924-938.

Malhotra, A. (1975). *Design criteria for a knowledge-based English language system of management: An experimental analysis.* (MAC TR 146. Project MAC). Cambridge, MA: Massachusetts Institute of Technology.

Miller, L.A., Heidorn, G.E., & Jensen, K. (1981). Text-critiquing with the EPISTLE system: An author's aid to better syntax. *AFIPS Conference Proceedings, 1981 NCC,* (pp. 649-655). Montvale, NJ: AFIPS Press.

Ramshaw, L.A., & Weischedel, R.M. (1984). Problem localization strategies for pragmatics in natural language front ends. *Proceedings of COLING 84.* Stanford, CA.

Robinson, J.J. (1982). DIAGRAM: A grammar for dialogues. *Communications of the ACM, 25,* 27-46.

Ross, J.R. (1979). Where's English. In C.J. Fillmore, D. Kempler, & W.S-Y Wang. (Eds.), *Individual Differences in Language Ability and Language Behavior,* (pp. 127-163). New York: Academic Press.

Sager, N. (1981). *Natural language information processing: A computer grammar of English and its applications.*

Reading, MA: Addison-Wesley.

Schank, R.C., Lebowitz, M., & Birnbaum, L. (1980). An integrated understander. *American Journal of Computational Linguistics, 6,* 13-30.

Sondheimer, N.K., Weischedel, R.M., & Bobrow, R.J. (1984). A knowledge of representation for semantic interpretation. *Proceedings of COLING 84,* (pp. 101-107). Sandford, CA.

Thompson, B.H. (1980). Linguistic analysis of natural language communication with computers. *Proceedings of the English International Conference on Computational Linguistics,* (pp. 190-201). Tokyo, Japan.

Trawick, D.J. (1983). *Robust sentence analysis and habitability.* (Tech. Rep. No. 5074: TR:83). Pasadena, CA: Computer Science Department, California Institute of Technology.

Walker, D.E. (1978). *Understanding Spoken Language.* New York: North-Holland.

Waltz, D.L. (1978). An English language question answering system for a large relational database. *Communications of the ACM, 21,* 526-539.

Warren, D.H.D., Pereira, L.M., & Pereira, F. (1977). PROLOG — The language and its implementation compared to LISP. *SIGPLAN Notices, 12,* 109-115.

Weischedel, R.M., & Black, J. (1980). Responding intelligently to unparsable inputs. *American Journal of Computational Linguistics, 6,* 97-109.

Weischedel, R.M., & Sondheimer, N.K. (1982). An improved heuristic for ellipsis processing. In *Proceedings of the 20th Annual Meeting of the Association for Computational Linguistics.* Cambridge, MA.

Weischedel, R.M., Voge, W.M., & James, M. (1978). An artificial intelligence approach to language instruction. *Artificial Intelligence, 10,* 225-240.

Wilks, Y.A. (1975). A preferential pattern-seeking semantics for natural language inference. *Artificial Intelligence, 6,* 53-74.

Wilks, Y. (1976). Natural language understanding systems within the AI paradigm — A survey and some comparisons. *American Journal of Computational Linguistics,* Microfiche 40.

Winograd, T. (1972). *Understanding natural language.* New York: Academic Press.

Woods, W.A. (1970). Transition network grammars for natural language analysis. *Communications of the ACM, 13,* 591-606.

Woods, W.A. (1973). Progress in natural language understanding — An application to lunar geology. In *AFIPS Conference Proceedings* (vol. 42). Montvale, NJ: AFIPS Press.

Woods, W.A., Kaplan, R.M., & Nash-Webber, B. (1972). *The lunar sciences natural language information system: Final report* (BBN Report No. 2378). Cambridge, MA: Bolt Beranek and Newman Inc.

Woods, W.A., Bates, M., Brown, G., Cook, C., Klovstad, J., Makhoul, J., Nash-Webber, B., Schwartz, R., Wolf, J., & Zue, V. (1976). *Speech understanding systems: Final report* (vols. 1-5). Cambridge, MA: Bolt Beranek and Newman Inc.

3
REFERENCE FAILURE

REPAIRING REFERENCE IDENTIFICATION FAILURES BY RELAXATION

Bradley A. Goodman

BBN Laboratories
10 Moulton Street
Cambridge, MA 02238

The goal of this work is the enrichment of human-machine interactions in a natural language environment.[1] We want to provide a framework less restrictive than earlier ones by allowing a speaker leeway in forming an utterance about a task and in determining the conversational vehicle to deliver it. A speaker and listener cannot be assumed to have the same beliefs, contexts, backgrounds or goals at each point in a conversation. As a result, difficulties and mistakes arise when a listener interprets a speaker's utterance. These mistakes can lead to various kinds of misunderstandings between speaker and listener, including reference failures or failure to understand the speaker's intention. We call these misunderstandings miscommunication. Such mistakes constitute a kind of "ill-formed" input that can slow down and possibly break down communication. Our goal is to recognize and isolate such miscommunications and circumvent them. This paper will highlight a particular class of miscommunication — reference problems — by describing a case study, including techniques for avoiding failures of reference.

1. Introduction

Cohen, Perrault and Allen (1981) showed in their paper "Beyond Question Answering" that "... users of question-answering systems expect them to do more than just answer isolated questions — they expect systems to engage in conversation. In doing so, the system is expected to allow users to be less than meticulously literal in conveying their intentions, and it is expected to make linguistic and pragmatic use of the previous discourse." Following in their footsteps, we want to build robust natural language processing systems that can detect and recover from miscommunication. The development of such systems requires a study on how people communicate and how they recover from problems in communication. This paper summarizes the results of a dissertation (Goodman, 1984) that investigated the kinds of miscommunication that occur in human communication with a special emphasis on *reference problems,* i.e., problems a listener has determining about whom or what a speaker is talking. We have written computer programs and algorithms that demonstrate how one could handle such problems in the context of a natural language understanding system. The study of miscommunication is a necessary task within such a context since any computer capable of communicating with humans in natural language must be tolerant of the imprecise, ill-devised or complex utterances that people use.

Our current research (Sidner, et al., 1981; Sidner, et al., 1983) views most dialogues as being cooperative and goal directed, i.e., a speaker and listener work together to achieve a common

---

[1] This research was supported in part by the Defense Advanced Research Project Agency under contract N00014-77-C-0378.

Reprinted with permission of the Association of Computational Linguistics.

goal. The interpretation of an utterance involves identifying the underlying plan or goal that the utterance reflects (Cohen, 1978; Allen, 1979; Sidner & Israel, 1981). This plan, however, is rarely, if ever, obvious at the surface sentence level. A central issue in the interpretation of utterances is the transformation of sequences of imprecise, ill-devised or complex utterances into *well-specified* plans that might be carried out by dialogue participants. Within this context, miscommunication can occur.

We are particularly concerned with cases of miscommunication from the hearer's viewpoint, such as when the hearer is inattentive to, confused about, or misled about the intentions of the speaker. In ordinary exchanges speakers usually make assumptions regarding what their listeners know about a topic of discussion. They will leave out details thought to be superfluous (Appelt, 1981; McKeown, 1983). Since the speaker really does not know exactly what a listener knows about a topic, it is easy to make statements that can be misinterpreted or not understood by the listener because not enough details were presented. One principal source of trouble is the description constructed by the speaker to refer to an actual object in the world. The description can be imprecise, confused, ambiguous or overly specific; it might be interpreted under the wrong context. This leads to difficulty for the listener when figuring out what object is being described, that is, reference identification errors. Such descriptions are "ill-formed" input. The blame for ill-formedness may lie partly with the speaker and partly with the listener. The speaker may have been sloppy or not taken the hearer into consideration; the listener may be either remiss or unwilling to admit he can't understand the speaker and to ask the speaker for clarification, or may simply feel that he has understood what he in fact has not.

This work is part of an on-going effort to develop a reference identification and plan recognition mechanism that can exhibit more "human-like" tolerance of such utterances. Our goal is to build a more robust system that can handle errorful utterances, and that can be incorporated in existing systems. As a start, we have concentrated on reference identification. In conversation people use imperfect descriptions to communicate about objects; sometimes their partners succeed in understanding and occasionally they fail. Any computer hoping to play the part of a listener must be capable of taking what the speaker says and either deleting, adapting or clarifying it. We are developing a theory of the use of extensional descriptions that will help explain how people successfully use such imperfect descriptions. We call this the theory of reference miscommunication.

Section 2 of this chapter highlights some aspects of normal communication and then provides a general discussion on the types of miscommunication that occur in conversation, concentrating primarily on reference problems and motivating many of them with illustrative protocols. Section 3 presents possible ways around some of the problems of miscommunication in reference. Finally there is a description of a partial implementation of a reference mechanism that attempts to overcome many reference problems.

We are following the task-oriented paradigm of Grosz (1977) since it is easy to study (through videotapes): it places the world in front of you (a primarily extensional world), and it limits the discussion while still providing a rich environment for complex descriptions. The task chosen as the target for the system is the assembly of a toy water pump. The water pump is reasonably complex, containing four subassemblies that are built from plastic tubes nozzles, valves, plungers, and caps that can be screwed or pushed together. A large corpus of dialogues concerning this task was collected by Cohen (see Cohen, 1981; 1982; 1984). These dialogues contained instructions from an "expert" to an "apprentice" that explain the assembly of the toy water pump. Both participants were working to achieve a common goal — the successful assembly of the pump. This domain is rich in perceptual information, allowing for complex descriptions of elements in it. The data provide examples of imprecision, confusion, and ambiguity as well as attempts to correct these problems.

The following exchange exemplifies one such situation. Here A is instructing J to assemble part of the water pump. Refer to Figure 1(a) for a picture of the pump. A and J are communicating verbally but neither can see the other. (The bracketed text in the excerpt tells what was actually occurring while each utterance was spoken.) Notice the complexity of the speaker's descriptions and the resultant processing required by the listener. This dialogue illustrates when listeners repair the speaker's description in order to find a referent, when they repair their initial reference choice once they are given more information, and when they fail to choose a proper referent. In Line 7, A describes the two holes in the *BASEVALVE* as "the little hole." J must repair the description, realizing that A doesn't really mean "one" hole but is referring to the "two" holes. J apparently does this since he doesn't complain about A's description and correctly attaches the *BASEVALVE* to the *TUBEBASE*. Figure 1(b) shows the configuration of the pump after the *TUBEBASE* is attached to the *MAINTUBE* in Line 10. In Line 13, J interprets "a red plastic piece" to refer to the *NOZZLE*. When A adds the relative clause "that has four gizmos on it," J is forced to drop the *NOZZLE* as the referent and to select the *SLIDEVALVE*. In Lines 17 and 18, A's description "the other – the open part of the main tube, the lower valve" is ambiguous, and J selects the wrong site, namely the *TUBEBASE*, in which to insert the *SLIDE-VALVE*. Since the *SLIDEVALVE* fits, J doesn't detect any trouble. Lines 20 and 21 keep J from thinking that something is wrong because the part fits loosely. In Lines 27 and 28, J indicates that A did not give him enough information to perform the requested action. In Line 30, J further compounds the error in Line 18 by putting the *SPOUT* on the *TUBEBASE*.

Excerpt 1 (Telephone)

A: 1 Now there's a blue cap
[J grabs the TUBEBASE]
       2. that has two little teeth sticking
       3. out of the bottom of it.

J: 4. Yeah.

A: 5. Okay. On that take the
       6. bright shocking pink piece of plastic
[J takes BASEVALVE]
       7. and stick the little hole over the teeth.
[J starts to install the BASEVALVE, backs off, looks at it again and then goes ahead and installs it]

J: 8. Okay.

A: 9. Now screw that blue cap onto
      10. the bottom of the main tube.
[J screws TUBEBASE onto MAINTUBE]

J: 11. Okay.

A: 12. Now, there's a –
      13. a red plastic piece
[J starts for NOZZLE]
      14. that has four gizmos on it.
[J switches to SLIDEVALVE]

J: 15. Yes.

A: 16. Okay. Put the ungizmoed end in the uh
      17. the other – the open
      18. part of the main tube, the lower valve.
[J puts SLIDEVALVE into hole in TUBEBASE, but A meant OUTLET2 of MAINTUBE]

J: 19. All right.

A: 20. It just fits loosely. It doesn't
      21. have to fit right. Okay, then take

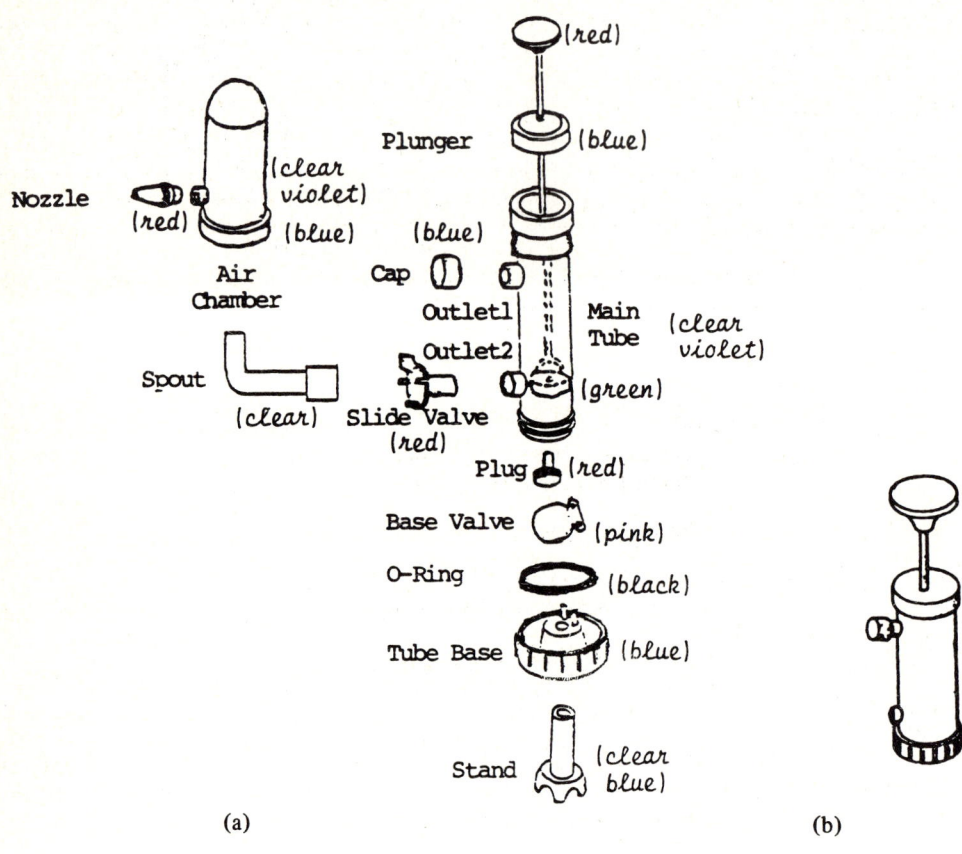

Figure 1. The Toy Water Pump.

```
                    22. the clear plastic elbow joint.
[J takes SPOUT]

        J:      23. All right.

        A:      24. And put it over the bottom opening, too
[J tries installing SPOUT on TUBEBASE]

        J:      25. Okay.

        A:      26. Okay. Now, take the —

        J:      27. Which end am I supposed to put it over?
                28. Do you know?

        A:      29. Put the — put the — the big end —
                30. the big end over it.
[J pushes big end of SPOUT on TUBEBASE, twisting it to force it on]
```

2. Miscommunication

People must and do manage to resolve lots of (potential) miscommunication in everyday conversation. Much of it is resolved subconsciously — with the listener unaware that anything is wrong. Other miscommunication is resolved with the listener *actively* deleting or replacing information in the speaker's utterance until it fits the current context. Sometimes this resolution is postponed until the questionable part of the utterance is actually needed. Still, when all these fail, the listener can ask the speaker to clarify what was said.[2]

There are many aspects of an utterance that the listener can become confused about and that can lead to miscommunication. The listener can become confused about what the speaker intends for the referents, the actions, and the goals described by the utterance. Confusions often appear to result from conflict between the current state of the conversation, the overall goal of the speaker, or the manner in which the speaker presented the information. However, when the listener steps back and is able to discover what kind of confusion is occurring, then the confusion can quite possibly be resolved.

2.1. Causes of miscommunication

This section attempts to motivate a paradigm for the kinds of conversation that we studied and tries to point out places in the paradigm that leave room for miscommunication.

2.1.1. Effects of the structure of task-oriented dialogues

Task-oriented conversations have a specific goal to be achieved: the performance of a task (e.g., Grosz, 1977). The participants in the dialogue can have the same skill level and they can simply work together to accomplish the task; or one of them, the expert, could know more and could direct the other, the apprentice, to perform the task. We have concentrated primarily on the latter case — due to the protocols that we examined — but many of our observations can be generalized to the former case, too. We will refer to this as the apprentice-expert domain.

The viewpoints of the expert and apprentice differ greatly in apprentice-expert exchanges. The expert, having an understanding of the functionality of the elements in the task, has more of a feel for how the elements work together, how they go together, and how the individual elements can be used. The apprentice normally has no such knowledge and must base his decisions on perceptual features such as shape (Grosz, 1981).

The structure of the task affects the structure of the dialogue (Grosz, 1977), particularly through the center of attention of the expert and apprentice. This is the phenomenon called focus (Grosz, 1977; Reichman, 1978; Sidner, 1979), which, in task-oriented dialogues is a very real and operational thing (e.g., focus is used in resolving anaphoric references). Shifts in focus correspond directly to the task, its subtasks, the objects in a task and the subpieces of each object. Focus and focus shifts are governed by many rules (Grosz, 1977; Reichman, 1978; Sidner, 1979). Confusion may result when expected shifts do not take place. For example, if the expert changes focus to an object but never discusses its subpieces (such as an obvious attachment surface) or never bothers to talk about the object reasonably soon after its introduction (i.e., between the time of its introduction and its use, without digressing in a well-structured way in between (see Reichman, 1978)), then the apprentice may become confused, leaving him ripe for miscommunication. The reverse influence between focus and objects can lead to trouble, too. A shift in focus by the expert that does not have a manifestation in the apprentice's world will also perplex the apprentice.

Focus also influences how descriptions are formed (Grosz, 1981; Appelt, 1981). The level of detail required in a description depends directly on the elements currently highlighted by the

---

[2] An analysis of clarification subdialogues can be found in Litman and Allen, 1984.

focus. If the object to be described is similar to other elements in focus, the expert must be more specific in the formulation of the description or may consider shifting focus away from the possibly ambiguous objects to one where the ambiguity won't occur.

2.2 Consequences of miscommunication

In this section we will make it clear that people do miscommunicate and yet they often manage to fix things. We will look at specific forms of miscommunication and describe ways to detect them. We will highlight relationships between different miscommunication problems but won't necessarily demonstrate ways to resolve each of them.

2.2.1 Instances of miscommunication

There are many ways hearers can get confused during a conversation. Figure 2 outlines some of them that were derived from analyzing the water pump protocols. This section defines and illustrates many of them through numerous excerpts. Each excerpt is marked in parentheses to show what modality of communication was used (see Cohen, 1984 for a description about the collection of these excerpts). Each bracketed portion of the excerpt explains what was occurring at that point in the dialogue. The confusions themselves, coupled with the description at the end of this section on how to recognize when one of them is occurring, provides motivation for the use of the algorithm outlined in Section 3 as a means for repairing communication problems. We will only discuss referent confusion in this paper. The other forms of confusion — Action, Goal, and Cognitive Load — are described in Goodman, 1982, 1984. Another categorization of confusions that lead to conversation failure can be found in Ringle and Bruce, 1981.

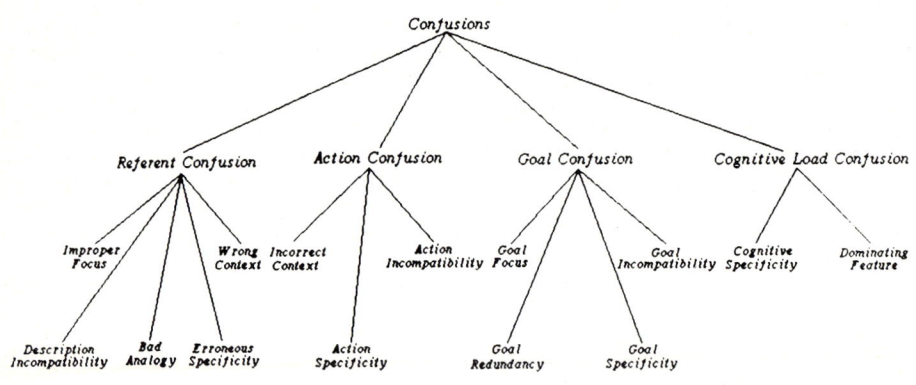

Figure 2. A taxonomy of confusions

Referent confusion occurs when the listener is unable to correctly determine what the speaker is referring to with a particular description. It occurs when the descriptions in the utterance are ambiguous or imprecise, when there is confusion between the speaker and listener about what the current focus or context is, or when the descriptions in the utterance are either incorrect or incompatible with the current or global context.

Erroneous Specificity

Ambiguous (and, thus imprecise) descriptions can cause confusion about the referent. Excerpt 2 below illustrates a case where the speaker's description is underspecified — it does not provide enough detail to prune the set of possible referents down to one.

Excerpt 2 (Face-to-Face)

S: 1. And now take the little red
2. peg.
[P takes PLUG]
3. Yes.

4. and place it in the hole at the
5. green end,
[P starts to PLUG into OUTLET2 of MAINTUBE]
6. no

7. the — in the green thing
[P puts PLUG into green part of PLUNGER]

P: 8. Okay.

In Lines 4 and 5, S describes the location to place a peg into a hole by giving spatial information. Since the location is given relative to another location by "in the hole *at the green end*", it defines a region where the peg might go instead of a specific location. In this particular case, there are three possible holes to choose from that are near the green end. The listener chooses one — the wrong one — and inserts the peg into it. Because this dialogue took place face to face, S is able to correct the ambiguity in Lines 6 and 7.

A speaker's description can be imprecise in several possible ways. (1) It may contain features that do not readily apply in the domain. In Line 3, Excerpt 3, the feature "funny" has no relevance to the listener. It is not until A provides a fuller description in Lines 5 to 8 that E is able to select the proper piece. (2) It may use a vague head noun coupled with few or no feature values (and context alone does not necessarily suffice to distinguish the object). In Excerpt 4, Line 9, "attachment" is vague because all objects in the domain are attachable parts. The expert's use of "attachment" was most likely to signal the action the apprentice can expect to take next. The use of the feature value "clear" provides little benefit either, because three clear, unused parts exist. The size descriptor "little" prunes this set of possible referents down to two contenders. (3) Enough feature values are provided but at least one value is too vague leading to trouble. In Excerpt 5, Line 3, the use of the attribute value "rounded" to describe the shape does not sufficiently reduce the set of four possible referents (though, in this particular instance, A correctly identifies it) because the term is applicable to numerous parts in the domain. A more precise shape descriptor such as "bell-shaped" or "cylindrical" would have been more beneficial to the listener.

Excerpt 3 (Telephone)

E: 1. All right.

2. Now.

3. There's another funny little
4. red thing, a
[A is confused, examines both NOZZLE and SLIDEVALVE]
5. little teeny red thing that's
6. some — should be somewhere on
7. the desk, that has um — there's
8. like teeth on one end.
[E takes SLIDEVALVE]

A: 9. Okay.

E: 10. It's a funny-loo — hollow,
    11. hollow projection on one end
    12. and then teeth on the other.

Excerpt 4 (Teletype)

A: 1. take the red thing with the
    2. prongs on it

    3. and fit it onto the other hole
    4. of the cylinder

    5. so that the prongs are
    6. sticking out

R: 7. Ok

A: 8. now take the clear little
    9. attachment

    10. and put on the hole where you
    11. just put the red cap on

    12. make sure it points
    13. upward

R: 14. Ok

Excerpt 5 (Teletype)

S: 1. Ok,

    2. put the red nozzle on the outlet
    3. of the rounded clear chamber

    4. Ok?

Improper Focus

Focus confusion can occur when the speaker sets up one focus and then proceeds with another one without letting the listener know of the switch (i.e., a focus shift occurs without any indication). An opposite phenomenon can also happen — the listener may feel that a focus shift has taken place when the speaker actually never intended one. These really are very similar — one is viewed more strongly from the perspective of the speaker and the other from the listener.

Excerpt 6 below illustrates an instance of the first type of focus confusion. In the excerpt, the speaker (S) shifts focus without notifying the listener (P) of the switch. As the excerpt begins, P is holding the *TUBEBASE*. S provides in Lines 1 to 16 instruction for P to attach the *CAP* and the *SPOUT* to outlets *OUTLET1* and *OUTLET2*, respectively, on the *MAINTUBE*. Upon P's successful completion of these attachments, S switches focus in Lines 17 to 20 to the *TUBE-BASE* assembly and requests P to screw it on to the bottom of the *MAINTUBE*. While P completes the task, S realizes she left out a step in the assembly — the placement of the *SLIDE-VALVE* into *OUTLET2* of the *MAINTUBE* before the *SPOUT* is placed over the same outlet.

S attempts to correct her mistake by requesting P to remove "the plas"[3] piece in Lines 22 and 23. Since S never indicated a shift in focus from the *TUBEBASE* back to the *SPOUT*, P interprets "the plas" to refer to the *TUBEBASE*.

Excerpt 6 (Face-to-Face)

S: 1. And place
    2. the blue cap that's left
[P takes CAP]
    3. on the side holes that are
    4. on the cylinder.
[P lays down TUBEBASE]
    5. the side hole that is farthest
    6. from the green end
[P puts CAP on OUTLET1 of MAINTUBE]

P: 7. Okay

S: 8. And take the nozzle-looking
    9. piece,
[P grabs NOZZLE]

    10. no

    11. I mean the clear plastic one.
[P takes SPOUT]

    12. and place it on the other hole
[P identifies OUTLET 2 of MAINTUBE]
    13. that's left,

    14. so that nozzle points away
    15. from the
[P installs SPOUT on OUTLET2 of MAINTUBE]

    16. right.

P: 17. Okay.

S: 18. Now

    19. take the

    20. cap base thing
[P takes TUBEBASE]
    21. and screw it onto the bottom.
[P screws TUBEBASE on MAINTUBE]
    22. ooops,
[S realizes she has forgotten to have P put SLIDEVALVE into OUTLET2 of MAINTUBE]
    23. un-undo the plas
[P starts to take TUBEBASE off MAINTUBE]

    24. no

---

[3] The whole word here is "plastic." People in general tend to be good at proceeding before hearing the whole utterance or even the whole word.

                    25. the clear plastic thing that I
                    26. told you to put on
[P removes SPOUT]

                    27. sorry.

                    28. And place the little red thing
[P takes SLIDEVALVE]
                    29. in there first,
[P inserts SLIDEVALVE into OUTLET2 of MAINTUBE]
                    30. it fits loosely in there.

Excerpt 7 below demonstrates the latter type of focus confusion that occurs when the speaker (S) sets up one focus — the *MAINTUBE*, which is the correct focus in this case — but then proceeds in such a manner that the listener (J) thinks a focus shift to another piece, the *TUBEBASE*, has occurred. Thus, Line 15 refers to "the lower side hole in the *MAINTUBE*" for S and "the hole in the *TUBEBASE*" for J. J has no way of realizing that he has focused incorrectly unless the description as he interprets it doesn't have a real world correlate (here something does satisfy the description so J doesn't sense any problem) or if, later in the exchange, a conflict arises due to the mistake (e.g., a requested action cannot be performed). In Line 31, J inserts a piece into the wrong hole because of the misunderstanding in Line 15. Line 31 hints that J may have become suspicious that an ambiguity existed but since the task was successfully completed (i.e., the red piece fitted into the hole in the base), and since S did not provide any clarification, he assumed he was correct.

                    Excerpt 7 (Telephone)

        S: 1. Um now.
                2. Now we're getting a little
                3. more difficult.

        J: 4. (laughs)

        S: 5. Pick out the large air tube
[J picks up STAND]
                6. that has the plunger in it.
[J puts down STAND, takes PLUNGER/MAINTUBE assembly]

        J: 7. Okay.

        S: 8. And set it on its base,
[J puts down MAINTUBE, standing vertically, on the TABLE]
                9. which is blue now,
                10. right?
[J has shifted focus to the TUBEBASE]

        J: 11. Yeah

        S: 12. Base is blue.
                13. Okay,
                14. Now
                15. You've got a bottom hole still
                16. to be filled,
                17. correct?

        J: 18. Yeah.
[J answers this with MAINTUBE still sitting on the TABLE; he shows no indication of what hole he thinks is meant — the one on the MAINTUBE, OUTLET2, or the one in the TUBEBASE]

S: 19. Okay.
   20. You have one red piece
   21. remaining?
[J picks up MAINTUBE assembly and looks at TUBEBASE, rotating the MAINTUBE so that TUBEBASE is pointed up, and sees the hole in it; he then looks at the SLIDEVALVE]

J: 22. Yeah.

S: 23. Okay.
   24. Take that red piece.
[J takes SLIDEVALVE]
   25. It's got four little feet on
   26. it?

J: 27. Yeah?

S: 28. And put the small end into
   29. that hole on the air tube —
   30. on the big tube.

J: 31. On the very bottom?
[J starts to put it into the bottom hole of TUBEBASE — though he indicates he is unsure of himself]

S: 32. On the bottom,
   33. Yes.

Misfocus can also occur when the speaker inadvertently fails to distinguish the proper focus because he did not notice a possible ambiguity, or when, through no fault of the speaker, the listener just fails to recognize a switch in focus indicated by the speaker. Excerpt 7 above is an example of the first type because S failed to notice that an ambiguity existed since he never explicitly brought the *TUBEBASE* either into or out of focus. He just assumed that J had the same perspective as him — a perspective in which no ambiguity occurred.

Wrong Context

Context differs from focus. The context of a portion of a conversation is concerned with the point of the discussion in that fragment and with the set of objects relevant to that discussion, though not attended to currently. Focus pertains to the elements which are currently being attended to in the context. For example, two people can share the same context but have different focus assignments within it — we're both talking about the water pump but you're describing the *MAINTUBE* and I'm describing the *AIRCHAMBER*. Alternatively, we could just be using different contexts — I think you're talking about taking the pump with new parts — in both cases we may be sharing the same focus — the pump — but our contexts are totally off from one another.[4] The kinds of misunderstandings that can occur because of context problems are similar to those for focus problems: (1) the speaker might set up or be in one context for a discussion and then proceed in another one without effectively letting the listener know of the change, (2) the listener may feel a change in context has taken place when in fact the speaker never intended one, or (3) the listener fails to recognize an indicated context switch by the speaker. Context affects reference because it helps define the set of available objects that are possible contenders for the referent of the speaker's descriptions. If the contexts of the speaker and listener differ, then misreference might might result.

---

[4] Grosz (1977, 1981) would describe this as a difference in "task plans" while Reichman (1978, 1981) would say that the "communicative goals" differed.

Bad Analogy

An analogy (see Gentner, 1980 for a discussion on analogies) is a useful way to help describe an object by attempting to be *more* precise by using shared past experience and knowledge — especially shape and functional information. If that past experience or knowledge doesn't contain the information the speaker assumes it does or isn't there, then trouble occurs. Thus, one more way referent confusion can occur is by describing an object using a poor analogy. An analogy used to describe an object might not be specific enough — confusing the listener because several pieces might conform to the analogy or, in fact, none at all appear to fit because discovering a mapping between the analogous object and some piece in the environment is too difficult. In Excerpt 8, J at first has trouble correctly satisfying A's functional analogy "stopper" in "the big blue stopper", but finally selects what he considers to be the closest match to "stopper".

Excerpt 8 (Telephone)

A: 1. Okay. Now.

2. take the big blue
3. stopper that's laying around

[J grabs AIRCHAMBER]

4. ... and take the black
5. ring —

J: 6. The big blue stopper?

[J is confused and tries to communicate it to A; he is holding the AIRCHAMBER here]

A: 7. Yeah,

8. the big blue stopper

9. and the black ring.

[J drops AIRCHAMBER and takes the O-RING and the TUBEBASE]

In other cases it might be too specific — confusing the listener because none of the available referents appear to fit it. In Line 8 of Excerpt 6, "nozzle-looking" forms a poor shape analogy because the object being referred to is actually an elbow-shaped spout. The "nozzle-looking" part of the description convinced the listener that what he was looking for was something specific like a nozzle (which is a small spout). Sometimes, when an object is a clear representative of a specific analogy class, the apprentice may become confused, wondering why the expert bothered to form an analogy instead of just directly describing the object as a member of the class. Hence, it would not be surprising if the apprentice ignored the best representative of the class for some less obvious exemplar. Thus, for example, it is better to say "nozzle" instead of "nozzle-looking." In Excerpt 9, the description "hippopotamus face shape" (a shape analogy) in Lines 2 and 3, and "champagne top" (a shape analogy) in Line 9, are too specific and the listener is unable to easily find something close enough to match either of them. He can't discover a mapping between the object in the analogy and one in the real world.

Excerpt 9 (Audiotape)

M: 1. take the bright pink flat
2. piece of hippopotamus face
3. shape piece of plastic
4. and you notice that the two
5. holes on it

[M is trying to refer to BASEVALVE]

6. match

7. along with the two
8. peg holes on the
9. champagne top sort of
10. looking bottom that had
11. threads on it

[M is trying to refer to TUBEBASE]

Description Incompatibility

Incompatible descriptions can lead to confusion also. A description is incompatible when (1) one or more of the specific conditions, i.e., the feature values, do not satisfy *any* of the pieces; (2) when one or more specified constraints do not hold (e.g., saying "the *loose* one" when all objects are tightly attached); or (3) if no *one* object satisfied *all* of the features specified in the description. In Lines 7 and 8 of Excerpt 9 above, M's use of "the two peg holes" leads to bewilderment for the listener because the described object has no holes in it. M actually meant "two pegs".

2.2.2 Detecting miscommunication

Part of our research has been to examine how a listener discovers the need for a repair of an utterance or a description during communication. The incompatibility of a referent or action is one signal of possible trouble. The appearance of an obstacle that blocks one from achieving a goal is another indication of a problem.

Incompatibility

Two kinds of incompatibility, action or referent, appear in the taxonomy of confusions. The strongest hint that there is a reference problem occurs when the listener finds *no* real world object to correspond to the speaker's description. This can occur when (1) one or more of the specified feature values in the description are not satisfied by *any* of the pieces (e.g., saying "the orange cap" when none of the objects are orange); (2) when one or more specified constraints do not hold (e.g., saying "the red plug that fits *loosely*" when all the red plugs attach tightly); or (3) if no *one* object satisfies *all* of the features specified in the description (i.e., there is, for each feature, an object that exhibits the specified feature value, but no one object exhibits all of the values). An action problem is likely if (1) the listener cannot perform the action specified by the speaker because of some obstacle; (2) the listener performs the action but does not arrive at its intended effect (i.e., a specified or default constraint isn't satisfied); or (3) the current action affects a previous action in an adverse way, yet the speaker has given no sign of any importance to this side-effect.

Goal obstacle

A goal obstacle occurs when a goal (or subgoal) one is trying to achieve is blocked. This blockage can result in confusion for the listener because he did not expect the speaker to give him tasks that could not be achieved. Often, though, it points out for the listener that some miscommunication (such as misreference) has occurred.

Goal redundancy

Goal redundancy occurs when the requested goal (or subgoal) is already satisfied. In some sense, it is a special kind of goal obstacle where the goal to be fulfilled is blocked because it is already satisfied. It is a simple goal obstacle because nothing has to be done to get around it. However, it can lead to confusion on the part of listeners because they may suspect they misunderstood what the speaker has requested since they wouldn't expect a reasonable speaker to request the performance of an already completed action. It provides a hint that miscommunication has occurred.

3. Repairing Reference Failures

3.1 Introduction

The previous section illustrated how task-oriented natural language interactions in the real world can induce contextually poor utterances. Given all the possibilities for confusion, when confusions do occur, they must be resolved if the task is to be performed. This section explores the problem of fixing reference failures.

Reference identification is a search process where a listener looks for something in the world that satisfies a speaker's uttered description. A computational scheme for performing reference identification has evolved from work by other artificial intelligence researchers (e.g., see Grosz, 1977). That traditional approach succeeds if a referent is found, or fails if no referent is found (see Figure 3(a)). However, a reference identification component must be more versatile than those constructed in the traditional manner. The excerpts provided in the previous section show that the traditional approach is wrong because people's real behavior is much more elaborate. In particular, listeners often find the correct referent even when the speaker's description does not describe any object in the world. For example, a speaker could describe a blue block as the "turquoise block." Most listeners would go ahead and assume that the blue block was the one the speaker meant.

A key feature to reference identification is "negotiation." Negotiation in reference identification comes in two forms. First, it can occur between the listener and the speaker. The listener can step back, expand greatly on the speaker's description of a plausible referent, and ask for confirmation that he has indeed found the correct referent. For example, a listener could initiate negotiation with "I'm confused. Are you talking about the thing that is kind of flared at the top? Couple inches long. It's kind of blue." Second, negotiation can be with oneself. This type of negotiation, called self-negotiation, is the one that we are most concerned with in this research. The listener considers aspects of the speaker's description, the context of the communication, and the listener's own abilities. He then applies that deliberation to determine whether one referent candidate is better than another or, if no candidate is found, what are the most likely places for error or confusion. Such negotiation can result in the listener testing whether or not a particular referent works. For example, linguistic descriptions can influence a listener's perception of the world. The listener must ask himself whether he can perceive one of the objects in the world the way the speaker described it. In some cases, the listener's perception may overrule the description because the listener can't perceive it the way the speaker described it.

To repair the traditional approach we have developed an algorithm that captures for certain cases the listener's ability to negotiate with himself for a referent. It can look for a referent and, if it doesn't find one, it can try to find possible referent candidates that might work, and then loosen the speaker's description using knowledge about the speaker, the conversation, and the listener himself. Thus, the reference process becomes multi-step and resumable. This computational model, which I call "FWIM" for "Find What I Mean", is more faithful to the data than the traditional model (see Figure 3(b)).

One means of making sense of an approximate description is to delete or replace portions of it that don't match objects in the hearer's world. In our program we are using "relaxation" techniques to capture this behavior. Our reference identification module treats descriptions as approximate. It relaxes a description in order to find a referent when the literal content of the description fails to provide the needed information. Relaxation, however, is not performed blindly on the description. We try to model a person's behavior by drawing on sources of knowledge used by people. We have developed a computational model that can relax aspects of a description using many of these sources of knowledge. Relaxation then becomes a form of communication repair (Brown & VanLehn, 1980) that hearers can use.

3.2 The relaxation component

When a description fails to denote a referent in the real world properly, it is possible to repair

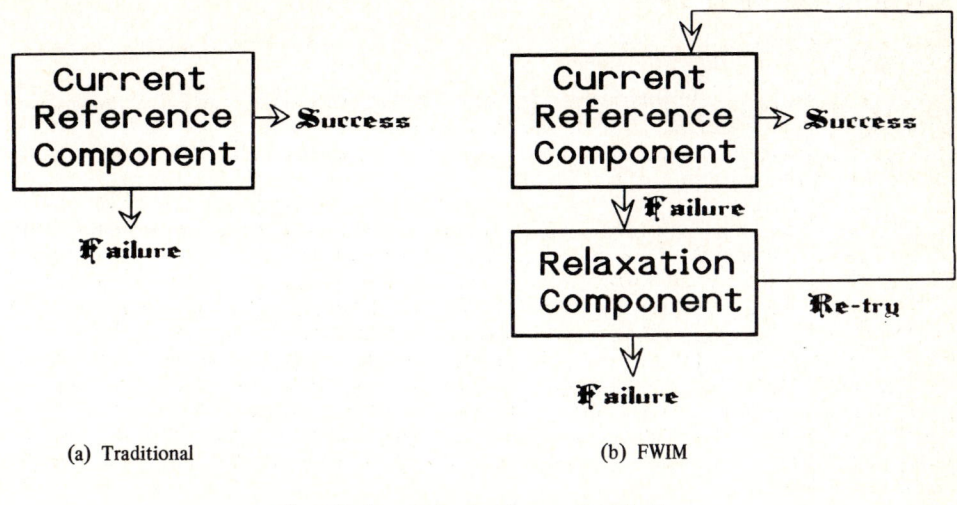

(a) Traditional   (b) FWIM

Figure 3. Approaches to reference identification.

it by a relaxation process that ignores or modifies parts of the description. Since a description can specify many features of an object, the order in which parts of it are relaxed is crucial (i.e., relaxing in different order could yield matches to different objects). There are several kinds of relaxation possible. One can ignore a constituent, replace it with something close, replace it with a related value, or change focus (i.e., consider a different group of objects.). This section describes the overall relaxation component that draws on knowledge sources about descriptions and the real world as it tries to relax an errorful description to one for which a referent can be identified.

3.2.1 Find a referent using a reference mechanism

Identifying the referent of a description requires finding an element in the world that corresponds to the speaker's description (where every feature specified in the description is present in the element in the world but not necessarily vice versa). The initial task of our reference mechanism is to determine whether or not a search of the (taxonomic) knowledge base that we use to model the world is necessary. For example, the reference component should not bother searching – unless specifically requested to do so – for a referent for indefinite noun phrases (which usually describe new or hypothetical objects) or extremely vague descriptions (which do not clearly describe an object because they are composed of imprecise feature values). A number of aspects of discourse pragmatics can be used in that determination (e.g., the use of a deictic in a definite noun phrase, such as "this X" or "the last X", hints that the object was either mentioned previously or that it probably was evoked by some previous reference, and that it is searchable) but we will not examine them here.

The knowledge base contains linguistic descriptions and a description of the listener's visual scene itself. In our implementation and algorithms, we assume it is represented in KL-One (Brachman, 1977), a system for describing taxonomic knowledge. KL-One is composed of CONCEPTs, ROLEs on concepts, and links between them. A CONCEPT is like a set, representing those elements described by it. A SUPERC link ("= =>") is used between concepts to show set inclusions. For example, consider Figure 4. The SuperC from Concept B to Concept A is like stating B⊆A for two sets A and B. An INDIVIDUAL CONCEPT is used to guarantee that the subset specified by a concept is unique. The Individual Concept D shown in the figure is defined to be a unique member of the subset specified by Concept C. ROLEs on concepts are like

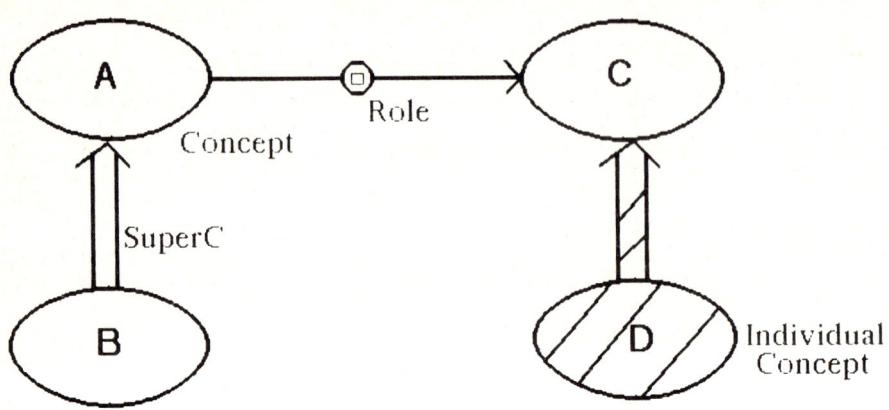

Figure 4. A KL-ONE Taxonomy

normal attributes and slot fillers in other knowledge representation languages. They define a functional relationship between the concept and other concepts.

Assuming that a search of the knowledge base is considered necessary, then a reference search mechanism is invoked. The search mechanism uses the KL-One Classifier (Lipkis, 1981) to search the knowledge base taxonomy. This search is constrained by a focus mechanism based on the one developed by Grosz (1977). The Classifier's purpose is to discover all appropriate subsumption relationships between a newly formed description and all other descriptions in a given taxonomy. With respect to reference, this means that all possible (descriptions of) referents of the description will be subsumed by it after it has been classified into the knowledge base taxonomy. If more than one candidate referent is below (when a description A is subsumed by B, we say A is "below" B) the classified description, then, unless a quantifier in the description specified more than one element, the speaker's description is ambiguous. If exactly one description is below it, then the intended referent is assumed to have been found. Finally, if no referent is found below the classified description, the relaxation component is invoked. We will only consider the last case in the rest of the paper.

3.2.2 Collect votes for or against relaxing the description

It is necessary to determine whether or not the lack of a referent for a description has to do with the description itself (i.e., reference failure) or outside forces that are causing reference confusion. For example, the problem may be with the flow of the conversation and the speaker's and listener's perspectives on it; it may be due to incorrect attachment of a modifier, it may be due to the action requested; and so on. Pragmatic rules are invoked to decide whether or not the description should be relaxed. These rules will not be discussed here so we will assume that the problem lies in the speaker's description.

3.2.3 Perform the relaxation of the description

If relaxation is demanded, then the system must (1) find potential referent candidates, (2) determine which features in the speaker's description to relax and in what order, and use those ordered features to order the potential candidates with respect to the preferred ordering of features, and (3) determine the proper relaxation techniques to use and apply them to the description.

Find potential referent candidates

Before relaxation can take place, potential candidates for referents (which denote elements in the listener's visual scene) must first be found. These candidates are discovered by performing a "walk" in the knowledge base taxonomy in the general vicinity of the speaker's classified description. A KL-One partial matcher is used to determine how close the candidate descriptions found during the walk are to the speaker's description. The partial matcher generates a numerical score to represent how well the descriptions match (after first generating scores at the feature level to help determine how the features are to be aligned and how well they match). This score is based on information about KL-One and does not take into account any information about the task domain. The ordering of features and candidates for relaxation described below takes into account the task domain. The set of best descriptions returned by the matcher (as determined by some cutoff score) is selected as referent candidates.

Order the features and candidates for relaxation

At this point the reference system inspects the speaker's description and the candidates, decides which features to relax and in what order,[5] and generates a master ordering of features for relaxation. Once the feature order is created, the reference system uses that ordering to determine the order in which to try relaxing the candidates.

We draw primarily on sources of linguistic knowledge, pragmatic knowledge, discourse knowledge, domain knowledge, perceptual knowledge, hierarchical knowledge, and trial and error knowledge during this repair process. A detailed treatment of all of them can be found in (Goodman, 1983, 1984; Sidner et al., 1984). These knowledge sources are consulted to determine the feature ordering for relaxation. We represent information from each knowledge source as a set of *relaxation rules*. These rules are written in a PROLOG-like language. Figure 5 illustrates one such linguistic knowledge relaxation rule. This rule is motivated by the observation in the excerpts that speakers typically add more important information at the end of a description (where they are separated from the main part of the description and thus provided more emphasis). Since the syntactic constituents often at the end are relative clauses or predicate complements, we created this more specific relaxation rule. However, a more general and more applicable rule is that information presented at the end of a description is usually more prominent.

Relax the features in the speaker's description in the order:
adjectives, then prepositional phrases, and finally relative clauses and predicate complements.

E.g.,
    Relax-Feature-Before (v1, v2)
    <— ObjectDescr(d),
        FeatureDescriptor(v1),
        FeatureDescriptor(v2),
        FeatureInDescription(v1,d),
        FeatureInDescription(v2,d),
        Equal(syntactic-form(v1,d),"ADJ"),
        Equal (syntactic-form(v2,d),"REL-CLS")

Figure 5. A sample relaxation rule

---

[5] Of course, once one particular candidate is selected, then deciding which features to relax is relatively trivial — one simply compares feature by feature between the candidate description (the target) and the speaker's description (the pattern) and notes any discrepancies.

Each knowledge source produces its own partial ordering of features. The partial orderings are then integrated to form a directed graph. For example, perceptual knowledge may say to relax color. However, if the color value was asserted in a relative clause, linguistic knowledge would rank color lower, i.e., placing it later in the list of things to relax.

Since different knowledge sources generally have different partial orderings of features, these differences can lead to a conflict over which features to relax. It is the job of the best candidate algorithm to resolve the disagreements among knowledge sources. It's goal is to order the referent candidates, C_i, so that relaxation is attempted on the best candidates first. Those candidates are the ones that conform best to a proposed feature ordering. To start, the algorithm examines pairs of candidates and the feature orderings from each knowledge source. For each candidate C_i, the algorithm scores the effect of relaxing the speaker's original description to C_i, using feature ordering from one knowledge source. The score reflects the goal of minimizing the number of features relaxed while trying to relax the features that are "earliest" in the feature ordering. It repeats its scoring of C_i for each knowledge source, and sums up its scores to form C_i's total score. The C_i's are then ordered by that score.

Figure 6 provides a graphic description of this process. A set of objects in the real world is selected by the partial matcher as potential candidates for the referent. These candidates are shown across the top of the figure. The lines on the right side of each box correspond to the set of features that describe that object. The speaker's description is represented in the center of the figure. The set of specified features and their assigned feature value (e.g., the pair Color-Maroon) is also shown there. A set of partial orderings is generated that suggest which features in the speaker's description should be relaxed first — one ordering for each knowledge source (shown as "Linguistic," "Perceptual," and "Hierarchical" in the figure). These are put together to form a directed graph that represents the possible, reasonable ways to relax the features specified in the speaker's description. Finally, the referent candidates are reordered using the information expressed in the speaker's description and in the directed graph of features.

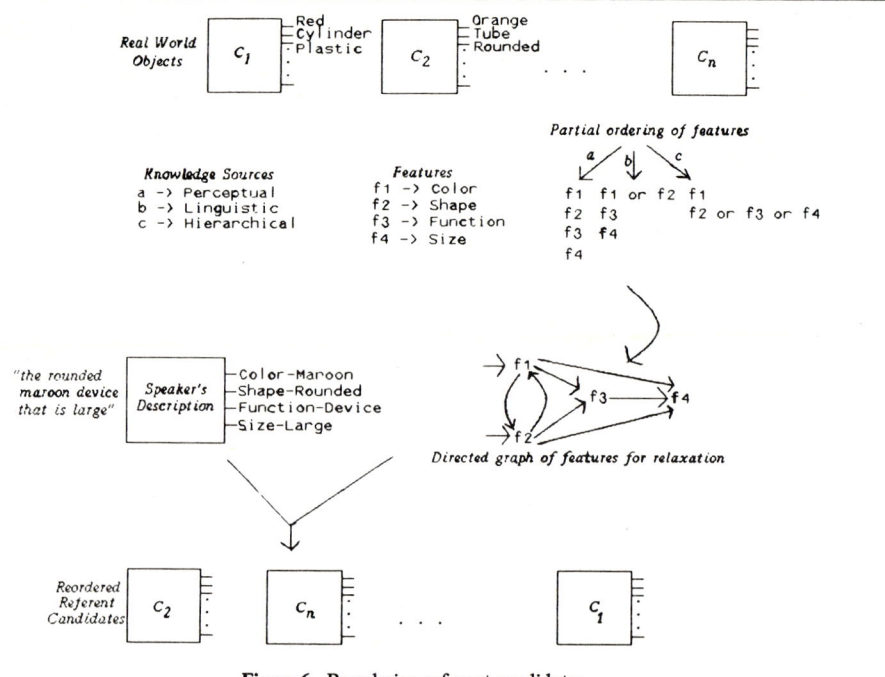

Figure 6. Reordering referent candidates

Once a set of ordered potential candidates is selected, the relaxation mechanism begins step 3 of relaxation; it tries to find proper relaxation methods to relax the features that have just been ordered (success in finding such methods "justifies" relaxing the description). It stops at the first candidate which is reasonable.

Determine which relaxation methods to apply

Relaxation can take place with many aspects of a speaker's description: with complex relations specified in the description, with individual features of a referent specified by the description, and with the focus of attention in the real world where one attempts to find a match. Complex relations specified in a speaker's description include spatial relations (e.g., "the outlet *near* the *top* of the tube"), comparatives (e.g., "the *larger* tube") and superlatives (e.g., "the *longest* tube"). These can be relaxed. The simpler features of an object (such as size or color) that are specified in the speaker's description are also open to relaxation.

Often the objects in focus in the real world implicitly cause other objects to be in focus (Grosz, 1977; Webber, 1978). The subparts of an object in focus, for example, are reasonable candidates for the referent of a failing description and should be checked. At other times, the speaker might attribute features of a subpart of an object to the whole object (e.g., describing a plunger that is composed of a red handle, a metal rod, a blue cap, and a green cup as "the green plunger"). In these cases, the relaxation mechanism utilizes the part-whole relation in object descriptions to suggest a way to relax the speaker's description.

Relaxation of a description has a few global strategies that can be followed for each part of the description: (1) drop the errorful feature value from the description altogether, (2) weaken or tighten the feature value but keep its new value *close* to the specified one, or (3) try some other feature value.

These strategies are realized through a set of procedures (or *relaxation methods*) that are organized hierarchically. Each procedure is an expert at relaxing its particular type of feature. For example, a Generate-Similar-Feature-Values procedure is composed of procedures like Generate-Similar-Shape-Values, Generate-Similar-Color-Values and Generate-Similar-Size-Values. Each of those procedures are specialists that attempt to first relax the feature value to one "near" the current one (e.g., one would prefer to first relax the color "red" to "pink" before relaxing it to "blue") and then, if that fails, to try relaxing it to any of the other possible values. If those fail, the feature would simply be ignored.

3.3 An example on handling a misreference

This section describes how a referent identification system can handle a misreference using the scheme outlined in the previous section. For the purposes of this example, assume that the water pump objects currently in focus include the *CAP*, the *MAINTUBE*, the *AIRCHAMBER* and the *STAND* (see Figure 1(a) for a picture of these parts). Assume also that the speaker tries to describe two of the objects: "... two devices that are clear plastic. One of them has two openings on the outside with threads on the end, and it's about five inches long. The other one is a rounded piece with a turquoise base on it. Both are tubular. The rounded piece fits loosely over ...". The reference system can find a unique referent for the first object but not for the second. The relaxation algorithm will be shown below to reduce the set of referent candidates for the second description down to two. It, then, requires the system/listener to try out those candidates to determine if one, or both, fits loosely. The protocols exhibit a similar result when the listener uses "fits loosely" to get the correct referent (e.g., Excerpt 6 exemplifies where the "fit" can confirm that the proper referent was found).

Figure 7 provides a simplified and linearized view of the actual KL-One representation of the speaker's descriptions after they have been parsed and semantically interpreted. A representation of each of the water pump objects that are currently under consideration is presented in Figure 8. Each provides a physical description of the object – in terms of its dimensions, the basic 3-D shapes composing it, and its physical features – and a basic functional description of the object.

The first entry in each representation in Figure 8 (that entry is shown in uppercase) defines the basic kind of entity being described (e.g., "TUBE" means that the object being described is some kind of tube). The words in mixed case refer to the names of features and the words in uppercase refer to possible fillers of those features from things in the water pump world. The "Subpart" feature provides a place for an embedded description of an object that is a subpart of a parent object. Such subparts can be referred to on their own or as part of the parent object. The "Orientation" feature, used in the representations in Figure 8, provides a rotation and translation of the object from some standard orientation to the object's current orientation in 3-D space. The standard orientation provides a way to define relative positions such as "top," "bottom," or "side."

Descr1:
    (DEVICE (Transparency CLEAR)
        (Composition PLASTIC)
        (Subpart (OPENING))
        (Subpart (OPENING))
        (Subpart (THREADS (Rel-Position END)))
        (Dimensions (Length 5.0))
        (Analogical-Shape TUBULAR))

Descr2:
    (FIT-INTO (Outer (DEVICE(Transparency CLEAR)
                      (Composition PLASTIC)
                      (Shape ROUND)
                      (Analogical-Shape TUBULAR)
                      (Subpart (BASE (Color TURQUOISE)))))
          (Inner ....)
          (FitCondition LOOSE))

Figure 7. The speaker's descriptions

The first step in the reference process is the actual search for a referent in the knowledge base. The reference identification process is incremental in nature, i.e., the listener can begin the search process *before* he hears the complete description. This was observed throughout the videotape excerpts and the algorithm presented here is actually designed to be incremental. The KL-One Classifier compares the features specified in the speaker's descriptions (Descr1 and the "Outer" feature of Descr2 in Figure 7) with the features specified for each element in the KL-One taxonomy that corresponds to one of the current objects of interest in the real world. Notice that some features are directly comparable. For example, the "Transparency" feature of Descr1 and the "Transparency" feature of *MAINTUBE* are both equal to "CLEAR." Other features require further processing before they can be compared. The OPENING value of "Subpart" in Descr1 is thought of primarily as a 2-D cross-section (such as a "hole"), while two CYLINDER subparts of *MAINTUBE* are viewed as (3-D) cylinders that have the "Function" of being outlets, i.e., OUTLET-ATTACHMENT-POINTS. To compare OPENING and CYLINDER, the inference must be made that both things can describe the same thing (similar inferences are developed in Mark, 1982). One way this inference can occur is by recursively examining the subparts of *MAINTUBE* with the partial matcher until the cylinders are examined at the 2-D level. At that level, an end of the cylinder will be defined as an OPENING. With that examination, the *MAINTUBE* can be seen as described by Descr1.

Descr2 presents different problems. Descr2 refers to an object that is supposed to have a subpart that is TURQUOISE. The Classifier determines that Descr2 could not describe either the *CAP* or *STAND* because both are BLUE. It also could not describe the *MAINTUBE*[6] or *AIR*

---

[6] Since Descr1 refers to *MAINTUBE*, *MAINTUBE* could be dropped as a potential referent candidate for Descr2. We will, however, leave it as a potential candidate to make this example more complex.

```
             (CAP      (Color BLUE)
                       (Composition PLASTIC)
      CAP              (Transparency OPAQUE)
                       (Dimensions (Length .25) (Diameter .5))
                       (Orientation (Rotation (0.0 0.0 90.0))
                                    (Translation (0.0 0.0 0.0))))


             (TUBE     (Color VIOLET)
                       (Composition PLASTIC)
                       (Transparency CLEAR)
                       (Dimensions (Length 4.125))
                       (Subpart (CYLINDER (Dimensions (Length .25) (Diameter 1.125))
                                          (Orientation (Rotation (0.0 0.0 0.0))
                         Lip                            (Translation (0.0 0.0 3.75)))
                                          (Function OUTLET-ATTACHMENT-POINT)))
                       (Subpart (CYLINDER (Dimensions (Length 3.5) (Diameter 1.0))
      MAIN               TubeBody         (Orientation (Rotation (0.0 0.0 0.0))
      TUBE                                              (Translation (0.0 0.0 .25)))))
                       (Subpart (CYLINDER (Dimensions (Length .25) (Diameter 1.125))
                                          (Orientation (Rotation (0.0 0.0 0.0))
                         Threads                        (Translation (0.0 0.0 0.0)))
                                          (Function THREADED-ATTACHMENT-POINT)))
                       (Subpart (CYLINDER (Dimensions (Length .375) (Diameter .5))
                                          (Orientation (Rotation (0.0 0.0 90.0))
                         Outlet1                        (Translation (0.0 .5 3.00)))
                                          (Function OUTLET-ATTACHMENT-POINT)))
                       (Subpart (CYLINDER (Dimensions (Length .375) (Diameter .5))
                                          (Orientation (Rotation (0.0 0.0 90.0))
                         Outlet2                        (Translation (0.0 .5 .625)))
                                          (Function OUTLET-ATTACHMENT-POINT)))


         (CONTAINER  (Dimensions (LENGTH 2.75))
                     (Composition PLASTIC)
                     (Subpart (HEMISPHERE (Color VIOLET)
                                          (Transparency CLEAR)
                         Chamber          (Dimensions (Diameter 1.0))
                         Top              (Orientation (Rotation (0.0 0.0 0.0))
                                                       (Translation (0.0 0.0 2.25)))))
                     (Subpart  (CYLINDER  (Color VIOLET)
                                          (Transparency CLEAR)
                         Chamber          (Dimensions (Length 1.0) (Diameter 2.25))
                         Body             (Orientation (Rotation (0.0 0.0 0.0))
                                                       (Translation (0.0 0.0 .375)))))
                     (Subpart  (CLYINDER  (Color BLUE)
                                          (Transparency OPAQUE)
                                          (Dimensions (Length .375) (Diameter 1.25))
       AIR                                (Orientation (Rotation (0.0 0.0 0.0))
       CHAMBER         Chamber                         (Translation (0.0 0.0 0.0)))
                         Bottom           (Function CAP OUTLET-ATTACHMENT-POINT)
                                          (Subpart (CYLINDER (Color BLUE)
                                                             (Dimensions (Length .375) (Diameter .5))
                                                             (Orientation (Rotation (0.0 0.0 0.0))
                                                             (Translation (0.0 0.0 0.0)))
                                                             (Function OUTLET-ATTACHMENT-POINT)))))
                     (Subpart  (CYLINDER  (Color VIOLET)
                                          (Transparency CLEAR)
                         Chamber          (Dimensions (Length .5) (Diameter .375))
                         Outlet           (Orientation (Rotation (0.0 0.0 90.0))
                                                       (Translation (.625 .625 .625)))
                                          (Function OUTLET-ATTACHMENT-POINT))))


            (TUBE    (Dimensions (Length 2.75))
                     (Composition PLASTIC)
                     (Subpart  (CYLINDER  (Color BLUE)
                                          (Transparency CLEAR)
                         Top              (Dimensions (Length 2.25) (Diameter .375))
                                          (Orientation (Rotation (0.0 0.0 0.0))
      STAND                                            (Translation (.5 0.0 .375)))
                                          (Function OUTLET-ATTACHMENT-POINT)))
                     (Subpart  (CYLINDER  (Color BLUE)
                                          (Transparency CLEAR)
                         Base             (Dimensions (Length .375) (Diameter 1.0))
                                          (Orientation (Rotation (0.0 0.0 0.0))
                                                       (Translation (0.0 0.0 0.0)))
                                          (Function OUTLET-ATTACHMENT-POINT))))
```

Figure 8. The Objects in focus

CHAMBER since each has subparts that are either VIOLET or BLUE. The Classifier places DESCR2 as best it can in the taxonomy, showing no connections between it and any of the objects currently in focus. At this point, a probable misreference is noted. The reference mechanism now tries to find potential referent candidates, using the taxonomy exploration routine described in Section 3.2.3, by examining the elements closest to Descr2 in the taxonomy and using the partial matcher to score how close each element is to Descr2.[7] The matcher determines *MAINTUBE, STAND,* and *AIR CHAMBER* as reasonable candidates by aligning and comparing their features to Descr2.

Scoring Descr2 to *MAINTUBE:*

o   a TUBE is a kind of DEVICE; (>)
o   the Transparency of each is CLEAR; (+)
o   the Composition of each is PLASTIC; (+)
o   a Tube implies Analogical-Shape TUBULAR, which implies Shape CYLINDRICAL, which is a kind of Shape ROUND; (>)
o   the recursive partial matching of subparts: A BASE is viewed as a kind of BOTTOM. Therefore, BASE in Descr2 could match to the subpart in *MAINTUBE* that has a Translation of (0.0 0.0 0.0) − i.e., *Threads* of *MAINTUBE*. However, they mismatch since color TURQUOISE in Descr2 differs from color VIOLET of *MAINTUBE*. (−)

Scoring Descr2 to *STAND:*

o   a TUBE is a kind of DEVICE; (>)
o   the Transparency of each is CLEAR; (+)
o   the Composition of each is PLASTIC; (+)
o   a TUBE implies Analogical-Shape TUBULAR, which implies Shape CYLINDRICAL, which is a kind of Shape ROUND; (>)
o   the recursive partial matching of subparts: BASE in Descr2 could match to the subpart in *STAND* that has a Translation of (0.0 0.0 0.0) − i.e., *Base* of *STAND*. However they mismatch since color TURQUOISE in Descr2 differs from color BLUE of *STAND*. (−)

Scoring Descr2 to *AIR CHAMBER:*

o   a CONTAINER is a kind of DEVICE; (>)
o   the Transparency of Descr2, CLEAR, matches the Transparency of *ChamberTop, ChamberOutlet* and *ChamberBody* of *AIR CHAMBER* but mismatches the Transparency of *ChamberBottom* of *AIR CHAMBER*. Therefore, the partial match is uncertain; (?)
o   the Composition of each is PLASTIC; (+)
o   the subparts of *AIR CHAMBER* have Shape HEMISPHERICAL and CYLINDRICAL which are each a kind of Shape ROUND; (>)
o   the recursive partial matching of subparts: BASE in Descr2 could match to the subpart in *AIR CHAMBER* that has a translation of (0.0 0.0 0.0) − i.e., *ChamberBottom* of *AIR CHAMBER*. However, they mismatch since color TURQUOISE in Descr2 differs from color BLUE of *AIR CHAMBER*. (−)

The above analysis using the partial matcher provides no *clear* winner since the differences are so close causing the scores generated for the candidates to be almost exactly the same (i.e., the only difference was in the score for Transparency). All candidates, hence, will be retained for now.

---

[7] The partial matcher scores are numerical scores computed from a set of role scores that indicate how well each feature of the two descriptions match. Those feature scores are represented as a scale: HIGHEST (+), (>◁), (=), (?), − LOWEST.

At this point, the knowledge sources and their associated rules that were mentioned earlier apply. These rules attempt to order the feature values in the speaker's description for relaxation. First, we'll order the features in Descr2 using linguistic knowledge. Linguistic analysis of Descr2, "... are clear plastic ... a rounded piece with a turquoise base ... Both are tubular ... fits loosely over ...," tells us that the features were specified using the following modifiers.

- o Adjective: (Shape ROUND)
- o Prepositional Phrase: (Subpart (BASE (Color TURQUOISE)))
- o Predicate Complement: (Transparency CLEAR), (Composition PLASTIC), (Analogical-Shape TUBULAR), (Fit LOOSE)

Observations from the protocols (as described by the rules developed in Goodman, 1984) has shown that people tend to relax first features specified as adjectives, then as prepositional phrases and finally as relative clauses or predicate complements. This suggests relaxation of Descr2 in the order:

[Shape] < [Color, Subpart]
    < [Transparency, Composition, Analogical-Shape, Fit].

The set of features on the left side of a "<" symbol is relaxed before the set on the right side. The order that the features inside the braces, "[...]" are relaxed is left unspecified (i.e., any order of relaxation is alright). Perceptual information about the domain also provides suggestions. Whenever a feature has feature values that are close, then one should be prepared to relax any of them to any of the others (we call this the "clustered feature value rule"). In this example, since the colors are all very close — BLUE, TURQUOISE, and VIOLET — then Color may be a reasonable thing to relax. Hierarchical information about how closely related one feature is to another can also be used to determine what to relax. The Shape values are a good example. A CYLINDRICAL shape is also a CONICAL shape, which is also a 3-D ROUND shape. Hence, it is very reasonable to match ROUNDED to CYLINDRICAL. All of these suggestions can be put together to form the order.

[Shape, Color] < [Subpart]
    < [Transparency, Composition, Analogical-Shape, Fit].

The referent candidates *MAINTUBE*, *STAND*, and *AIR CHAMBER* can be examined and possibly ordered for relaxation using the above feature ordering. For this example, the relaxation of Descr2 to any of the candidates requires relaxing their SHAPE and COLOR features. Since they each require relaxing the same features, the candidates cannot be ordered with respect to each other (i.e., none of the possible feature orders is better for relaxing the candidates). Hence, no one candidate stands out as the most likely referent.

While no ordering of the candidates was possible, the order generated to relax the features in the speaker's description can be used to guide the relaxation of each candidate. The relaxation methods mentioned at the end of the last section come into use here. Generate-Similar-Shape-Values can determine that HEMISPHERICAL and CYLINDRICAL shapes of the *AIR CHAMBER* are close to the 3D-ROUND shape. This holds equally true for the cylindrical shapes of the *MAINTUBE* and the *STAND*. Generate-Similar-Color-Values next tries relaxing the Color TURQUOISE. It determines the colors BLUE and GREEN as the best alternates. Here only two clear winners exist — the *AIR CHAMBER* and the *STAND* — while the *MAINTUBE* is dropped as a candidate since it is reasonable to relax TURQUOISE to BLUE or to GREEN but not to VIOLET. Subpart, Transparency, Analogical-Shape, and Composition provide no further help (though, the fact that the *AIR CHAMBER* has both CLEAR and OPAQUE subparts might put it slightly lower than the *STAND* whose subparts are all CLEAR. This difference, however, is not significant.). This leaves trial and error attempts to try to complete the FIT action. The one (if any) that fits — and fits loosely — is selected as the referent. The protocols showed that people often do just that — reducing their set of choices down as best they can and then taking each of the remaining choices and trying out the requested action on them.

4 Conclusion

Our goal in this work is to build robust natural language understanding systems, allowing them to detect and avoid miscommunication. The goal is *not* to make a perfect listener but a more tolerant one that could avoid many mistakes, though still wrong on occasion. In Section 2, we introduced a taxonomy of miscommunication problems that occur in expert-apprentice dialogues. We showed that reference mistakes are one kind of obstacle to robust communication. To tackle reference problems, we described how to extend the succeed/fail paradigm followed by previous natural language researchers.

We represented real world objects hierarchically in a knowledge base using a representation language, KL-One, that follows in the tradition of semantic networks and frames. In such a representation framework, the reference identification task looks for a referent by comparing the representation of the speaker's input to elements in the knowledge base by using a matching procedure. Failure to find a referent in previous reference identification systems resulted in the unsuccessful termination of the reference task. We claim that people behave better than this and explicitly illustrated such cases in an expert-apprentice domain about toy water pumps.

We developed a theory of relaxation for recovering from reference failures that provides a much better model for human performance. When people are asked to identify objects, they go about it in a certain way, find candidates, adjust as necessary, re-try, and, if necessary, give up and ask for help. We claim that relaxation is an integral part of this process and that the particular parameters of relaxation differ from task to task and person to person. Our work models the relaxation process and provides a computational model for experimenting with the different parameters. The theory incorporates the same language and physical knowledge that people use in performing reference identification to guide the relaxation process. This knowledge is represented as a set of rules and as data in a hierarchical knowledge base. Rule-based relaxation provided a methodical way to use knowledge about language and the world to find a referent. The hierarchical representation made it possible to tackle issues of imprecision and over-specification in a speaker's description. It allows one to check the position of a description in the hierarchy and to use that position to judge imprecision and over-specification and to suggest possible repairs to the description.

Interestingly, one would expect that "closest" match would suffice to solve the problem of finding a referent. We showed, however, that it doesn't usually provide you with the correct referent. Closest match isn't sufficient because there are many features associated with an object and, thus, determining which of those features to keep and which to drop is a difficult problem due to the combinatories and the effects of context. The relaxation method described circumvents the problem by using the knowledge that people have about language and the physical world to prune down the search space.

Acknowledgements

I want to thank especially Candy Sidner for her insightful comments and suggestions during the course of this work. I'd also like to acknowledge the helpful comments of George Hadden, Diane Litman, Marc Vilain, Dave Waltz, Bonnie Webber and Bill Woods on this paper. Many thanks also to Phil Cohen, Scott Fertig and Kathy Starr for providing me with their water pump dialogues and for their invaluable observations on them.

References

Allen, J.F. (1979). *A plan-based approach to speech act recognition*. Ph.D. thesis, University of Toronto.
Appelt, D.E. (1981). *Planning natural language utterances to satisfy multiple goals*. Ph.D. thesis, Stanford University.
Brachman, R.J. (1977). *A structural paradigm for representing knowledge*. Ph.D. thesis, Harvard University.
Brown, J.S., & VanLehn, K. (1980). Repair theory: A generative theory of bugs in procedural skills. *Cognitive Science, 4,* 379-426.
Cohen, P.R. (1978). *On knowing what to say: Planning speech acts*. Ph.D. thesis, University of Toronto.

Cohen, P., Perrault, C., & Allen, J. (1981). Beyond question answering. In W. Lehnert & M. Ringle, (Eds.), *Knowledge representation and natural language processing.* Hillsdale, NJ: Lawrence Erlbaum Associates.

Cohen, P.R. (1981). The need for referent identification as a planned action. *Proceedings of IJCAI-81,* (pp. 31-35). Vancouver, British Columbia, Canada.

Cohen, P.R., Fertig, S., & Starr, K. (1982). Dependencies of discourse structure on the modality of communication: Telephone vs. teletype. *Proceedings of ACL,* (pp. 28-35). Toronto, Ontario, Canada.

Cohen, P.R. (1984) The pragmatics of referring and the modality of communication. *Computational Linguistics, 10,* 97-146.

Gentner, D. (1980). *The structure of analogical models in science.* Bolt Beranek and Newman Inc.

Goodman, B.A. (1982). Miscommunication in task-oriented dialogues. *KRNL Group Working Paper.* Bolt Beranek and Newman Inc.

Goodman, B.A. (1983). Repairing miscommunication: Relaxation in reference. *Proceedings of AAAI-83,* (pp. 134-138). Washington DC.

Goodman, B.A. (1984). *Communication and miscommunication.* Ph.D. thesis, University of Illinois, Urbana.

Grosz, B.J. (1977). *The representation and use of focus in dialogue understanding.* Ph.D. thesis, University of California, Berkeley.

Grosz, B.J. (1981). Focusing and descriptions in natural language dialogues. In A. Joshi, B. Webber & I. Sag, (Eds.), *Elements of discourse understanding* (pp. 84-105). Cambridge England: Cambridge University Press.

Lipkis, T. (1981). A KL-One Classifier. *Proceedings of the 1981 KL-One Workshop,* (pp. 128-145). (Tech. Rep. No. 4942). Bolt Beranek and Newman Inc.

Litman, D.J., & Allen, J.F. (1984). A plan recognition model for clarification subdialogues. *Proceedings of Coling 84,* (pp. 302-311). Stanford, CA: Stanford University.

Mark, W. (1982). Realization. *Proceedings of the 1981 KL-One Workshop,* (pp. 78-89). (Tech. Rep. No. 4942). Bolt Baranek and Newman Inc.

McKeown, K.R. (1983). Recursion in text and its use in language generation. *Proceedings of AAAI-83,* (pp. 270-273). Washington DC.

Reichman, R. (1978). Conversational coherency. *Cognitive Science, 2,* 283-327.

Reichman, R. (1981). *Plain speaking: A theory and grammar of spontaneous discourse.* Ph.D. thesis, Harvard University.

Ringle, M., & Bruce, B. (1981). Conversation failure. In W. Lehnert & M. Ringle, (Eds.), *Knowledge representation and natural language processing.* Hillsdale, NJ: Lawrence Erlbaum Associates.

Sidner, C.L., & Israel, D.J. (1981). Recognizing intended meaning and speaker's plans. *Proceedings of the International Joint Conference on Artificial Intelligence,* (pp. 203-208). Vancouver, BC: The International Joint Conference on Artificial Intelligence.

Sidner, C.L. (1979). *Towards a computational theory of definite anaphora comprehension in English discourse.* Ph.D. thesis, Massachusetts Institute of Technology.

Sidner, C.L., Bates, M., Bobrow, R.J., Brachman, R.J., Cohen, P.R., Israel, D.J., Schmoize, J., Webber, B.L., & Woods, W.A. (1981). *Research in knowledge representation for natural language understanding.* (Tech. Rep. No. 4786). Bolt Beranek and Newman Inc.

Sidner, C.L., Bates, M., Bobrow, R., Goodman, B., Haas, A., Ingria, R., Israel, D., McAllester, D., Moser, M., Schmoize, J., & Vilain, M. (1983). *Research in knowledge representation for natural language understanding.* Annual Report, 1 September 1982 – 31 August 1983. (Tech. Rep. No. 5421). Cambridge, MA: BBN Laboratories Inc.

Sidner, C., Goodman, B., Haas, A., Moser, M., Stallard, D., Vilain, M. (1984). *Research in knowledge representation for natural language understanding.* Annual Report, 1 September 1983 – 31 August 1984. (Tech. Rep. No. 5694). Cambridge, MA: BBN Laboratories Inc.

Webber, B.L. (1976). *A formal approach to discourse anaphora.* Ph.D. thesis, Harvard University.

GENERATING RESPONSES TO PROPERTY MISCONCEPTIONS USING PERSPECTIVE[1]

Kathleen F. McCoy

Dept. of Computer and Information Sciences
University of Delaware
Newark, De. 19716

In a dialogue between a database or expert system and a user, it is likely that the user will exhibit a property misconception by attributing a property or property value to an object that the object does not have. This paper discusses a method for responding to such misconceptions so as to avoid further confusion on the part of the user. The method involves reasoning on a model of the user to determine possible sources of the error. A response refuting the user's erroneous beliefs supporting the error can then be given. The process of generating a response to a property misconception is made context sensitive by working on a model of the user which has been augmented with contextual information provided by a new notion of object perspective. Object perspective serves to highlight certain aspects of the user model which have been made important by the previous discourse. This new notion of object perspective is introduced and its implications on correcting misconceptions are discussed.

1. Introduction

Miscommunication abounds in many kinds of dialogues. In this paper I concentrate on one particular kind of dialogue: that which takes place between a human information seeker and a database or expert system. In such a situation it is likely that a form of miscommunication may occur between the user and the system. In particular, the user may exhibit a *property misconception*. That is, s/he may attribute a property or property value to an object that that object does not have. Such a misconception may leave the system unable to directly respond to the user's query. For instance suppose the user enters the following query:

U. Give me the HULL-NO of all DESTROYERS whose MAST-HEIGHT is above 190.

A system faced with such a query has a problem. DESTROYERS with such a large MAST-HEIGHT do not exist, yet the user is clearly trying to ask something of the system. If the system simply responds "there are none", then the user may be misled into believing that such ships could possibly exist, but that there are simply none in the database at the present time (see Kaplan, 1982; Mays, 1980). In order to avoid such confusion which would perpetuate the miscommunication, the system must have the ability to effectively respond to property misconceptions revealed by the user.

The subject of this paper is a methodology for generating effective responses to property misconceptions. This methodology has been implemented as part of the ROMPER (Responding to Object-related Misconceptions using PERspective) system, reported in McCoy (1985). The methodology works as follows. Rather than storing *a priori* a list of property misconceptions

---

[1] The work reported here was done while the author was at the University of Pennsylvania and was partially supported by the NFS grant No MCS81-07290 and by the ARO grant DAA20-84-K-0061.

along with some canned response, the methodology calls for an analysis of the user model upon encountering a misconception. This analysis looks for structural configurations that have been found to support particular kinds of misconceptions. Each identified configuration has a response strategy associated with it which may then be instantiated. The whole process is made context sensitive by a new notion of *object perspective* which acts to filter the user model, highlighting those aspects which are made important by previous dialogue, while suppressing others.

This paper first investigates response strategies used by human experts and identifies structural configurations of the user model which would suggest the use of such strategies. It next discusses contextual effects on responses and introduces the notion of object perspective. Next an application of object perspective, the assessment of object similarity, is presented; this assessment is particularly important in the user model analysis. Finally, the paper closes with a discussion of how one particular misconception can be responded to differently under two different perspectives.

2. Corrective Strategies

An analysis of transcripts of human conversational partners revealed that responses to misconceptions very often included more than a simple denial of the wrong information. This was particularly true in circumstances where the misconception was about something important to the current goals of the conversation. In addition to denying the information involved in the misconception, many of the misconception response pairs included not only the corresponding correct information, but also additional information to back up the denial-correction pair. The additional information often involved refuting the faulty reasoning which may have led the user to the misconception.

While it may seem that the kinds of faulty reasoning that the user may have been using to come up with the misconception are limitless, the transcript analysis revealed a surprisingly small number of misconception support relations that were refuted by the human experts. In addition, these few misconception support relations could be couched in terms of a knowledge-base structure rather than its content. Thus a system, reasoning on a model of the user, might look for such relations in a domain independent fashion. If one were found, information refuting the found misconception support might be included in the corrective response.

The misconception support that was refuted by human experts in corrective responses to property misconceptions took two forms: either the user was seen as having confused the *object* involved in the misconception with a similar object (or to have made a bad analogy from that similar object), or the user was seen as having confused the *attribute* being discussed with a similar attribute. These two different kinds of misconception support gave rise to two distinct response strategies used by human experts to respond to property misconceptions.

2.1. Wrong Object

The first of these is exemplified by the following exchange:

U.   Give me the HULL-NO of all DESTROYERS whose MAST-HEIGHT is above 190.

S.   There are no DESTROYERS in the database having MAST-HEIGHT above 190. All DESTROYERS in the database have a MAST-HEIGHT between 85 and 90. Were you thinking of an AIRCRAFT-CARRIER?

The user's statement reveals a property misconception: s/he has attributed the property of MAST-HEIGHT above 190 to the object DESTROYERS when MAST-HEIGHT cannot take on such values for that object. But suppose, however, that an analysis of the user model reveals that the user believes that DESTROYERS and AIRCRAFT-CARRIERS are very similar objects, and that the MAST-HEIGHT of AIRCRAFT-CARRIERS can be above 190. Because of the similarity the expert may conclude that the user's misconception may have been the result of the user either confusing the DESTROYER with the AIRCRAFT-CARRIER or making a bad analogy from the AIRCRAFT-CARRIER. In such a situation a helpful corrective response might contain a reference to the object that the user might have confused with the object in the query, to help stem any further trouble. In this instance the reference takes the form of an alternative correct query containing the object of possible confusion.

This example is typical of responses used in such situations. It may be analyzed in three parts. The expert has:

1. Denied that a DESTROYER can have the property MAST-HEIGHT above 190.

2. Stated the correct value for the property MAST-HEIGHT of object DESTROYER.

3. Offered an alternative correct query containing the object AIRCRAFT-CARRIER — an object for which the property MAST-HEIGHT may take on the value above 190 and an object that is seen as similar to the DESTROYER.

In general the schema followed may be stated as:

Denying that the object involved in the misconception has the property stated in the misconception.

State the corresponding correct property that the object involved in the misconception does have.

Offer an alternative correct query containing the property involved in the misconception and an object that is similar to the object involved in the misconception.

This schema may be followed in the case of a property misconception when (1) there is another object which has the attribute involved in the misconception, and (2) the system believes that the user believes that that object is similar to the misconception object.

2.2 Wrong Attribute

The second kind of response strategy containing support for the misconception correction is exemplified by the following exchange:

U. Give me the HULL-NO of all DESTROYERS whose MAST-HEIGHT is above 3500.

S. All DESTROYERS in the database have a MAST-HEIGHT between 85 and 90. Were you thinking of the DISPLACEMENT?

The user's statement again reveals a property misconception: s/he has attributed the property of MAST-HEIGHT above 3500 to the object DESTROYERS when MAST-HEIGHT cannot take on such large values for DESTROYERS. But suppose, however, that an analysis of the user model reveals that the user believes that MAST-HEIGHT and DISPLACEMENT are similar attributes. Further suppose that DISPLACEMENT can take on the value given in the user's statement. Because of the similarity, the expert may conclude that the user's misconception may have been the result of the user confusing the attributes MAST-HEIGHT and DISPLACEMENT. In such a situation a helpful corrective response might contain a reference to the attribute the user may be confusing with the attribute in the query to help stem any further trouble. In this instance, like above, the reference takes the form of an alternative correct query. This time, however, the correct query contains the *attribute* of possible confusion.

The response in the last example is typical of responses used in this kind of situation. It may be analyzed in three parts. The expert has:

1. Denied that a DESTROYER can have the property MAST-HEIGHT above 3500.

2. Stated the correct value for the property MAST-HEIGHT of object DESTROYER.

3. Offered an alternative correct query containing the attribute DISPLACEMENT — an attribute which may take on a value above 3500 and which is seen as being similar to the attribute MAST-HEIGHT.

In general the schema followed may be stated as:

> Deny that the object involved in the misconception has the property stated in the misconception.
>
> State the corresponding correct property that the object involved in the misconception does have.
>
> Offer an alternative correct query containing the object involved in the misconception and a property that is similar to the property involved in the misconception.

This schema may be followed in the case of a property misconception when (1) there is another property that the object involved in the misconception does have, and (2) the system believes the user believes that property is similar to the misconception property.

2.3 No Support

Of course there were cases found in the transcripts when the expert apparently could find no reason why the user came up with the misconception. In this case, no additional information was included in the correction to refute the user's reasoning in arriving at the misconception. The following exchange is an example of this:

> U. Give me the HULL-NO of all DESTROYERS whose MAST-HEIGHT is above 9200.
>
> S. There are no DESTROYERS in the database with a MAST-HEIGHT above 9200. All DESTROYERS have a MAST-HEIGHT between 85 and 90.

Just as in the above two examples, the user's statement reveals a property misconception. This time, however, the expert can find no reason for the misconception. Because of this, a minimal response is given.

In general the schema followed in such cases may be stated as:

> Deny that the object involved in the misconception has the property stated in the misconception.
>
> State the corresponding correct property that the object involved in the misconception does have.

This schema may be followed in the case of a property misconception when neither of the two kinds of misconception support identified above can be found in the user model.

3. Role of Contextual Information

In the above section I introduced a number of correction strategies which were found to be used by human experts in responding to property misconceptions. In addition, I briefly indicated the structural analysis of the user model that a system might use in order to choose one of the given strategies in a particular situation. However, if this structural analysis were to take place on a flat standard semantic network representation of the user, it would seem unable to take contextual affects into account since flat semantic network representations do not change with context. However, context seems to play a significant role in an expert's response to a particular misconception. As an example consider the following two exchanges where the same misconception is responded to differently under different circumstances.

U. I am interested in investing in some securities to use as savings instruments. I want something short-term and I don't have a lot of money to invest so the instrument must have small denominations. I am a bit concerned about the penalties for early withdrawal. What is the penalty on a T-bill?

E. T-bills don't have a penalty. Were you thinking of a money market certificate?

Contrast the above interaction with the one below with a different preceding context:

U. I am interested in investing in some securities. Safety is very important to me, so I would probably like to get something from the government. I am a bit concerned about the penalties for early withdrawal. What is the penalty on a T-bill?

E. T-bills don't have a penalty. Were you thinking of a T-bond?

Notice that in both cases a similar object is offered as a possible object of confusion. However, the established context influences which object is chosen. In the first case the context's focus on attributes such as maturity and denominations has influenced the choice of the money market certificate as the object offered in the response. In contrast, the second example highlights attributes such as issuer and safety and therefore causes the T-bond to be offered.

In the previous section I described a method for generating responses to property misconceptions which involved looking for certain structural configurations in the user model. However, if this model were to be represented as a flat semantic network, then there would be no way to account for the influence of previous discourse. In fact, it is likely that the flat representation would be underspecified with respect to correcting misconceptions since it would contain *all* of the information that the system believes the user knows about the misconception object. It is the case, however, that the discourse in which the misconception is revealed can serve to highlight certain aspects of the user model. This highlighted user model contains less extraneous information and thus is more specified with respect to matching a particular user model configuration and instantiating the corresponding response strategy. *Perspective* is a notion which can be used to model this contextual affect.

In sum, contextual information must be taken into account when correcting misconceptions. I claim that the process described above is not affected by context, but rather what that process works on is affected by context. The user model analysis should not be done on a flat semantic network representation of the user's beliefs. Rather it should be done on a network enriched with perspective which causes certain aspects of the user model to be highlighted while others are suppressed. The following sections will show how the analysis done on such an enriched model will be sufficient to explain the contextual affects described above.

4. Object Perspective

In this section I introduce a notion of object perspective as an augmentation to the standard semantic network representation. Perspective accounts for different highlighting of attributes in different situations. This highlighting is reflected in the salience values given to the attributes of the domain objects. I claim that this difference in attribute salience values can be used to calculate different highlighting in the user model and system knowledge base. This highlighting can, in turn, account for the choice of different responses being given to the same misconception in different situations.

4.1 Previous Notions

The notion of object perspective or point of view has been discussed in the AI literature. It is this notion that is used to explain how different aspects of the same object may be highlighted in different circumstances. For instance, a building may be viewed as someone's home or as an architectural work. These two different "perspectives" cause different attribute highlighting.

This phenomenon has been explained and used by a number of researchers (Grosz, 1977; Grosz, 1981; Bobrow & Winograd, 1977; Tou et al., 1982)[2] who maintain that an object is viewed through a particular perspective by viewing it as a member of one particular superordinate when, in fact, the object may have many superordinates. The object inherits only attributes through the superordinate in perspective. Therefore different perspectives on the same object cause different properties to be inherited (and therefore highlighted).

Although this notion of object perspective is intuitively appealing, in practice its use is rather difficult. Notice that it hinges on the use of a limited inheritance mechanism. That is, objects are seen as inheriting properties from only one of their superordinates. One problem with this notion of object perspective is that attributes are inherited from the top of the generalization hierarchy, not just from immediate superordinates. So, an object's perspective may therefore involve not just one superordinate but a chain of superordinates. In such cases, one must not only determine the perspective from which a particular object is being viewed, but also the perspective from which each member of this chain of superordinates is viewed.

For instance, suppose we are trying to determine what attributes should be inherited by a particular building. We know that buildings may be looked at from at least two perspectives: that of architectural work and that of someone's home. We may pick the perspective of architectural work as the current perspective. So, our building should inherit those attributes contributed by the concept "architectural works" in the hierarchy. However, what if our hierarchy has more than one superordinate of "architectural works". For instance, there might be architectural works from a particular period (which would emphasize those aspects which were important during that period) and architectural works from a particular architect (which would emphasize the more specific aspects of the architecture that that architect was famous for). Thus, to determine what attributes should be inherited by our building, we must not only determine what perspective should be taken on the building, but also what perspective should be taken on the chosen perspective, and so on. In general, there is no *a priori* limit to the number of perspectives that must be determined.

Another problem with the notion of object perspective as defined above is that it gives us no way of determining which objects in the hierarchy are more important than others (that is, beyond those that are ruled out when choosing a perspective). At an intuitive level it seems that the highlighting of other objects should be affected by the point of view taken on the main object under discussion. There is no way for the notion of object perspective as previously defined to account for this intuition.

A final problem with the notion of object perspective is that it is unable to capture any regularity in the highlighting of the attributes of a *group of objects* (unless, of course, every member of the group has the same immediate superordinates). Yet, it is very common in a conversation to talk about several objects (not just one). In doing so it is likely that the same (or at least similar) attributes will be discussed for each. Again, the previous notions of object perspective cannot handle this phenomenon.

4.2 Perspective: Definition and Representation

So far we have noted that a different highlighting of the user model could account for different responses being given to the same misconception in different situations. We turned to previous notions of object perspective to account for this highlighting. However, the previous notions were found inadequate for our purposes. In this section I formulate a new notion of

---

[2] (McKeown et al., 1985) uses a slightly different notion of object perspective. In their work, perspective carves out a predetermined piece of the generalization hierarchy to be highlighted. This notion of object perspective avoids some of the problems I will discuss below and in fact may be a more specific instantiation of the notion of object perspective that I will introduce in the next section.

object perspective which overcomes the problems associated with the previous notions of object perspective. Our representation of this new notion of object perspective must meet the following criteria. It must:

1. account for the highlighting of systematic groups of attributes of an object and the suppression of others;

2. account for a particular highlighting of attributes being evident for an entire group of objects;

3. account for the heightened importance of some objects;

4. not require an arbitrary number of assessments to calculate;

5. be triggered by preceding discourse;

I claim that all of the above criteria can be met by a simple notion of object perspective which can be defined in the following way:

First, the perspectives which can be taken on the domain objects will sit in some structure which is completely orthogonal to the generalization hierarchy.

Second, the number of such perspectives that need be defined for the objects in a given domain of discourse is small and finite. Moreover, *any* given domain object may be viewed from any one of several perspectives defined for that domain.

Third, each perspective comprises a set of attributes with associated salience values. It is these salience values which dictate which attributes are highlighted and which are suppressed.

Fourth, one such perspective is designated *active* at a particular point in the discourse.

This notion of object perspective works as follows. An object or group of objects is still said to be viewed through a perspective. In particular any object which is accessed by the system is viewed through the *active* perspective. However, rather than dictating which attributes an object inherits, the active perspective affects the salience values of the attributes that an object posesses (either directly or inherited through the generalization hierarchy). The active perspective essentially acts as a filter on an object's attributes — raising the salience of and thus highlighting those attributes which have a high salience rating in the active perspective, and lowering the salience of and thus suppressing those attributes which are either given a low salience value or do not appear in the active perspective.

To sum up this new notion of object perspective let us take an example. We could imagine the world of investment containing the fragment of knowledge representation shown in figure 1. We could imagine taking two different perspectives on the object in this domain.[3] One might be called the "Savings Instruments" perspective which roughly corresponds to the way one looks at the objects in which one is thinking of investing money for savings purposes. The second might be called the "Issuing Company" perspective which corresponds to how the objects are looked at when the issuer involved is very important.

Now in building a model of the securities domain, we would construct a taxonomy of the objects as usual. In addition to this taxonomy, we must build a separate structure containing the perspectives which can be taken on these objects. Each perspective would list attributes with salience values indicating their importance in the perspective. Two elements in the perspective

---

[3] There may be other perspectives which may be taken on these objects. These two are chosen to illustrate the mechanism.

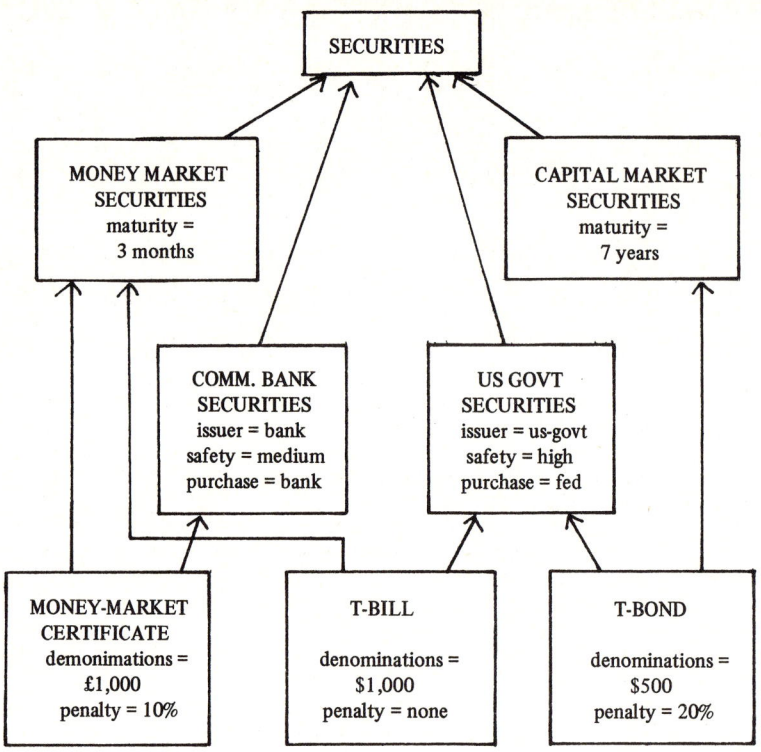

Figure 1. "Flat" Representation of Securities

structure might look like this (here we are assuming salience values of 1 = high, .5 = medium, and 0 = low):

<pre>
    Savings Instruments   Issuing Company
      maturity — 1.0   issuer — 1.0
   denominations — 1.0   safety — 1.0
       safety — 0.5   purchase — 0.5
</pre>

These perspectives indicate what attributes are most important (and thus should perhaps be accessed first) in the two different views of the domain objects. Thus, if a particular security is accessed when the active perspective is "Savings Instruments", then that security's maturity and denominations are highlighted while its safety is somewhat highlighted. Other attributes of that security will be suppressed.

Notice how this notion of object perspective allows different highlighting than could be achieved using the previous notions of object perspective. This can be seen by the highlighting of the attributes of object T-Bond under perspective "Savings Instruments". This perspective would cause the attributes denominations and maturity to be given a very high salience rating, while attribute safety would receive a somewhat high salience rating. Notice that the attribute denominations is directly associated with T-Bonds, maturity is associated with one of its superordinates, and safety is associated with another superordinate. In the previous notion of object perspective, only attributes associated with one superordinate could be highlighted.

Filtering the knowledge base through the perspective of "Issuing Company" makes it easy to see how object importance might be affected by the active perspective. Essentially the idea is that the whole becomes highlighted because its parts are highlighted. In this case, the parts are the attributes associated with objects. Notice that in the perspective of "Issuing Company" all of the highlighted attributes are contributed by objects Commercial Bank Securities and US Govt Securities. Thus, these objects would be highlighted by the perspective, while objects which have none of the highlighted attributes (such as Money Market Securities and Capital Market Securities) would be suppressed.

5. Object Similarity

This section discusses one way in which object perspective affects misconception responses. Recall that one strategy used in correcting property misconceptions was termed the *wrong object* strategy because it included an alternative correct query which contained an object similar to the "wrong object" appearing in the user's original statement. Recall that this strategy is applicable when the user erroneously attributes a property to one object but a similar object does have that property. The strategy specifies mentioning this second object as a possible source of confusion. In order to determine whether or not the strategy is applicable we need a measure of object similarity.

Object similarity has previously been shown to be important in tasks such as organizing explanations (McKeown, 1982), offering cooperative responses to pragmatically ill-formed queries (Carberry, 1984), and identifying metaphors (Weiner, 1984). In the above systems the similarity of two objects is based on the distance between the objects in the generalization hierarchy. One problem with this approach is that it is *context invariant*[4]. It has been shown in the psychological literature, however, that the perception of how similar two objects are is highly context dependent. In order to reflect this, our similarity metric must have the ability to take certain contextual information into account.

Such a metric was introduced by Tversky (Tversky, 1977) and has been adopted for use in this work. Tversky's metric is based on *feature mapping* as opposed to distance on some dimension. He assumes that each object can be represented as a set of features (properties) that that object possess. He defines an interval similarity scale which is based on common and disjoint features of the objects involved. Suppose that we have two objects, a and b, where A is the set of properties associated with object a and B is the set of properties associated with object b. Tversky's measure can be expressed as:

$$s(a,b) = \theta f(A \cup B) - \alpha f(A - B) - \beta f(B - A)$$

fo. ome $\theta, \alpha, \beta > = 0$.

Essentially, the equation states that the similarity of two objects is some function of their common features minus some function of their disjoint features. The importance of each particular feature involved (determined by the function f) and the importance of each piece of the equation (determined by θ, α, and β) may change with context. Setting the values of θ, α, an β seems to be dependent on the importance of the objects involved. Thus, work on centering ar focusing which addresses the prominence of an object at a point in the discourse (see Grosz, 1981; Sidner, 1983; Joshi & Weinstein, 1981; Grosz et al., 1983) will probably be useful for fixing these values.

---

[4] See (McCoy, 1985) for additional problems and discussion of this point.

This work has concentrated on the way that the f function can be changed with context, and the affects of such a change on misconception responses. I claim that this function can be set using the new notion of object perspective to explain different responses to the same misconception under different circumstances. That is, the salience values found in the knowledge base after it has been filtered through perspective are the ones that get used in the similarity judgements.

6. Responding using Perspective

To see how the change in perspective can change the misconception response consider the misconception which was discussed earlier.

U. I am interested in investing in some securities to use as savings instruments. I want something short-term and I don't have a lot of money to invest so the instrument must have small denominations. I am a bit concerned about the penalties for early withdrawal. What is the penalty on a T-bill?

Given the preceding discourse it is reasonable to assume that the perspective of "Savings Instruments" is the active perspective at the time of the misconception utterance. A system attempting to respond to this misconception would proceed by first attempting to instantiate the wrong object schema described above. Recall that this schema is applicable when there is a similar object which has the property involved in the misconception. The system might collect all objects which have the attribute in question and then test their similarity to the object involved in the misconception. In our knowledge base there are two objects which have the attribute involved in the misconception: Money Market Certificates and T-Bonds. The following is a list of the attributes possessed by the misconception object, T-Bill, and the two objects, Money Market Certificates and T-Bonds, which the user may be confusing with T-Bill, along with the salience values attached under perspective "Savings Instruments".

T-Bill	MM Cert	T-Bond
1 — maturity = 3 months	1 — maturity = 3 months	1 — maturity = 7 years
1 — denom = $1,000	1 — denom = $1,000	1 — denom = $500
.5 — safety = high	.5 — safety = med	.5 — safety = high
.	.	.
.	.	.
.	.	.

Applying the Tversky metric using the salience values attached by perspective (and assuming a value of 1 for θ, α, and β) we get:

s(T-Bill, MM-Cert) = f(maturity, denom) - f(safety)
    = 2 - .5 = 1.5 = = = > high similarity

s(T-Bill, T-Bond) = f(safety) - f(maturity, denom)
    = .5 -2 = -1.5 = = = > low similarity

With these calculations the system would choose the Money Market Certificate as the possible object of confusion and respond:

S. Treasury Bills don't have a penalty. Were you thinking of a Money Market Certificate?

Contrast the above interaction with the one below with a different preceding context:

U. I am interested in investing in some securities. Safety is very important to me, so I would probably like to get something from the government. I am a bit concerned about the penalties for early withdrawal. What is the penalty on a T-bill?

With this preceding discourse the active perspective is more likely to be the "Issuing Company" perspective. The attributes and salience values dictated for the objects under this perspective would be:

T-Bill	MM Cert	T-Bond
1 — issuer = us govt	1 — issuer = bank	1 — issuer = us govt
1 — safety = high	1 — safety = medium	1 — safety = high
.5 — purchase = fed	.5 — purchase = bank	.5 — purchase = fed
.	.	.
.	.	.

Using these values the similarity metric would produce the following calculations:

s(T-Bill, MM Cert) = f() - f(issuer, safety, purchase)
= 0 - 2.5 = -2.5 = = = > low similarity

s(T-Bill, T-Bond) = f(issuer, safety, purchase) -f()
= 2.5 - 0 = 2.5 = = = > high similarity

Using these calculations a reasonable response by the system would be:

S. Treasury Bills don't have a penalty. Were you thinking of a Treasury Bond?

As the examples show, changes in the active perspective can account for the same misconception being responded to in two different ways.

7. Conclusion

If we want our natural-language front-ends to database or expert systems to mimic human behavior, they must have the ability to handle misconceptions. This paper has described a methodology for handling object-related misconceptions and has illustrated this methodology on misconceptions involving object properties.

The proposed method for responding to property misconceptions requires associating response schemas with certain structural configurations of the user model. The response schemas described in this paper were derived from a corpus of transcripts and were associated with user model configurations which would explain their use by an expert in responding to a misconception.

A system might use the pairing of strategies to configurations upon encountering an object-related misconception by searching the user model for one of the identified configurations. If one was found, the associated schema could be instantiated to generate a corrective response.

One apparent draw-back of this method is that although misconception responses found in transcripts of human conversational partners appear to be context sensitive, this method seems to be insensitive to previous context. It was proposed, however, that the context-dependent nature could be explained not by having the *process* of correcting misconceptions change with context, but rather by having *what the process worked on* change with context. A new notion of object perspective was introduced as an augmentation to a flat semantic network representation of the user. Object perspective provides a highlighting of the user model due to previous discourse. This resulting user model was shown to be sufficient to account for different responses being given to the same misconception in different situations.

References

Bobrow, D.G., & Winograd, T. (1977). An overview of KRL, a knowledge representation language. *Cognitive Science, 1,* 3-46.
Carberry, S.M. (1984). Understanding pragmatically ill-formed input. In *Proceedings of Coling '84,* Stanford University.
Grosz, B. (1977). *The representation and use of focus in dialogue understanding.* Technical Report 151, SRI

International, Menlo Park CA.
Grosz, B. (1981). Focusing and description in natural language dialogues. In A. Joshi, B. Webber, and I. Sag (Eds.), *Elements of discourse understanding.* New York: Cambridge University Press.
Grosz, B., Joshi, A.K., & Weinstein, S. (1983). Providing a unified account of definite noun phrases in discourse. In *Proceedings of ACL '83,* Cambridge, MA.
Joshi, A.K., & Weinstein, S. (1981). Control of inference: Role of some aspects of discourse structure — centering. In *Proceedings of IJCAI '81,* Vancouver, Canada.
Kaplan, S.J. (1982). Cooperative responses from a portable natural language query system. *Artificial Intelligence, 19,* 165-187.
Mays, E. (1980). Correcting misconceptions about database structure. In *Proceedings of CSCSI,* Victoria, BC.
McCoy, K.F. (1985). *Correcting object-related misconceptions.* PhD thesis, University of Pennsylvania.
McKeown, K. (1982). *Generating natural language text in response to questions about database structure.* PhD thesis, University of Pennsylvania.
McKeown, K., Wish, M., & Matthews, K. (1985). Tailoring explanations for the user. In *Proceedings of IJCAI '85.* Los Angeles, CA.
Sidner, C.L. (1983). Focusing in the comprehension of definite anaphora. In M.Brady and R.C. Berwick (Eds.), *Computational works of discourse.* Cambridge, MA: MIT Press.
Tou, F., Williams, M., Fikes, R., Henderson, A., & Malone, T. (1982). RABBIT: An intelligent database assistant. In *Proceedings of AAAI,* Carnegie-Mellon University.
Tversky, A. (1977). Features of similarity. *Psychological Review, 84,* 327-352.
Weiner, E. (1984). A knowledge representation approach to understanding metaphors. *Computational Linguistics, 19,* 1-14.

THE DYNAMICS OF REFERENTIAL MEANING IN SPONTANEOUS CONVERSATION: SOME PRELIMINARY STUDIES[1]

Anthony Anderson

MRC Applied Psychology Unit
15 Chaucer Road
Cambridge CB2 2EF
England

Simon C. Garrod

Department of Psychology
University of Glasgow
Adam Smith Building
Glasgow G12 BRT

This chapter examines in detail the mechanisms by which the participants in an ongoing dialogue circumvent the failure of referring expressions to uniquely identify the intended referents. A computer-controlled game task was used which required pairs of subjects to cooperate verbally in order to move position markers (one for each player) through the spatial network of a maze from start positions to goal positions. The nature of the game is such that correct reference to particular locations in the maze is critical to the successful completion of the task. This requirement results in many examples of failures of reference in the data. From our results we argue that: (a) the initial failures of reference are caused by the intrinsic imprecision of the referring expressions used, and (b) the initial failures are circumvented by means of the subjects tacitly negotiating local conventions regarding what is meant by what is said.

1. Introduction

1.1 Theoretical Background

In the set of studies reported in the present chapter, we examine the mechanisms by means of which the participants in an ongoing dialogue circumvent one particular type of failure of communication, namely, the failure of the referring expressions that they use to correctly and unambiguously designate the intended referents. The task employed in the present set of studies requires subjects to refer to particular locations within a spatial network, and correct reference is critical to the successful completion of the task. This constraint allows us to study the mechanisms by means of which these difficulties are resolved. We will argue that the difficulties arise in the first instance from the intrinsic imprecision of the referring expressions. Accordingly, we will discuss the relevance of the present set of studies for some of the theories of meaning mooted in recent years, and we will begin by briefly (and in a highly selective fashion) reviewing some influential theories of meaning.

The problem of arriving at an adequate characterization of the meanings of expressions in natural language is one which has received much attention in recent years. A wide variety of theories concerning the processes by which we comprehend meaning have been mooted, and this wide variety of theories embody a correspondingly wide range of philosophical presuppositions. A full review of the various theories of meaning would occupy more space than is available in the present chapter; accordingly, we will focus upon particular theories which illustrate certain key issues. For a fuller review of such theories see Johnson-Laird (1983).

---

[1] This work was funded by a Science and Engineering Research Council studentship granted to the first author, and by an ESRC project grant to the second author.

One important issue, from the point of view of the present set of studies, is the extent to which meaning is (explicitly or tacitly) viewed by the various theorists as a relatively static or a relatively dynamic entity. This key issue provides a perspective from which to view the various theories of meaning, and is important in relation to the empirical work which is described later in the present paper.

Some semantic theories (e.g., Katz & Fodor, 1963) presuppose that the meaning (e.g., of a lexical item) is essentially a predictable function of its semantic subcomponents: semantic 'primitives' or 'markers' combine together in an essentially additive fashion to result in a particular meaning for the lexical item. This view of meaning, which we will call the *fixed meaning assumption* (see Garrod & Anderson, in preparation) is currently the dominant view amongst theorists.

On the other hand, some other theorists presuppose that meaning is a fairly 'flexible' entity: that is, there is a certain amount of ambiguity inherent in natural language. Woods (1981), for example, argues that ambiguity is an inherent and indeed essential feature of natural language. He postulates the existence of an inner 'mental language' in terms of which external (natural) language is interpreted. This postulated internal 'mental language' is, he argues, capable of vastly greater discriminative subtlety than is external, natural language. Natural language is highly ambiguous, and when a particular natural language string is interpreted in terms of the inner language, the correct sense of the ambiguous natural language string is selected by context. The high degree of ambiguity of natural language allows it to be parsimonious vis-a-vis the internal mental language. This constitutes an economic and elegant solution to the problem of expressing the variety of distinctions of which our mental language is capable: a very large number of lexical items would be required in natural language if it were to be capable of encoding the myriad distinctions on a one-to-one basis.

Rommetveit (1974, 1983) takes a similar view with respect to the inherent flexibility of meaning. Rommetveit discusses the meaning potentials of a word or expression, that is, the variety of meanings which a word or expression may assume, and is critical of the notion of expressions having unequivocal 'literal meanings'. Given that expressions are open to multiple interpretations, Rommetveit suggests that when we are involved in a conversation, we engage in a tacit contract with our interlocuter to adopt particular interpretations of the expressions we use, as opposed to their other possible interpretations. This contractual view of language use emphasises the multiplicity of possible interpretations of expressions.

Rommetveit's emphasis on the contractual nature of language use finds an echo in the views of the philosopher David Lewis (1969). Lewis discusses the existence of conventions in language. These conventions are similar to other conventional regularities in behaviour, such as driving on the left-hand-side of the road, or wearing certain clothing for weddings.

Lewis's analysis of language conventions originates in a more general treatment of conventional regularities which relies heavily on the idea of a co-ordination game, in which interdependent decisions are made by members of a population of players (Schelling, 1960). Co-ordination games are defined as decision making games in which particular conjoint decisions yield optimal payoffs for all participants in just those circumstances where all those participants make the same decision. Such games present players with co-ordination problems in cases where more than one conjoint decision is optimal (a so-called co-ordination equilibrium). On what basis is any player to decide between supporting one co-ordination equilibrium or another?

Lewis argues that each player must base his/her decision on what is expected of the others, and this expectation will in turn depend upon higher order expectations about what the other players expect him to do, and so on *ad infinitum*. Hence any rational justification for a particular course of action in a proper co-ordination game will involve increasingly higher order replications of the other agents' practical reasoning. An illustration of such embedded reasoning is given in Figure 1, for a two person game for three levels of replication.

Of course, the most certain way of coming to a co-ordinated decision is through explicit

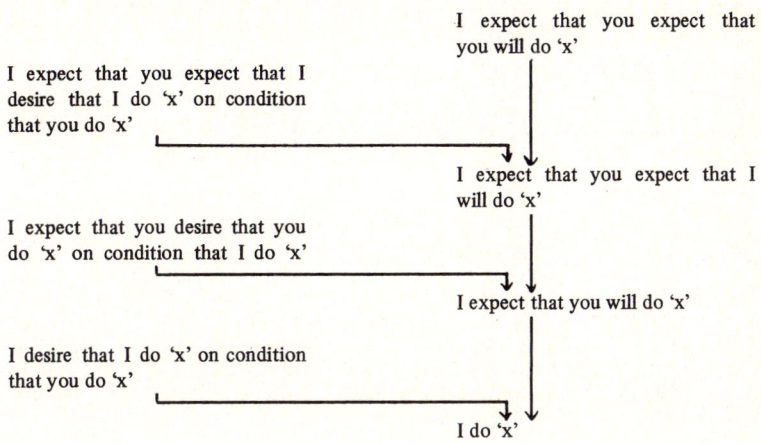

Figure 1. A schematic representation of the practical reasoning needed to justify a course of action 'x' to yield co-ordination. The arrows indicate how one expectation may be used to justify another (modified from Lewis, 1969).

---

agreement between all parties, but this is rarely an option which is available. The alternative, at least for populations which have regular experience of similar co-ordination problems, is to rely on precedent. Decisions which are known to have regularly been followed before by the relevant parties will be ideal choice candidates since they justify both higher and lower order expectations about each other's likely course of action. This leads to the idea that conventions are regularities in action or belief which form part of a co-ordination problem and whose choice is justified through mutually known precedence. Thus more formally Lewis suggests that any regularity in action or belief is a convention when:

(1) Everyone within the relevant population whose members adhere to the convention does in fact conform to the convention.

(2) Everyone within the relevant population believes that the other members of the population conform to the convention.

(3) This belief that everyone else conforms gives each individual a good and decisive reason to conform to the convention.

(4) There is a general preference among members of the relevant population for general conformity to the convention (rather than less-than-general conformity).

(5) The particular regularity to which the population conforms is not the only possible regularity which meets conditions (3) and (4) above: that is, the regularity is arbitrary in character, another regularity being an equally good candidate for adoption as a convention.

(6) The conditions (1) – (5) above are matters of common or mutual knowledge among the members of the population.

That conventions exist in language use is, to quote Lewis, 'a platitude'. In his formulation, a particular language L (that is, a particular function which assigns meanings to linguistic strings) is used by, or is the language of, a given population P by virtue of a convention of truthfulness and trust in L which holds in P. Under suitably different conditions, a different language would be used by P.

The similarity between Lewis' notion of conventions and Rommetveit's notion of tacit contracts concerning options of interpretation endorsed by dialogue participants is noteworthy: Rommetveit's 'tacit contracts' could be viewed as 'micro-conventions' or local conventions which holds in a population of two.

Thus, one useful dimension with respect to which a variety of theories of meaning can be classified is the extent to which they presuppose that meaning is a relatively fixed or a relatively flexible entity. It is worth noting that the theories which presuppose that meaning is a relatively 'static' entity tend to be those couched in terms of the understanding of written language by a reader. The theories which presuppose that meaning is a relatively flexible entity, on the other hand, tend to focus upon language in use, whether in conversation (e.g., Rommetveit, 1974) or in terms of the problems posed by writing language-producing computer programs capable of interacting with human beings in a 'conversational' mode (Woods, 1981). Furthermore, the critique of the earlier semantic theories (e.g., Rommetveit's 1974 critique of the Harvard-MIT school of psycholinguistics) is centred on their failure to focus on natural language in use. This tendency for some theorists to theorize, by default, about the understanding of written language has been referred to as the 'written language bias' (Linell, 1982).

In the present paper, we will present analyses of, and excerpts from, actual spoken dialogues which emerge when speakers engage in a specially designed co-operative game. This game requires the dialogue participants to describe locations within the spatial network of a maze. The design of the maze game is such that a semantic analysis of the referring expressions generated by the speakers is possible. We will argue that the data are best interpreted in terms of the dialogue participants being masters of their language to the extent that they are able to establish tacit local conventions regarding the interpretations of the referring expressions that they use. This involves the interlocutors, we contend, in a certain amount of tacit negotiation of the referential meaning of those expressions.

In the next section of the paper, we will describe in some detail the experimental maze game task before proceeding to a description of the data and the analyses undertaken.

1.2 The Experimental Task: The Dialogue Maze Game

The dialogue maze game was designed for the purpose of eliciting natural dialogues which contained spontaneously-generated descriptions of physical locations (within a predefined spatial network) whose exact position is known to the investigator. Such descriptions can then be analyzed against what is actually being described.

The essence of the game is as follows. First, each player is seated in a different room in front of a computer visual display terminal linked to a main computer (the terminals being Imlac corporation PDS1-G and Intelligent systems corporation Intecolor 8001 terminals, and the main computer being a Data General Nova 210 computer). On the screen of each terminal a simple maze is displayed. The maze consists of a configuration of *nodes* (small box-like structures placed at equi-distant locations, in which players' position markers can be located between moves), and these nodes are connected by *paths* (which are links between adjacent nodes along which players can move): see figure 2. Both players see a version of the same maze. The actual structure of the mazes (the configuration of nodes and the network of pathways) is identical for both players; however, there are certain differences of detail, which are explained below. Both subjects have the task of moving their respective position markers through the maze from a start position to a goal. The locations of start and of goal positions are different for the two players; furthermore, each player sees only his or her own start position (marked by an x) and goal position (marked by an asterisk), and the game is so designed that the players have to make alternate moves.

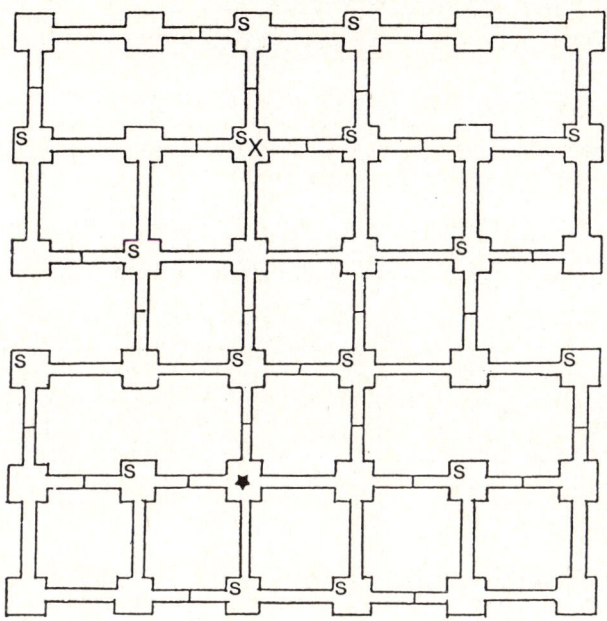

Figure 2. Typical game; state of the maze at the start of the game for player A. Key: X = start position, Asterisk = goal position, S = switch node, and the small barriers across some of the pathways represent initial gate positions. Notice that player A is unable to see his partner's position. Notice also that the maze is exactly symmetrical, in terms of its configuration of paths and in terms of the locations of gates and switch nodes, about a vertical centre line.

However, each player's maze also contains obstacles in the form of *gates* (see Figure 2) which cannot be traversed and which are present in approximately half of the available pathways in the maze. It is in overcoming these obstacles that verbal co-operation is required and this can only be achieved by means of a given player persuading his partner to use one of the *switch positions*. Switch positions, like gates, are distributed differently for each player, and the fundamental rule of the game is as follows. If a given player (e.g., player A) requires his or her configuration of barriers to be changed, he or she must enlist the co-operation of player B, find out where in the maze player B is currently located, and guide player B into one of the switch nodes visible to A on A's screen. If the players achieve this successfully, then the entire configuration of barriers is changed for player A such that all paths which were previously gated are now open, whilst all previously open paths are now blocked.

Since the position of barriers and switch nodes is different for both players (cf. Figures 2 and 3), verbal co-operation is required in order to effect appropriate barrier switching and avoid it occurring accidentally (the switching mechanism operates even if a player moves into one of his partner's switch nodes unintentionally).

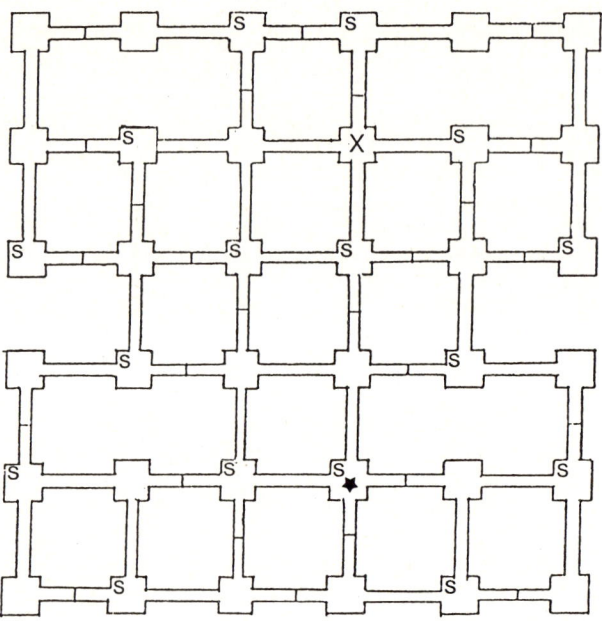

Figure 3. Typical game; state of the maze at the start of the game for player B. Key: X = start position, Asterisk = goal position, S = switch node, and the small barriers across some of the pathways represent initial gate positions. Player B is unable to see his partner's position, like player A, and in the particular game depicted here, the two players are set a task in which the positions of their start and goal locations are such that the players have to move identical distances within the maze in a 'parallel' fashion (cf. Figures 2 and 3). Both of these features are optional in the maze game program.

Attempts by a player to move across a barrier result in the player's position marker 'rebounding' on the barrier back to the node from which the move began; the player is thus returned to the node from which the illegal move was initiated, and a number of penalty points (arbitrarily set at two in the present studies) is registered against both players.

Typically, a game will consist in the players attempting to move towards their respective goals with dialogue intervening between moves. The dialogue will contain descriptions of players' current positions, switch node locations, goal positions and so on. All speech which occurs between players is recorded using a TEAC A-3340 4-channel reel tape recorder, each subject's speech being recorded on a separate channel. A tone is simultaneously recorded whenever a move takes place, a slightly different tone being used for each player, and the tones are each allotted a single recording channel. Thus it is possible to analyse any description of a player's current position against a record of that position printed out at the end of each session.

The software for the maze game was written by Jim Mullin, computer manager in the Department of Psychology, University of Glasgow.

1.3 The Location-Descriptions Generated by Subjects

In this section we will give an informal description of the variety of methods used by subjects to generate descriptions of maze location. In the present paper, our main purpose is to examine the interactive aspects of maze location description; consequently, our classification of the location descriptions is summarized in an informal fashion. It should be noted, however, that the informal four-fold classification of the descriptions given here was reinforced by a more formal statistical analysis which is described in detail elsewhere, and is summarized briefly in the present section.

An informal examination of the transcripts of the 48 games studied suggests that subjects employ a comparatively small number of description-generation 'strategies'. In fact four basic ways of generating location descriptions were observed, and all four take the form of procedures: they constitute a set of instructions to the listener as to how and where to search to find the desired location. These four generic types of description are each illustrated in turn.

(1) PATH or TOUR type descriptions

In these descriptions, the speaker generates a description of a location by first identifying a perceptually prominent point in the maze network, such as a corner, and then specifying the relevant location in terms of a series of movements along available pathways from the previously-identified prominent point to the location in question. For example (see Figure 3 for an illustration of the referent):

I am: if you go to the top right corner, and go down one and along two to the left, that's where I am.

In this type of description, the speaker is, so to speak, inviting the listener on a tour of the maze, the destination of the tour being the location in question.

(2) LINE type descriptions

This class of descriptions involves describing the maze as if it were an array of lines, for example (see Figure 3 for an illustration of the referent):

I'm on the second top line, third from the right.

The listener is first tacitly instructed to locate a particular line of nodes (in this case a row, or horizontal line) and is then instructed to look along that row until the relevant location is reached.

(3) FIGURAL type descriptions

In this type of description, the speaker makes reference to a shape or figure formed by part of the path structure of the maze. The listener is typically directed to locate a specified figure, and then the relevant point is described relative to that figure. For example (see Figure 4 for an illustration of the referent)[2]:

B   Where's the asterisk?
A   Er: mine is: if you take the uppermost left hand side: right?
A   Right.
B   See where the L is?
A   Yes.

---

[2] In the excerpts of dialogue which we discuss in this chapter, the following conventions hold. Colons (:) denote short pauses (i.e., pauses of less than 1 second's duration), and figures quoted in parentheses, for example (2 secs), are approximate durations of longer pauses.

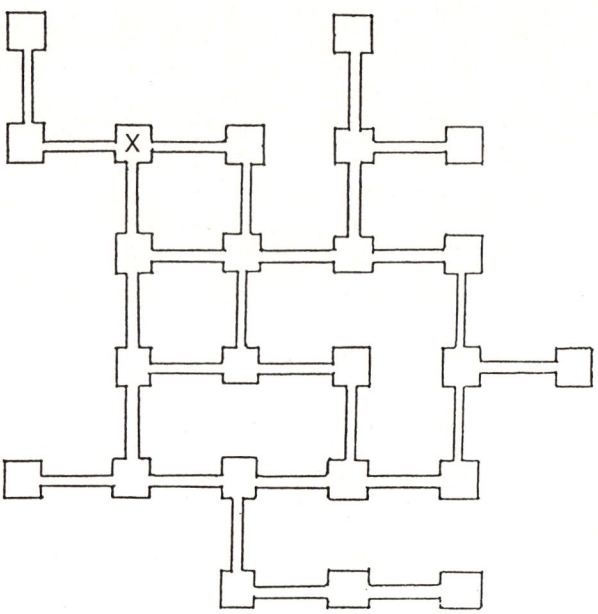

Figure 4. The location (X) described as the "last part of an L (shape)... in the middle square of that line". In this diagram, the locations of gates and switch nodes has been omitted for clarity.

B Right: ehm: my asterisk is at the: (1 sec) sort of: last part of the L: see what I mean?
   In the middle square of that line:
A OK.

Once again, this type of description is implicitly procedural: the listener is tacitly instructed to look for a particular shape (in this case an L shape) and is then instructed to search in the area in which the specified shape is located to arrive at the relevant point.

(4) MATRIX or COORDINATE type

In this type of description, the speaker generates descriptions using an implicit set of procedures which are very similar to the Cartesian coordinates of mathematics. The interlocutors assume a coding system for denoting the horizontal and vertical lines of nodes in the maze, and they then describe locations by giving two-place coordinates. For example, the columns of nodes might be lettered A to F from left to right, and the horizontal lines of nodes might be numbered 1 to 6 from bottom to top. The point illustrated in Figure 3 might then be described as 'D5'. The listener's task on hearing such a description would be to locate the relevant column (D, the fourth from the left) and the relevant row (row 5, the fifth row from the bottom) and to seek their intersection.

A more extensive semantic analysis of each of these types of description is given in Garrod, Anderson and Sanford (1984), where it is argued that each depends upon the speakers adopting some particular mental model of the maze being described. Such models embody quite different assumptions about the spatial organization of the maze, both in terms of what entities are describable and how these entities relate to each other in space. So, for instance, with a 'horizontal line' type of model, vertical relations are only expressible with respect to the higher order units (e.g., lines) and horizontal relations with respect to lower order units (e.g., boxes). On the other hand, in path type models spatial relations are reflected in terms of the possibilities of movement between boxes along actual paths in the maze. Consequently, relations like "next to" may be interpreted quite differently given the two models, since in the line type euclidean distance is roughly maintained while in the path type distance depends on the number of paths which have to be traversed to get from one point to another. Similar contrasts can be made between models which underly the other types of description.

According to the Gerrod et al. analysis, a particular description language may be characterised as a combination of (1) some set of terms, such as 'row', 'level', 'right-turn indicator' which are mapped extensionally onto sets of elements in (2) some particular mental model of the maze, where the relationship between the terms and the model constitutes a definition of a particular signalling system (see Lewis, 1969).

Another informal observation which was made during a perusal of the transcripts of dialogue and which was subsequently supported by a more formal analysis was that within a given dialogue, the interlocutors predominantly use only one of the above four types of location description. Thus, each participant within a given dialogue tends to use the same type of location description as does his or her partner. This observation is particularly pertinent to our hypothesis (see below) that interlocutors within a given dialogue tacitly negotiate local conventions regarding the choice of, and the referential semantics of, a particular 'language' in which to couch location descriptions. As noted above, the informal four-fold classification given above was reinforced by the results from a classification study using hierarchical cluster analysis (Anderson, 1983). In this analysis, the lexical items comprising all of each individual speaker's location descriptions were classified in terms of their being instances of each of 38 lexical categories. The 38 lexical categories were chosen in such a fashion that a suitably general description of the data was achieved using a comparatively small number of categories. For example, there were cateogries for instances of usage of edge names (TOP, BOTTOM, LEFT, RIGHT), a general category of 'SHAPE' for a variety of different instances of figural-type descriptions (T shapes, L shapes, rectangular shapes), etc. When this was done for all descriptions generated by all 96 speakers (48 dyads), the lexical categories were cluster analysed using three different clustering algorithms, and the results from these three different analyses were then compared to ascertain which clusters of lexical categories emerged consistently across all three analyses. Eight such consistently-occurring clusters were observed. Each cluster corresponded to either a complete description type (e.g., one cluster comprised those lexical categories corresponding to the lexical items used in line type descriptions) or some sub-category of a description type (e.g., lexical categories corresponding to the lexical items in expressions like '... left hand corner ...' were observed to consistently cluster together).

A second cluster analysis was undertaken, in which the data corresponding to individual speakers were cluster analyzed. This was achieved by rotating the data matrix (38 lexical categories x 96 subjects) through 90 degrees to yield a data matrix of 96 subjects x 38 lexical categories. A cluster analysis of individual subjects was then undertaken, yielding clusters of subjects such that all members of a cluster were classified together on the basis of their having used similar lexical categories (and hence, similar location descriptions). Again, three different clustering algorithms were used and the results yielded by all three were compared to ascertain which groupings of subjects were emerging consistently. This analysis yielded results which lent support to the above-noted observation that the two partners within a particular dialogue were likely to use the same description format as one another. In only two games out of 48 were the two speakers within a particular dialogue consistently classified as belonging to different clusters of subjects by all three clustering criteria. In contrast, 29 of the remaining games showed the pattern that both participants within a given dialogue were consistently grouped together within one or other of the several clusters of subjects by all three clustering criteria. These data thus support the view that a fairly high degree of inter-speaker entrainment is occurring within the dialogues.

To summarize, the above data suggest the following conclusions. Firstly, there are a number of different ways of describing locations within the maze networks used in the present experiments. Secondly, speakers within a given dialogue would appear to use only one of these description types predominantly within that dialogue. Thirdly, the speakers within a given dialogue would appear to constrain one another's choice of description.

1.4 Hypotheses

Given the above observations, an important question arises, namely, what processes govern the selection of a particular type of description to be used by both partners within a given dialogue? A corollary to this question is the following: since any description is referentially ambiguous (this point is explored further below), and the task requires that locations are specified with great precision, how do the interlocutors achieve successful, unambiguous reference?

In the next sections of the paper, we will present analyses of the location descriptions which, we argue, will provide an answer to the above questions. In particular, we wish to propose that the interlocutors within a given dyad tacitly negotiate local conventions regarding the choice of, and the precise referential semantics for, the particular type of referring expression that they use. To illustrate what we mean by this, consider the following excerpt of dialogue (see Figure 2 for an illustration of the referent):

1. B: So where are you aiming for?
2. A: I'm aiming for: right. If you take each wee box as a coordinate position, three, four.
3. B: Oh.
4. A: Three, five rather.
5. B: What's three? Three is - (speaker A interrupts)
6. A: Right, O.K., say the top left and you go along two and down one.
7. B: Yeah – oh yes.
8. A: Right? And you're just in the one to the right of that.
9. B: Have you got – wait a minute, you're going along two –
10. A: Right – top left.
11. B: Right.
12. A: Go along two.
13. B: One, two.
14. A: To the right, and down one.
15. B: One. Yeah, I'm just to – (A interrupts)
16. A: You're just to the right of that.

This extract of dialogue is interesting. In utterances 1 – 4, player B asks player A where her goal is located, and player A generates a matrix – or coordinate-type description which has as its origin the bottom left-hand corner node of the maze, but the description as given does not specify the location of this origin. In utterance 5, player B attempts to clarify for herself the nature of the procedures implicit in player A's description. Player B seems willing to accept this form of description provided its extensional semantics are made clear. In the next utterance, however (utterance 6), player A does not take up player B's implicit offer to negotiate, and instead tries an entirely different type of description, a path-type description. (Note that player A's first description involved the counting of node positions, whilst the path description generated by the same individual counts movements between nodes; this highlights the ambiguity of the descriptions, because all four types of location are subject to this potential ambiguity.) In utterance 7, player B seems to understand this path description ("yeah – oh yes") and player A follows up the description with a further confirming and clarifying description (utterance 8) in terms of player B's (previously specified) position. However, as noted above, player A's path description is ambiguous with respect to whether node positions or path movements are being counted, and consequently player B commences a further clarifying description (utterance 9). In the subsequent utterances of the above extract of dialogue, a decomposed path description is generated by player B, in an attempt to resolve the ambiguity problem. Note that the 'decomposed' description allows player B to acknowledge her understanding (or otherwise) of the various subcomponents of the description in turn, as they are uttered. Subsequently, these two players used path-type descriptions predominantly.

This extract of dialogue illustrates how the choice of a type of description to be used throughout the dialogue is negotiated. Initially there is a failure of reference. In this particular case, player A's initial description is not preceded or accompanied by appropriate clarification of the referential meaning of the coordinates stated in the description, but rather than clarifying the semantics of the initially proffered description, these subjects try using another type of description, a path description. In other cases, reference fails not because of unintentional vagueness on the part of the speaker who proffers the initial description as in the example discussed above, but because the two participants unwittingly attach different interpretations to the same referring expression and conflict occurs; an example of this type of failure will be considered in the discussion section. In general terms, following the failure, other alternative types of description are employed by the subjects, one of these types is selected for use and the details of its extensional semantics are clarified, and thereafter the negotiated description type is used predominantly throughout the game.

The chosen description type is only used predominantly, rather than exclusively, because players often describe their locations, particularly towards the end of a game, in terms of path movements away from their previously specified goal positions. This type of description, relative to a previously specified position, is used in the case of all types of game and all predominant description types.

The above extract of dialogue has been discussed at some length in order to illustrate what we mean by 'tacitly negotiated local conventions' concerning which referring expressions are to be used by the interlocutors and how they are to be interpreted.

It is clear, therefore, that we use the term 'negotiation' in a sense closely related to its commonsense interpretation: by negotiation, we mean the arriving at a mutually acceptable agreement concerning the choice of, and the extensional-semantic interpretation of, the referring expressions used by a pair of interlocutors, by means of discussion between the interlocutors.

In general, negotiation is aimed at establishing mutual knowledge between the interlocutors in order to secure the relevant local language conventions for that stretch of dialogue. It always involves in some way both the communicator (i.e., the one initiating a description) and his or her audience . In rare instances promoting mutual knowledge about semantic conventions seems to be the primary function of an utterance (as in the second part of utterance 2 above). More often, however, it seems to be secondary and depends upon treating a description as a precedent for subsequent descriptions by both parties. The basic negotiation paradigm is therefore one where the communicator proffers some type of description of a position and the audience offers some response. So long as the description is understood (in some way) by the audience it is then taken as a precedent for further descriptions, that is, the 'language' of description in the sense given above is now established through precedence. Of course there is a range of devices available to both communicator and audience to promote or even block this process of proffering and accepting descriptions, all of which constitute negotiation. We wish to draw attention to some of these:

1. The 'prolongation' or 'decomposition' of a description by the speaker to allow his listener to signal his understanding, or the lack of it, at the time of utterance. This was illustrated in the 'decomposed' path description generated by player A in the extract of dialogue discussed above.

2. The use of explicit clarification questions. This mechanism is commonly employed by subjects to clarify the extensional semantics of a count procedure (i.e., whether node

positions or path movements are being scanned). A typical example of this would be where a speaker might ask "Do you count the bottom row as one or zero?" (an answer to which would effectively resolve this type of ambiguity; the counting of node positions would involve designating the bottom as 'one', whereas the counting of path positions would involve designating the bottom as zero).

3. The listener's response to a proffered description. In the ideal case of perfect understanding, the listener need only acknowledge his or her understanding (by saying 'OK', 'Uh-huh', or some such expression). Failure to completely understand a proffered description would be signalled by repeating all or part of the description, or restating the description in slightly different terms as a 'check'.

Whilst these are unlikely to be the only operative mechanisms underpinning tacit negotiation of the type discussed here, for the purposes of the present paper we will concentrate on these three mechanisms as a first step. The devices by means of which the negotiation is achieved (i.e., those sketched briefly above) are clearly rather subtle ones; we wish to stress that the negotiation we refer to in this paper is very much a tacit process.

Returning to our first question above (i.e., what processes govern the selection of a particular type of description to be used within a particular game), one relevant variable may well be the nature of the actual maze on which the game is played. In the experiments reported here, one variable which was manipulated was the shape of the mazes on which games were played. In some cases, the mazes were symmetrically shaped with respect to a vertical centre line, and they were regularly (essentially square) shaped; for an example of this type of maze, see Figure 2. On the other hand, other mazes were specially constructed so that they were asymmetrical in both the horizontal and vertical planes, and irregularly shaped (see Figure 4). This variable was expected to influence the choice of a description type so that symmetrical, regularly-shaped mazes would, we hypothesize, be more consonant with description types which abstract from the actual maze structure and treat the maze as if it were a regular array of parallel lines or a regular two-dimensional space of nodes. In contrast, we might expect the irregular, asymmetrical maze shapes to be more amenable to description in terms of those description types which involve explicit attention to the actual maze structure. We therefore hypothesize that there will be an asymmetry in the choice of a description type so that regular, symmetrical mazes will more often be described using line- and coordinate-type descriptions, whereas irregular, asymmetrical mazes will more often be described in terms of path- and figural-type descriptions.

Having chosen a type of description for subsequent predominant use, there remains a considerable degree of ambiguity which requires to be eliminated. We hypothesize that tacit negotiation is also used to eliminate ambiguity. Because the maze game requires great precision of reference, we hypothesize that the normal state of affairs in these dialogues will be one in which the tacit negotiation takes place during the early stages of the dialogue, and, having thus eliminated semantic ambiguity during the early stages of the game, 'negotiation' should be less in evidence during the later stages.

We therefore hypothesize that: (a) there should be more repetitions, clarifications, restatements, and 'decomposed' descriptions (i.e., those which allow listener feedback following each component of the description) during the first 50% of the location descriptions within a given game as opposed to the second 50% of descriptions. Conversely, (b) one might expect greater prevalence of merely acknowledging the understanding of a proffered description by a listener during the later 50% of descriptions than in the earlier descriptions, since one would expect the later descriptions to be those in which negotiation is absent, and the mere acknowledgement of understanding by the listener to be sufficient. As a corollary to (a) and (b) above, and the hypothesis of early negotiation followed by conformity to the negotiated solution, (c) one would expect the early descriptions within a given game to be longer in terms of the total number of utterances spoken per description.

2. Method

Three experimental studies were undertaken: however, the principal analyses detailed below (the analyses concerning the establishment of interlocutors of local conventions regarding the location descriptions that they generate) were undertaken on all data from all three studies. Consequently, in much of what follows, the three studies outlined below will be treated as one large database.

The three studies were as follows:

(1) 'Baseline' study. In this study, 16 subjects (eight males and eight females, comprising four pairs of males and four pairs of females) took part in an exploratory study. Subjects firstly played a practice game on a small, symmetrical maze based on a 4 x 4 matrix of nodes. This allowed subjects to familiarize themselves with the rules of the game. They then played two games on the larger, symmetrical mazes based on a 6 x 6 matrix of nodes (of which the maze shown in Figures 2 and 3 is an example). In the analyses to be described below, only the data from the first unseen-partner condition game on the larger maze will be considered, because it is during the first game that we would expect any negotiation to take place.

(2) 'Monster' games. In this study, 10 pairs of subjects (four pairs of males and six pairs of females) took part in a total of eleven practice games on the smaller (4 x 4) maze. These games were then followed by a total of 12 games on the same larger, symmetrical mazes used in study 1. The games were played such that each pair of subjects played one practice game followed by at least one game on a larger maze. The game on the larger mazes involved the use of an option available in the maze game program which was not employed in the game in study 1. This is the maze 'monster'. This is a semi-intelligent computer-controlled third player, marked by a letter M, which is programmed to 'pursue' one or other player in the maze. It is located in the same position in both players' mazes, and it moves once for every three moves of both subjects. It moves towards whichever player is currently nearest, but it obeys the pursued player's gate configuration, that is, it will not traverse his or her barriers. If the 'monster' manages to occupy the same node as a given player, that player is considered to be 'eaten' and the game is thenceforth terminated. The presence of the monster complicates the game strategically, and results in a significantly greater amount of speech between moves and a slower rate of moving as compared with the, otherwise equivalent, 'baseline' games (Anderson, 1983).

In two cases, one or other subject in a dyad was 'eaten' in a very short space of time; in such cases, the subjects were asked to play a second 'monster' game.

(3) In study 3, eight pairs of subjects (three male and five female dyads) took part in a total of 14 games played on mazes based on 6 x 6 matrices of nodes, but which had been specially constructed so that in one case the maze was radically asymmetrical (see Figure 4), whilst in the other, bilateral symmetry about a vertical centre line was retained, but with the special property that the vertical lines of pathways were more complete than the horizontal lines. These two mazes were designed to be less amenable to 'abstract', vector-based description types (i.e., 'row' and 'matrix' types) than were those mazes used in the previous studies. Most of the subjects in study 3 played two games, one on each of the two mazes. The order of presentation of the two mazes was conterbalanced across subjects. No practice games were played before the first main game; this was an attempt to obviate any possible effect of the square, symmetrical practice maze on the subjects' choice of a description type. Instead, the subjects were instructed with particular care.

This study was an attempt to ascertain whether there was an influence of maze shape upon the choice of a description type used to describe that maze.

The subjects in all cases were undergraduate students of the University of Glasgow. They were paid at the rate of £1.00 per hour for participating in the experiments. In all three studies, only same-sex pairs of subjects played the game.

3. Procedure

All subjects were carefully instructed regarding the rules of the game. The necessity of the players describing their locations to one another in order to effect appropriate gate-switching, and of mutual cooperation, were stressed. Care was taken to ensure that subjects within a dyad did not attempt to compete with one another in moving towards goal, but instead they were instructed to attempt to solve the problem together. Subjects were in fact given a (hypothetical) norm of number of moves to game solution and instructed to attempt to better this figure. It was hoped that this would help prevent subjects competing with one another. Subjects were instructed to try to solve the game in as few moves as possible, whilst incurring as few penalty points as possible.

In general, the subjects appeared to enjoy participating in the game task, becoming very task-involved, and the resulting dialogues have a spontaneous and natural quality.

4. Selection of Data and the Analyses Employed

4.1 Analysis 1: The influence of maze shape on choice of description

In order to test the hypothesis that the shape of the maze on which a game is played will influence the choice of a type of location description, it was decided to use an objective criterion to select examples of games employing the chosen categories of types of description (the chosen categories being vector-based, i.e., line and matrix-type descriptions, and non-vector-based, i.e., path and figural-type descriptions). The dialogues which were selected for use in the present analysis were selected on the basis of the cluster analysis of subjects described in Anderson (1983); only those games in which both subjects predominantly employed the same single type of description as one another were classified as having used either of the two categories of description. Comparing the distribution of these two broad categories of description type across the two different types of maze (i.e., regular, symmetrical versus 'specially-constructed') resulted in a 2 x 2 analysis which was analyzed using the Fisher exact probability test.

4.2 Analysis 2: The tacit negotiation of local conventions

In order to test the hypotheses regarding the tacit 'negotiation' by subjects of the choice of, and the referential semantics of, a type of location description, it was decided to examine the data from those games played on the larger mazes which were played first by each dyad. This permitted an analysis of a total of 18 games. (In the case of the two dyads who had played very short 'monster' games initially and who had then played a second 'monster' game, the data from the two monster games were considered together, the descriptions from the two games being concatenated to form one long sequence.)

Each sequence of dialogue in which a location description occurred was extracted from each transcript. The sequence would typically begin with either a wh-question ("Where are you?") or a description being spontaneously generated, and ended with either an acknowledgement of understanding ("Uh-huh", "OK", or some such expression) or, in come cases, with no such response. As noted above, the response to a proffered description could take several forms (a restatement, a repetition, etc.). Consequently, the length of description sequences varied considerably, from as little as one utterance (in cases where no wh-question preceded the description and no acknowledgement or other response followed it) to prolonged sequences as exemplified by the segment of dialogue discussed in the introduction.

To test the negotiation hypothesis, it was decided to conduct a within-subjects comparison between the first 50% of all location descriptions generated within a given game with the second 50%. If negotiation does take place during the early stages of the games, with a later predominant use of one particular description type (its referential semantics also having been clarified during the early stages of the game) one would expect significant differences between those sequences in the first half of the game as compared to those in the second half. This test is somewhat crude, since some dyads will be able to complete their negotiation during the very early stages of the game, whereas others may continue to negotiate during the second half of the game. Nevertheless, it was felt that an objective division criterion was preferable to attempting to subjectively define a hypothetical point in each transcript at which the putative negotiation ceased, and comparing those descriptions generated before this point with those generated after it.

The following is a description of the measures used:

(1) The amount of discussion of the descriptions in utterances spoken by both players was measured. In addition to comparing the mean amount of discussion of the descriptions in terms of the number of utterances spoken per description in the first 50% as opposed to the second 50% of the descriptions, a plot was made of the mean amount of discussion of each description in utterances as a function of its serial position in the sequence of descriptions. The number of utterances per description was averaged out across all 18 dialogues for each serial position. We would predict an early peak, followed by a stabilization of the curve, if negotiation is taking place during the early stages of the games.

Note that this measure concerns the overall amount of verbal interaction within a segment of dialogue in which a location is being described. This is distinct from ellipsis, in which the descriptions themselves are shortened. For example, we might expect a more elaborate description such as the following one:

I'm er: on the second bottom: row: er: third along from the left.

to be generated during the early part of the game, whereas the later descriptions within the same game might be somewhat more brief, for example:

I'm er: third row, third box.

This latter description is a highly elliptical row description in which the direction in which counting of nodes (or of path movements) should take place is no longer specified. In fact, a comparison of the lengths, in words, of those utterances which describe locations yielded, as would be expected, a significant difference. The data were scored across all 18 games as were the analyses of tacit negotiation considered in detail in this paper. It was found that later descriptions, i.e., those comprising the second 50% of all descriptions, were significantly shorter, with a mean length of 8.8 words overall, than the earlier descriptions, which had a mean length of 10.6 words. Thus the location-describing utterances are subject to the normal process of ellipsis.

The measure with which we are concerned here, however, concerns the number of utterances which both speakers contribute to the dialogue sequences in which locations are described. We are thus interested in both the description itself and also all reactions and comments by the listener. A large score on this measure reflects a greater number of utterances being generated by both speakers, i.e., a greater amount of discussion of a proffered description taking place. Hence the prediction of greater values on this measure during the earlier stages of a given game.

(2) The percent of descriptions within each half of the total in which the initial response to a proffered description is either a simple acknowledgement or no response at all was the second measure. The listener's first reaction was examined in this measure. If the listener responded immediately with a simple acknowledgement or gave no response, then this was scored. Null responses were scored within this category because any misunderstanding resulted in a response by the listener; no reply would be taken by the speaker to indicate understanding by the listener. Any other type of response was not scored in this category; for example, cases in which a speaker generated a 'decomposed' type of description, that is, one in which pauses were inserted by the speaker after each subcomponent of the description to allow the listener to give feedback (see below), and where the listener gave more than one simple acknowledgement, were not included in this category. This category of response, therefore, included only 'ideal' cases in which a speaker generated an entire description in one utterance and the listener indicated his or her understanding by means of either a simple acknowledgement or a null response.

(3) The third measure was the percentage of descriptions in each half of the sequence of location descriptions in which a proffered description was 'decomposed' by the speaker, that is, spoken by the speaker with pauses inserted by the speaker after each subcomponent of the description was spoken, in order to allow the listener to give feedback to indicate understanding (or otherwise). If the description was generated with pauses inserted after each subcomponent, this was recorded as one instance of this category. If, on the other hand, no such pauses were employed by the speaker, that is, the proffered description was spoken fluently, no score was registered on this variable.

(4) The number of restatements of the description was the fourth measure. In this case, restatements occurring anywhere in the description sequences were counted.

(5) The number of repetitions of the description was the fifth measure. Also, the score on this variable was incremented by 1 if a repetition of the whole description, or any part of it, was made by either subject anywhere in the description sequence.

(6) The number of clarifications was the final measure. This variable was again scored to reflect the presence in any part of a description sequence of a question or statement which clarified the description given (e.g., "Do you count the bottom as one?" or "Are you starting from the left?").

All of these variables were scored to reflect the proportion (expressed as a percentage) of the first 50% of all descriptions spoken versus the proportion of the second 50% of descriptions containing these features. A within-subjects comparison was drawn for each variable.

5. Results

5.1 Analysis 1: The influence of maze shape on choice of description

This analysis involved comparing the proportion of games played on the two types of maze (regular-symmetrical versus 'specially-constructed') involving the use of the two broad categories of description type (vector-based, i.e., line and matrix, versus non-vector-based, i.e., path and figural, types) with what one might expect by chance. If the proportions are uneven in such a way that a particular category of maze type was described more often by one form of description rather than another, then one might conclude that maze shape does influence the choice of a description type. The results were as follows:

	TYPE OF DESCRIPTION		
Type of Maze	Row Matrix	Path/Figural	Total
Regular	20	3	23
Specially-constructed	2	5	7
Total	22	8	30

Table 1. The distribution of the two classes of description type across the two classes of maze type (N=30). The distribution is significantly different from what would be expected by chance (p=0.0064, Fisher exact probability test), indicating that particular types of mazes are, to some extent, associated with particular types of location description.

On first inspection, the distribution of the two types of description across the two types of maze do indeed appear asymmetrical in nature, especially in the case of the line- and matrix-type descriptions, the vast majority of which are used to describe locations on regular, symmetrical mazes. When these data are subjected to analysis using a Fisher exact probability test, the results are significant with $p = 0.0064$. One can conclude, therefore, that there is some evidence of an asymmetrical distribution of description types across mazes, lending some support to the argument that the nature of the maze constrains the type of description to be used in describing it to some extent.

5.2 Analysis 2: The tacit negotiation of local conventions

The following table shows the mean values of the different measures in the first half as opposed to the second half of all descriptions spoken in each of the 17 games analyzed. All variables conform to the patterns predicted.

Measure	First 50% of all descriptions	Second 50% of all descriptions	Significance
Length of descriptions (Utterances)	3.86	2.94	$T = 23, N = 18$ $p < 0.005$
Proportion of listeners' responses which are simple acknowledgements*	45.8%	59.8%	$T = 36, N = 17$ $p < 0.05$
Description 'decomposed'*	25.44%	13.8%	$T = 30, N = 16$ $p < 0.01$
Repetitions*	39.2%	25.8%	$T = 30, N = 16$ $p < 0.025$
Restatements*	13.3%	10.3%	$T = 62.5, N = 17$ NS
Clarifications*	13.35%	2.53%	$T = 0, N = 14$ $p < 0.005$

Table 2. A comparison of the prevalence of each of the six measures which were hypothesized to reflect the existence of tacit negotiation in the first half as opposed to the second half of the total location descriptions uttered (N = 18). In the case of those measures which are marked with an asterisk, the values quoted are average proportions (percentages) for the first half, as opposed to the second half, of all descriptions generated by both subjects participating in a given dialogue which contain examples of the measure in question. Significant differences in the predicted directions are obtained in five of the six comparisons.

Each measure will be discussed in turn.

(1) The amount of discussion of the descriptions. As noted earlier, one statistic which was computed was the amount of discussion (in utterances spoken by both players) of the descriptions. We hypothesized that, if negotiation (as defined in the introduction section) does take place during certain location-describing verbal interactions, it should involve greater amounts of speech as compared with descriptions in which no negotiation takes place. Consequently, one

should expect significant differences in the length (in utterances spoken by both players) of those segments of dialogue which contain descriptions involving negotiation as compared to those segments of dialogue containing descriptions which do not involve negotiation. Because the successful playing of the game requires precision of reference, we further hypothesized that negotiation should take place early in any given game.

Figure 5 shows a plot of the mean amounts of discussion of the location descriptions as a function of their serial position, for the first 20 descriptions, averaged over the data from 18 games. The graph displays clearly the predicted pattern: there is an early peak during the first 4 descriptions, which are 4 to 6 utterances long on average, followed by a decline in description length, until descriptions 7 and 14 (inclusive) show a relatively stable amount of speech per description of around 3 utterances. From the fifteenth to the eighteenth description there is another, much less pronounced, peak. This peak reflects the change in description type which often occurs towards the end of a game, in which there is often a tendency to describe one's position in terms of path movements away from the, previously specified, goal position. This change in description format is often accompanied by the generation of 'checking' descriptions, hence the slight lengthening of the verbal interactions. (The reason that this second 'peak' appears spread out across several descriptions is that different dyads of subjects reach this point after different numbers of moves, hence the apparent 'spreading-out' of this lengthening over several descriptions.) For example:

  B: Whereabouts are you?
  A: I'm in the second: top line now: second top er: third in from the left.
  B: Third in.
  A: So I'm – I'm only two away from my base.
  B: Aye, aye.

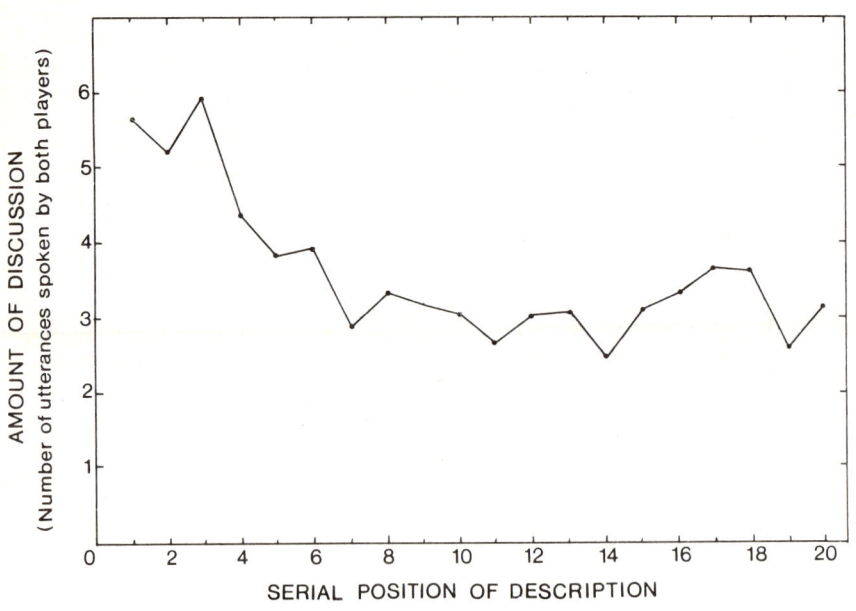

Figure 5. Amount of discussion (mean number of utterances spoken by both players) of a proffered location description as a function of that description's serial position in the dialogue (N = 18).

The speaker generating the description first generates a complete description, and then restates it in terms of distance away from his (previously specified) goal position (which he describes as his 'base'). Towards the end of a game, subjects often change from the established description format to describe themselves in terms of path movements away from their goal positions, generating such descriptions as "I'm two away from my goal now". The change from a previously established description format to this path-type description often involves the generation of 'checking' descriptions by one or other of the interlocutors, thus lengthening the overall verbal interaction.

Overall, the pattern is very much as one would predict: an early peak which, we argue, reflects the tacit negotiation taking place, and later (approximate) stabilization at a level (of approximately 3 utterances) which is economical in terms of the amount of speech output per description.

When the amount of discussion of the first 50% and the second 50% of the total location descriptions uttered are compared statistically (see Table 2) a highly significant difference emerges: the descriptions in the first half, with a mean length of 3.9 utterances, are significantly longer than the descriptions in the second half, which have a mean length of 2.94 utterances (Wilcoxon $T = 23, n = 18, p < 0.005$, 1 tail).

(2) The percentage of each half of the total descriptions uttered in which the listener's initial response was a simple acknowledgement (or in which no response was given) were compared. In this case, we predicted that early negotiation would be accompanied by little use of the simple acknowledgement of a proffered description, whereas later descriptions should involve the simple acknowledgement of the proffered description to a much greater extent. The percentages involved conformed to the prediction, and the difference was statistically significant: 45.8% of the first half of the descriptions involved the use of simple acknowledgement as a response to a proffered description, whereas 59.8% of the second half of the descriptions were initially responded to in this fashion (Wilcoxon $T = 36, n = 17, p < 0.05$, 1 tail).

(3) The percentages of each half of the total location descriptions uttered in which the speaker paused after each subcomponent of the description to allow listener feedback. The values again lay in the predicted direction. Of the first half of all descriptions generated, a mean of 25.4% involved such 'decomposition' of a location description, whereas only 13.8% of the second half of the location descriptions involved such 'decomposition'. This difference was significant (Wilcoxon $T = 19, n = 15, p < 0.01$, 1 tail).

(4) The number of restatements of a proffered description in each half of the total descriptions generated. In this case, we predicted a greater use of restatements in the first half of the descriptions as opposed to the second half. The difference found was small, in the predicted direction, but non-significant (with means of 13.3% for the first half and 10.3% for the second half; see Table 2. (Wilcoxon $T = 62.5, n = 17$, NS).

(5) The number of repetitions in each half of the total descriptions generated. The obtained values again conformed to our prediction, with a mean of 39.2% of the first half of all descriptions spoken involving repetition of all or part of the proffered description, as compared to a mean of 25.8% of the second half of all descriptions (see Table 2). This difference was significant (Wilcoxon $T = 30, n = 16, p < 0.025$, 1 tail).

(6) The number of clarification questions/statements in each half of all descriptions spoken. The means conformed to prediction (see Table 2), with 13.35% of the first half involving the use of clarifications as compared with 2.53% of the second half of the descriptions on average. This difference was highly significant (Wilcoxon $T = 1, n = 14, p < 0.005$, 1 tail).

Overall, the hypothesis that a certain amount of tacit negotiation of local conventions will take place during the early stages of the games has received some support from these analyses; the mechanisms identified in the introduction are indeed more often in evidence during the early stages of the games. The plot of the amount of discussion of the descriptions as a function of their serial position would seem to indicate that the tacit negotiation takes place during the first few description sequences; this accords well with what we expected given the nature of the maze game. One of the measures which we argued would reflect tacit negotiation taking place failed to yield a significant difference, however. These findings are discussed in more detail below.

6. Discussion

The hypothesis that a certain amount of tacit negotiation of local conventions takes place during the early stages of the dialogues rather than during the later stages has received a substantial degree of support from the results of this study: all but one of the mechanisms which were identified as possible vehicles for tacit negotiation showed significant differences in the predicted directions. This was despite the fact that, as noted earlier, subjects often change from their previously-established description type during the later stages of the game. As a result, there is a certain amount of re-negotiation. This phenomenon would explain why one of our measures, that of the number of restatements during the first half as opposed to the second half of all descriptions, failed to yield a significant difference. Restatements are commonly used as 'checking' descriptions during the 're-negotiation' alluded to above, hence the non-significant result.

In the present paper we have argued that failures of reference are due to the inherent imprecision of natural language, and we have examined some of the mechanisms which subjects spontaneously employ in order to circumvent such failures. We would therefore argue that the present data are less compatible with those theories of meaning which embody the 'fixed meaning assumption', and instead are more consonant with the view that meaning is a less fixed, more open entity which may be as much an 'emergent' property of the social interaction in which the dialogue is embedded as it is an 'inherent' property of the language itself.

In the introduction section, we noted that the negotiation of meaning which we examine in the present paper is initially motivated by a failure of reference, and we distinguished between two types of failure of reference, namely failures because of the unintentional vagueness of a proffered description, and failures because the two dialogue participants spontaneously adopt different, conflicting interpretations of the expressions that they use. We would argue that the same mechanisms of tacit negotiation would be employed by the subjects in both cases.

A good example of a failure occurring because the two participants unwittingly adopted systematically different interpretations of the same referring expressions came from a practice game played by two female subjects in study 2 (the study involving games in which the maze 'monster' was present in the maze). This example is worth considering in some detail, because it illustrates the serious problems that a failure of this type can cause. The two subjects employed an infrequently-used variant of the coordinate type of description, which involved the specification of counted distances from two edges of the maze which are perpendicular to one another – for example, 'second from the left and second from the top.' This scheme specifies an intersection of two lines of nodes to yield a unique location, involving the same implicit scanning procedures as the other more elliptical coordinate type descriptions discussed earlier. In the case of the particular dialogue under consideration here, a conflict (in the sense of systematically different interpretations of the same expression being adopted by the two speakers) arose, in that player A used ordinal count values to refer to path movements, whereas player B used ordinals to refer to node positions when generating and interpreting descriptions. This conflict resulted in serious misunderstandings, as for example in the following exchange (see Figure 6 for an illustration of the referent):

1. A: ... and where are you then?
2. B: I'm in the: second from the l: left and I'm the: second from the top.
3. A: Second from the left, second from the top: right: that's one of my Ss.
4. B: Is it? (1 sec.) It's one of mine.
5. A: Wait a minute. Second: second from the top?
6. B: Uh-huh.
7. A: No you're not: second ...

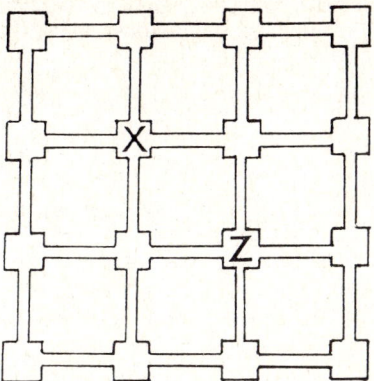

Figure 6. The location (X) described by one subject as 'second from the left and second from the top'. This subject uses the term 'second' to refer to the second node position. Her partner, however, used ordinal numbers to refer to path movements, and consequently misunderstands the description to be referring to location (Z). This systematic difference in interpretation resulted in repeated misunderstandings between these two subjects, the only solution to which was the adoption of an entirely different type of description (the path type) for subsequent use.

---

8. A: So you're still at the bottom: left-right hand corner, the corner, the corner of it?
9. B: No: yeah
10. A: Oh: yeah
11. B: I'm in the bottom right-hand corner of the top left hand square: (laughs) Is that what you mean?
12. A: No. (laughs)
13. B: (laughs).
14. A: Right: Take it: (1 sec.) take it from the left, right? Like — (speaker B interrupts)
15. B: Right I'm: one along: and and two up.
16. A; One along: and two up. One a-: Right. So you're really diagonally opposite the asterisk?
17. B: I'm: well: your asterisk, aye.

In this protracted exchange, player B initally generates a description which player A checks by means of reference to the locations of switch nodes (which were discussed previously and at the point in the dialogue from which this excerpt comes their locations were a matter of mutual knowledge). This leads to player A suspecting (in utterance 5) that a misunderstanding is occurring. In utterance 8, player A again attempts a check, this time generating a figural-type description which refers to the bottom right-hand corner of the maze, her phrase 'the corner of it' referring to that junction of pathways which is immediately diagonally opposite to the bottom right-hand corner proper; from this it is evident that player A is understanding player B to be located at location Z in Figure 6. Player B then responds with another figural description, one which treats the structure of the maze pathways as a set of 9 adjacent square shapes (utterance 11; see also Figure 6). The difficulty is resolved in utterances 14 to 17 by the generation of a decomposed path-type description. Perhaps not surprisingly, these subjects eventually used path-type descriptions predominantly, since these were much less problematic for them than the type of description they initially used. In fact, these same subjects had run into exactly the same difficulty slightly earlier in the same game, except that, in the earlier case, player B was describing the location Z in Figure 6, whereas player A understood her to be referring to location X in Figure 6. The nature of the misunderstanding is identical in both cases. This example clearly illustrates the difficulties which the more serious failures of reference can cause, and again points to negotiation as the solution to the difficulty.

Whilst our hypothesis that the tacit negotiation of local conventions takes place has received some support from the results reported in the present chapter, it is necessary to note here two caveats.

Firstly, it is undoubtedly the case that other phenomena are occurring which we have not yet examined empirically. For example, a good deal of inference regarding what is meant by what is said must be taking place; otherwise, the occasional novel coinage (like 'double box' or 'right turn indicator' which are terms used spontaneously by subjects to refer to particular configurations of pathways in the maze network, the meanings of which are rarely discussed explicitly) would cause problems for the interlocutors.

Secondly, it is necessary to note the differences between the present experimental game task and everyday, spontaneous conversation before attempting to generalise from the present set of results to conversation in general. The present maze game task differs from the normal situation in which everyday spontaneous conversations take place in two important respects:

(1) In the game task, subjects are only in contact with one another auditorially. It is not possible for them to exploit eye contact, facial expression, posture changes or any other such paralinguistic cues to signal understanding (or the lack of it).

(2) In the present task, there are explicit penalties for misunderstanding – if subjects fail to understand one another, this has immediate deleterious effects on their performance in the game. There may be no such feedback in the case of misunderstanding during an ongoing face-to-face conversation; indeed, Rommetveit (1974) argues that (partial) misunderstanding is the norm rather than the exception.

Both of these differences would tend to put pressure on the subjects in the maze game task to negotiate to a greater extent than would, perhaps, otherwise be the case.

However, it would seem not implausible to argue that, sometimes at least, the participants in an ordinary, spontaneous, face-to-face conversation will find it necessary to engage in the sort of tacit negotiation which we have described in the present paper. This situation would arise when, for example, great precision of reference is required. We would argue, therefore, that although the game task does differ in some ways from everyday conversation, there will be cases of everyday conversation to which the present set of results do indeed generalise.

The possibility that not only are speakers of a language bound by the global conventions of the linguistic community at large, but that they are also capable of establishing local conventions regarding meaning, highlights the inherently social nature of language use. The possibility of local conventions is consonant with Rommetveit's theoretical position; his (1974) notion of 'contextually appropriate optional elaborations' of meaning can be readily described in terms of local conventions, as can his notions concerning the 'contractual' nature of language.

In conclusion, the research reported in the present paper is consonant with the view that meaning is by no means a static, fixed entity, but rather, there does indeed appear to be a degree of flexibility inherent in meaning, which speakers are capable of exploiting.

References

Anderson, A. (1983). *Semantic and social-pragmatic aspects of meaning in task-oriented dialogue.* Unpublished PhD thesis, University of Glasgow.

Garrod, S., Anderson, A., & Sanford, A. (1984). *Semantic negotiation and the dynamics of conversational meaning.* Glasgow Psychology Technical Report No. 1, Department of Psychology, University of Glasgow.

Garrod, S., & Anderson, A. (in preparation). Saying what you mean in conversational dialogue: A study in semantic coordination.

Johnson-Laird, P.N. (1983). *Mental models.* Cambridge: Cambridge University Press.

Katz, J.J., & Fodor, J.A. (1963). The structure of a semantic theory. *Language, 39,* 170-210.

Lewis, D.K. (1969). *Convention: A philosophical study.* Cambridge, MA: Harvard University Press.

Linell, P. (1982). *The written language bias in linguistics.* SIC 2, University of Linkoping, Studies in Communication.

Rommetveit, R. (1974). *On message structure.* London: Wiley.

Rommetveit, R. (1983). In search of a truly interdisciplinary semantics. A sermon on hopes of salvation from hereditary sins. *Journal of Semantics, 2,* 1-28.

Schelling, T.C. (1960). *The strategy of conflict.* Cambridge, MA: Harvard University Press.

Woods, W.A. (1981). Procedural semantics as a theory of meaning. In A. Joshi, B. Webber, and I. Sag (Eds.), *Elements of discourse understanding.* Cambridge: Cambridge University Press.

4

INFERENCE AND EXPECTATION FAILURE

THE USE OF INFERRED KNOWLEDGE IN UNDERSTANDING PRAGMATICALLY ILL-FORMED QUERIES[1]

M. Sandra Carberry

Department of Computer and Information Sciences
University of Delaware
Newark, Delaware 1 97 16

Utterances that are syntactically and semantically well-formed but violate the rules of the underlying world model exhibit pragmatic ill-formedness. If the speaker's utterance does not conform to the listener's world model, then the speaker either holds beliefs that contradict the listener's beliefs or the speaker has improperly formulated the language representation for the intended query. This paper presents a context-based strategy for constructing a cooperative but limited response to pragmatically ill-formed queries that are generally interpreted correctly by human listeners. The strategy presumes that the system assimilates the preceding information-seeking dialogue by inferring the task-related plan motivating the information-seeker's queries. When the system is presented with a pragmatically ill-formed query, this inferred knowledge is used to suggest substitutions for the erroneous proposition causing the ill-formedness; these suggestions must then be evaluated on the basis of relevance and semantic criteria.

1. Introduction

An ill-formed utterance is usually viewed as one that violates the syntactic or semantic rules of language. Techniques for handling ill-formedness have been proposed by a number of researchers and range from special components for processing such utterances (Harris, 1977) to meta-rules for relaxing the rules of language (Weischedel and Sondheimer, 1983, and also this volume).

However, an utterance can also exhibit pragmatic ill-formedness. Any information-processing system, whether human or machine, must interpret utterances relative to a model of the world. If the speaker's utterance does not conform to the listener's world model, then two possibilities exist:

(1) the speaker holds erroneous beliefs about the world, or at least beliefs that contradict the listener's beliefs

(2) the speaker has improperly formulated the language representation for the intended query.

Consider for example the query

"Which apartments are for sale?"

---

[1] This work has been partially supported by a grant from the National Science Foundation, IST-8311400, and a subcontract from Bolt Beranek and Newman Inc. of a grant from the National Science Foundation, IST-8419162.

In a real estate world model, single apartments are rented, not sold. However, apartment buildings, condominiums, townhouses, and houses are for sale. Thus the above query contains the erroneous proposition

$$\text{Sale-Status (x:\&APARTMENTS, FOR-SALE)}$$

If the speaker believes that single apartments are sold, then failure is due to incorrect beliefs about the relations in which members of the entity set Apartment can participate. On the other hand, if the speaker knows that single apartments are not sold, then the error is caused by improper formulation of the intended query's English language representation; perhaps the speaker really wants to know, depending upon the context in which the utterance occurs, which condomimiums are for sale, which apartment buildings are for sale, or even which apartments are for rent.

Utterances that are syntactically and semantically well-formed but violate the rules of the underlying world model exhibit pragmatic ill-formedness. A natural language system will view these as ill-formed even if a native speaker finds some such utterances perfectly normal. This phenomenon has been termed "pragmatic overshoot" (Weischedel and Sondheimer, 1983).

When such failures occur, a human listener often constructs and responds to a modified query that is well-formed with respect to the listener's world model and is believed to represent the speaker's intent in making the utterance or at least satisfy his perceived needs. Alternative ways of responding depend upon whether the listener believes there is a discrepancy between his beliefs and those of the speaker, the seriousness with which the speaker views those discrepancies, and how much faith the speaker has in the correctness of his own beliefs.

(1) If the listener detects no discrepancies between his beliefs and those of the speaker regarding the world model but merely attributes the erroneous query to an improper formulation of the intended query's English language representation, then the listener may continue the dialogue without notifying the speaker of his language error.

(2) If the discrepancies are important to the speaker's current goals and the listener considers himself an expert in the domain under discussion, he must explicitly correct the speaker's misconceptions. Joshi's extension of Grice's maxim of quality (Joshi, 1983) dictates that a respondent must not say anything which will lead the questioner to make a false inference. Thus the respondent must correct detected misconceptions lest the questioner infer that the respondent's knowledge supports them. The problem of generating such misconception responses is under investigation by McCoy (1985).

(3) If the listener is uncertain of his knowledge in the domain under discussion, he may choose to enter a negotiation dialogue during which the disparate beliefs of the two participants are "squared away".[2]

(4) If the listener does not feel that the speaker's erroneous beliefs are significant enough to warrant correction, he may continue the dialogue without explicitly correcting the speaker's misconceptions. Although Joshi's revised maxim of quality would suggest that all misconceptions should be corrected, Grice's maxim of relevance (Grice, 1975) supports the argument that listeners are not expected to correct minor, unimportant misconceptions. An informal poll of a dozen human subjects indicates that the overwhelming majority would be unhappy with a system that constantly corrected minor errors, both because this correction process would slow down and distract from their information-seeking goals and because it would cause them to concentrate upon greater precision in their queries, at the expense of a truly natural dialogue.

---

[2] The term "square away" is used by Joshi (1983) to describe the process by which dialogue participants resolve discrepancies in their mutual beliefs.

A robust natural language interface should have the ability to handle pragmatically ill-formed utterances much as people do. This would require that they not merely inform the speaker that the query is incorrect but that they attempt to deduce what the speaker meant to say and respond accordingly. Otherwise the communication could not be called "natural" because whenever the speaker's utterance violated the system's view of the world, the system would be unable to understand utterances to which human listeners respond with relative ease.

This paper addresses the problem of handling intensional failures that result from pragmatically ill-formed queries during an information-seeking dialogue. It is concerned with those cases in which a human listener's response would typically address the speaker's intent in making the utterance but would not necessarily explicitly correct the speaker's improper use of language or potential erroneous beliefs.

2. Information-Seeking Dialogues

An information-seeking dialogue generally contains two participants, one seeking information and the other attempting to provide that information. Underlying such a dialogue is a task which the information-seeker wants to perform, often at some time in the future; such tasks include treating a patient (medical information system), pursuing a degree at a university (student advising system), or taking a business trip (travel reservation system). The information-seeker's purpose in engaging in the dialogue is to obtain the information necessary for him to construct a plan for accomplishing his task.

In a cooperative information-seeking dialogue, the information-provider is engaged in helping the information-seeker construct a plan for his underlying task. However, naturally occurring communication is both incomplete and imperfect. Not only does the information-seeker fail to explicitly communicate all aspects of his/her goals and partially constructed plan for achieving these goals, but also the utterances are often imperfectly formulated. They may be ill-formed in the strict syntactic or semantic sense or they may present pragmatic problems in understanding. Such pragmatic difficulties include violation of the listener's view of the world.

However, the information-seeker expects the information-provider to facilitate better and more helpful communication by assimilating the dialogue, inferring from it much of what the information-seeker wants to accomplish, and using these inferences to remedy many of the information-seeker's faulty utterances.

3. What is Pragmatic Overshoot?

A world model is a representation of knowledge about the world, such as entities and their attributes, relationships between entities, and constraints upon these relationships. In order to understand an utterance, a listener must relate it to his world model and interpret it with respect to his beliefs about that world.

However, the speaker's view of the world may differ from that of the listener. As a result, the speaker's utterance may be syntactically and semantically correct yet violate the pragmatic rules of the listener's world model. This phenomenon has been termed "pragmatic overshoot" (Weischedel and Sondheimer, 1983).

Discrepancies of this type may be classified as extensional or intensional failures. An extensional failure occurs when the information-seeker's utterance presumes the existence of data items which, although theoretically possible, do not currently exist in the world model. Extensional failure represents a discrepancy in the beliefs of speaker and listener about the current contents of the database. Thus extensional failures are transient since adding new information to the database may prevent the failure from recurring.

An intensional failure occurs when the information-seeker's utterance presumes a structure for the underlying world model that differs from that of the information-provider. Thus for a given knowledge representation scheme, intensional failures are permanent since no addition of data to the database will prevent them from recurring. This paper is concerned with one subclass of intensional failures, those which specify a relationship that does not exist in the world model. Consider the previous example

"Which apartments are for sale?"

The above query contains the erroneous proposition that there is a sale relation operating upon an entity set of type "Apartment". In a real estate world model, single apartments are rented, but apartment buildings, condominiums, houses, shopping centers, and movie theaters are sold. Even if the system's world knowledge is expanded to include more data or encompass a larger domain, the erroneously presumed relationship will still fail to exist.

4. Review of Related Work

Previous researchers have attempted to treat pragmatically ill-formed utterances in isolation, without recourse to the preceding dialogue context.

Mays (Mays, 1980; Webber & Mays, 1983) proposed a mechanism for detecting the occurrence of pragmatic overshoot by using a robust data model that incorporated a generalization dimension into an entity-relationship model. Generalization arcs could be marked as mutually exclusive; thus the entity sets "Faculty" and "Student" would be mutually exclusive sub-entity sets of the entity set "People" whereas "Men" and "Faculty" would be non-mutually exclusive. This allowed Mays to differentiate between an intensionally valid presupposition such as "a faculty member who is male" and the intensional failure "a faculty member who is a student".

Upon detecting an instance of pragmatic overshoot, Mays suggests providing the information-seeker with related knowledge about the underlying data model, thereby indicating the cause of the failure. For example, if failure is due to the presumption of a nonexistent relationship R between entity sets X and Y, May's response would contain information about each possible relation R between some entity set Z1 and entity set X, each possible relation R between some entity set Z2 and entity set Y, and each possible relation S between entity sets X and Y. However, he does not use a model of whether these possibilities are applicable to the information-seeker's underlying task. In a large world model, such responses will be too lengthy and include too many irrelevant alternatives.

Kaplan (1979), Chang (1978), and Sowa (1976) have investigated pragmatic overshoot resulting from missing joins in the logical representation of the information-seeker's query. Consider, for example, a university world model in which phone numbers are associated with homes, faculty offices, and research labs. Then the query

"What is Dr. Bell's phone number?"

exhibits a missing join since it fails to fully specify the relations to be joined; the above utterance might be an abbreviated version of any of the following queries:

"What is the phone number of Dr. Bell's home?"
"What is the phone number of Dr. Bell's office?"
"What is the phone number of Dr. Bell's research lab?"

Kaplan (1979) proposed using the shortest relational path connecting the entity sets. Chang (1978) proposed an algorithm based on minimal spanning trees, using an *a priori* weighting of the arcs to indicate the likelihood of their intended join. Sowa (1976) defined a formalism called conceptual graphs for describing the semantics of stored data and its interrelationships and suggested using these structures to construct the expanded relation when the speaker's query incompletely specified the necessary relations and join elements.

While each of these strategies produces a join which seems reasonable given the current database, none of these researchers presents a model of whether the proposed path is relevant to the speaker's intentions. Thus each of the models will give unreasonable results in certain discourse situations.

5. Plan-Based Responses to Pragmatic Overshoot

My model for constructing a response to utterances involving pragmatic overshoot is based on the Gricean theory of meaning (Grice, 1969; Grice, 1957). Grice claims that the meaning of an utterance on a particular occasion is the effect that the speaker intends to produce in his listener by means of the listener's recognition of that intent. He furthermore notes that the speaker, in making the utterance, must believe that the listener will be able to deduce these intentions.

According to Grice's theory, a listener must infer the speaker's intent in making an utterance. Now consider an utterance involving pragmatic overshoot. Such an utterance violates the listener's world model. However, the listener, in attempting to deduce the speaker's intended meaning, will be guided by the Gricean maxim of relevance (Grice, 1975). Since the information-seeker is unlikely to flout this maxim in an information-seeking dialogue, the listener will believe, in the absence of contradictory indications such as clue words suggesting topic shift, that the speaker's utterance is relevant to the task at hand. A cooperative information-provider will have used the information exchanged during the dialogue and his knowledge of the domain to hypothesize the information-seeker's goals and plans for achieving these goals. This context model of goals and plans provides clues for interpreting utterances. When pragmatic overshoot occurs, a human listener can often modify the information-seeker's ill-formed query to form a similar query X that is both meaningful and relevant to the current dialogue context. Thus the information-provider may infer that this revised query was the utterance intended by the speaker and respond accordingly.

Analysis of naturally occurring dialogues shows that pragmatically ill-formed utterances are handled with relative ease by human information-providers and that the context within which each utterance occurs affects interpretation. Thus I claim that a natural language system must build a context model representing the information-seeker's plans and goals as inferred from the ongoing dialogue and use this context model to interpret pragmatically ill-formed utterances. Only alternative queries which might represent the information-seeker's intent or at least satisfy his perceived needs should be considered; thus consideration should be limited to queries relevant to instantiating and expanding the partial plan represented in the context model.

My hypothesis is that the context model suggests substitutions for the erroneous proposition causing the pragmatic overshoot; these substitutions produce revised queries, all of which are relevant to the information-seeker's underlying task-related plan and therefore meaningful to the overall dialogue context. In the event that multiple substitutions are suggested, a choice from among these should be made using criteria such as relevance of the revised query to the current focus of attention in the dialogue and semantic similarity of the substitution to the ill-formed proposition employed by the information-seeker.

This paper is mainly concerned with the suggestion mechanism, the component that proposes revisions to the information-seeker's erroneous utterance on the basis of knowledge acquired from the preceding dialogue.

6. Pragmatic Overshoot Processing

6.1 Knowledge Representation

My system for handling utterances involving pragmatic overshoot requires a representation for each of the following:

(1) the relationships among attributes and entity sets in the underlying world model
(2) a generalization hierarchy relating attributes, relations, entity sets, and functions

(3) a context model of the speaker's underlying task-related plan as inferred from the preceding dialogue.

An entity-relationship model states the possible primitive relationships among entity sets. The world model also includes a generalization hierarchy of entity sets, attributes, relations, and functions and specifies the types of attributes and the domains of functions.

The plan construction component is described in Carberry (1983). It requires a representation of the set of domain-dependent plans and goals which an information-seeker might pursue. This component hypothesizes and tracks the changing task-level goals of an information-seeker during the course of a dialogue. My approach is to infer a lower-level task-related goal from the speaker's explicitly communicated goal, relate it to potential higher-level plans, and build the complete plan context as the dialogue progresses.

6.2 The Suggestion Mechanism

It is assumed that a separate mechanism, such as that devised by Mays, detects the occurrence of pragmatic overshoot and identifies the erroneous proposition. This proposition contains either a non-existent attribute or entity relationship or a function applied to an inappropriate domain. The pragmatic overshoot processor must attempt to infer what the speaker intended to communicate in making the ill-formed utterance. The current context model and the possible expansions of its constituent goals and actions suggest relevant queries which the information-seeker might be expected to utter. The suggestion mechanism applies to this model a set of heuristics that propose substitutions for the proposition causing the pragmatic overshoot. Each of these substitutions produces a revised query which is relevant to the current dialogue context and might represent the information-seeker's intent in making the ill-formed utterance.

The suggestion mechanism's heuristics consist of two classes of rules. The first proposes a simple substitution for an attribute, entity set, relation, or function appearing in the erroneous proposition. The second proposes a conjunction of propositions representing an expanded relationship path as a substitution for the proposition specified in the information-seeker's query. These two classes of rules may be used together to propose a substitution consisting of both an expanded relationship path and an attribute or entity set substitution.

6.3 Simple Substitutions

Suppose a student wants to pursue an independent study project in a university world model. Such projects can be directed by full-time faculty but not by faculty who are "extension" or "on sabbatical". The student might erroneously enter the query

"What is the classification of Dr. Bell?"

Only students have a classification attribute in a university world model; this attribute can have values such as Nursing — 1989, Business — 1987, or Arts & Science — 1988 indicating the student's college and anticipated graduation date. Faculty have attributes such as rank, status, title, tenure, and age. Pursuing an independent study project under the direction of Dr. Bell requires that Dr. Bell's status be "full-time" or "part-time". If the information-provider knows the student wants to pursue independent study, then he might infer that the student needs the value of this status attribute and answer the revised query

"What is the status of Dr. Bell?"

Errors such as the above occur when the information-seeker's beliefs about how to reference a particular attribute or entity set differ from that of the information-provider. Such errors occur even more often in spoken dialogues when the information-seeker is distracted and inadvertently uses incorrect terminology. In both cases, the underlying task-related plan inferred from a preceding dialogue appears to provide the primary clues as to the speaker's intentions, with relevance and semantic criteria differentiating among multiple possibilities.

The substitution mechanism contains five simple substitution rules for suggesting revisions of the information-seeker's erroneous proposition. The following is one such exemplary rule:

Rule-S1

If the information-seeker's proposition erroneously presumes that a member Ent1 of entity set Ent-Set has an attribute Att1, then replace attribute Att1 with attribute Att2 if

(1) a proposition specifying attribute Att2 on a member Ent2 of entity set Ent-Set appears in an expansion of the context model
(2) entities Ent1 and Ent2 unify; (one is a variable or both are the same constant)

The context model represents the speaker's partially constructed plan as inferred from the preceding dialogue. This plan can be expanded by substituting a plan for each constituent goal and action. Since each plan itself contains constituent goals and actions, this expansion can be continued to any desired degree of detail, until eventually the plan contains only primitive actions.

Intuitively, the information provider anticipates that the information-seeker might need to know each attribute value in an expansion of the goals and actions in the plan. Suppose this inferred plan contains an attribute Att2 of a member of entity set Ent-Set, namely

$$Att2 (Ent2:\&ENT\text{-}SET, att\text{-}val:\&ATT2\text{-}DOMAIN)$$

and that the information-seeker requests the value of attribute Att1 for a member Ent1 of entity set Ent-Set. Then if Ent1 and Ent2 unify (one is a variable or both are the same constant), a cooperative listener might infer that the information-seeker wants the value of Att1 for the entity Ent1 specified in his utterance, especially if attributes Att1 and Att2 are semantically similar.

The substitution mechanism searches an expansion of the information-seeker's inferred plan for propositions whose arguments unify with the arguments in the erroneous proposition causing the pragmatic overshoot. RULE-S1 then suggests substituting the attribute from the plan's proposition for the attribute specified in the information-seeker's query. This substitution produces a query relevant to the current dialogue and may capture the speaker's intent or at least satisfy his needs.

Consider for example the query

"What is the area of the special weapons magazine of the Alamo?"

which appears in a dialogue transcript of an information-seeker attempting to load cargo onto ships using the REL natural language interface (Thompson, 1980). The semantic representation of this query contains the erroneous proposition

$$Area(SPEC\text{-}WEAPONS\text{-}MAGAZINE, \_areaval:\&SQUARE\text{-}FEET).$$

The special weapons magazine is a member of the entity set STORAGE-AREA; members of this entity set have attributes such as Remaining-Capacity and Total-Capacity, but no Area attribute. A plan for loading an item of cargo into a storage area of a ship would contain the precondition

$$Greater (\_rem\text{-}cap:\&CUBIC\text{-}FEET, \_item\text{-}sz:\&CUBIC\text{-}FEET)$$
$$\text{such that}$$
$$Volume(\_item:\&CARGO, \_item\text{-}sz:\&CUBIC\text{-}FEET)$$
$$Remaining\text{-}Capacity(\_stor\text{-}area:\&STORAGE\text{-}AREA, \_rem\text{-}cap:\&CUBIC\text{-}FEET)$$

specifying that the volume of the item to be loaded must be less than the remaining capacity of the storage area under consideration. If the preceding dialogue indicates that the information-seeker wants to load an item of cargo onto the Alamo, then the substitution mechanism would suggest substituting the attribute Remaining-Capacity for the attribute Area in the information-seeker's erroneous proposition; note that this also results in substituting for the variable _areaval a variable of type &CUBIC-FEET since this parameter represents the value of the particular attribute applied to the entity in question.

6.4 Expanded Path Substitutions

Suppose a student wants to contact Dr. Bell to discuss the appropriate background for a new seminar. Then the student might utter the query

"What is Dr. Bell's phone number?"

Phone numbers are associated with homes, faculty offices, and research labs. Course discussions with professors may be handled in person or by phone; contacting a professor by phone requires that the student dial the phone number of Dr. Bell's office. Thus the listener might infer that the student needs the phone number of the office occupied by Dr. Bell.

Although the above query appears quite natural, it incompletely specifies the desired path between entities in the world model; this was referred to earlier as the "missing joins" problem. One cause of such incompleteness appears to be a desire on the part of the information-seeker to use abbreviated, terse queries under the assumption that the existing context will insure proper interpretation. Carbonell (1983) demonstrated this phenomenon in an experiment during which subjects were found to persist in using abbreviated statements and queries, even in the presence of explicit and repeated instructions to adhere to syntactically and semantically complete sentences. A second cause of "missing joins" in the semantic representation of natural langauge queries is a difference in the world model structures presumed by information-seeker and information-provider.

Analysis of naturally occurring dialogues indicates that utterances with "missing joins" occur frequently and are generally interpreted correctly by human listeners. The underlying task-related plan inferred from the preceding dialogue appears to provide significant clues as to the speaker's intentions by indicating those paths in the world model that are relevant to the current context.

The substitution mechanism contains two expanded path rules for handling missing logical joins in the semantic representation of the information-seeker's query. These rules apply when the attributes and entity sets are not directly related by the relationship $R1u$ specified by the information-seeker — but there is a path R in the entity relationship model between the attribute and entity set or between the entity sets. This is referred to as path expansion since by finding the missing joins, an expanded relational path is constructed. The following is an exemplary path expansion rule:

Rule-S2

Suppose the information-seeker's proposition erroneously presumes a relation $R1u$ between a member of entity set Ent-SetA and a member of entity set Ent-SetB,
$\qquad$ R1u(Enta:&ENT-SETZ, Entb:&ENT-SETB)
If
(1) a path in an expansion of the information-seeker's inferred plan contains the propositions
$\qquad$ R0(x : &ENT-SETA, y1 : &ENT-SET1)
$\qquad$ R1(y1 : &ENT-SET1, y2 : &ENT-SET2)
$\qquad$ R2(y2 : &ENT-SET2, y3 : &ENT-SET3)
$\qquad\qquad$ . . . .
$\qquad\qquad$ . . . .
$\qquad\qquad$ . . . .
$\qquad$ Rn(yn : &ENT-SETn, z : &ENT-SETB)

(2) R1u is one of the relations R0, R1, ..., Rn

(3) the terms Enta and x and the terms Entb and z unify,

then replace the erroneous proposition
        R1u(Enta : &ENT-SETA, Entb : &ENT-SETB)
with the expanded path
        R0(Enta : &ENT-SETA, y1 : &ENT-SET1)
        R1(y1 : &ENT-SET1, y2 : &ENT-SET2)
        R2(y2 : &ENT-SET2, y3 : &ENT-SET3)
        . . . . .
        . . .
        Rn(yn : &ENT-SETn, Entb : &ENT-SETB).

The entities in a model of a task-related plan are determined in part by their relationship to other entities. Since the information-seeker is attempting to instantiate and expand his partially constructed plan, the information-provider anticipates that the information-seeker might need to know which entities are related according to the relationships appearing in an expansion of the goals and actions in this plan. Suppose this inferred plan includes a composition of relations R1, R2, and R3 relating members of entity sets Entity-SetA and Entity-SetB. If the information-seeker erroneously requests those members of Entity-SetA that are related by R1 (or alternatively R2 or R3), to members of Entity-SetB, then a cooperative listener might infer that the information-seeker wants the members of Entity-SetA that are related by the composite relation

$$R1 * R2 * R3$$

to a member of Entity-SetB.

The substitution mechanism searches an expansion of the information-seeker's inferred plan for a single path containing a sequence of propositions connecting the entity sets presumed related by the relation in the information-seeker's erroneous proposition. RULE-S2 suggests substituting the composite relation represented by the sequence of propositions appearing in the information-seeker's inferred plan for the erroneously specified relation specified in the information-seeker's query, thus constructing an expanded path. This substitution produces a query that is relevant to the overall task which the information-seeker is pursuing and may capture the information-seeker's intended meaning.

For example, consider the pragmatically ill-formed query

"Which faculty does Data-Logic give money to?"

In a University world model, faculty are not given money by industries; faculty do consult for industries and earn consulting fees, and research projects have government or industrial money supporting them. Therefore the above query contains the erroneous proposition

Gives-Funds(DATA-LOGIC,_fac1:&FACULTY)

If the preceding dialogue indicates that the information-seeker is evaluating individual faculty contributions to the university, then Figure 1 presents a portion of this inferred task-related plan.

Upon analyzing this plan, the substitution mechanism finds that the propositions

    Principal-Investigator(_proj:&RESEARCH-PROJECT,_fac:&FACULTY)
    Gives-Funds(_cmp1:&COMPANY,_proj:&RESEARCH-PROJECT)

appear along a single path and that the terms _fac1:&FACULTY and DATA-LOGIC from the

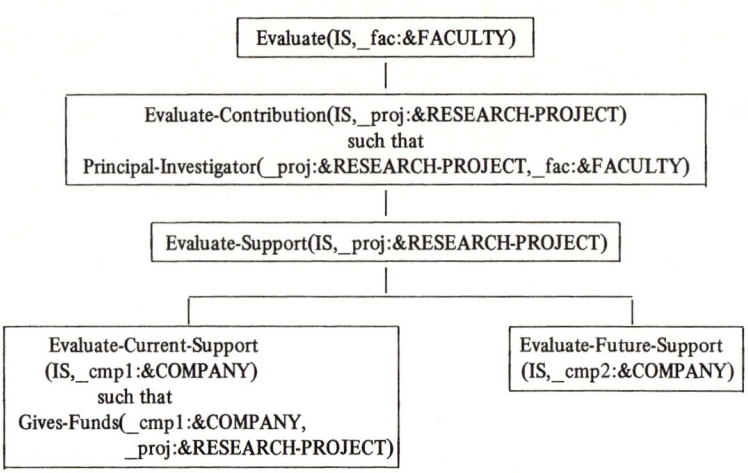

Figure 1. A Portion of a Plan for Evaluating Faculty

information-seeker's erroneous proposition unify with the terms _fac:&FACULTY and _cmp1: &COMPANY respectively from the above propositions. Thus RULE-S2 suggests replacing the erroneous proposition appearing in the information-seeker's query with the expanded path represented by the conjunction of the propositions

>Gives-Funds(DATA-LOGIC, _proj:&RESEARCH-PROJECT)
>Principal-Investigator(_proj:&RESEARCH-PROJECT, _fac1:&FACULTY)

This revised query is semantically equivalent to a request for the faculty who are the principal investigators of projects supported by DATA-LOGIC.

To limit the components of path expressions to those relations which can be meaningfully combined in a given context, RULE-S2 makes a strong assumption: that the relations comprising the relevant expansion appear on a single path within the context model representing the information-seeker's inferred plan. Further research is needed to identify the conditions under which this restriction can be relaxed.

6.5 Selection Mechanism

The substitution and path expansion rules propose substitutions for the erroneous proposition that caused the pragmatic overshoot. Each substitution produces a revised query that is relevant to the information-seeker's overall task. If only one revised query is suggested, then the information-provider is justified in believing that it is the utterance most likely to represent the information-seeker's intentions, since it is the only suggested variant of the information-seeker's query relevant to his inferred task-related plan.

However, if multiple revised queries are suggested, then the information-provider must decide if any of these is significantly more likely than the others to represent the information-seeker's intended utterance. Two criteria appear important in selecting the most appropriate interpretation. The first is relevance to the current focus of attention in the information-seeker's task-related plan.

Grosz (1977) introduced the concept of focusing in her work on determining the referents of definite noun phrases and Sidner (1981) extended this in her study of anaphora resolution. McKeown (1982) built upon Sidner's work in her investigation of focusing constraints for natural language generation. She claims that when faced with a choice of whether to move to a new topic or continue with a recently introduced topic, the speaker will remain in the recently introduced topic if he has something further to say about it; otherwise the speaker must reintroduce this topic at a later time. In addition, a speaker will choose to continue with the current topic before switching back to a previous one.

These considerations lead to the working assumption that the information-seeker is most likely to continue with aspects of the current focused task and the most recently considered subaction of this task before considering other subtasks of the overall plan. Thus the more relevant a revised query to that aspect of the task upon which the information-seeker's attention is currently focused, the more anticipated is that query at this point in the dialogue.

The second criterion is similarity of the revised query and the infromation-seeker's utterance. The more the information-seeker's utterance is altered, the less closely it resembles what he actually said, and therefore the less likely it is to represent what he meant to say. The revision operations alter the information-seeker's actual utterance in two ways:

(1) by expanding the relational path
(2) by substituting a new term for that employed by the information-seeker

Thus two metrics are relevant to measuring the semantic difference between the revised query and that uttered by the information-seeker:

(1) the length of the path expansion
(2) the semantic resemblance between a substituted term and that employed by the information-seeker

These two metrics will not be described in this paper, except to say that a generalization hierarchy is used to semantically compare substitutions with the items for which they are substituted; details can be found in (Carberry, 1985).

The criteria of relevance and semantic difference are used to select the revised query most likely to represent the information-seeker's intent in making the utterance. The selection algorithm attempts to weigh relevance against semantics. It employs an evaluation function

$$E(\text{revised-query}) = w1 * \text{FocusShift}(\text{revised-query}) + w2 * \text{SemanticDifference}(\text{revised-query})$$

to evaluate each suggested revised query, where FocusShift(revised-query) measures how much the revised query would change the current focus of attention in the dialogue and SemanticDifference(revised-query) measures the semantic difference between the revised query and the information-seeker's actual utterance. Relevance to the current focus of attention in the dialogue seems to be more significant than semantic difference since highly relevant queries are strongly anticipated by the listener. Thus $w1$ should exceed $w2$. However, precisely how much semantic difference is required before a less relevant revised query is preferred over a more relevant but semantically less similar one and precisely how less relevant the query may be, are both matters that require further investigation. The relative values of the weights $w1$ and $w2$ determine the emphasis upon relevance versus semantics.

Since this evaluation function is imperfect, two suggested revisions with similar, although unequal, evaluations must be regarded as equally plausible. Therefore the selection mechanism evaluates each suggested revised query and selects as appropriate interpretations those whose evaluation is within an arbitrarily set tolerance of the suggestion whose evaluation is smallest.

7. Example

The following example illustrates the suggestion and selection strategies and demonstrates how the information-seeker's underlying task, as inferred from the preceding dialogue, influences interpretation.

> IS: "I am the manager of a real estate investment trust."
> "We'd like to invest between 30 and 50 million dollars."
> "Which apartments are for sale?"

The last utterance in the above dialogue contains the erroneous proposition

Sale-Status(_x:&APARTMENT, FOR-SALE)

The preceding dialogue indicates that IS, the information-seeker, represents a real estate investment trust interested in expanding its holdings by a large sum of money. Figure 2 presents a portion of an expanded plan for achieving this goal. The substitution mechanism proposes substituting the entity set APART-COMPLX for the entity set APARTMENT appearing in the information-seeker's erroneous proposition, producing a semantic representation equivalent to the query

"Which apartment complexes are for sale?"

The substitution mechanism will also propose substituting the entity sets SHOP-CENTER and OFFICE-PLAZA for the entity set APARTMENT; other substitutions will also be suggested by components of the plan for the information-seeker; in other contexts, such as the example at the beginning of the chapter, they might be suggested. Thus the suggestion mechanism captures the perspective from which the information-seeker has made his utterance.

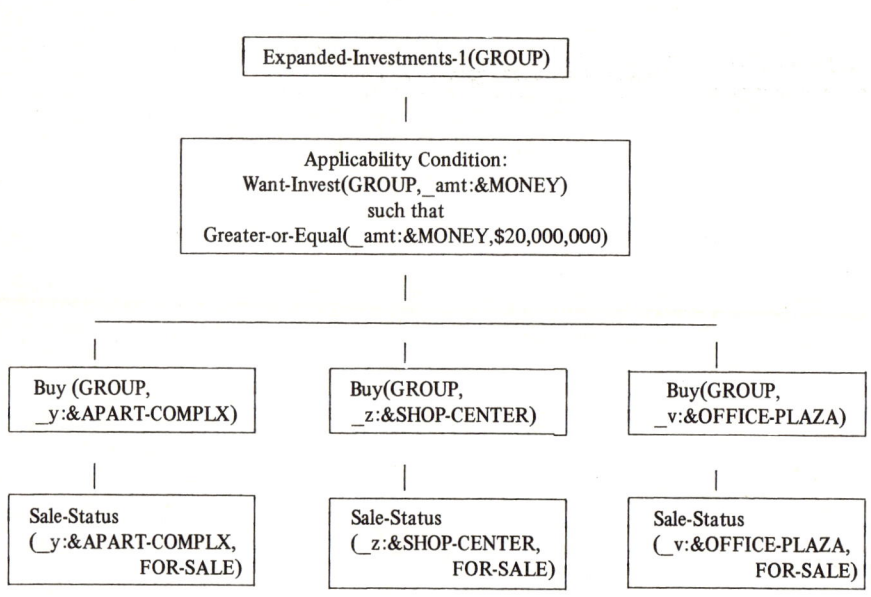

Figure 2. Portion of an Expanded Plan for Investments

The selection mechanism must evaluate these proposed revised queries. Its metrics find that the revised query resulting from the substitution of APART-COMPLX for the entity set APART-MENT is highly relevant to the current focus of attention in the underlying plan and that APART-COMPLX and APARTMENT are semantically very similar. The other suggestions are not evaluated nearly as highly and thus the selection mechanism selects the revised query equivalent to

"Which apartment complexes are for sale?"

as the most appropriate interpretation of the ill-formed utterance.

8. Conclusions

If we want systems that engage in truly natural dialogue, then they must be able to understand pragmatically ill-formed queries that are handled easily by human listeners. Such understanding requires the use of knowledge gleaned from assimilations of the preceding dialogue. Previous strategies for handling pragmatic overshoot analyzed the utterance in isolation from the dialogue context and thus were unable to model whether the proposed response was relevant to the speaker's intentions.

This paper has argued that the information-seeker's underlying task-related plan as inferred from the preceding dialogue should be the primary mechanism for suggesting potential interpretations of such queries, since this plan captures the perspective from which the user has made the ill-formed utterance. The methodology proposed in this paper is based on the hypothesis that the information-seeker's inferred task-related plan, represented by a context model, suggests a substitution for the proposition causing the pragmatic overshoot and that such suggestions then must be evaluated on the basis of relevance and semantic criteria. This approach is superior to previous strategies because it uses a model of the established dialogue context to identify and address the speaker's perceived intentions in making the utterance.

References

Carberry, M.S. (1985). Pragmatic modeling in information system interfaces. Ph.D. Dissertation, Dept. of Computer and Information Sciences, University of Delaware, Newark, Delaware.
Carberry, M.S. (1983). Tracking user goals in an information-seeking dialogue. In *Proceedings of the National Conference on Artificial Intelligence.*
Carbonell, J.G. (1983). Discourse pragmatics and ellipsis resolution in task-oriented natural language interfaces. In *Proceedings of the 21st Annual Meeting of the Association for Computational Linguistics.*
Chang, C.L. (1978). Finding missing joins for incomplete queries in relational data bases. Technical Report RJ2145, IBM Research Laboratory.
Grice, H.P. (1975). Logic and conversation. In P. Cole and J.L. Morgan (Eds.), *Syntax and semantics: Speech acts, Volume 3.* New York: Academic Press.
Grice, H.P. (1969). Utterer's meaning and intentions. *Philosophical Review, 68,* 147-177.
Grice, H.P. (1957). Meaning. *Philosophical Review, 56,* 377-388
Grosz, B.J. (1977). The representation and use of focus in a system for understanding dialogs. In *Proceedings of the International Joint Conference on Artificial Intelligence.*
Harris, L.R. (1977). User oriented data base query with the ROBOT natural language query system. *International Journal for Man-Machine Studies, 9,* 697-713.
Joshi, A.K. (1983). Mutual beliefs in question-answer systems. In N. Smith (Ed.), *Mutual beliefs.* New York: Academic Press.
Kaplan, S.J. (1979). Cooperative responses from a portable natural language data base query system. Ph.D. Dissertation. Department of Computer Science, University of Pennsylvania, Philadelphia, Pennsylvania.
Mays, E. (1980). Failures in natural language systems: Applications to data base query systems. In *Proceedings of the National Conference on Artificial Intelligence.*
McCoy, K.F. (1985). The role of perspective in responding to property misconceptions. In *Proceedings of the International Joint Conference on Artificial Intelligence.*
McKeown, K.R. (1982). The TEXT system for natural language generation. In *Proceedings of the 20th Annual Meeting of the Association for Computational Linguistics.*

Sidner, C.L. (1981). Focusing for interpretation of pronouns. *American Journal of Computational Linguistics, 7*, 217-231.

Sowa, J.F. (1976). Conceptual graphs for a data base interface. *IBM Journal of Research and Development,* 336-357.

Thompson, B.H. (1980). Linguistic analysis of natural language communication with computers. In *Proceedings of International Joint Conference on Artificial Intelligence.*

Webber, B.L., & Mays, E. (1983). Varieties of user misconceptions: Detection and correction. In *Proceedings of International Joint Conference on Artificial Intelligence.*

Weischedel, R.M., & Sondheimer, N.K. (1983). Meta-rules as a basis for processing ill-formed input. *American Journal of Computational Linguistics, 9.*

UNDERSTANDING BY EXPLAINING EXPECTATION FAILURES

Roger C. Schank and Christopher C. Owens

Yale University

> Our original models of natural language processing by computer dealt primarily with issues of how to represent meaning in the machine and how to extract meanings from text. Today, our work is concentrated primarily on how people learn, how they organize their memories, and how they recover from failed expectations. This chapter explains this change of emphasis by focusing on something which is or should be at the center of Artificial Intelligence — the role of explanation in predictive understanding.

1. Introduction

An exercise we at Yale often employ when teaching a class about the task of natural language understanding is to read a newspaper story aloud, stopping after each word and asking the students in the class to try to predict what will come next. In fact the reading aloud part is irrelevant to the exercise; anyone can try this on his own by covering the printed text of a newspaper story with a piece of blank paper and uncovering it one word at a time.

The reason that this is such an important exercise is because it dramatically focuses our attention upon certain aspects of the text understanding process that would not otherwise be available to introspective examination. It forces us to pay attention to certain kinds of processing and certain kinds of knowledge that are used in text understanding. In particular, this exercise teaches us four lessons that are of importance in the context of this chapter.

The first lesson from the exercise is that we can and actually do make predictions about what is likely to follow a given text fragment. Often we can do so with surprising accuracy. Obviously this depends to a great deal on what kind of story we are reading and how far into it we have read at the time we start trying to make predictions. After hearing only one or two words of a story we remain at a loss to predict what will come next; once we have read further into the story the range of predictions we can comfortably make becomes narrower and narrower. We only make predictions that would make sense as continuations of the story we have read so far.

A tempting idea is to say that these predictions are based on some implicit knowledge we have of language use patterns. The idea of transitional probabilities in sentences partially describes this. The words: "Hammered out an ..." are almost always followed by "agreement", so if we have in our heads a notion of what words are likely to follow others, when we see the phrase we can predict its conclusion with good accuracy. But this misses a great deal of the point we wish to make about sentence understanding. Transitional probabilities are merely a description of what language patterns are actually used in the sentences we utter; they are not a theory of how we can understand sentences.

We learn a second lesson from our exercise by observing that the predictions are much easier to make at certain points during the text than at others, and that this is more than just a function of how much of the story has been read. The transitional probability argument of the preceding paragraph might be invoked again here to claim that cases in which one word has a much higher transitional probability than do any others are easy to predict while cases in which there are a large number of choices with approximately equal transitional probabilities will be much harder to predict. But that just doesn't say anything about what kind of process might account for these differences. Clearly, making predictions about what will follow depends to a great extent upon how well we understand what the sentence is about. Consider the sentence fragment:

... the driver of the second vehicle was taken by ambulance to St. Raphael's Hospital, where doctors determined that he was ...

The range of predictions we can make here is limited not only because of the way language is used, not only because we have some knowledge about what kinds of words are likely to follow but instead because we know what the story is about and because we have some knowledge about what kinds of ideas are likely to follow as well.

On the other hand, the sentence fragment:

... At a news conference yesterday afternoon at the White House President Reagan announced that he would soon ...

gives us much more room to make a variety of predictions because our idea of what the story is about is in some sense much more vague than was our idea about what the first story is about. Although it may be more difficult to predict the conclusion of the sentence in this case than it was for the first example, it is nevertheless evident that the variety of predictions we are able to make is constrained by what little we do know about the story thus far and by what we know about what is going on in the real world.

The point of this second lesson is that the predictions we make when reading a story one word at a time come from a partial model we have built about what the story is about. We make predictions based upon the model, and we only make predictions that would make sense in the context of that model as possible conclusions to the story.

Our third relevant lesson from the exercise is that many different kinds of knowledge must be used to successfully make predictions in the course of reading and understanding a piece of text. Real-world knowledge, like the kind we have been discussing above, is not alone sufficient to make these kinds of predictions. We are emphatically not claiming that there is no purely linguistic knowledge involved in text understanding. Our point is that several kinds of knowledge, traditionally viewed as separate and distinct mental entities, are in fact integrated in the text understanding process. Neither knowledge about what happens in the real world, nor knowledge about which kinds of sentence — use patterns are likely and which are not, nor knowledge about what kinds of facts people are likely to put into stories is alone sufficient to account for all the predictions we might make. We must combine all these kinds of knowledge and apply them all to the task.

In Schank and Birnbaum (1980) we discuss the use of these different types of knowledge to understand the sentence fragment:

Yesterday seven Libyan gunmen shot their way into ...

We would be much more likely to predict, for example:

... a government building.

than we would:

... a paper bag.

What allows us to be confident about the likelihood of the former sentence compared with the latter is our knowledge about gunmen and their goals, their plans and their typical actions, our knowledge about government buildings and paper bags, our knowledge about how sentences might be put together, our knowledge about which types of stories are likely to be found in newspapers and which are not, and our knowledge of what goals someone might have in telling this story. All of these kinds of knowledge must be used, and any process model which attempts to account for story understanding must be able to use all of them.

This idea about the role of knowledge in understanding has had a profound effect upon the nature of our work in Artificial Intelligence. If we have succeeded in convincing ourselves that these kinds of knowledge are in fact present in people and that they are necessarily used quite early in the understanding process, then we must first examine some more general problems. We must discover how this knowledge might be acquired, how it might be represented for use by the various processing elements of an intelligent system, how it can be organized, stored in a memory, indexed, accessed, retrieved and applied to the understanding process. Later in this chapter we will discuss the history of some of our attempts to answer these questions.

The final observation from our exercise is an obvious one, but one that is essential to the development of our ideas about the text understanding process. It is that our predictions are often wrong. At first this hardly seems important: no understanding process is going to be perfect. If we could always correctly predict the remainder of a sentence, story or paragraph then books would be a great deal shorter than they are. The relevant questions here are: how do the situations in which predictions are correct differ from the situations in which they are wrong and, how do each of those situations relate to the understanding process? What exactly happens to us when we have built up certain expectations about what will follow in a piece of text and those expectations are violated?

2. Expectation Failure

The general idea of failure-based understanding is that examining how we make comparisons between our expectations and what actually occurs is the key to our knowledge of the understanding process itself. Expectations that turn out to be correct are in some sense uninteresting; all we do when an expectation is correct is to continue reading. It is the failed expectations that control the understanding process. An expectation failure causes us to reconsider our partial model of the story, to modify it slightly in order to make it compatible with the new input responsible for the failed expectation, or even to discard the partial model completely in favor of some other model that fits the new input better. As we discuss various types of memory organization strategies and their applicability to understanding, we will consider how each deals with recovery from expectation failure and how each uses the information gained from failures to direct the understanding process.

Consider again the exercise described in the introduction. Is reading a story one word at a time and stopping to make predictions in any way comparable to real understanding tasks? This exercise may not be as artificial as it seems: deliberate prediction and substantiation is far more than just an interesting way to see what kinds of knowledge structures we might have in our minds. The process closely mimics something that is happening during real understanding tasks. People really do build up a partial model of a story as they read or hear it; they use the partial model to set up expectations as to what will follow; they use the cases where their expectations are confirmed to strengthen their model; they use the cases where the expectations fail to be confirmed to modify their model or to cast their model out in favor of another. It is very difficult to account for the way people behave on real understanding tasks without resorting to something like this process.

There are of course differences between reading one word at a time and real reading tasks. We do not deny that the long pauses present in the former example gives us the chance to make our predictions conscious and explicit, whereas they would not be so during the course of normal reading. Furthermore, the long pauses give us a chance to make many more predictions than we would in a real understanding task, thereby complicating the job of picking among competing candidate predictions. The slow process used in the exercise biases us away from a shallow understanding mode and towards deeper understanding.

None of these differences are relevant to our tasks here: to demonstrate the functionality of expectations and the expectation failures and to explore the ways memory can be organized to support the processes of expectation generation, prediction and substantiation, and model building and modification in response to input. Expectations and expectation failures control the processes of understanding and learning.

So if we do really use a method of setting up expectations and testing them when we are engaged in an understanding task, it pays to consider why this is the case. What can a system that generates expectations and tests them do that we cannot do otherwise?

The answer is that expectations provide a reduced set of possibilities against which to test input. A system that did not use this kind of prediction/substantiation mechanism would have to decide, for each new fragment of input, which of all the structures in its entire memory could account for the input. But we want to be able to do fast disambiguation of ambiguous episodes without recourse to exhaustive analysis. Setting up some small number of expectations accomplishes this: a piece of input text might, under an expectation-based system, only be compared against a few memory structures: those that were suggested as possible candidates by an expectation-generation process. Since we are looking for the best match among a few memory structures rather than among all structures in the entire memory, disambiguation can proceed much more quickly. This captures our intuition that "to see something, one must be looking for it."

3. Static Memory Structures

Our first attempts at natural language understanding systems contained parsers whose job it was to translate text into an internal representation of the meaning of that text, and other modules whose job it was to operate upon that representation in order to perform the other kinds of tasks associated with understanding (like the application of knowledge structures to the input representation in an attempt to answer questions or summarize stories). Examples of these parsers were ELI (Schank, 1975) and the Conceptual Analyzer (Riesbeck, 1975; Birnbaum & Selfridge, 1979).

The representation we used was Conceptual Dependency (Schank, 1972). In these parsers, expectations were used to process the incoming text; the expectations came from demons attached to particular words. A demon is a piece of procedural code whose function it is to set up expectations for other kinds of words or concepts and to build memory structures based upon the presence or absence of those other words or concepts. Demons could handle limited forms of lexical ambiguity and could, also to a limited extent, deal with failed expectations.

A typical demon might be the one for the word "gave" (presented here in a highly reduced form). It could have rules like:

Look to the left for a human.
Look to the right for a physical object.
   If you find one, build an ATRANS (transfer of possession) concept
      and fill in the human as the actor and the object as the object.
   If you find the word "to", look to its right for a human. Fill in that
      object as the recipient of the ATRANS. ("John gave a book to Mary")
Or, if, to the right of the word "gave" you find the name of a person and
   the name of an object in sequence, fill in the person as the recipient and
   the object as the object. ("John gave Mary a book").
But, if you find that you are trying to fill the object slot with an action,
   e.g., "a kiss", then throw out the ATRANS, because this is not a transfer
   of possession at all. Build a structure representing kissing instead.

The problems rapidly become obvious. Tremendous amounts of knowledge must be encoded procedurally and in a non-perspicuous fashion. Furthermore, extensibility becomes nearly impossible due to the interaction of demons for various words. Also, although some kinds of expectations are being set up and subsequently used, there is a large amount of world knowledge involved in making our predictions that intuitively we don't want to see represented at the lexical level. Consider:

John was really upset at the way Mary had botched the sales meeting. When they got back to the office he let her have it.

versus

John knew that Mary deserved to receive the sales achievement award. When they got back to the office he let her have it.

In the first case "let her have it" is preferred in its colloquial sense meaning to berate strenuously. In the second case the more compositional meaning of the phrase is preferred. We obviously know this because of what has come before, but trying to write the correct demons to perform this kind of disambiguation is an unimaginably difficult task. What we need is a kind of knowledge structure to provide expectations at a more episodic level. If, in the first case, we know enough to be expecting some kind of angry criticism, then the phrase "let her have it" will be recognized as satisfying the expectation. Similarly in the second case if we are expecting the presentation of a trophy, the same phrase will satisfy the expectation as well. Memory and knowledge about real-world events simply must be activated and allowed to provide this kind of expectation, while no amount of manipulating the code for word-based demons will ever be particularly convincing at this task.

If we had somehow known how to activate the correct knowledge structure about patterns of behavior in either of the two cases above, and if the knowledge structure could drive the language understanding part of the system, then the expectation of either a criticism or an award could come from the knowledge structure and that would handle the ambituity of the phrase "let her have it".

One of our early knowledge structures proposed to address the problem of allowing memory to provide some top-down control over the understanding process was the script (Schank & Abelson, 1977). Scripts were an attempt to capture the fact that for a great many activities, specific events tend to occur in a fairly standard temporal sequence. Scripts were designed to provide a story understander with a framework of expectations relative to these kinds of stereotypical activities such as eating a meal in a restaurant, buying something, travelling someplace by public transportation and the like.

Theoretically, a script represented a fossilized set of inferences of the kind possibly made by a planner: one idea behind scripts was that the connection between one event in a script and the next, such as between going to a restaurant and sitting down to eat, could be recovered by a plan and goal understander. Going to a restaurant, for example, satisfies the goal of being there, which is itself a necessary precondition to eating a meal there, which in turn satisfies the goal of alleviating hunger. A plan/goal analyzer has enough knowledge at its disposal to be able to understand this type of event sequence without resource to scripts. What it lacks, however, is any mechanism to organize this information so that it is usable in practical situations. The problem with getting a system to understand a story using plans and goals alone was one of combinatorial explosion: an understander would quickly bog down trying to chain together all its atomic inference rules in search of an overall pattern. Within the example of a story about going to a restaurant, for example, the first step might be going to one's car. There are a great number of plans we can infer from someone going to their car, and without some overall framework to impose expectations on the understanding process we would be forced to explore all the possible things that might be going on. Scripts narrowed this range of expectations. With the restaurant script active, we make only certain permitted inferences. If a person is getting into his or her car we assume that it is because that person wants to go to the restaurant.

In fact the plan/goal connection between the various events in a script sequence was never explicitly represented within the script structure. All that a script-based understander could do was, given some part of a script sequence, predict the remainder. A script-based understander could also fill in missing input: if we have heard that someone went to a restaurant and that, upon leaving, he left a big tip, we can assume that he ate his dinner.

So scripts provide expectations by virtue of stereotyped temporal sequence. Expectations can succeed, in which case we assume any elided intervening steps. When expectations fail we do not do anything interesting. Anomalous input is in fact ignored. The problem of changing one's mind about which script one is in is ignored as well.

For example, if we learn that John went into a restaurant, asked a waiter for help changing a flat tire outside, received the help, gave the waiter a tip and then left, we don't want to assume that John ate a meal. Script-based understanders did not have a principled way to determine whether this is a story about fixing flat tires or about eating meals in restaurants. The information about flat tires will simply be ignored by a script based understander when in fact it should clue us in to the fact that we are not dealing with a restaurant story at all but with something else altogether.

Furthermore, scripts are rather large and unwieldly. Representing any reasonable level of detail seems to require an unreasonable level of specificity. Variations on a theme, such as eating in a sit-down restaurant versus eating in a fast-food restaurant are difficult to handle. Tracks within a script were proposed to handle variations, but they failed to fully capture the shared knowledge underlying the multiple tracks of a script.

The problems with these systems center around the fact that they could not easily be extended. Like parsers based upon lexical demons, they became unwieldly when programmed to handle many variations or exceptions; they break down under the weight of added information and added specificity.

The questions of learning and generalization are somewhat moot relative to a script-based understander because there is no good representation of the story in memory after it has been read. We simply pay attention to the features that match the script and ignore everything else. All we can say after having read the story is: "another instance of the restaurant script." It is not clear from these systems how one would in practice learn a script from experience. Script-based understanders relied upon hand-coded scripts; they could not generalize new scripts based upon their input stories.

4. Dynamic Memory Structures

Fitting a piece of text into the appropriate knowledge structure is not all there is to understanding. Where, for example, does one draw the line between understanding the meaning of a piece of text, incorporating that meaning into one's memory or acting upon that meaning. For a somewhat frivolous example, if you told someone that the building in which he was standing was on fire, you would in some sense not consider him to have understood the sentence until you saw him leave the building. In Schank (1982), we pointed out that if human readers are repeatedly given the same story to read, they will not only recognize it as the same story, they will eventually get bored of having to read it and will complain. When they are presented with unexpected facts or unexpected outcomes, human readers become surprised and devote greater effort to the task of trying to understand the unusual facts. People spontaneously notice thematic similarities between different stories. They are able to be reminded of one event by another, to tell the difference between stories that are variations on a common theme and stories which differ in a more fundamental way, and to form reasonable abstract generalizations of a story. These are all aspects of behavior which we should consider as part of the larger context of understanding.

This kind of behavior requires that a program's knowledge or memory somehow change in response to what it has read. This is the point made in Schank (1982). This kind of change is nothing other than a realistic memory, and it is essential to support the varieties of realistic text understanding we are talking about. A program understands a piece of text at a reasonably deep level if it can change its memory as a result of having read the text. It might find a memory structure that covers the text perfectly and simply add the episode as another example, it might modify an existing memory structure to cover the new input, it might try to see the input in a different light to fit an existing memory structure, or it might make up a new memory structure from scratch.

The theory of Dynamic Memory (Schank, 1982) was developed in response to some of the shortcomings of understanding systems based upon scripts and other static memory structures. Our results from trying to program systems based on scripts, as well as important psychological evidence (Bower, Black, & Turner, 1979) indicated a more modular, hierarchically organized memory was necessary to account for the types of understanding that people do. Among other new structures, Dynamic Memory introduced Memory Organization Packages (MOPs) to answer some of these problems. MOPs also began to address some of the issues of generalization and learning that we discussed above.

Briefly, a MOP organizes pieces of information together in a fashion not dissimilar to a script, except with greater modularity, greater flexibility and with more reliance on hierarchical memory. MOP memory has Packaging and Abstraction relationships. Scenes are packaged together into MOPs; while each scene itself is a specialization of more general classes of scenes and an abstraction of less general classes.

The scenes within a MOP are themselves memory structures, which exist within the abstraction net in memory independent of the MOP itself. In the restaurant MOP, for example, the read-menu scene is marked as being a specialization of the more general get-info. Likewise, the restaurant MOP itself is a specialization of a more general MOP covering the purchase of some consumable item: with scenes roughly corresponding to choosing, requesting, receiving, consuming, paying. Examples of systems based upon this theory were IPP (Lebowitz, 1980), Cyrus (Kolodner, 1980), Boris (Dyer, 1983) and MOPtrans (Lytinen, 1984).

MOPs are amenable to a more principled treatment of expectation failures than are scripts. With MOPs, failed expectations can serve as indices from one MOP to another. For example, within the Restaurant MOP, if no headwaiter appears to greet us when we arrive, we can use that failure as a link to the Informal Restaurant MOP, which will from now on guide our expectations as we continue to process the episode. This is superior to choosing a track in a script because informal-restaurant is not a totally different structure than restaurant; it contains only the differences.

In theory, MOPs also had a way of dealing with anomalous input for which there was no particular failure indexed. MOPs theoretically allow for the processing of anomalous input by the means of causal equivalency substitution. What this means is that, if a given MOP was active and a piece of anomalous input was encountered (anomalous input, in this case, means input that is not expected by any of the active expectations generated by the MOP), the system can try to see if the anomalous input is equivalent in any way to any of the facts currently expected. For example, in a restaurant episode we expect the waiter to bring a menu to the table. If in fact this does not happen and instead the waiter recites a list of the day's specials, our understanding system should be able to perform the substitution based upon the hierarchical memory. The read-menu part of the restaurant MOP is an instance of get-information, as is listening to the waiter recite a list of the day's specials. Since these share a common parent, one can be substituted for the other in an understanding situation.

The problem with this in practice is that causal links are not explicitly specified in MOPs. Get-information is just one of many possible abstractions of getting a menu from the waiter, and nowhere are we told that it is the relevant one (as opposed to, for example, get-physical-object). If the waiter brought us a rock instead of a menu, MOP-based understanders would be perfectly happy with the substitution, since rocks and menus share a common abstraction, namely physical-object. MOPs share with scripts the problem that, although the causal nature of the packaging links is implictly there, it is not represented explicitly enough to be useful to the understanding process.

MOPs, therefore, deal with anomalous input and expectation failures by indexing from one

MOP to the next based upon a specific type of failure. This is a principled improvement over scripts, but it still is inadequate. The problem is that for this kind of indexing to work, specific types of failure must be known in advance and set up as indices. It is not clear how this indexing scheme could be acquired from experience.

MOP memory can be used for certain types of inductive learning; this was developed in the IPP program. Because this program was not limited to saving only the features of an episode that corresponded to an existing script, it could create new memory structures by examining the features common to more than one episode and combining those features into a new MOP. IPP, for example, was able to form the new generalization that the victims of kidnappings in Italy are often wealthy industrialists, because that pattern occurred several times in the stories that it had processed. Using the same method, however, the program also formed the generalization that terrorist attacks in Central America always kill exactly two people. The program simply had no way of knowing why the first generalizations, to it, were summarizations of patterns that had appeared in more than one story.

The problem of forming useful generalizations can be related to the problem of handling expectation failures: How do we take a new or unexpected piece of information and connect it to existing memory structures? In the case of expectation failures, we are simply looking for a new knowledge structure that, had it been active, would have provided a correct expectation instead of the failed one. In the case of generalization, we are looking to create a memory structure, from two or more episodes, that would provide expectations suitable to each of the episodes individually. We want to make up a category that in some meaningful way accounts for the episodes. What is needed here is *not* just inductive generalization of the kind done by IPP. The key is that the generalization must capture the *meaningful, interesting* commonalities among the episodes.

5. Explanation Pattern Memory

For an idea on how to handle expectation failures and generalization in a more principled way, consider again human behavior. Consider what happens when people encounter surprising input, when they encounter input that is either not predicted by any of their active expectations or that actively contradicts one of their active expectations. They recover from the resulting expectation failures by *explaining* the input. An explanation is something that causes previously anomalous input to make sense. To explain anomalous input, for us, is to recover from an expectation failure by connecting the anomalous input to the mental model of the story, either by seeing the anomalous input in a different light, by switching to a different mental model or by making up a new mental model altogether. An explanation is something that causes the input and a mental model of the story to match each other.

To explain something is also to create a coherent causal structure corresponding to the episode. There are many different ways of doing this. One possible way is just to access some library of causal rules and to chain them together into a complete explanation. Like the plan-based understanders discussed before, however, this suffers from bad combinatorics. People do better than this because a well-organized memory helps to construct explanations. When people are trying to recover from failed expectations by explaining what is happening, they can be *reminded* of related experiences. Presumably the explanation that was used to make sense out of the related experience will be useful in explaining the current episode. Even if it cannot be applied directly, it can perhaps be modified slightly to fit the current situation. Barring even this, it will almost certainly be a useful source of causal connections or of further remindings.

To make use of past explanations as a help in explaining new situations, we must have a memory full of explanations and a way to retrieve explanations from that memory. We need some kind of structure to hold fossilized explanations the way a script holds fossilized plans. We need a way to index these structures in a memory, and we need a way of examining an episode and selecting features that are likely to be useful as indices in finding relevant explanations. We also need some strategies to modify near-miss explanations into useful ones, and a way to decide, based upon a near miss, which of these modification strategies to apply.

The new memory structure is an Explanation Pattern (XP). XP-memory is not designed to replace MOPs as the central organizing structure of a dynamic memory. Instead, an XP contains additional information about memory structures that would not be present in a MOP. An XP, for example, contains causal annotation: information about why the MOP makes sense, how the pieces of the MOP connect together, and what features of the class of episodes represented by the MOP are essential to the believability of this XP versus those which are peripheral. An XP is a *view* of a MOP in this sense; there can be more than one XP related to the same MOP, each XP containing a different set of causal connections among the same scenes.

A program is currently being developed at Yale that explains anomalous events by applying XPs. For a detailed description of the program's operation, see Schank (in press). The program produces novel explanations by retrieving old explanations and modifying them to fit new situations. Thus it is both an understanding program and a learning program. Its goal is to discover explanations that will help it understand an event that has caused an expectation failure. In doing so it learns new explanations and stores them for future use. The general algorithm employed by the program is as follows:

1: ANOMALY DETECTION
   Attempt to fit story into memory.
   If successful DONE; otherwise an anomaly has been detected.

2: XP SEARCH
   Search for an XP that can be applied to explain the anomaly.

3: XP ACCEPTING
   Attempting to apply XPs.
   If successful then skip to step 5.

4: XP TWEAKING
   If unable to apply XPs directly then attempt to revise them into XPs that might apply better.
   If successful send these tweaked XPs back to step 3.

5: XP INTEGRATION AND GENERALIZATION
   If any results accepted, integrate results back into memory, making appropriate generalizations.

An example for which the program provides explanations is the unexpected death of the young race horse Swale, one week after having won the Belmont Stakes horse race. Although the death is an expected part of the race-horse-life script, this death has come too early. Using this failure and rule-based specialization much as in the same manner as MOPtrans, the program begins considering the death as an example of early-death. Possible routine explanations for early-death include, for example, death from sickness. This, however, can be ruled out because Swale's good health can be inferred from his recent victory at a major racing event.

The program is faced with a failure in that no active memory structure could have generated the expectation of Swale's death. It must be reminded of some similar episode that it has explained in the past and re-use that explanation in the current scene.

One retrieval strategy that can be used to find explanations is coordination of anomalies. If two unusual things cooccur, there may be an explanation pattern that relates them. One feature of Swale that might be selected this way is his excellent athletic condition, compared with other race horses. Looking in XP memory under the indices Death and Excellent Athletic Condition can cause a retrieval of the XP originally created to explain the death of Jim Fixx, an author who advocated the health benefits of jogging, but who nevertheless died of a hidden congenital heart defect while jogging:

The Jim Fixx XP
   Joggers jog a lot.

Jogging results in physical exhaustion because jogging is a kind of exertion and exertion results in exhaustion.
Physical exhaustion coupled with a heart defect can cause a heart attack.
A heart attack can cause death.

There is a failure in trying to apply this XP however, since it requires that the actor be a jogger and because a jogger is a kind of human. The program can characterize this failure as an inapplicable theme, and can use this characterization to index a repair strategy, namely substituting another theme for the failed one. Since an XP contains causal information, we know *why* it is relevant that Jim Fixx was a jogger: namely that jogging is a form of physical exertion. We are not faced with an unconstrained generalization problem. The program need only look among Swale's known themes to find one that represents some kind of physical exertion. Running in horse races satisfies this constraint, resulting in the explanation: "Swale had a congenital heart defect. The exertion of running in horse races strained his heart and brought out the latent defect. He had a heart attack and died." This tweak also results in a generalized XP that doing activity involving physical exertion can result in heart attacks; this new XP can be accessed directly in the future without need for search and tweaking.

Another feature of Swale was his success at an early age. Indexing into XP memory with death and early-success as keys can retrieve the explanation pattern originally generated to explain the death of the rock star Janis Joplin:

The Janis Joplin XP:
Being a star performer can result in stress because it is lonely at the top.
Being stressed-out can result in a need to escape and relax.
Needing to escape and relax can result in taking recreational drugs.
Taking recreational drugs can result in an overdose.
A drug overdose can result in death.

On the face this is silly when applied to Swale. Many of the steps do not apply to horses, and the failure does not seem repairable by trying alternate themes. But, as a source of other remindings, for example re-indexing into the XP memory using Drug Overdose and Death as keys, this XP can be a source of a useful explanation, namely that Swale's owner was giving him drugs to improve his performance and that he accidentally gave him an overdose, killing him.

By writing programs that can develop these and other explanations, we are trying to deal with expectation failures during the understanding process by learning new knowledge structures that, had they been active originally, would have generated correct and useful expectations that would not have failed. In the process of retrieving and modifying XPs we are dealing with a new kind of failure as well: the failure of a retrieved XP to fit the explanation against which it is being tried. It is this failure that drives the tweaking process; a program must characterize this type of failure at a suitable level of abstraction and use that characterization to suggest repair strategies.

When an episode is processed two things happen: The original retrieved XP, plus the modifications suggested in response to the XP's failure to fit the current situation, yield both an explanation or categorization of the current situation and a new generalized memory structure.

The concept of explanation-based learning has been discussed elsewhere, notably by Segre and DeJong (1985), who use explanation construction as a means to select the causally significant features of an episode. Within their explanation methodology, features of an episode that participate in the explanation are important, and should be included in the generalized description of the episode; those that do not can safely be ignored.

While we also use explanations to select features of an episode, we differ from Segre and DeJong's approach in that we keep an explanation around after it has been constructed. We store these explanations in a memory where they can be accessed later, and tweaked to explain new events. Consequently, we deal with the new kind of failure discussed above: the failure of a retrieved explanation to fit a new event, and with the concomitant repair strategies indexed under those failure types.

One particularly noticeable thing about script-based and, to a lesser extent, MOP based understanders is their dullness and predictability. Although these understanders can recognize instances of patterns, they will never surprise anyone with their insight or creativity. They will never come up with an interpretation of a story or episode that would not be perfectly obvious to any person with remotely relevant experience. The ability to adapt an old explanation to a new situation; to see, in effect, the analogy between two events that are on the surface unrelated yet that share common causal patterns, is something that we generally accept as evidence of creative, intelligent behavior. We would like explanation-based understanders to exhibit some of this kind of intelligence.

We believe that the ability to make creative explanations comes from the way in which explanatory knowledge is indexed in the system and from the sophistication of explanation modification strategies. Good indexing is essential to any kind of explanatory success at all; XPs are useless to an understander unless they can be found and applied to the understanding process at the appropriate time. In our example above we treated rather lightly the reminding method by which an appropriate XP can be found, while in point of fact this is the key to having the whole process work. Deciding what features of an episode are useful as indices into XP memory and deciding under what features to index a new XP are current areas of research on this program.

Another area of development lies in the application of opportunism to the XP retrieval process. When we are retrieving patterns from memory, we must be attuned to the fact that the misses are as important as the hits. Whatever search mechanism we use to find candidate explanation patterns must miss gracefully. When an explanation pattern shares some, but not all of the indices derived from the episode we are trying to explain, chances are good that the explanation pattern, although not directly applicable to the current episode, will be useful in some ways: either as a candidate for tweaking or as a pointer to other explanation patterns via the features that cause the episode to fail to match this explanation pattern. In the example above of explaining Swale's death by using a pattern retrieved from memory that doesn't quite match the current episode, the difference between the features present in the episode (race horse) and the features required by the pattern (jogger) are the source of the new generalization that physical exertion coupled with a heart defect can cause the sudden death of a person or animal in seemingly top physical condition.

6. Summary

In summary, then, what does explanation say about the role of failures in discourse processing? We have shown how expectations are set up and tested as part of the routine kinds of understanding that people use all the time. We have argued that it is when expectations fail that something interesting happens. At the simplest level, which we have called failure-based understanding, expectation failures are a clue to help us select the appropriate knowledge structure into which to try to fit an episode. By paying attention to the information obtained from expectation failures, a system can refine or change its model of what an episode is about.

At a more complex level, namely failure-based learning, people can deal with a less tractable type of expectation failure and do so creatively. By constructing an explanation of a failed expectation, we can develop new knowledge structures to handle similar situations in the future. We learn new ways to categorize episodes based upon the causal connections uncovered during the explanation process. Within the explanation task itself failures play a role; characterizing the failure of an XP to fit is the key to finding useful tweaking strategies.

Just as the text understanding task itself must make use of knowledge structures to constrain search and to apply some top-down direction to the process, so must the explanation subtask. The relevant knowledge structure to this task is the explanation pattern, the details of which we are only beginning to work out at the implementation level. Explanation patterns, with their fossilized traces of causal reasoning, can be retrieved by a reminding process and modified to fit a situation which is causally equivalent but different in surface features. This is a kind of single-trial learning and generalization of which we know people to be capable, and one which we would like to see in our programs.

We have seen that a major problem with explanation patterns is how to find the appropriate one to retrieve for a given episode. Issues of what features of an episode for use for search keys, which features of an XP to use as indices, and what kind of search strategy might we employ so that "near miss" candidates are useful in constructing a good explanation are the key to understanding certain aspects of creative reasoning. The task of constructing explanations is not the first place in Artificial Intelligence that these particular problems have appeared, but it is one in which they can be well-defined and one in which work on them can progress.

Acknowledgement

We would like to thank Chris Riesbeck and David Leake for their helpful comments on drafts of this paper. The explanation-based understander described above is being developed by the authors, Chris Riesbeck, Alex Kass and David Leake and is being supported in part by the Advanced Research Projects Agency of the Department of Defense and monitored under the Office of Naval Research under contract N00014-82-K-0149.

References

Birnbaum, L., & Selfridge, M. (1970). *Problems in conceptual analysis of natural language.* Technical report 168, Yale University.
Bower, G.H., Black, J.B., & Turner, T.J. (1979). Scripts in memory for text. *Cognitive Psychology, 11,* 177-220.
Dyer, M.G. (1983). *In-depth understanding.* Cambridge, MA: MIT Press.
Kolodner, J.L. (1980). *Retrieval and organizational strategies in conceptual memory: A computer model.* Ph.D. Thesis, Yale University.
Lebowitz, M. (1980). *Generalization and memory in an integrated understanding system.* Ph.D. Thesis, Yale University.
Lytinen, S. (1984). *The organization of knowledge in a multi-lingual, integrated parser.* Ph.D. Thesis, Yale University.
Riesbeck, C. (1975). Conceptual analysis. In R.C. Schank, *Conceptual information processing.* Amsterdam: North-Holland.
Schank, R.C. (1972). Conceptual dependency: A theory of natural language understanding. *Cognitive Psychology, 3,* 552-631.
Schank, R.C. (1975). *Conceptual information processing.* Amsterdam: North-Holland.
Schank, R.C. (1982). *Dynamic memory: A theory of learning in computers and people.* Cambridge: Cambridge University Press.
Schank, R.C., & Abelson, R. (1977). *Scripts, plans, goals and understanding.* Hillsdale, NJ: Lawrence Erlbaum Associates.
Schank, R.C., & Birnbaum, L. (1980). *Memory, meaning and syntax.* Technical report 189, Yale University.
Schank, R.C. (in press). *Explanation patterns: Understanding mechanically and creatively.* Hillsdale, NJ: Erlbaum.
Segre, A.M., & DeJong, G.F. (1985). Explanation based manipulator learning: Acquisition of planning ability through observation. In *Proceedings of the IEEE International Conference on Robotics and Automation,* St. Louis, Missouri.

SOME ASPECTS OF DEFAULT REASONING IN INTERACTIVE DISCOURSE*†

Aravind K. Joshi and Bonnie L. Webber

Department of Computer and
Information Science
Moore School/6389
University of Pennsylvania
Philadelphia, PA 19104

Ralph M. Weischedel

BBN Laboratories Inc.
50 Moulton Street
Cambridge, MA 02238

We are concerned with interaction between two agents: one agent is the user and the other agent is a system which plays the role of a helpful cooperative agent. Cooperativeness in interaction has many dimensions. In this chapter, we will be particularly concerned with the role of the cooperative agent in preventing the user coming to false conclusions. In cooperative man-machine interaction, it is taken as *necessary* that a system respond truthfully to a user's question. It is not, however, *sufficient*. In particular, if the system has reason to believe that its response might lead the user to draw an inference that it knows to be false, then it must block this inference by modifying or adding to its response. We describe two kinds of false conclusions we are attempting to block by modifying otherwise true responses: (1) false conclusions drawn by standard default reasoning, and (2) false conclusions drawn in a task-oriented context on the basis of the user's expectations about the way a cooperative agent will respond. Finally, we discuss constraints limiting the cooperative agent's responsibilities with respect to anticipating the false conclusion that the user may draw from its response.

1. Introduction

We are concerned with interaction between two agents: one agent is the user (U) and the other agent is a system (S) which plays the role of a helpful cooperative agent. For example, S can be a knowledge base system, an expert system, or even a robot interacting with the physical environment carrying out tasks and reporting about the state of the world. Cooperativeness in interaction has many dimensions, for example, S provides U with some requested information, S helps U to formulate an appropriate question, S explains to U the structure of S's knowledge base, S demonstrates to U how to carry out a certain procedure, S corrects possible misconceptions of U, S prevents U from coming to false conclusions, etc. In this chapter, we will be particularly concerned with the last aspect, i.e., preventing false conclusions.

In cooperative man-machine interaction, it is taken as *necessary* that a system truthfully and informatively respond to a user's question. It is not, however, *sufficient*. In particular, if the system has reason to believe that its planned response might lead the user to draw an inference that it knows to be false, then it must block it by modifying or adding to its response. The problem is that a system neither can nor should explore all conclusions a user might possibly

---

* Acknowledgement: This work was supported by U.S. Army grants DAA6-29-84-K-0061, DAAB07-84-K-FO77, DAA29-84-9-0027, DAAG29-85-K-0061, DARPA grant number N00014-85-K-0018 and NSF grants MCS-8219116-CER, MCS-82-07294, MCS-83-05221, DCR-84-10543, and IST-8419162.
† This work is based on some joint work carried out by the three authors, which has been reported in part in Joshi, Webber, and Weischedel (1984b,c). Some related work also appears in Joshi, Webber and Weischedel (1984a).

draw: its reasoning must be constrained in some systematic and well-motivated way.

Such cooperative behavior was investigated in Joshi (1982) in which a modification of Grice's *Maxim of Quality* is proposed:

Grice's *Maxim of Quality* –
Do not say what you believe to be false or for which
you lack adequate evidence.

A modification of Grice's *Maxim of Quality* was proposed by Joshi (1982)
as follows:
If you, the speaker, plan to say anything which may
imply for the hearer something that you believe to be
false, then provide further information to block it.

This behavior was studied in the context of interpreting certain definite noun phrases. In this paper, we investigate this revised principle as applied to question answering. In particular the goals of the research described here are to:

1. characterize tractable cases in which the system as respondent (R) can anticipate the possibility of the user/questioner (Q) drawing false conclusions from its response and can hence alter or expand its response so as to prevent it happening;

2. develop a formal method for computing the projected inferences that Q may draw from a particular response, identifying those factors whose presence or absence catalyzes the inferences;

3. enable the system to generate modifications of its response that can defuse possible false inferences and that may provide additional useful information as well.

Before we begin, it is important to see how this work differs from our related work on responding when the system notices a discrepancy between its beliefs and those of its user (Kaplan, 1982; Mays, 1980; McCoy, 1983; Webber & Mays, 1983). For example, if a user asks "How many French students failed CSE121 last term?", he shows that he believes *inter alia* that the set of French students is non-empty, that there is a course CSE121, and that it was given last term. If the system simply answers "None", he will assume the system concurs with these beliefs since the answer is consistent with them. Furthermore, he may conclude that French students do rather well in a difficult course. But this may be a false conclusion if the system *doesn't* hold to all of those beliefs (e.g., it doesn't know of any French students). Thus while the system's assertion "No French students failed CSE121 last term" is true, it has misled the user (1) into believing it concurs with the user's beliefs and (2) into drawing additional false conclusions from its response.[1] The differences between that related work and the current enterprise are that:

1. It is *not* assumed in the current enterprise that there is any *overt indication* that the domain beliefs of the user are in any way at odds with those of the system.

2. In that related work, the user draws a false conclusion from what is said because the presuppositions of the response are not in accord with the system's beliefs (following a nice analysis in Mercer and Rosenberg, 1984). In the current enterprise, the user draws a false conclusion from what is said because the system's response behavior is not in accord with the user's expectations. It may or may not also involve false domain beliefs that the system attributes to the user.

In this chapter, we describe two kinds of false conclusions we are attempting to block by

---

[1] It is a feature of Kaplan's CO-OP system (1982) that it points out the discrepancy by saying "I don't know of any French students".

modifying otherwise true response:

> false conclusions drawn by standard default reasoning — i.e., by the user/listener concluding (incorrectly) that there is nothing special about this case

> false conclusions drawn in a task-oriented context on the basis of the user's expectations about the way a cooperative expert will respond.

In Section 2, we discuss examples of the first type, where the respondent (R) can reason that the questioner (Q) may inappropriately apply a default rule to the (true) information conveyed in R's response and hence draw a false conclusion. We characterize appropriate information for R to include in his response to block that false conclusion. In Section 3, we describe examples of the second type. Finally, in Section 4, we discuss our claim regarding the primary constraint posed here on limiting R's responsibilities with respect to anticipating false conclusions that Q may draw from its response: that is, it is only that part of R's knowledge base that is already in focus (given the interaction up to that point, including R's formulating a direct answer to Q's query) that will be involved in anticipating the conclusions that Q may draw from R's response.

2. Blocking Potential Misapplication of Default Rules

Default reasoning is usually studied in the context of a logical system in its own right or an agent who reasons about the world from partial information and hence may draw conclusions unsupported by traditional logic. However, one can also look at it in the context of interacting agents. An agent's reasoning depends not only on his perceptions of the world but also on the information he receives in interacting with other agents. This information is partial, in that another agent neither will nor can make everything explicit. Knowing this, the first agent (Q) will seek to derive information *implicit* in the interaction, in part by contrasting what the other agent (R) *has made* explicit with what Q assumes *would have been* made explicit, were something else the case. Because of this, R must be careful to forestall inappropriate derivations that Q might draw. The question is on what basis R should reason that Q may assume that some piece of information (P) would have been made explicit in the interaction, were P true.

One basis, we contend, is the likelihood that Q will apply some *standard default rule* of the type discussed by Reiter (1980) if R doesn't make it explicit that the rule is not applicable. Reiter introduced the idea of default rules in the stand-alone context of an agent or logical system filling in its own partial information. Most standard default rules embody the sense that "given no reason to suspect otherwise, *there's nothing special about the current case*". For example, for a bird what would be special is that it can't fly — i.e., "Most birds fly". Knowing only that Tweety is a bird and no reason to suspect otherwise, an agent may conclude by default that there's nothing special about Tweety and so he can fly.

This kind of default reasoning can lead to false conclusions in a single agent situation, but also in an interaction. That is, in a question-answer interaction, if the respondent (R) has reason for knowing or suspecting that the situation goes counter to the standard default, it seems to be common practice to convey this information to the questioner (Q), to block his potentially assuming the default. To see this, consider the following example.

2.1 Example

Suppose it's the case that most associated professors are tenured and most of them have PhDs. Consider the following interchange

Q: Is Sam an associate professor?
R: Yes, but he doesn't have tenure.

There are two things to account for here: (1) Given the information was not requested, why did R include the "but" clause, and (2) why this clause and not another one? We claim that the answer to the second question has to do with that part of R's knowledge base that is currently in focus. This we discuss more in Section 4. In the meantime, we will just refer to this subset as "RBc".

Assume RBc contains at least the following information:

(a) Sam is an associate professor.
(b) Most associate professors are tenured.
(c) Sam is not tenured.

(b) may be in RBc because the question of tenure may be in context. Based on RBc, R's direct response is clearly "Yes". This direct response however could lead Q to conclude falsely, by default reasoning, that Sam is tenured. That is, R can reason that, given just (b) and his planned response "Yes" (i.e., if (c) is not in Q's knowledge base), Q could infer by default reasoning that *Sam is tenured,* which R knows with respect to RBc is false. Hence, R will modify that planned response to block this false inference, as in the response above.

In general, we can represent R's reasoning about Q's reaction to a simple direct response "Yes, B(a)", given Q believes "Most Bs F", in terms of the following default schema, using the notation introduced in Reiter (1980).

$$\frac{\text{told}(R,Q,B(c)) \:\&\: (\text{Most } x)\,[B(x) \Rightarrow F(x)] \:\&\: \neg\text{told}(R,Q,\neg F(c)): M(F(c))}{F(c)}$$

As in Reiter's discussion, "M(P)" means it is consistent to assume that P. In the associate professor example, B corresponds to the predicate "is an associate professor", F, to the predicate "has tenure", and c, to Sam. Using such an instantiated rule schema, R will recognize that Q is likely to conclude F(c) – "Sam has tenure" – which is false with respect to RBc (and hence, with respect to all of R's knowledge base). Thus R will modify his direct response so as to block this false conclusion.

3. Blocking False Conclusions in Expert Interactions

The situations we are concerned with here are ones in which the system is explicitly tasked with providing help and expertise to the user. In such circumstances, the user has a strong expectation that the system has both the experience and motivation to provide the most appropriate help towards achieving the user's goals. The user does not expect behavior like:

Q: How can I get to Camden?
R: You can't.

As many studies have shown (Allen, 1982), an advice seeker (Q) expects that an expert (R) will attempt to recognize the plan Q is attempting to follow in pursuit of a goal and respond to Q's question accordingly. Further studies (Pollack, Hirschberg, & Webber, 1982; Pollack, 1984a, 1984b) show that Q may also expect that R will respond in terms of a better plan if the recognized one is either sub-optimal or unsuitable for attaining Q's perceived goal. Thus because of this principle of "expert cooperative behavior", Q may expect a response to a more general question than the one he has actually asked. That is, in asking an expert "How do I do X?" or "Can I do X?", Q is anticipating a response to "How can I achieve my goal?"

3.1 Example

Consider a student (Q) asking the following question, near the end of the term.

Q: Can I drop CIS577?

Since it is already too late to drop a course, the only direct answer the expert (R) can give is

"No". Of course, part of an expert's knowledge concerns the typical states users get into and the possible actions that permit transitions between them. Moreover it is also part of this expertise to infer such states from the current state of the interaction, Q's query, some shared knowledge of Q's goals and expectations and the shared assumption that an expert is expected to attend to these higher goals. How the system should go about inferring these states is a difficult task that others are examining (Carberry, 1983; Pollack, 1984a, 1984b). We assume that such an inference has been made. We also assume for simplicity that the states are uniquely determined. For example, we assume that the system has inferred that Q is in state Sb (student is doing badly in the course) and wants to be in a state Sg (student is in a position to do better in this course or another one later), and that the action a (dropping the course) will take him from Sb to Sg.

Given this, a response purely of "No" may lead Q to draw some conclusions that R knows to be false. For example, R can reason that since a principle of cooperative behavior for an expert is to tell Q the best way to go from Sb to Sg, Q is likely to conclude from R's response that there is no way to go from Sb to Sg. This conclusion however would be false if R knows some other ways of going from Sb to Sg. To avoid potentially misleading Q, R must provide additional information, such as

R: No, but you can take an incomplete and ask for more time to finish the work.

As we noted earlier, an important question is how much reasoning R should do to block false conclusions on Q's part. Again, we assume that R should only concern itself with those false conclusions that Q is likely to draw that involve the part of R's knowledge base currently in focus (RBc). RBc includes, of course, the subset which R needs in order to answer the query in the first place. (See section 4 for further details).

We will make this a little more precise by considering several cases corresponding to the different states of R's knowledge base with respect to Sb, Sg, and transitions between them. For convenience, we will give an appropriate response in terms of Sb, Sg and the actions. Clearly, it should be given in terms of descriptions of states and actions understandable to Q. (Moreover, by making further assumptions about Q's beliefs, R may be able to validly trim some of its response.)

1. Suppose that it is possible to go from Sb to Sg by dropping the course and that this is the only action that will take one from Sb to Sg.

$$Sb \xrightarrow{\alpha} Sg$$

   In the this case, the response is
   R: Yes. α is the only action that will take you from Sb to Sg.

2. Suppose that in addition to going from Sb to Sg by dropping the course, there is a better way, say β, of doing so.[2]

$$Sb \overset{\alpha}{\underset{\beta}{\rightrightarrows}} Sg$$

   In this case, the response is
   R: Yes, but there is a better action β that will take you from Sb to Sg.

3. Suppose that dropping the course does not take you from Sb to Sg, but another action β will. This is the situation we considered in our earlier discussion.

$$Sb \xrightarrow{\beta} Sg$$

---

2 "Betterness" is yet another area for future research.

In this case the response is
R: No, but there is an action β that will take you from Sb to Sg.

4. Suppose that there is no action that will take one from Sb to Sg.

      Sb   Sg

In this the response is
R: No. There is no action that will take you from Sb to Sg.

Of course, other situations are possible. The point, however, is that the additional information that R provides to prevent Q from drawing false conclusions is limited to just that part of R's knowledge base that R is focussed on in answering Q's query.

4. Constraining the Respondent's Obligations

As many people have observed — from studies across a range of linguistic phenomena, including co-referring expressions (Grosz, 1977; Grosz, Joshi, & Weinstein, 1983; Sidner, 1982), left dislocations (Prince, 1981), epitomization (Ward, 1982), etc. — a speaker (R) normally focuses on a particular part of its knowledge base. What he focuses on depends in part on (1) context, (2) R's partial knowledge of Q's overall goals, as well as what W knows already as a result of the interaction up to that point, and (3) Q's particular query, etc. The precise nature of how these various factors affect focusing is complex and is receiving much attention (Grosz, 1977; Grosz, Joshi, & Weinstein, 1983; Sidner, 1982). However, no matter how these various factors contribute to focusing, we can certainly assume that R comes to focus on a subset of its knowledge base in order to provide a *direct answer* to Q's query (at some level of interpretation). Let us call this subset RBc for "R's current beliefs". Our claim is that one important constraint on cooperative behavior is that it is determined by RBc only. Clearly the information needed for a direct response is contained in RBc, as is the information needed for many types of helpful responses. In other words, RBc — that part of R's knowledge base that R decides to focus on in order to give a direct response to Q's query — also has the information needed to generate several classes of helpful responses. The simplest case is presupposition failure (Kaplan, 1982), as in the following

Q; How many Q's were given in CIS 500?

where Q presumes the CIS 500 was offered. In trying to formulate a direct response, R will have to ascertain that CIS 500 was offered. If it was (Q's presumption is true), then R can go ahead and give a direct response. If not, then R can indicate that CIS 500 was not offered and thereby avoid misleading Q. All of this is straightforward. The point here is that the information needed to provide this extra response is already there in that part of R's knowledge base which R had to look up anyway in order to try to give the direct response.

In the above example, it is clear how the response can be localized to RBc. We would like to claim that this approach has a wider applicability: that RBc alone is the basis for responses that anticipate and attempt to block interactional defaults as well. Since RBc contains the information for a direct response, R can plan one (r). From r, R can reason whether it is possible for Q to infer some conclusion (g) which R knows to be false because — g is in RBc. If so, then R should modify r so as to eliminate this possibility. The point is that the only false inferences that R will attempt to block are those whose falsity can be checked in RBc. There may be other false inferences that Q may draw, whose falsity cannot be determined solely with respect to RBc (although it might be possible with respect to R's entire knowledge base). While intuitively this may not seem enough of a constraint on the amount of anticipatory reasoning that the modified maxim of quality imposes on R, it does constrain things a lot by only considering a (relatively small) subset of the knowledge base. Factors such as context may further delimit S's responses, but they will all be relative to RBc.

5. Conclusion

Since the behavior of expert systems will be interpreted in terms of the behavior users expect of cooperative human experts, we (as system designers) must understand such behavior patterns so as to implement them in our systems. If such systems are to be truly cooperative, it is not sufficient for them to be simply truthful. Additionally, they must be able to predict limited classes of false inferences that users might draw from dialogue with them and also to respond in a way to prevent those false inferences. The current enterprise is a small but non-trivial step in this direction. We are investigating other cases where a cooperative expert should prevent false inferences by another agent, including preventing inappropriate default reasoning (Joshi, Webber, & Weischedel, 1984a, 1984b) and preventing false reasoning about achieving goals.

References

Allen, J. (1982). Recognizing intentions from natural language utterances. In M. Brady & R.C. Berwick (Eds.), *Computational models of discourse.* Cambridge, MA: MIT Press.

Carberry, S. (1983). Tracking user goals in an information-seeking environment. In *Proceedings of the national conference on artificial* (AAAI).

Grosz, B. (1977). *The representation and use of focus in dialogue understanding.* Technical Report 151, SRI International, Menlo Park, CA.

Grosz, B., Joshi, A.K., & Weinstein, S. (1983). Providing a unified account of definite noun phrases in discourse. In *Proceedings 21st annual meeting* (Association for Computational Linguistics), Cambridge, MA.

Joshi, A.K. (1982). Mutual beliefs in question-answering systems. In N. Smith (Ed.), *Mutual belief.* New York: Academic Press.

Joshi, A.K., Webber, B., & Weischedel, R. (1984a). Living up to expectations: Computing expert responses. In *Proceedings of AAAI-84,* Austin TX.

Joshi, A.K., Webber, B., & Weischedel, R. (1984b). Preventing false inferences. In *Proceedings of Coling84,* Stanford, CA.

Joshi, A.K., Webber, B., & Weischedel, R. (1984c). Default reasoning in interaction. In *Proceedings of AAAI workshop on non-monotonic reasoning,* New Paltz, NY.

Kaplan, J. (1982). Cooperative responses from a portable natural language database query system. In M. Brady & R.C. Berwick (Eds.), *Computational models of discourse.* Cambridge, MA: MIT Press.

Mays, E. (1980). Failures in natural language systems: Applications to database query systems. In *Proceedings of the first national conference on artificial intelligence* (AAAI), Stanford, CA.

McCoy, K. (1983). Correcting misconceptions: What to say. In *Proceedings of CHI'83 conference on human factors in computing systems,* Cambridge, MA.

Mercer, R., & Rosenberg, R. (1984). Generating corrective answers by computing presuppositions of answers, not of questions. In *Proceedings of the 1984 conference* (Canadian Society for Computational Studies of Intelligence), University of Western Ontario, London, Ontario.

Pollack, M., Hirschberg, J., & Webber, B. (1982). User participation in the reasoning processes of expert systems. In *Proceedings of AAAI-82,* Carnegie-Mellon University, Pittsburgh, PA.

Pollack, M. (1984). *Goal inference in expert systems.* Technical Report MS-CIS-84-07, University of Pennsylvania, Philadelphia, PA.

Pollack, M. (1984). Good answers to bad questions. In *Proceedings of the 1984 conference* (Canadian Society for Computational Studies of Intelligence), University of Western Ontario, London, Ontario.

Prince, E. (1981). Topicalization, focus movement and Yiddish movement: A pragmatic differentiation. In *Proceedings of the 7th annual meeting* (Berkeley Linguistics Society), Berkeley, CA.

Reiter, R. (1980). A logic for default reasoning. *Artificial Intelligence, 13,* 81-132.

Sidner, C.L. (1982). Focusing in the comprehension of definite anaphora. In M. Brady (Ed.), *Computational models of discourse.* Cambridge, MA: MIT Press.

Ward, G. (1982). A pragmatic analysis of epitomization: Topicalization it's not. In *Proceedings of the Summer meeting 1982* (Linguistics Society of America), College Park, MD.

Webber, B. & Mays, E. (1983). Varieties of user misconceptions: Detection and correction. In *Proceedings of IJCAI-83,* Karlsruhe, West Germany.

INDUCTION AND DECISION MAKING
IN THE VICINITY OF DIALOGUE FAILURE

Owen Egan

Linguistics Institute of Ireland

An inductive approach to dialogue processing is adopted. Communication is considered as a probabilistic outcome, and failure likewise as a matter of degree, to be described not in absolute terms but only in relation to the probability of error, its cost, and the comparable cost of dialogue intervention aimed at lessening the probability of error. A dialogue might be said to be approaching failure when the relative costs of miscommunication and intervention to seek clarification are such that one of the participants is obliged, by minimal norms of rational decision-making, to intervene. The implications of this viewpoint are discussed, particularly the notions of coherence and cooperation in dialogue which are implicit in it. Applications are made to sample dialogues.

1. Introduction

The idea of dialogue as decision-making under uncertainty, or a "game" in the technical sense which this term acquires in game theory, has been put forward from time to time (Herdan, 1966; Saarinen, 1979; Suppes, 1984, Ch. 7) but has never attracted sustained attention in discourse research. This is not because there is anything wildly implausible about it. One can hardly disagree with the general view, expressed, for example by Suppes that "it is a probabilistic question whether a speaker expresses what he intended, and it is equally a matter of probability how what is expressed is received by the listener" (Suppes, 1984, p. 152). Neither is there anything inherently controversial in the application of game-theoretic ideas to dialogue and discourse generally. The notion of meaning has strong links with the perceived likelihood of falsification. The speech-act of assertion, for example, can be taken "as a kind of gamble that the speaker will not be proved wrong" (Dummett, 1976, p. 126). There is an even stronger case for the decision-theoretic viewpoint when we think of the formulation of "good" questions in terms of the expected yield of information, or the selection of likeliest interpretations in the case of ambiguous or ill-formed input. Thus Zaefferer (1977) suggests that we sometimes choose between alternative interpretations of ambiguous statements on the basis of "utility", i.e., the product of a subjective prior probability and the cost of error, whole or partial. But while there are no *a priori* objections to these approaches, in practice they have rarely been adopted with any persistence in discourse analysis. The principal reason is that dialogue is more naturally thought of as a discrete process operating on deductive principles rather than as an inductive process which accumulates evidence from different sources and arrives at a probabilistic judgment.

I want to consider the inductive approach once more. One reason is the increasing attention being given to pragmatic and illocutionary features of discourse, which, psychologically considered, are inductions based on multiple cues. Another is the shift to larger units of discourse, which also argues for integrative mechanisms of an inductive nature, such as the "principle of accumulation" advocated by Pause (1984) for the resolution of anaphora, or the sequenced "relaxations" of reference used by Goodman (1983, and also this volume) in the repair of reference failures.

The formalisms I have in mind are those of decision theory, based on expected values or utilities, and the inductive models of the "judgmental" tradition in cognitive psychology. Among the latter we can distinguish at least the following three main types: (1) Bayesian formalisms (Edwards, Lindman, & Savage, 1963; Fischoff & Beyth-Marom, 1983), which model the revision of subjective probabilities in the light of new evidence by methods which are optimal in objective probability theory, (2) measures of dissimilarity, or badness-of-fit, including statistical indices such as chi square used to quantify imperfect or "fuzzy" matching between concepts, and (3) regression equations (Einhorn, Kleinmuntz, & Kleinmuntz, 1979), which express uncertain inference as an optimal computation on the values of observed cues. For a review of current efforts to incorporate such mechanisms into cognitive models in artificial intelligence see Mamdani, Efstathiou, and Pang (1985).

First I outline the essentials of the decision-making perspective as they apply to dialogue, giving special attention to the topic of dialogue failure. I argue that we cannot talk about failure in any absolute sense but only about degrees of failure. We cannot identify the "point" at which failure occurs. The best we can do is to say that dialogue is approaching failure when the relative costs of present intervention and the anticipated future loss due to failure are such that a rational listener is obliged, by norms of rational decision-making, to seek clarification. The notion of cooperation in dialogue is discussed. It implies that participants must model the values and anticipated losses of the other person and adjust their own values and costs accordingly. The problem of segmenting dialogue into "moves" or "options" is raised, in the hope of finding an empirically-based "grammar" of dialogue which would be used to infer ill-formedness at the level of speech acts. The paper concludes by returning to the notion of rationality and its importance for our treatment of dialogue failure. Norms of rationality are found to apply not only to the management of dialogue but also to the internalized models which the participants have of each other. In many cases it is solely in view of the assumption that participants behave in a rational way generally, and in particular in the management of dialogue, that individual moves can be perceived as incongruent, and thus taken as further evidence for approaching failure.

2. The Decision-making Perspective

The principal elements of the decision-theoretic perspective are familiar from the common-sense view of dialogue as a pragmatic undertaking, i.e., as a means of getting something done. Dialogue, from this point of view, is a form of cooperative, goal-directed activity, in which the participants attempt to achieve something by the most efficient means. Accordingly, some of the well-formedness of dialogue should be captured in axioms of rationality. The model is particularly appropriate if we think of the kind of dialogue which takes place when people ask for directions or work cooperatively on some task. The plan-based approach to dialogue in artificial intelligence (e.g., Allen & Perrault, 1980; Grosz, 1981; Hobbs & Evans, 1980) is concerned with such cases. Dialogues of this sort are attractive from the computation perspective because they have the advantage that the dialogue is embedded in an extra-linguistic task on which the participants will generally have a good deal of shared, prior knowledge. When the intending traveller asks the attendant in the train station about the times of departing trains, the participants know the task which has to be accomplished (to get the traveller onto the right train), and the various subtasks which must be performed first (going to the correct platform). Knowledge of the structure of the external task can then be used to "drive" the dialogue along to its conclusion. A good account of some real-life dialogue can be given with this model.

There are difficulties with the plan-based approach, not the least of which is its low generalizability. Most dialogue is not so firmly mapped onto an external task as these examples suggest. And even when it is, a good many of the goals generated in the course of dialogue are internal to the dialogue itself, "communication goals" as we might call them to contrast with "action goals". In reality the traveller must get on the train *and* manage a piece of dialogue, and the interaction of these goals, in on-line processing, does not reduce to a simple task-subtask relationship, as the plan-based approach might suggest. We will return to some of these difficulties later when we apply the concepts of action, outcome, cost, value, and efficiency, to dialogues with goals which are largely internal.

Our intention of dealing with ordinary dialogue of all kinds, not just the plan-based variety, forces us to go back a little further into the conceptual primitives of the decision-theoretic viewpoint. In particular, some minimal degree of clarity and plausibility must be obtained on the following notions, as they might apply to dialogue:

1 Desired outcome of dialogue, i.e., (successful) communication
2 Communication failure
3 Value of communication
4 Cost of failure
5 Current estimate of probability of failure
6 Expected cost of failure, i.e., the product of 4 and 5

Actually our interest in the decision-theoretic point of view has nothing to do with the quantification of costs, which is usually quite impossible in the case of dialogue. Rather we introduce the idea of cost in order to bring into focus the structure of dialogue as a form of rational decision-making under uncertainty. Even the notion of failure, which suggests something decisive and identifiable, to which costs might be assigned in some objective way, will also lack a precise reference in most dialogues. Instead we must think of the desired outcome of dialogue in quite general terms as "communication", namely a certain congruence of meaning structures, those encoded in the utterance by the speaker, and those elicited in the listener. How much congruence we require is a matter we need not go into here, except to say that "meaning structures" need not be given a mentalistic interpretation. "Result-congruence", in Suppes's (1984, p. 156) sense, will do just as well. It is true that in the type of dialogue studied in the plan-based approach we might well be able to give an external definition to congruence. In general, however, we must settle for a measure which is internal.

3. Proximal and Distal Failure

Communication failure is the other side of the coin: lack of congruence. We need to distinguish between proximal and distal failure. Brunswik's terms (Hammond, 1966) are suitable here because they help to introduce an inductive paradigm. His idea was that our perception of distal objects, such as a motor car approaching, or unfriendliness in an informant, is mediated by a computation on proximal cues, namely the more elementary units of visual and auditory perception. It is not clear how decisively natural dialogue can be segmented into elementary and higher-order units. (We generally beg the question in the layout of the written version.) It would be surprising, nonetheless, if the transaction, the speech act, and the word, to mention but three units, do not turn out to be "natural classes" which can be mapped onto lower-level cues with a good deal of certainty. A transaction is a distal object of dialogue, such as obtaining information, buying something, introducing people to each other. Some transactions are "set pieces" for which we can supply a "script" and a "discourse chart", but of course many others will have a general name like "having a conversation" and will pose more serious problems of segmentation. I think of speech acts (assertion, interrogation, exclamation, instruction, directive, promise, and the like) as the middle-distance objects of dialogue, corresponding to the tables and chairs of the perceptual world. They are normally the smallest meaningful "move" or "action" in dialogue. The decision-theoretic model depends crucially on our being able to identify them, since they are the individual actions of a dialogue and the bearers of costs and values. I will take words as given proximal cues in the recognition of speech acts, ignoring the notorious problems of word-recognition in speech decoding.

Distal failure is failure in larger units of discourse, which might be the entire transaction, or part of it. It may not be evident at the time. There can be failure to correctly identify speech acts, as when we do not realise that we are being asked something. Very often this kind of failure is evident only in retrospection. Proximal failure, such as failure to hear individual words, or failure to decode some piece of input, is more likely to be noticed immediately. For our purposes the important relationship between proximal and distal failure is that the former does not necessarily lead to the latter. Normal conversation can cope with a certain amount of proximal failure. In the area of second language learning considerable work has been done on the "strategies" by which the learner can handle very large amounts of proximal failure while still preserving reasonable odds of a successful distal communication (Faerch & Kasper, 1983). But if proximal failure does not entail distal failure, it does of course make it more likely. There is an inductive relation between them.

The value of communication and the cost of failure are inversely related, just as communication and failure are. If there is a lot to be gained by successful communication in a particular dialogue, then there is also a lot to be lost by failure. We will have to think of gain and loss too in somewhat general terms, since it is only in very exceptional circumstances that the value of communication could be quantified with any plausibility. An educational example which comes to mind is that of the oral examination in a second language, where specific penalties are attached to dialogue failures of different kinds. In other words failures are preclassified and assigned a rating on a scale of "seriousness". Another is in experimental paradigms of the "twenty questions" or the "concept attainment" variety, including professional diagnosis from limited information by means of question asking. Here it is understood that a cost-unit attaches to every request for information, and that optimal strategies can be defined in decision-theoretic terms. The assumption is borne out by the facts. Good interrogators maximize the expected informational yield of their questions, taking into account the informational value of possible answers (the number of hypotheses they falsify) and the likelihood of obtaining such answers under the obtaining model of the informant (Bruner, Goodnow, & Austin, 1956; Kleinmuntz, 1968). For the most part, however, we cannot be so precise about the reasons we have for the moves we make in a dialogue. The most we can do is allude to the "value" of communication in some very general sense, which is partly social, partly cognitive, partly pragmatic.

4 Relativity of the Notion of Failure

It is only when the participants have lost something through miscommunication that we can talk about "failure" or "breakdown" in dialogue in something like the ordinary sense of these terms. Some dialogues, for example the sort that takes place between an egocentric talkative person and a captive, uninterested listener, might be said to be in a state of permanent failure. Congruence of meaning scarcely exists, and efforts to ensure it or check on it may be few and far between. (If there is no effort at all to maintain congruence we may wish to withdraw the term "dialogue".) But in another sense nothing is failing, nothing is breaking down. In fact as far as the ordinary meaning of the terms "failure" and "breakdown" is concerned we could just as well say that the conversation is immune from failure, since neither party has much to lose through miscommunication. Conversely, when we stand to suffer heavy losses due to misunderstanding, we will often consider the communication unclear, to the point of asking for repetition and clarification, even though the probability of miscommunication, viewed by a third party, is very slight. One thinks also of legal and philosophical dialogue, in which there may be a considerable outlay of resources in order to achieve slight gains in clarity.

If we except the kind of dialogue which is closely mapped onto an external task, we cannot talk in any absolute sense, therefore, about successful communication or its opposite, failure, but only about successes and failures relative to the resources of the participants and the special purposes being served by the dialogue. The point is partly acknowledged in the distinction between "input failure" and "model failure" (Ringle & Bruce, 1980) in which the latter is relative to the capacity of the listener. But even with a fixed capacity, failure is relative to costs and some model of rational decision making. Pressed for a definition of failure, the best that can be done is to say that dialogue is approaching failure when the costs of present intervention relative to distal failure are such that a rational listener is obliged, by norms of rationality, to admit failure.

The following extract from an oral examination in a second language (Irish) will illustrate some of the points being made. It is an instance of partial communication. The examiner is trying to elicit "normal" conversation from the student, consisting of small-talk about the student's interests and such like. It is understood that students will get credit for being able to sustain the conversation, and that the examiner will adapt the level of the conversation to the ability of the student. The examiner speaks first. I have translated errors in the student's Irish with English errors which are similar in type and gravity.

1a Now when you're coming to school, are you alone or is there somebody with you?
1b I come with my friend.
2a What's her name? Or is it a boy? Or a girl?
2b A girl.
3a A girl? What's her name?
3b Maighréad.
4a Is she in your class?
4b It is.
5a Were you talking to her about the exam?
5b She is very nice, and she lives down the road from me.
6a Yes ...
6b And we went, in the summer we went to Dublin.
7a How did you go? By train? Or ...
7b By train, yes.
8a How much did it cost? How much money did you have to pay?
8b It was cheap, but my mother gave me the money.
9a Yes. Where did you go?
9b We went to the shops, to the Phoenix Park, to the Zoo.
10a Yes. What did you see there?
10b Monkey, zebra, and a lot more.
11a Were you allowed give food to the animals?
11b Yes, we ate our lunch in the Zoo, and bought sweets and lemonade in the shops.
12a Very good.

The dialogue clearly fails in places. But we cannot say that it has failed completely at any point. Turn-taking norms are broadly respected, and the topic-shifts which result from the student's incomplete understanding of the previous utterance are not as serious as they might have been. One might argue that the dialogue fails in absolute terms in the last exchange, and that a more honest examiner would have signalled the fact. But other cases could be found to make the point that any distinction between (a) dialogue which involves a large amount of incomprehension on one side, and is sustained only by massive amounts of cooperation from the other side, and (b) mutually incomprehensible dialogue sustained by insincere "go-ahead" signals, must, in the end, be a difference of degree rather than one of kind. This is the point made earlier about the relativity of the notion of failure to dialogue goals and agreed standards of coherence.

In attempting to enumerate the different levels at which dialogue can fail, Riley (1980) recommends the traditional fourfold division: pragmatic, illocutionary, syntactic, and semantic. Communication can be successful at one level but fail at another. We can understand the literal meaning of an utterance and still wonder why it has been directed at us by the speaker. Here communication is achieved at the semantic and syntactic levels, but falters at the illocutionary and pragmatic levels. Conversely, as in the case of the student above, communication can falter at the semantic and syntactic levels and still be maintained by appropriate turn-taking, by appropriate illocution (replying to an interrogative with a declarative), and by partial successes in topic development. Whatever marks we deduct from the student for the lapses in 5b and 11b, where the student has not fully understood the question and continues with a related topic instead of asking for clarification, it is still true that even more would have to deducted if the student changed topics entirely, or didn't know that he was being asked a question, or didn't even know that his turn in the dialogue had arrived. So we cannot say, in any absolute sense of the term, that the dialogue above "fails" at 5b, or 6b or 11b, but only that it falters, or fails to a degree.

There is a sound epistemological basis for retreating to relativistic notions of communication and failure, namely the relativity in the ascription of propositional attitudes to others (Dennett, 1971). As long as we remain within the idiom of propositional attitudes, of which the idiom of speech-acts is a special case, there can be no absolute criterion of synonymy, even when we paraphrase ourselves. As we make the criteria for synonymy more and more demanding, there is no point at which we move from inductive estimates of the other person's meaning to deductions and translations which are somehow "guaranteed by the very meanings of the words" (Quine, 1960, Ch. 6). In addition, the adequacy of our inductive synonymy-standards must be assessed relative to the task in hand. And sometimes we must wait quite a bit before deciding about adequacy, since an apparently successful communication can turn out later not to be so.

Riley (1980) argues that a dialogue which descends into total incomprehensibility, in the sense that the participants no longer understand a single word of what they are "saying" to each other, can still constitute communication of sorts if the exchanges are in accordance with some pragmatic norms, such as those governing turntaking, register, politeness, and so on. The claim is difficult to assess since we have no idea what kind of pragmatics might apply to incomprehensible utterances. I would say that we must know what the participants hope to achieve through the dialogue before we can talk about pragmatic norms, or decide whether they have been complied with. Here too the notions of communication and communication failure force us back to prior questions about the minimal axioms of rationality, and the minimal sharing of those axioms which must be assumed before we can talk about a valid piece of dialogue. The topics of rationality and cooperation are now introduced jointly.

5. Cooperative Dialogue

It is assumed in normal dialogue that the participants will cooperate with each other by correcting mistaken assumptions, by avoiding responses which though correct in themselves could lead to mistaken inferences, and by a variety of other means calculated to maximize the value of the exchange for the other participant. In the dialogue above the examiner tends to ask "multiple choice" questions (2a, 7a), speaking in menus, as it were, in order to make it easy for the student to respond in a natural way in spite of limited competence — a metalinguistic goal which in the unusual setting of a language examination is the principal goal of the student.

The important elements of a cooperative relationship in dialogue are outlined in Figure 1. The cost of failure was mentioned earlier. To this is now added a congruence estimate and the

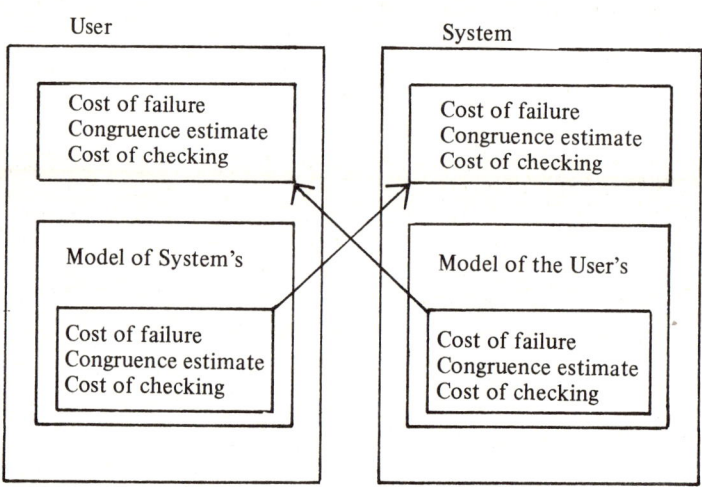

Figure 1. Elements of cooperative dialogue.

cost of checking. The congruence estimate may be thought of as the subjective probability of failure, the likelihood, in the view of each participant, that the dialogue will fail. The term "checking" is intended to cover all the actions which a speaker can take to make this outcome less likely, in other words, actions likely to improve the current index of congruence. The principal cost of checking will usually be disruption of the dialogue, although we can think of more formalized situations, diagnostic tasks with time limits, for example, where there is an extra-linguistic price to be paid for each check. In the dialogue above it is clear that the student considers requests for clarification more disruptive than illicit topic-shifts, probably as a result of the advice often given to candidates for oral examinations not to stop or to admit failure.

Cooperation in the ordinary sense also assumes some estimate of the costs of failure for the other and some willingness to make these costs one's own. In addition some estimate of the other's perplexity, or fear of failure is essential, including unrealistic fears. The listener must be prevented from thinking that somebody is wrong when in fact the dialogue is working. Some estimate of the cost to the other of checking is also needed, since it will be related to the base-rate frequency of perplexity and go-ahead signals.

The essential element of cooperation, however, is not any of these but an adjustment of the listener's cost of failure in the direction of the perceived cost of failure to the speaker. In the case of human dialogue we think in terms of some kind of averaging, "my neighbour as myself" as it were. In the case of machines, on the other hand, we expect a simple replacement of the system's costs with the user's costs. In practice we would be happy if the machine behaved like a moderately cooperative person.

The modelling of the other's costs is a more complex operation than it might seem at first. It involves hypotheses about the likely behaviour of the other, which in turn involves suppositions about what is reasonable for people to do in such situations. These are matters closely related to failure and the detection of failure in dialogue. In fact it is only because certain norms of rationality can be presupposed, as a kind of constant backdrop to dialogue, that the idea of congruence can be discussed.

6. Models of Rationality

The importance of the concept of rationality in the study of language and cognition is now generally conceded (Egan, in press). The central point is that cognitive attributions of all kinds, starting with the attribution of meaning to words and sentences, presuppose that certain norms of rationality are being complied with. The point is fairly obvious when it comes to the attribution of intentions to actions, including speech-acts during dialogue. We cannot believe or assert things at will. Requirements of consistency must be met or else assertions degenerate into word-strings, possibly interpretable at lexical and syntactic levels, but not at the level of speech acts. It follows too that the attribution of non-rationality to others is extremely difficult to defend on epistemological grounds (Quine, 1960, Ch. 6; Cohen, 1981). Faulty translation is not only the most charitable interpretation for seemingly irrational discourse, but also the most defensible explanation. In the context of dialogue this means that we should consider failure a more reasonable hypothesis than inconsistency in the respondent, and more importantly, that apparent inconsistency will be an important indicator of impending failure.

The full scope of the term "inconsistency", and its connection with the notion of failure in dialogue, may be seen in the following dialogue, taken from Faerch & Kasper (1983). It is very similar to the dialogue previously reported. A native speaker of English is speaking to a learner.

1a What do you read at home?
1b mmmm History.
2a What subjects do you read about?
2b I read history home and sometimes ...
3a Mhm.
3b In my school.
4a Do you like history. In school. Do you like learning history?

4b Yes.
5a Do you have history lessons in school?
5b Kings?
6a When you learn about, I don't know, old kings.
6b Oh yes, I have that. No.
7a Yes. Do you like it?
7b Not about history.
8a But you like reading books about history?
8b Err. This history, you know, young histories.
9a Aha. For example, 1930 or so.
9b Yes.
10a Do you mean recent, in more recent years?
10b Err. A history is, maybe, on a boy, girl, ...
10b ...
11a Young people, life, ...
12a Oh you mean a story, maybe.
12b Yes
13a Just a story, about people. Not ...
13b No.
14a Not necessarily in the past.
14b No.
15a Now I understand you.

A less persistent interrogator might not have detected the "faux ami", the Danish word "histoire" which can mean a story or a novel as well as history. If the error was not detected the dialogue, like the previous one, could still have limped along with a partial overlap of meanings. In fact, however, the interrogator forces a certain level of articulation and coherence on the student, in cooperation with the student's goal to learn English. As a result, an apparent inconsistency points the way to some faulty translation.

What categories of incongruence signals can be proposed in cases such as this? Let us set aside for the moment the many direct signals of distress, in the form of gestures, laughter, facial expressions, and so on, which would be an integral part of the dialogue in a fuller transcript. Confining ourselves to the present transcript, we first note a good many proximal failures, silences, expressions like "young history" which are ill-formed in the sense that they violate grammatical selection restrictions (Weischedel & Sondheimer, 1983). There are also some violations of what may be called the etiquette of dialogue. The pattern of turn-taking is not very distinct, and in 6b, "Oh yes, I have that. No", the respondent appears to reply to different questions in the same response. Finally there is inconsistency of content, when the student appears to be contradicting herself, saying that she doesn't like history when previously she said that she did. In dialogues relating to external tasks contradictions of this sort will sometimes take the form of goal obstacles, which point to failure in view of the assumption that the listener will not be expected to do the impossible (Goodman, 1983, p. 210).

The following, therefore, might all contribute, negatively, to a congruence index:

1 Non-verbal signals of perplexity
2 Proximal failures
3 Mismanagement of dialogue
4 Apparent inconsistency in the content of dialogue

Proximal failures can be identified roughly with perceptual, lexical, and syntactic failures (Ringle & Bruce, 1980). Violations of the "grammar" of dialogue occur at the illocutionary or speech-act level. Inconsistency in the content of dialogue may be said to occur at the level of "plans", since it is the term of "plan", and related terms such as "goal", "goal compatibility", "obstacle", and such like, which introduce, in an indirect way, the notion of rationality which I want to bring into focus here.

7. Failure Detection and the Classification of Speech Acts

In decision-theoretic and plan-based approaches to dialogue the speech-act is the primary unit of analysis, corresponding as it does to a single action in the execution of some non-linguistic plan, or a move in a game. Given some suitable classification of the speech-acts in a dialogue, and assuming, as seems reasonable, that there are certain constraints operating on the sequence of moves, it should be possible to establish norms, which in turn could be used to measure the deviations of actual sequences from the norm, thus quantifying the extent of dialogue mismanagement.

"Grammars" of dialogue, operating on a lexicon of speech-acts, have been proposed by Sinclair and Coulthard (1975), Burton (1981), and Riley (1980). There are serious problems with grammars of dialogue. Of these the most serious is the problem of circularity or empirical vacuity, since there is often enough vagueness in the definition of the speech-acts to allow us to find an admissable parse by reclassifying the deviant act (Levinson, 1983; Toolan, 1985). Nonetheless, grammars of dialogue can be used as rough guides to normative dialogue management, akin to rules of etiquette. In the model of dialogue proposed for the ESPRIT project on dialogue failure (Christie, et al., 1985) a modification of Burton's grammar is expressed as an ATN and is fitted to representative dialogue. Empirical transition probabilities are used as prior probabilities for incoming dialogue moves and yield a measure of "goodness-of-fit" of the ongoing dialogue to the grammar. Such a measure would be one contribution to an overall index of congruence.

Its contribution would have to take other factors into account. Firstly, it would be unrealistic to consider the classification of speech-acts as something which is done once and for all. It must be open to revision. In practice many utterances will admit more than one classification. This can be acknowledged by an initial provisional classification and possible alternative classifications. The alternatives might be determined on an *a priori* basis, or in the light of characteristics of the present utterance. Given an initial parsing of the dialogue, and its attendant measure of fit, we envisage an iterative process which now works on a passage of dialogue, substituting plausible alternatives locally in order to arrive at a classification which maximizes congruence over the passage as a whole. Such a procedure openly acknowledges the circularity in syntactic-like descriptions of dialogue and thus purges it of the "viciousness" it would have if it were not explicitly acknowledged.

Secondly, the computation of congruence at this level of the dialogue must interact with similar procedures operating at both higher and lower levels, at the lexical and syntactic levels, on the one hand, and at the plan-level on the other hand. It is difficult to say how the interaction should be modelled. According to the principle of charity mentioned earlier, congruence at plan-level must be conceded. It must be assumed that people will not request the impossible, that they will not knowingly contradict themselves, and so on. In other words, incongruence at plan level will be treated as apparent incongruence only – not, of course, because this is necessarily the case but because cooperative dialogue is unthinkable otherwise. In this context the axioms of "common sense" will take the form of some small set of rules from logic and decision theory, such as the "naturalistic" sets of axioms proposed respectively by Fitch (1952) and Edwards, Lindman, and Phillips (1967).

The importance of incongruence at plan level lies in the fact that it (1) indicates failure at lower levels, and (2) provides a criterion by which adequate correction can be recognized, namely the lessening of incongruence. In general the effect of incongruence at plan level is to relax the criteria at lower levels in order to find a more acceptable reading higher up. Naturally the process must be carried on in conjunction with the iterative classification of speech acts described above. Essentially it operates by reinstating some of the less plausible speech-act classifications, in the hope of trading off the loss of illocutionary congruence for a gain in congruence at plan level.

A similar mechanism might be tried at lexical and semantic levels. It is not clear, however, whether any priority can be recommended in the relaxation of local congruence indices. The crucial construct here is the rank ordering of features for relaxation (Goodman, 1983 and this volume) or the "weighting" of different forms of relaxation, for example those at the level of topic and those at the level of relevance (Carberry, 1984 and this volume), in order to arrive at an optimal repair. The ordering is certain to depend a good deal on the task and the capacities of the participants. In dialogue with non-native speakers, for example, the lexicon is always suspect, but not with native speakers.

8. Conclusion

When dialogue is looked at as decision-making under uncertainty our attention is drawn to the description of dialogue at the level of speech acts, since these are the dialogue equivalents of moves or actions. Failure appears as a continuous rather than a discrete feature, and the detection of failure is centred on inductive mechanisms which integrate signals of incongruence from various descriptive levels of the dialogue, ranging from the perceptual and lexical levels to the level of planning, at which incongruence can be detected in terms of incompatibility with basic laws of logic and decision theory. The model of failure detection is based on a provisional identification of the speech-act sequence, including likely alternative classification, which is then iterated to maximize its fit to a normative sequence. Further revisions may be necessary to reduce incongruence at plan level, including relaxations, i.e., introduction of alternatives which are locally less likely, at syntactic and lexical levels. Incongruence at plan level is obligatorily treated as illusory. Effectively it becomes a signal of failure at lower levels. It is removed at the cost of increased incongruence at lower levels. The model has the capacity to set up expectations at each point in the dialogue concerning the likeliest speech acts to follow, and to paraphrase previous speech acts in terms of dialogue goals. The most urgent requirement for developing such a system at present is a good empirical fit of the illocutionary "grammar" to corpora of dialogue gathered in standard settings, such as cooperative tasks or question-answering sessions.

References

Allen, J.F. & Perrault, C.F. (1980). Analyzing intentions in utterances. *Artificial Intelligence, 15*, 143-178.
Bruner, J.S., Goodnow, J.J., & Austin, G.A. (1956). *A study of thinking.* New York: Wiley.
Burton, D. (1981). Analysing spoken discourse. In M. Coulthard & M. Montgomery (Eds.), *Studies in discourse analysis.* London: Routledge & Kegan Paul.
Carberry, M.S. (1984). Understanding pragmatically ill-formed input. In *Proceedings of Coling '84.* Stanford, CA.
Christie, B., Egan, O., Ferrari, G., Gardiner, M., Harper, J., Reilly, R., & Sheehy, N. (1985). *Communication failure in dialogue: Techniques for detection and repair. Deliverable II: A dialogue classification system.* Commission of the European Communities: Esprit Project 527. Dublin: Educational Research Centre.
Cohen, L.J. (1981). Can human irrationality be experimentally demonstrated? *Behavioral and Brain Sciences, 4*, 317-370.
Cohen, P.R., & Perault, R.C. (1979). Elements of a plan-based theory of speech acts. *Cognitive Science, 3*, 177-212.
Dennett, D.C. (1971). Intentional systems. *Journal of Philosophy, 68*, 87-106.
Dummett, M. (1976). What is a theory of meaning? In G. Evans & J. McDowell (Eds.), *Truth and meaning.* Oxford: Clarendon Press.
Edwards, W., Lindman, H., & Phillips, L. (1967). Emerging technologies for making decisions. In F. Barron, W. Dement, W. Edwards, H. Lindman, L. Phillips, J. Olds, & M. Olds (Eds.), *New directions in psychology* (Vol. 2). New York: Holt, Rinehart, & Winston.
Edwards, W., Lindman, H., & Savage, L.J. (1963). Bayesian statistical inference for psychological research. *Psychological Review, 70*, 193-242.
Egan, O. (in press). The concept of belief in cognitive theory. *Annals of Theoretical Psychology, 4.*
Einhorn, H.J., Kleinmuntz, D.N., & Kleinmuntz, B. (1979). Linear regression and process-tracing models of judgment. *Psychological Review, 85*, 465-485.
Faerch, C., & Kasper, G. (1983). Plans and strategies in foreign language communication. In C. Faerch & C. Kasper (Eds.), *Strategies in interlanguage communication.* London: Longmans.
Fischoff, B., & Beyth-Marom, R. (1983). Hypothesis evaluation from a Bayesian perspective. *Psychological Review, 90*, 239-260.
Fitch, F.B. (1952). *Symbolic logic.* New York: Random House.
Goodman, B. (1983). Repairing miscommunication: Relaxation in reference. In *Proceedings of IJCAI '83,* Karlsruhe, West Germany.

Grosz, B. (1981). Focusing and description in noctural language dialogues. In A. Joshi, B. Webber, & I. Sag (Eds.), *Elements of discourse understanding*. Cambridge: University Press.
Hammond, K. (1966). The psychology of Egon Brunswik. New York: Holt, Rinehart & Winston.
Herdan, G. (1966). *The advanced theory of language as choice and change*. Berlin: Springer-Verlag.
Hobbs, J., & Evans, D.A. (1981). Conversation as planned behaviour. *Cognitive Science, 4*, 349-377.
Kleinmuntz, B. (Ed.). (1968). *Formal representation of human judgment*. New York: Wiley.
Levinson, S.C. (1983). *Pragmatics*. Cambridge: University Press.
Mamdani, A., Efstathiou, J., & Pang, D. (1985). Inference under uncertainty. *Commission of the European Communities: AIP Information Newsletter, 85-08*, 136-142.
Pause, P.E. (1984). Das Kumulationsprinzip – eine Grundlage fuer die Rekonstruktion von Textverstehen und Textverstaendlichkeit. *Zeitschrift fuer Literaturwissenschaft und Linguistik, 14*, 38-56.
Quine, W.V.O. (1960). *Word and object*. Cambridge MA: MIT Press.
Riley, P. (1980). When communication breaks down: Levels of coherence in discourse. *Applied Linguistics, 1*, 201-216.
Ringle, M.H., & Bruce, B.C. (1980). Conversation failure. In W.G. Lehnert, & M.H. Ringle (Eds.), *Strategies for natural language processing*. Hillsdale, NJ: Erlbaum.
Saarinen, E. (Ed.). (1979). *Game-theoretical semantics*. Dordrecht, Holland: Reidel.
Sinclair, J. McH., & Coulthard, R.M. (1975). *Towards an analysis of discourse*. London: Oxford University Press.
Suppes, P. (1984). *Probabilistic metaphysics*. New York: Basil Blackwell.
Toolan, M. (1985). Analysing fictional dialogue. *Language and Communication, 5*, 193-206.
Weischedel, R.M., & Sondheimer, N.K. (1983). Meta-rules as a basis for processing ill-formed input. *American Journal of Computational Linguistics, 9*, 161-177.
Zaefferer, D. (1977). Understanding misunderstanding: A proposal for an explanation of reading choices. *Journal of Pragmatics, 1*, 329-346.

PERSONAE: MODELS OF STEREOTYPICAL BEHAVIOR[1]

Henry B. McLoughlin

Department of Computer Science
University College Dublin

A dialogue system interacts with people and therefore it needs to represent knowledge about them and the situations in which they and the systems will interact. This knowledge should be sufficient to allow the system to understand the behavior of the user and also enable it to plan to influence them. Much of behavior is stereotypical. Having recognised persons' behavior as being in accordance with a particular type we "identify" them with that type and expect them to behave in a certain way. This paper reports on preliminary work to develop sufficiently rich models of individuals which incorporate this idea of stereotypes. Knowledge structures called *Personae* are proposed which contain information about a particular stereotype. The acquisition and use of Personae are discussed and some problems for future research are outlined.

1. Introduction

How do people organize the knowledge which they must have in order to understand the behavior of other people? How does one know how to behave towards strangers the first time one meets them?

In this paper we consider the type of knowledge of others which we seem to possess and which we use both in understanding how others behave and in planning to influence their behavior. We propose a structure called a Persona which is used to organize our information about people. We discuss the acquisition and use of Personae and we consider their generality.

The ideas discussed in this paper have arisen out of work which is being carried out in the Department of Computer Science at U.C.D. to develop a computer based system which will play the role of a telephone receptionist in typed simulations of telephone conversations. In particular, the ideas were developed as a means of solving a number of problems which we encountered in our attempts to represent the knowledge which the system needs to maintain about its users.

2. Knowledge Requirements of an AI System

An Artificial Intelligence (AI) system needs to have knowledge of the world in which it operates. It needs some way of representing what is the case in the world, and how changes can be brought about in that world. The SHRDLU system (Winograd, 1972), which operated in a table-top blocks world, needed to know what blocks existed, and what their identifying properties were. These properties included their colour, shape, size, and position. In addition, the system knew of a number of operations which it could perform to bring about changes in the world.

---

[1] This research project is being partly funded by the National Board of Science and Technology.

These operations had a set of conditions which had to be true when the operation was performed, and a set of results which would be true afterwards. Given a particular state of affairs in the world, and the goal of achieving a different state of affairs, the system could form a plan, consisting of a sequence of operations, to achieve this goal.

When dealing with a world as simple as the blocks world this knowledge was sufficient for the system to perform well. However, as we progress towards building larger systems, where the world in which the system operates is more complex, there is a corresponding increase in the complexity of the knowledge which the system's world model must contain. In addition, there is a decrease in the certainty that this world model corresponds with the actual world. Both of these factors are particularly apparent in systems whose world model contains knowledge of people.

Most models of cognition in AI are variants of what Konolige (1985) terms the belief-desire-intention paradigm. An agent with a set of beliefs about the world and a state of affairs which he desires to bring about will form intentions or plans to bring about that desired state of affairs. Thus, a system will have a set of goals which it wants to achieve, and a model of the world which will include both its beliefs about objects in the world and its knowledge of plans which cause change in the world. It seems reasonable that the system should model other agents as being structurally similar to itself. So, it will represent others as having beliefs and goals and a set of plans to achieve these goals.

It is not possible to directly observe the beliefs, goals or plans of another person. Our knowledge of them will usually be indirect. They may tell us about them or we may infer them from observing their actions, but in neither case can we be fully certain that our model conforms with reality.

When we engage in conversation with other persons we are acting according to a plan we have formulated which we believe will achieve some desired goal. We may be attempting to get the other persons to perform some action for us or we may be supplying them with some information. Either way we are seeking to bring about a change in their beliefs. How is it possible to plan to influence the beliefs of others? We would need to know about their beliefs and we would need to know how changes could be brought about in those beliefs. Here we run into the problem of what in speech act theory is referred to as perlocutionary effect (Searle, 1969; Austin, 1962).

A perlocutionary effect is an effect brought about as a result of an utterance. Thus, by arguing with you I may bring about the effect of convincing you. The arguing is what is known as an illocutionary act. There is never a guarantee that a given utterance will bring about a desired perlocutionary effect. Sometimes, and with some people, a particular utterance may produce the desired effect; other times it may fail.

Cohen and Perrault (1979) developed a system which could plan to produce illocutionary acts. Their system maintained a model of the world which included the beliefs of other people. These beliefs were nested to allow the system to reason about the users' beliefs about the system's beliefs. The system had knowledge of a number of simple operators which produced changes in the belief models of the system and the user. Given a desired state of affairs, such as wanting to know the time of arrival of a train at a railway station, the system would produce a plan to achieve the goal. The plan would contain operators which corresponded to illocutionary acts; when these were executed they would produce certain effects in the belief structure of the hearer. They were primarily interested in illocutionary acts and they used simple operators to account for the perlocutionary effects. An example of one of these was their Cause-to-want operator. If the hearer believed that he could perform a certain act, and if he believed that the speaker wanted him to perform the act, then he would want to perform it.

Clearly, persuading someone to perform some action is a more complex matter. Consider the following piece of dialogue. Mario is a member of the Mafia and Mr. Casey is the owner of a grocery store.

Mr. Casey: "I haven't got the money, you'll have to come back next week."

Mario: "Mr. Casey, your little daughter is so sweet. I hope you take good care of her."

Mr. Casey: "Look I'll find the money somewhere, give me another day."

Mario seems to persuade Mr. Casey to get the money, but, for us to understand what is happening we need to supply a lot of knowledge about the people concerned and the types of things which they are capable of doing. Mr. Casey is persuaded because he detects a threat to his daughter. He uses his knowledge of the Mafia and their tactics to decide how serious the threat is. From Mario's point of view, he uses his knowledge of Mr. Casey's goal of protecting his daughter and his knowledge of Mr. Casey's knowledge of him and his Mafia background to make an utterance which will be seen as a threat and will persuade Mr. Casey. While this example may be a bit contrived it does show rather clearly how both understanding the behavior of others and planning to influence their behavior depends upon having a lot of knowledge about their beliefs, plans and goals.

Goffman (1959) states that "Information about the individual helps define the situation, enabling others to know in advance what he will expect of them and what they may expect of him. Informed in these ways, others will know how best to act in order to call forth a desired response from him". In our daily lives, we encounter quite a number of people in different situations. Most adults are able to cope with these encounters; they can understand the behavior of the other person and make their ideas and wants known to them. Where the person with whom we are dealing is known to us it is reasonable to assume that we bring to bear the knowledge which we have gleaned from past dealings with that person. The more knowledge we have about them the more certain we are that we will be able to understand their behavior and draw from them the responses we desire. As people get to know each other more, the likelihood of their misunderstanding each other diminishes.

3. Stereotypes

There are, however, numerous times when we encounter someone whom we have never met before. In coping with such situations we seem to bring into play knowledge which we have gleaned from experiences with similar individuals. For example, when we meet a particular taxi driver for the first time we use our knowledge of taxi drivers in general in dealing with him. Similarly, as noted by Gahagan (1984), "We do not need to know anything about the personality of a bank clerk to be able to predict what he or she will do when we hand in a pay-in slip and some money."

It seems that we abstract from our experiences with a number of different people a stereotype which represents the common aspects of their behavior. Gahagan (1984) states that "Stereotypes are beliefs about the characteristics possessed by some group and possessed therefore by any member of that group." Once we possess a particular stereotype, we can use the knowledge which it contains to understand the behavior of members of the class which it represents. It acts like a filter through which we see others. The advantage of this is that we will have less work to do in understanding their actions; there are expectations about the sort of activities which they should be performing. This makes understanding easier for us.

Few of us would experience difficulty in understanding the references made in the following utterances:

"He was a typical salesman."

"If you're going to behave like a student then you shall be treated like a student."

"But what could one expect from a paid up member of the Communist Party?"

In the above, the speaker seems to presume that the hearer will understand what is being said by virtue of the hearer applying his knowledge of typical salesmen, students and Communist Party members. Mentioning the particular stereotype seems to call to mind the knowledge which we associate with that stereotype.

The amount and type of knowledge which is stored will vary but we can categorize the knowledge under a number of headings. There will frequently be a goal or set of goals which guide the person's behavior. Thus, a salesman will usually aim to sell a lot of goods, make a high commission, and get recognition within the company as the best salesman. A communist may want to practice the communist way of life as he sees it and to spread the doctrine of communism.

To achieve these goals there may be a number of strategies which the person will typically use. Often we can categorize a person by observing the strategies they are using. A person who seeks to influence us by complaining about us failing to perform certain acts is more likely to be classed as a typical mother-in-law (should the person be female) rather then as a salesperson.

Many stereotypes have particular props associated with them. During the late sixties and early seventies it was generally believed that all students were long-haired radicals who wore denim. Evidence to the contrary abounded, but such was the power of the stereotype that many people simply ignored it. This is one of the interesting things about stereotypes: they may be very inaccurate, but still be widely used. Consequently, behavior may be completely misunderstood by a person applying the wrong stereotype.

4. Personae

We propose a data structure which we call a *Persona* to contain and organize the knowledge of stereotypes and their behavior which we use when we interact with others. A Persona is a collection of information which we have about a particular type of person. It contains knowledge about what goals we can expect them to pursue, what means they will employ in achieving them, and what obligations they assume when playing the role. It outlines the beliefs which we can expect them to have and the beliefs about them which we may assume. It describes the props which they use in their performances and the qualifications which they must have in order to adopt the persona.

4.1 Elements of a Persona

A Persona is designed to contain and organize enough knowledge to be able to account for all the aspects of a stereotype which we have described in the previous section. This does not mean that all the elements which we discuss below will always be present in any given Persona. Instead, a Persona can cope with as much or as little information as is available.

4.1.1 Overall Goals

Associated with a Persona there may be a set of goals which the person assuming the Persona (henceforth the player) will have. These goals will often represent the reasons why the player is assuming the Persona. These are not the only goals that the player will have, they are the ones which characterize the Persona.

4.1.2 Plans

In our model intentional behavior results from an agent acting out a plan which has been formulated to achieve a particular goal. We have mentioned earlier that stereotypes often exhibit certain behavioral characteristics. These are modeled by a set of plans associated with the Persona. These plans are used by the player to achieve his goals. We note that the player is not limited to these and only these plans; what we are saying is that the player is likely to use these plans because they represent what we consider to be the typical behavior of this type of person.

These are the plans which result in the characteristic behavior associated with the Persona. For example, if we hear that Jake and his gang have held up the stagecoach and that the sheriff decides to round up a posse, then we can attribute this action to the sheriff performing one of the characteristic actions associated with the Persona he is playing. If we heard that he had gone to visit his aunt then we might be a bit confused.

Similarly, if we heard that the bank had been robbed and then heard that Fred had started to round up a posse then we would assume that Fred was the sheriff. We know that only a certain type of person is qualified to perform this action so if someone performs it then we must assume that he or she is actually qualified and thus we can bring to bear all the knowledge we have about that type of person if we notice him or her performing one of these acts.

4.1.3 Obligations

Obligations model the ways in which we are able to produce a desired response from the player of the Persona. One of the problems which we referred to earlier was how one can possibly plan to call forth a given response from someone. The fact that we do so is beyond question. We will confidently summon a cab and persuade the driver to bring us to our destination even though we may never before have met the driver. We ring the telephone operator and ask for a particular number and it never occurs to us that they might refuse or ask us why they should do so. We summon the fire brigade in an emergency without having to decide on exactly how we will go about convincing them to leave the fire station and drive to our house.

Airenti, Bara and Colombetti (1984), have proposed an elegant way of representing the type of obligation we are concerned with here. They make use of a theme which they call the Profession theme and a number of contracts which represent socially determined games such as taking a taxi. These contracts have a number of roles associated with them. The taxi contract has the roles of subject and client. The Profession theme will generate the motivation for persons to play their role in a contract in which they are the subjet, if others indicate that they wish to play the role of client.

Using this theme and a number of contracts, we can represent the obligations associated with the Persona. Each contract will represent a particular task which the player may be required to perform.

4.1.4 Situations

Most Personae have associated with them a number of situations in which we would expect to find the person playing a role. Some Personae are very strongly tied to situations. A waitress almost always plays her role in the restaurant situation. A salesman will have a number of typical situations in which he plays his role. Parents on the other hand, whilst they will have a number of characteristic situations, will be less tied to them and can perform in almost any setting. These situations can be represented by using the situational scripts proposed by Schank and Abelson (1977).

4.1.5 Beliefs

We can associate certain beliefs with those who are playing a particular Persona. We may have beliefs which we attribute to them and beliefs about them. These beliefs enable us to know what knowledge we can expect players to have and what we expect them not to know.

4.1.6 Props

Associated with most Personae is a set of physical props. These are used by persons both to perform their actions to achieve their goals and to confirm the role which is being played. Thus a doctor's stethoscope acts both as an instrument to examine the patient's respiration and as a prop to confirm that the person playing the role is playing the role of doctor. These props can be very helpful in recognizing the particular stereotype.

4.1.7 Qualifications

These are the qualifications which we believe the player must have if he is playing the role. These can be assumed unless we have evidence to the contrary. We presume that a doctor has actually been trained in medical school, that a salesman is actually selling for the company which he claims employs him, and that an electrician is capable of wiring our house.

5. Some examples of Personae

Some of the easiest Personae to recognize are those which are associated with specific occupations in society. Some examples are the Salesman, the Policeman, the Mailman and the Waitress. These correspond closely with what Schank and Abelson (1977) have called role themes.

We shall consider the salesman Persona in some detail. A salesman will usually have goals to sell as many goods as he can so as to get as much commission as possible. He will have a number of strategies which he uses. One of these is the "hard sell" where he attempts to sell by bombarding the potential buyer with information on the product to such an extent that the buyer may buy the product if only to get the salesman to go away.

On other occasions he may use the "soft sell" where he gently cajoles the buyer, attempting to convince him to buy the product without explicitly mentioning the purchase until he has sold the idea first. Door to door salesmen may try to sell by convincing the householder that unless they buy the product they will be the only family on the block without it.

When we think of a typical salesman we can call to mind those situations in which we would expect to see them playing their roles. These include the "door to door" situation, the "shop floor" situation and the "telephone sales" situation. Each of these is represented by situations within the Persona.

A salesman is obliged to provide information about the product which he is selling and to answer any queries which the buyer has. In addition, he is expected to help the buyer in choosing a particular product and if requested he should provide demonstrations.

In every Persona where the player is conscious of playing the role, we can attribute to them the belief that playing it will aid them in achieving their goals. For a particular Persona we can attribute certain other beliefs to the player. A salesman will be expected to know about the product which he is selling. A teacher is expected to understand his subject, and a policeman is expected to have a certain knowledge of the law.

We assume that the person playing the role is as he professes himself to be and that he is not conning us; and so we expect that he will have whatever qualifications that salesmen possess. A policeman should have successfully undergone a period of training and so too should a medical doctor.

There will be certain props which we shall expect to find in the situations where we find someone playing a role. The salesman is expected to dress neatly and to possess a briefcase. The policeman should have at least an identity card and probably should wear a uniform. The Scottish highlander should wear a kilt, the priest should wear vestments.

It is possible for some of these props to be missing but this may weaken the overall impression of the Persona. In many cases it is in the interest of the player to have the Persona recognized and so they will want to display as many of the props of the Persona as possible. A priest wearing denim may find it difficult to be taken seriously, a salesman wearing a dirty track-suit would have similar problems.

Props can play a vital role in the Persona. Having the right props can convince the audience that the player possesses all the other characteristics as well. This is a favourite ploy in the confidence trickster who attempts to convince his audience that he is what he pretends to be by skilfully using the props of that particular Persona.

6. Some uses of Personae

A Persona has many uses. An author can provide his readers with a description of a particular character simply by mentioning the Persona which the character plays. A hostess at a party will frequently introduce two strangers to each other by mentioning their occupations as well as their names. This provides the people concerned with a certain amount of knowledge about each other, and this will make it easier for them to have a conversation. Before a business meeting, if it is necessary to introduce those present, the introductions will generally include not only the names but also the job titles of those present.

In each of these examples the participants are being informed which Persona to use to understand another person. This Persona immediately forms the basis of the person's model of the other. Once this basis is established the story or the conversation can proceed. The author may go on to develop the characters in his story and this will provide the reader with information which he can use to build upon the Persona. The characters will no longer be just typical salesmen or doctors; they will have individual traits and beliefs.

Similarly, as a conversation proceeds we learn more about the other participants and we use this knowledge to build upon the Persona and arrive at a more accurate picture of them. As we get to know people we build up a more complex model of them. If we meet particular persons a number of times we will have built up a model of them which takes into account the knowledge which we have gleaned from the previous encounters; we will not have to start with the bare Persona on each occasion.

There are occasions when we are not explicitly told what Persona a person is playing. On these occasions we may choose a particular Persona to apply by recognizing aspects of the Persona in the other's behavior. We may recognize the Persona from the situation or from the props being used. We quickly recognize the waitress in the restaurant situation. We assume that the person wearing the white coat and carrying a stethoscope in the hospital ward is a doctor and that the man in uniform directing traffic at the cross-roads is indeed a policeman on traffic duty.

In many cases it is in the players' interest to communicate to us the Persona which they are playing. For a doctor to be able to play his role he may find it necessary that we recognize and acknowledge that he is a doctor. A policeman will indicate that he is playing his role by showing us his identification badge.

Another use for Personae is in humor. So many jokes are aimed at stereotypical examples of ethnic groups. We have Irish jokes, Mexican jokes, mother-in-law jokes. Many of these would fall flat and lack humor unless the audience was able to recognize the Persona being referred to.

7. How general are Personae?

We have mentioned a number of Personae which can be recognized by most adults in our culture. We now wish to consider a number of questions relating to Personae. In particular, we wish to consider how general they are. Are there Personae which only small groups of people share? Are people always aware when they are assuming a Persona? Does any model of the salesman Persona differ from yours? Are some Personae less well defined than others?

Personae are developed as a result of exposure to particular types of people. Some of us will have had different experiences and this will be reflected in the Personae which we form. Thus, my Persona for a policeman may describe him as being someone who protects citizens and as a nice helpful person. Someone with a criminal record, however, may see the police as constituting a threat to him.

Amongst the most widely recognized Personae are those which we can identify with occupations such as salesman, teacher, or doctor. This is only to be expected as most of us will have had sufficient exposure to members of these professions to enable us to abstract Personae from them.

Some people will have more detailed Personae for certain roles as a result of having had a lot of exposure to people of the particular class. If I work in a hospital it is likely that my doctor Persona will contain more knowledge than that of someone who rarely visits a doctor.

8. More Abstract Types of Personae

In addition to the Personae we have discussed so far, which are associated with well defined occupations, it is clear that other types exist. Because a Persona is employed to organize knowledge about a type of person we can imagine having Personae to deal with all types of categories. For example, a receptionist in a university department may in the course of her work build up models of the type of people who come into her office. She may use the "typical student" Persona, the "angry parent" Persona and the "one of those idiots who never read notices" Persona. These are models she uses to deal with her personal contacts in work, and it is unlikely that they will be shared by many others.

We may have a Persona for organizing knowledge about males and females. We may have ones for Communists and for French people. Some of these will have less information than others. Some will not have all the components which a common occupation Persona will have. This may arise out of our ignorance or simply because there are no corresponding props or situations. As stated before, a Persona is used to organize our information about people. It can contain a lot of knowledge but does not always have to.

There are some Personae which are even more abstract. They may correspond to an honest person, or a happy person. These are human qualities which can be found in many Personae. So one can refer to a person as an honest salesman. In itself, saying they were honest or saying that they were policemen both provide us with information. The combination of these will provide us with a picture of the person.

9. Interaction amongst Personae

Individuals will usually have at their disposal a number of Personae which they will use to achieve their aims. It seems inevitable that there will be some form of interaction amongst these Personae. We have identified three possible ways in which they can interact but we have not as yet explored these interactions in any detail. We note here some of our initial observations about them.

The first type of interaction is where someone uses a number of different Personae at different times. A man may appear to us as a lawyer at work, a member of the army reserve force at weekends, and a do-it-yourself enthusiast to his neighbors in the evenings. However, if we heard that a man worked for the local civil rights organization during the day and was a member of the Ku Klux Klan during the evenings we would be quite surprised. This is because the goals of these particular Personae seem to be contradictory. A person can get away with this sort of double life as long as they do not play to the same audience. Once they are found out the audience will stop believing in the player's sincerity when playing at least one of the Personae.

The second type is where one Persona is overlaid on another. This frequently happens. We have friendly policemen, French Communists, Irish priests, and enthusiastic lecturers. If we encounter such people frequently we will probably develop individual Personae for them, but in many cases the knowledge which we use in modelling them arises from a combination of the knowledge contained in two or more Personae. Once again problems can occur if the Personae have contradictory goals associated with them.

The third type is perhaps the most complex. If we accept that we have a number of Personae which we use in understanding others, then it is reasonable to assume that other people will use Personae to understand us. This means that we will have to represent within the beliefs which we attribute to them, the Persona which we believe they are using to understand us. The Persona which they are actually using and the one which we think they are using in understanding us may be quite different.

10. Conclusions and Further Research

Personae offer us a way of organizing our knowledge about types of people. This knowledge of people can be used both in making sense of the behavior of others and in formulating our own plans to influence them. They can be used when we meet someone whom we have not met before. They provide us with a basis upon which to build an accurate model of another person.

When we began developing the idea of Personae we were primarily interested in developing richer user models for our telephone conversation system. However, in the course of the work it became apparent that the ideas could be applied in some other areas of Natural Language Processing such as story understanding and story generation.

There are a number of problems which have become apparent to us and which we have not attempted to solve in this paper. We mention three of them here and conclude by suggesting that work on them may form the basis of some fruitful research. The first of these problems has been mentioned in the last section: we would like to know what are the rules for combining Personae and how we might detect and rectify clashes between them.

The second problem is to try to understand how we actually form Personae from our experiences with other people.

The third area involves exploring the differences between recognising a Persona which someone is playing, and actually playing it oneself.

References

Airenti, G., Bara, B.G., & Colombetti, M. (1984). Plan formation and failure in communicative acts. In *Proceedings of the Sixth European Conference on Artificial Intelligence*. Amsterdam: Elsevier Science.
Austin, J.L. (1962). *How to do things with words*. New York: Oxford University Press.
Cohen, P.R., & Perrault, C.R. (1979). Elements of a plan-based theory of speech acts. *Cognitive Science, 3*, 177-212.
Gahagan, J. (1984). *Social interaction and its management*. New York: Methuen.
Goffman, E. (1959). *The presentation of self in everyday life*. New York: Penguin.
Konolige, K. (1985). User modelling, common-sense reasoning and the belief-desire-intension paradigm. In *Proceedings of the Ninth International Joint Conference on Artificial Intelligence*. Los Angeles, CA: Morgan Kaufmann.
Schank, R., & Abelson, R. (1977). *Scripts, plans, goals, and understanding*. Hillsdale, NJ: Lawrence Erlbaum Associates.
Searle, J.R. (1969). *Speech acts*. Cambridge: Cambridge University Press.
Winograd, T. (1972). *Understanding natural language*. New York: Academic Press.

5

PLANS AND GOALS

SOME REQUIREMENTS FOR A MODEL OF THE PLAN INFERENCE PROCESS IN CONVERSATION[1]

Martha E. Pollack

Artificial Intelligence Center
SRI International

The study of naturally occurring discourse suggests that, to avoid communicative failure, computer systems that engage in answering questions will need to be able to reason about their questioners' plans, including those that may actually be invalid. Existing AI models of plan inference in communication are shown to rest on assumptions that preclude their inferring invalid plans. Examples are provided that demonstrate that successful communication requires an ability not only to infer invalid plans that underlie queries, but also to reason about the sources of the invalidity in such cases, and to distinguish between different types of invalidities.

1. Introduction

If you overheard the following conversation, you would probably find it quite unremarkable:

(1) Q: I want to talk to Kathy. Do you know the phone number at the hospital?
R: She's already been discharged. Her home number is 555-1238.

Yet, for designers of computer systems that answer questions, this conversation offers a serious challenge. R's response, *while wholly appropriate,* does not include the information explicitly requested by Q. In fact, given what R knows, a direct response stating the hospital's phone number would have been quite inappropriate. How can we account for this apparent paradox? A commonsense analysis is that Q has failed to ask for information that *is* appropriate — i.e., that will help Q achieve his goal. Furthermore, R has told Q why she believes that the information requested is not appropriate to his goal; she has explained why his plan is invalid. If R realizes the inappropriateness of the query, it is, of course, equally inappropriate for her to answer it directly and without comment.

To avoid communicative failure, computer systems that engage in answering questions, including expert systems and help systems, should be able to perform as well as R does. If the commonsense analysis of (1) is correct, these systems must be able to evaluate the questions asked of them so as to determine whether or not these queries are appropriate to the questioner's goals. To do this, the systems will need to be able to reason about their questioner's plans, including those that may actually be invalid.

---

[1] Much of the work discussed in this paper was done at the University of Pennsylvania, where the author was supported by an IBM Graduate Fellowship.

A number of AI researchers have similarly claimed that the ability to reason about plans is important in systems that communicate intelligently, and have developed models of the plan inference process in communication. In this paper I shall review these models, and shall argue that, while they are useful for explaining and automating a variety of communicative processes, they rest on assumptions that preclude the inferring of invalid plans. I shall then claim that examples taken from naturally occurring discourse, including the example given above, demonstrate that successful communication requires an ability not only to infer invalid plans, but also to be able to reason about the source of the invalidity. I shall suggest specific reasoning capabilities that a model of the plan inference process in conversation must support. Although I shall not describe such a model here, one can be found in my doctoral thesis (Pollack, in preparation).

2. Existing Models of Plan Inference in Conversation

Studies of human conversation have inspired claims that "(h)uman conversational participants depend upon the ability of their partners to recognize their intentions, so that those partners may be capable of responding appropriately" (Sidner, 1981, p.203), and that people who use question-answering systems "expect to engage in a conversation whose coherence is manifested in the interdependence of their often unstated plans and goals with those of the system" (Cohen, Perrault & Allen, 1982, p.245). The incorporation of plan inference capabilities has enabled question-answering systems to handle indirect speech acts (Perrault & Allen, 1980); provide more information than is requested in the query (Allen, 1983a); provide helpful information to a yes/no query answered "no" (Allen, 1983a); disambiguate requests (Sidner, 1983); recover certain forms of intersentential ellipsis (Carberry, 1985); and handle such discourse phenomena as clarification and correction subdialogues (Litman & Allen, 1984). In short, the inclusion of plan inference capabilities in systems that model human conversation has enabled such systems to display a wide range of observed communicative behaviors.

Yet, implicit in all these models is an important assumption, whose tenability, unfortunately, is belied by an analysis of naturally occurring conversation. This assumption, which I call the *appropriate-query assumption*, is that the query being analyzed accurately requests information that the questioner needs to achieve his goal.[2] To see the reliance of these plan inference models on the appropriate-query assumption, we need to consider more carefully how they are designed. To simplify discussion, I shall adopt the following terminology: I shall refer to the agent whose plan is being inferred as the *actor* (alternately "he") and to the agent who is inferring the plan as the *inferring agent* (alternately "she"). Thus, in the example above, Q is the actor and R the inferring agent.

2.1 The Representation of Actions

The representation of plans and actions used in most plan inference systems and, in particular, in those included in models of human conversation is a direct outgrowth of the representation first developed in the STRIPS system (Fikes & Nilsson, 1971) and later expanded in the NOAH system (Sacerdoti, 1977). In these representations, each action α is modeled by an operator, which may contain some or all of the following parts:

1. *a header* — which names α;
2. *a precondition list* — which describes what must be true for α to be performed;
3. *an effect list* — which describes what will be true after α is performed;

---

[2] Two exceptions can be found in Carberry's thesis (Carberry, 1985) and in the work on the Consul system (Mark, 1981). Each of these includes techniques for handling less than completely appropriate queries under certain restricted conditions. Carberry's system can process queries that presume relationships that do not exist in the underlying model; however, to do so successfully, it must already have inferred, from previous discourse, the questioner's plan. Consul can handle queries arising from plans that are inappropriate in one domain (electronic mail) provided they are appropriate in another (U.S. mail).

4. *a list of constraints* — which describes restrictions on legal instantiations of the operator[3]; and
5. *a body* — which may be a set of subactions whose performance constitutes performance of α, or a set of subgoals whose achievement constitutes performance of α.

Operators, which are also called action-schemas, may be parameterized; for example, one typical operator is:

Header: PICKUP(x)
Precondition: ONTABLE (x) $\land$ HANDEMPTY $\land$ CLEAR(x)
Effect List: $\neg$ONTABLE(x) $\land$ $\neg$HANDEMPTY $\land$ HOLDING(x)

Each operator can thus actually represent a class of actions, in this case, the class of actions including "picking up a ball," "picking up the red block," etc. A particular instance of an action in that class is represented by an *operator-instance,* which is an operator along with a list of parameter bindings and a time specification. There is thus a distinction between the particular instance of the action in the world and the operator-instance modeling it. A similar distinction should also be made between a *property* of the world at some specific time, and the *proposition* modeling that property. Plan inference systems have access to a set of operators representing actions that can be performed in a domain; we can call the set of such operators the *operator library*. Operator libraries typically contain only operators that model valid domain actions.

2.2. The Representation of Plans

Since plans, intuitively, are collections of actions whose performance leads to some goal,[4] it is not surprising that, in systems that model actions with operators, plans are modeled as collections of operators. More specifically, they are modeled as directed graphs whose nodes may be either operators or propositions; when a node is labeled with a proposition **p**, it should be interpreted as standing for any action that would achieve **p**. I shall draw a distinction between *plans* and *plan graphs* analogous to that made between *actions* and *operators:* in each case, the latter is a representation of the former.

2.3. Constructing Plan Graphs

The next question to ask is: how do the existing plan inference systems work? That is, how do they construct plan graphs that represent the inferred plan of some actor? Each of the major systems for plan inference in conversation (Allen, 1983a; Allen, 1983b; Perrault & Allen, 1980; Carberry, 1985; Litman & Allen, 1984; Sidner, 1983; Sidner, 1985) has a set of inference rules for building plan graphs. Each rule states the conditions under which a piece of a plan graph — a plan subgraph — can be constructed; constructing a plan subgraph corresponds to inferring part of a possible plan. The conditions for constructing a subgraph always refer both to plan subgraphs that have already been inferred as candidate representations of the actor's plan, and to operators in the operator library. A typical rule, for example, allows a plan subgraph that includes the node to be expanded by adding an arc from α to a new node β if there is an operator whose header is β and which includes α in its body. Thus, if the operator in the top part of Figure 1 were in an operator library, the expanded plan subgraph shown could be constructed from the initial subgraph shown.

Although various plan inference systems state their own inference rules slightly differently from one another, in fact there are only four basic conditions under which any of them will

---

[3] For example, restrictions on the types of parameters, on the relations between parameters, and on the ordering of the subactions into which an action is decomposed. The distinction between constraints and operators was introduced by Litman and Allen (1984).

[4] At least this is the conception that underlies AI research in plan synthesis; see, e.g., Nilsson (1980, p. 282), Fikes and Nilsson (1971, p. 190), Wilkins (1983, p. 733), and Waldinger (1981, p. 251).

Given: OPERATOR

Header: Display the current
message on the screen

Body: Type 'TYPE .'

Expand: INITIAL SUBGRAPH   To: EXPANDED SUBGRAPH

Display the current message
on the screen
↑
|
Type 'TYPE .'

Type 'TYPE .'

Figure 1. Constructing a Subgraph from an Operator

construct a plan subgraph. These are shown in Figure 2. Each of the systems mentioned above uses some subset of the relations shown there to construct subgraphs. Starting with some action or actions believed to be in the plan, the plan inference system repeatedly applies these rules to construct candidate plan graphs until one of the candidates satisfies some termination condition[5].

A plan subgraph with nodes α and β and an arc from α to β, i.e.,

$$\begin{array}{c}\beta\\\uparrow\\\alpha\end{array}$$

can be constructed provided either α or β is already in the subgraph, and

α **R** β holds, where **R** is one of the following relations:

1. **R** = *causes*, i.e., β is on the effect list of α, so there is an operator in the operator library of the form

   Header: α
   Effects: ... β ...[6]

2. **R** = *is-a-precondition-of*, i.e., α is on the precondition list of β, so there is an operator in the operator library of the form

   Header: β
   Preconditions: ... α ...

3. **R** = *is-a-way-to*, i.e., α is part of the body of β, so there is an operator in the operator library of the form

   Header: β
   Body: ... α ...

4. **R** = *enables*, i.e., α has on its effect list some γ that is also on the precondition list of β, so there are two operators in the operator library of the form

   Header: α Header: β
   Effects: ... γ ...   Preconditions: ... γ ...

Figure 2. Possible Relations in Plan Graphs

---

[5] Depending on the details of the particular system, the resulting plan graph may only represent part of the actor's plan. For instance, imagine that β is believed to be in the plan, and α is inferred using rule 3 (α *is-a-way-to* β). Further imagine that α has some sister actions on the body list of the β operator. Some plan inference systems will introduce those sister actions into the plan along with α, while others will not. In the latter case, however, the additional actions could easily be filled in by referring to the operators in the operator library.

[6] The ellipses denote the fact that β need not be the only proposition in the list.

While the plan synthesis literature often views plans as data structures, within plan inference work they are generally seen as mental attitudes. After all, inferring another agent's plan means figuring out what sequence of actions he "has in mind"; the fact that he plans some particular sequence of actions does not guarantee that executing that sequence of actions will actually lead to his goal, but only that he believes that it will.[7] To make sense of the plan inference problem, plans need to be seen as sequences of actions to which an agent stands in some particular relationship. So, in one of the earliest AI accounts of plan inference, Allen (1983a)[8] notes that:

> Goals and plans of agents are indicated by using an operator WANT, i.e.,
>   WANT(A,P) = A has a goal to achieve P.
> By this, we mean that the agent A actually intends to achieve P, not simply that A would find P a desirable state of affairs ... *The properties of the WANT operator will be specified completely by the planning and plan inference rules.* (p. 117; emphasis mine.)

He then uses this notation in plan inference rules; one typical rule is:

$$SBAW(ACT) = i => SBAW(E), \text{ if E is an effect of ACT}$$

which can be glossed as: if S believes that A has a goal ACT and ACT has as an effect E, then S may believe that A has as a goal achieving E (the "action-effect rule," p. 121). Notice that the rule leaves unspecified who it is that believes that E is an effect of A. If we take it to be a belief of the inferring agent, then it is not clear that she is inferring the actor's plan; if we take it instead to be a belief of the actor, then it is not clear how the inferring agent comes to have direct access to it. In fact, in Allen's system as well as those that have followed it, there is only a single set of operators – those in the operator library, which represent the system-as-inferring-agent's knowledge; a plan inference rule can be applied so long as some operator therein satisfies its *if*-clause. The "SBAW" context is largely transparent to the reasoning process that is performed by Allen's system: the reasoning is all performed directly on its object, and the B and W operators are carried directly from antecedent to consequent in each inference rule. Without any repercussions, the W (and B) operators can be omitted, resulting in rules that are completely equivalent to those in Figure 2. In fact, in his example plan graphs, Allen often leaves off the W and B operators entirely, and this practice has been continued in more recent work in plan inference (Litman & Allen, 1984; Carberry, 1985)[9].

All the plan inference rules shown in Figure 2 are similar to the sample rule given above: they operate on the members of the operator library. This means that a subgraph joining two nodes α and β will be constructed only if (i) α and β are both encoded in the operator library, and (ii) they are encoded, in the operator library, in one of the configurations shown in Figure 2 – i.e., they are related by one of the relations: *causes, is-a-precondition-of, is-a-way-to,* or *enables*. Recall too that the operator library typically contains only valid domain information. Thus there are two further conditions that will be satisfied whenever a subgraph joining nodes α and β is constructed: (iii) α and β both model valid domain actions or propositions, and (iv) the relation-

---

[7] And, amongst other things, that he intends it as a way of achieving his goal. For further discussion, see Pollack (in preparation).

[8] Allen (1983a) presents, in abridged form, his dissertation research, (Allen, 1979).

[9] Actually, there seems to be some tension as to what is really meant by the W operator. Allen, as discussed earlier, states that AW(P) means "A has a goal to achieve P," which seems to imply that P is a single action or property, not a whole plan. Consistent with this, he says that "SBAW(x) = i => SBAW(y)" should be taken to mean that "if S believes A has a goal of X, then S may infer that A has a goal of Y" (p. 120). But he uses these rules to infer not just that A has a goal of Y – i.e., that his plan contains Y – but that it contains X and Y related to one another in some particular way specified by the rule. So the action-effect rule considered above should probably be written "SBAW(ACT) = i => SBAW(ACT -->$_{enables}$ E), if E is an effect of ACT." Writing the rule this way would clarify his model, but it would not affect my claim here.

ship between the operators α and β that is encoded in the operator library actually holds between the actions or properties they model. Then, since an inferred plan graph is a graph all of whose subgraphs satisfy conditions (i) through (iv), (and which satisfies some additional conditions), the inferred plan itself will be composed only of valid actions and properties that are included in the operator library and that are actually related to one another in ways represented there. No plan which is invalid, either by virtue of containing some nonrealizable action or property, or by virtue of inproperly relating two actions or properties, can be inferred; nor can any plan — valid or not — be inferred if it contains actions or properties not in the operator library, or if it contains actions or properties related to one another in ways not encoded in the operator library.

3. Implicit Assumptions of Existing Models

Having seen how plans are inferred by existing plan inference systems, we can now ask the following question: what sort of assumptions are needed to guarantee that these systems' behavior is correct? Put otherwise, what are the conditions under which these systems will be able to infer correctly an actor's plan?

Before outlining the necessary assumptions it is important to reiterate one of the design principles that is typically adopted by these systems, to wit, the decision to encode in the operator library only valid domain knowledge. I shall call this the *principle of parsimony*. Since the operator library represents the system's knowledge, and the system is, in turn, the inferring agent, the principle of parsimony can be stated as follows:

Principle of Parsimony(PP)
: the inferring agent does not have knowledge of any invalid domain information that she knows to be invalid.

The PP should not be confounded with a second assumption, *correct knowledge,* which states that the inferring agent is never mistaken:

Correct-Knowledge Assumption (CKA)
: all the domain knowledge that the inferring agent has is valid.

Given that plan inference systems emulate domain experts, the CKA is not a wholly unreasonable assumption.[10] The PP, however, is obviously too strong: experts do have knowledge of many typical misconceptions that can arise in their domain of expertise, upon which they can draw during the plan inference process. I shall return to this point below. However, it is worthwhile first to ask what sort of assumptions are necessary to guarantee the behavior of systems observing PP.

It turns out that two such assumptions are necessary. The first is the *closed-world assumption,* which, in general, is that anything not inferrable from a system's knowledge is assumed to be false (Reiter, 1978). In terms of the problem at hand, it is more convenient to state the contrapositive of the closed-world assumption, i.e., that everything that is true is inferrable from the system's knowledge:

Closed-World Assumption (CWA)
: the inferring agent has knowledge of all valid domain information.

That is, the operator library must contain representations of all valid actions in the domain and, furthermore, must contain representations of all propositions and actions that can be related to

---

[10] Though see the discussion below for the consequences of weakening it.

any encoded action. It may seem confusing to claim that all such information needs to be explicitly encoded in the operator library, when all that the CWA demands is that it be implicitly encoded in – i.e., inferrable from – the system's knowledge. However, recall that what is inferred by existing plan inference systems is possible plans. There are no facilities for inferring new actions or new relations between actions and/or propositions. Hence all such information must be encoded explicitly.

Why is the CWA necessary? Without it, there can be cases in which the actor's plan, while valid, contains actions or relations between actions about which the system does not have knowledge; the system would necessarily fail to infer the plan in such cases. The CWA rules out such a possibility: if we assume that the system has knowledge of all valid domain information, then there are not valid plans that an actor can have that the system cannot, in principle, infer. However, while the CWA is enough to make the claim that the system will behave correctly whenever it is faced with a valid plan, it is still not enough to guarantee that it will always behave correctly. To make this claim, another assumption, which I shall call the *valid-plan assumption*, is also necessary:

Valid-Plan Assumption (VPA)
    the actor's plan is valid.

When this assumption is made, the system never needs to infer a plan that contains invalid domain information. Thus, since the system is guaranteed to perform correctly whenever faced with a valid plan by the CWA, making the VPA guarantees that the system will always be able to perform correctly, i.e., be able, in principle, to infer the actor's plan.[11]

In question answering, one important corollary of the VPA is the *appropriate-query assumption* – the assumption that

Appropriate-Query Assumption (AQA)
    the query being analyzed accurately requests information that the questioner needs to achieve his goal.

If, by the VPA, the questioner's plan is assumed to be valid, i.e., it is assumed that the questioner knows what he needs to do to achieve his goal, then it follows that his query can be assumed to be appropriate, i.e., to reflect a need for information he truly requires to perform his plan and, consequently, achieve his goal.

However, evidence from naturally occurring data – some of which will be presented below – shows that the AQA is unwarranted; people seeking advice often ask for information that is not, in fact, appropriate to their goals. Thus, to construct a cooperative question-answering system, one needs to give up the AQA, and, since the AQA is a direct corollary of the VPA, one needs also to give that up. Thus systems must be designed that can infer plans even when those plans may not be valid.

One approach to doing this might be to encode some typical invalid plans – i.e., to give up the PP. Several CAI systems (Brown & Burton, 1978; Stevens, Collins, & Goldin, 1979; Woolf & McDonald, 1983) have done just this, encoding sets of erroneous beliefs that their users are likely to have and marking them as erroneous. While this seems to be a useful strategy, it is necessarily incomplete. It is impossible for any person to have complete knowledge of the potential beliefs of other people, since the range of beliefs is, in principle, infinite. This means that system designers cannot anticipate *a priori* all potential misconceptions the users of their system may have. It also means that the intelligent agent that such systems emulate – the human being – cannot know *a priori* all potential misconceptions that people asking her questions might have.

---

[11] If the CKA is dropped, then it becomes possible that an invalid plan can be inferred if the invalidity happens to correspond directly to an invalid belief of the inferring agent; but in this case the fact that the plan is invalid goes undetected.

Sometimes she will have to deal with a novel (to her) belief. Fortunately, human beings seem to be quite "robust" – good at dealing with novel situations; in fact it has been argued (Dennett, 1984) that such robustness is a definitive feature of high-level intelligence.

The alternative approach is to maintain the PP and develop an account of robustness in plan inference. Said otherwise, the approach is to develop methods of reasoning from valid information to determine novel invalid plans.[12] Such strategies are developed in Pollack (in preparation). Of course, in any complete reasoning system, the most reasonable approach would be to encode typical invalid knowledge, and fall back onto strategies for manipulating valid information whenever some invalidity not explicitly encoded – i.e., not anticipated *a priori* – is encountered.

The CWA is also overly strong. Once it is abandoned, then whenever the inferring agent infers some novel-to-her plan, she needs to consider whether it is invalid or whether it is simply a valid plan that contains information she hasn't previously known. A similar consequence results from abandoning the CKA. If we allow the possibility that the inferring agent mistakenly believes that some invalid plans are valid, then, again, whenever she infers a novel plan, she will have to consider whether it is invalid or whether her own beliefs are invalid.

Before concluding this discussion, it may be enlightening to consider things from a slightly different perspective. An analogy can be drawn between the CKA and the VPA and claims of soundness, and between the CWA and a claim of completeness. On this view, the CKA can be seen to assert that the inferring agent's knowledge is sound (with respect to "truth" in the "real world"), while the VPA can be seen to assert that the actor's knowledge is sound (again, using the "real world" as a model). The CWA can be seen to assert that the inferring agent's encoded knowledge is complete (once more, with respect to the "real world"). Then since the inferring agent's knowledge is sound and complete, and the actor's is also sound (though not necessarily complete), it follows that the actor's knowledge is a subset of the inferring agent's. Hence, a search through the inferring agent's space of possible plans is guaranteed to find the actor's plan. As we shall now see, however, the actor's knowledge is not necessarily sound.

4. Evidence against the Appropriate-Query Assumption

Discourse fragment 1, presented above in Section 1, is typical of naturally occurring conversations. While existing systems that perform plan inference in question answering have constrained the queries that are analyzed to be single requests for a piece of domain information, an examination of conversation between experts and people consulting them for advice reveals that advice-seekers do not just ask a simple question; instead they often include a great deal of extra information. The following four examples, taken verbatim from a set of questions asked of the designer of a computer-mail system, illustrate this phenomenon:[13]

(2) Q: Is there any way to tell a bboard to load only new messages (I would like to speed up entry into infomac)?

(3) Q: How can I scroll up and down in mail? I usually forget what number mail I'm looking at and if the info I need is off the screen, I can't get to it easily.

---

[12] Of course, having such reasoning principles encoded means that, in some sense, the invalid plans are "implicitly" encoded, in the same way that a sentence of some language L is implicitly encoded by a grammar generating L. This is inevitable. What PP requires is that the invalid information not be explicitly encoded. To continue the analogy with the grammar, this is akin to requiring that all sentences of L not be listed explicitly.

[13] These examples come from a set of transcripts collected in July and August 1984, in which users of an electronic mail system were asked to submit questions that they had to the designer of the system, in order to assist her in the task of writing a user manual. Both the questions sent to the designer and her responses to them were automatically archived, resulting in a permanent record of the conversations. My thanks to Sharon Perl, the mail-system designer, for giving me this set of transcripts.

(4) Q: Is there any way to dynamically load a file of definitions and commands and have them interpreted by mail? I've wanted to do this so I could have a large file of net addresses and aliases that I would manually load if I wanted to send mail to one of them. I figgure (*sic*) this would decrease the startup time for mail, since I would take all ths symbol defs out of my mailinit file.

(5) Q: From Mail I enter Emacs to write a message, using send. In Emacs, I realize I would like to see a message whose number I know, maybe even copy it into the message I am composing. What is the easiest way to do this? Thanks.

What examples like these suggest is that advice-seekers are aware of the necessity of, and inherent difficulty in, inferring their plans; they can be seen as attempting to assist in this process by providing information that they believe relevant to figuring out their plan. One of the ways they do this is by explicitly mentioning their goal. In 2, for example, Q not only asks how to load only new messages, but also lets it be known that his goal in doing this is to speed up entry into the infomac bboard.

The fact that existing plan inference systems at least implicitly make the AQA explains why they have not needed to be concerned with explicitly mentioned goals. Under the AQA the only truly interesting plans to infer are those that support goals not explicitly mentioned in the query. This is because, if one assumes that the underlying plan is valid and one is given an explicitly mentioned goal, one can simply provide an answer that addresses directly that explicitly mentioned goal. Notice, however, that such a claim is predicated on the assumption that the goal is not only mentioned in the query, but also explicitly marked as the goal. In fact, queries may mention several actions without making explicit how those actions are meant to fit together, or which of them is the questioner's goal. Consider, for example, the following two queries:

(6) Q: I want to talk to Kathy. I need to find out the phone number at the hospital.

(7) Q: I want to talk to Kathy. I need to find out whether she wants to go to Maryland.

In 6, if Q believes that Kathy is at the hospital, then he probably intends to enable the action "talking to Kathy" by "finding out the phone number in the hospital". On the other hand, if Q believes that Kathy is the only one in the office with a phone book, then he may intend "talking to Kathy" to enable "finding out the phone number at the hospital." Example 7 is similarly ambiguous, its meaning potentially depending upon whether Q intends to have his talk on the way to Maryland, or to have his talk consist in finding out whether Kathy wants to go to Maryland.

The linguistic form of the query does not, by itself, reveal the form of Q's plan. Yet, to provide an appropriate answer, Q's conversational partner must evaluate the plan. For instance, if she interprets Q as intending to "talk to Kathy" by "calling the hospital" and she knows that Kathy is no longer in the hospital, and, consequently, that Q's plan is not well-formed, then she should so inform him. Further, if she knows an alternative action that will support the desired goal, she may also suggest that. So, for example, as was seen with example 1, she may tell the questioner Kathy's home phone number. Without first inferring the Q's plan, there is no way for Q's conversational partner to evaluate it, determine whether it is well-formed, and plan correctly for the desired goal.

When the VPA and its corollary, the AQA, are abandoned, then, even if the query both explicitly mentions the goal action and marks it as such, it may still be essential to infer the questioner's plan. Without doing this, it will be impossible to ensure that the response given does not mislead the questioner.[14] To see this, consider the following two queries (the first is a variation of 2):

---

[14] And not misleading one's interlocutor seems to be a requirement in cooperative conversation; see Joshi (1982).

(8) Q1: Can I have mail load only new messages? I figure this would speed up entry into it.

(9) Q2: Can I have my username changed? I figure this would speed up entry into mail.

Assume that, in the mail system under discussion, entry time depends upon the number of messages loaded, and not upon username. Assume further that neither of the queried actions is achievable, i.e., there is no way to have mail load only new messages, nor is there a way to have one's username changed. It seems that, intuitively at least, a coherent plan behind 8 can nonetheless be inferred: if the queried action were achievable, it would result in fewer messages being loaded to mail, and, since entry time depends on the number of messages loaded, this would result in faster entry to mail. Thus an answer "You can't load only new messages, but you can clean out your mail file" is a valid one. In contrast, there is no reason to believe that changing one's username would lead to faster entry into mail. An answer similar to that suggested for 8 would be unsuitable for 9, since it would mislead Q2 into believing that entry time does depend upon username. If, at a later date, Q2 learned that it had become possible to change his username, he might do so, intending to speed entry to mail. Similarly Q1 might later try to get mail to load only new messages, should that option become available. Q1, who later has mail load only new messages, will succeed in his goal, while Q2, who has his username changed, will fail. The appropriate answer to question 9 is something like "Why do you think that will speed entry to mail?" (i.e., "I can't figure out your plan").

The claim then is that, at least in answering questions, successful, cooperative communication depends not only upon the ability to determine whether the plan underlying a query is valid or invalid, but also upon the ability, when encountering an invalid plan, to locate the source of the invalidity. Consider, as further evidence of this claim, the following examples:[15]

(10) Q: Is there some way to set the protection on my files so that only members of my group have permission to read them? I don't want Tom to be able to snoop around in them anymore."
R: Well, the command is "set protection = (g:r,w)", but it won't keep Tom out. He's the system manager: he can override file protections.

(11) Q: Is there some way to set the protection on my files so that only faculty have permission to read them? I don't want Tom to be able to snoop around in them anymore.
R: Sorry, there's no way to do that, and even if you could it wouldn't keep Tom out. He's the system manager: he can override file protections.

(12) Q: How can I tell the bboard to load only new messages? I figure this would speed up entry to INFOMAC.
R: There's no way to do that. But you can have the bboard administrator clean out the bboard file so that it will load faster.

(13) Q: I'd like to take advantage of the discount offer you sent me, and subscribe to your magazine. How can I specify that I don't want the magazines to start arriving until October, since I'm going to be moving before then.
R: Sorry, you can't do that. To get the discount rate you have to take the subscription before August 31st, and once you take it, the magazines will start arriving automatically within 3 weeks.

(14) Q: How do I get to train track 11?
R: Do you work around here?
Q: No.
R: Then why do you want to go to track 11?

---

[15] As will be obvious, example 12 is the same as 2 above. Example 14 is adapted from one found in Horrigan's train-station transcripts (Horrigan, 1977). The others are based upon naturally occurring tokens I observed outside the context of the mail transcripts mentioned above.

(15) Q: How can I change my password? I'm running out of room and need more storage.
R: Why do you think that one has anything to do with the other?

If we were to attempt an analysis of R's responses in these examples, what would we say? In 10 and 11, as in example 1 above, R seems to locate the source of the problem with Q's plan in some erroneous belief which she then refutes in her response: to wit, that Kathy is at the hospital (in 1), and that by protecting a file one prevents the system manager from reading it (in 10 and 11). Each of these erroneous beliefs in turn supports a second erroneous belief that some intended course of action will result in some particular goal. That is, as a result of his belief that Kathy is at the hospital, Q mistakenly believes that calling the hospital will lead to his talking with her; as a result of his belief about protecting a file, he mistakenly believes that, by doing that, he can keep Tom from reading it. Of course, there are also differences between these three cases that reveal themselves in R's responses: in 1, but not in 10 and 11, Q's goal can be achieved by some other means than that anticipated by Q — Q can call Kathy at home, but there is no way for him to prevent Tom from reading his files; in 1 and 10, but not 11, the information Q requests is (or at least may be) available, despite the fact that it cannot be put to a use that will facilitate Q's goal — so R may know the phone number at the hospital, and does know how to set the protection for group-read only, but, as far as R knows, there is no way to set the protection so that only faculty can read the file.

The next two examples are of a different sort. In these, R seems to determine that Q's plan is reasonable as far as it goes, but is nonetheless invalid, since it contains an action that cannot be performed: there is no way to get the bboard to load only new messages, and there is no way to postpone the delivery of the magazine subscription. So 12 and 13 are like 11 in that R cannot provide the requested information, but differ from it in that, if she could, Q's queries in 12 and 13 would be appropriate. 12 and 13 differ from one another in that in 12, but not 13, there is a way for Q to achieve his goal.

Finally, in 14 and 15, the explanation of R's response seems to be that she is unable to determine what Q's plan consists in. These two examples seem to be like 1 and 10 in that R probably could provide the information Q requests. However, she is unable to determine whether Q's plan is valid, and, if not, what the source of the invalidity is; as a result she requests further information to help her decide.

Taken together, the analyses of examples 1 and 10 through 15 illustrate the sort of reasoning that can support the generation of appropriate responses to queries. They demonstrate the importance, first, of being able to distinguish between plans that are destined to fail because the actions they contain do not relate to one another in the way they were intended (as in 1 through 11), and those that will fail simply because there is no way to perform them (12 and 13); second, of being able, in the former case, to isolate the erroneous belief that is the source of the mistake about the way the actions fit together; and third, of being able, in both cases, to reason about actions that cannot be performed (as in 11 through 13) as well as about those that can. A system that can perform such reasoning will be well on its way to being able to generate appropriate answers to queries, and thereby avoid communicative failure.

5. Conclusion

I have argued that existing models of plan inference in communication have rested on some very strong assumptions and I have presented examples, taken from naturally occurring discourse, that show that these assumptions are not always warranted. Such examples suggest that strict adherence to the Appropriate-Query Assumption by systems that answer questions will result in communicative failure in many instances. To avoid this, systems will have to be designed so that they can perform plan inference without assuming the appropriateness of the queries they are asked or of the plans underlying those queries; they will need to have the capability to reason about the source of such invalid plans, and to distinguish between different types of invalidities. Techniques for providing systems with these capabilities are presented in Pollack (in preparation).

References

Allen, J. (1979). *A plan based approach to speech act recognition.* Technical Report TR 121/79, University of Toronto.

Allen, J. (1983). Recognizing intentions from natural language utterances. In M. Brady and R.C. Berwik (Eds.), *Computational Models of Discourse*. Cambridge, MA: MIT Press.

Allen, J. (1983). ARGOT: A system overview. *International Journal of Computers and Mathematics, 9,* 97-110.

Brown, J.S., & Burton, R.R. (1978). Diagnostic models for procedural bugs in basic mathematical skills. *Cognitive Science, 21,* 155-192.

Carberry, M. (1985). *Pragmatic modeling in information system interfaces.* PhD thesis, University of Delaware.

Cohen, P., Perrault, C.R., & Allen, J. (1982). Beyond question answering. In W. Lehnert and M. Ringle (Eds.), *Strategies for Natural Language Processing*. Hillsdale, NJ: Lawrence Erlbaum Associates.

Dennett, D. (1984). Cognitive wheels: The frame problem of AI. In C. Hookway (Ed.), *Minds, machines and evolution*. New York: Cambridge University Press.

Fikes, R.E., & Nilsson, N. (1971). STRIPS: A new approach to the application of theorem proving to problem solving. *Artificial Intelligence, 21,* 189-208.

Horrigan, M.K. (1977). *Modelling simple dialogs.* Technical Report TR 108, Dept. of Computer Science, University of Toronto.

Joshi, A.K. (1982). Mutual belief in question answering systems. In N. Smith (Ed.), *Mutual Belief.* New York: Academic Press.

Litman, D., & Allen, J. (1984). *A plan recognition model for subdialogues in conversation.* Technical Report TR 141, University of Rochester.

Mark, W. (1981). Representation and inference in the Consul system. In *Proceedings of the 7th International Joint Conference on Artificial Intelligence,* Vancouver, BC.

Nilsson, N.J. (1980). *Principles of Artificial Intelligence.* Palo Alto, CA: Tioga Publishing Co.

Perrault, C.R., & Allen, J.F. (1980). A plan-based analysis of indirect speech acts. *American Journal of Computational Linguistics, 6,* 167-182.

Pollack, M.E. (In preparation). Inferring domain plans in question answering. University of Pennsylvania doctoral thesis.

Reiter, R. (1978). On closed world data bases. In H. Gallaire and J. Minker (Eds.), *Logic and data bases.* New York: Plenum Press.

Sacerdoti, E.D. (1977). *A structure for plans and behavior.* New York: American Elsevier.

Sidner, C.L., & Israel, D. (1981). Recognizing intended meaning and speaker's plans. In *Proceedings of the 7th International Joint Conference on Artificial Intelligence,* Vancouver, BC.

Sidner, C.L. (1983). What the speaker means: The recognition of speakers' plans in discourse. *International Journal of Computers and Mathematics, 9,* 71-82.

Sidner, C.L. (1985). Plan parsing for intended response recognition in discourse. *Computational Intelligence, 1,*

Stevens, A., Collins, A., & Goldin, S.E. (1979). Misconceptions in student's understanding. *International Journal of Man-Machine Studies, 11,* 145-156.

Waldinger, R. (1981). Achieving several goals simultaneously. In B.L. Webber and N.J. Nilsson (Eds.), *Readings in Artificial Intelligence.* Palo Alto, CA: Tioga Press.

Wilkins, D.E. (1983). Representation in a domain-independent planner. In *Proceedings of the 8th International Joint Conference on Artificial Intelligence,* Karlsruhe, West Germany.

Woolf, B., & McDonald, D. (1983). Human-computer discourse in the design of a PASCAL tutor. In *Proceedings of the CHI'83 Conference on Human Factors in Computing Systems.* ACM SIGCHI.

PLANS AND GOALS IN STORY COMPREHENSION[*]

A. Cahill and D.C. Mitchell

University of Exeter

Successful communication in dialogue may depend on the participants' capacity to fill in unstated or presupposed information. In the case of a story this may involve inferring a *plan* to achieve some *goal* specified in the setting. This chapter reports a series of six experiments which were carried out to determine whether people actually draw such inferences in the course of interpreting stories. An attempt was also made to discover something about the preconditions for making the inferences. The results suggest that people *do* draw plan inferences and that they do this while they are actually reading the text. This occurs even if the passage specifies no more than the protagonists' goals or the actions that they subsequently carry out to achieve these goals. The paper ends with a brief discussion of the relevance of these findings for work on man-machine interaction.

1. Introduction

An important source of model failure in dialogue is likely to be that associated with the failure to cope with ellipsis. The form in which new information is both requested and provided by participants in a dialogue is likely to be influenced by the working assumptions they have about the aims and intentions and about the knowledge possessed by the other person. Information which is assumed to be familiar or which can easily be inferred by the listener will often either be expressed in some distinctive manner — marking the fact that it is presupposed — or, in extreme cases, it might be excluded from the utterance entirely. The listener is left with the task of filling in the missing information and making sense of the communication as a whole. When both of the participants in the dialogue are human this might prove to be relatively straightforward.

People share unstated assumptions about each other's goals, motives and plans and they will often be able to cope with the ellipsis by using this information together with general information about the context of the dialogue. If human-computer interaction is to work effectively it may be necessary to furnish the machine with some comparable knowledge about goals and plans and about how to use them.

As a first step in making such provisions it would be useful to know more about the way in which people cope with unstated information. This is the aim of the present chapter. The work concentrates in the interpretation of *stories* because this is a textual form in which ellipsis occurs very naturally and in which it is used uniformly enough to draw reasonably general conclusions from the data.

2. A Model of Story Comprehension

In an A.I. approach to the comprehension of narrative text Schank and Abelson (1977) and

---

[*] This work was carried out as part of a PhD programme by the first author under the supervision of the second.

Wilensky (1978a,b) have developed a model which places considerable emphasis on the goals and plans which readers might attribute to the characters in a story in order to understand their actions in the course of interpreting the passage.

For the purposes of this work a "plan" may be viewed either as a course of action which can lead to the satisfaction of a goal or as a piece of knowledge that can be used to organize actions so that goals can be satisfied.

According to the model one of the major aspects of understanding a story is the problem of finding an explanation for each event mentioned in the text. In many stories this information is not provided explicitly and so it has to be inferred. Often this problem of computing a satisfactory explanation depends on a detailed knowledge first of the kinds of *goals* people are likely to have in different circumstances and secondly on the kinds of *plan* they might adopt and actions they might carry out to achieve such goals.

The model elaborated by Wilensky specifies a sequence of operations that might be used to settle upon a reasonable explanation for an event. The first step involves checking to see whether the event is part of a known plan. If it is, then the plan can be taken as the explanation and there is no further work to do (Step 1). If not, the system tries to infer some kind of plan which could be used to explain the event (Step 2). If it is unable to do this the whole process fails. If it succeeds then it checks to see whether the plan is one that can be used to satisfy some goal introduced earlier in the passage (Step 3), in which case the new *inferred* plan is taken as the explanation of the event. If the inferred plan is not one which could satisfy an existing goal, the system proceeds by guessing what kind of goal *might* be achieved using this plan (Step 4) and then tries to ascertain whether this hypothesised goal could be instrumental in achieving a known plan (Step 5). If the answer is positive, then the plan-goal sequence is taken as the explanation. If not, the system iterates between inferring new plans and new goals either until some kind of link is established or until no further inferences can be made, in which case the event remains unexplained.

The process is perhaps best illustrated by applying it to a particular example – say "Mary felt hungry so she reached for her copy of Spanish for Beginners".

In this sentence, the event to be explained is the reaching for the copy of the book. At the outset the only plan that can be inferred is "obtain food to satisfy hunger". Working through the sequence of operations outlined above, the system first checks whether the event (reaching for the book) is part of the known plan (satisfying hunger). Clearly it is not since books are not edible, and so the system moves on inferring that the intended plan is to *read* the book – presumably in order to be able to speak Spanish. Step 3 involves checking to see whether this plan can be used to satisfy the goal of alleviating hunger. Since speaking Spanish does not achieve this directly, the system goes on to infer what it *might* accomplish – perhaps coming up with something like "communicate with the locals". Finally, it infers that this new goal might be instrumental in achieving the original goal (food could be obtained by begging, threatening, bargaining etc.). Thus, in the end, the event is interpreted as part of a plan to achieve an indirect goal which is itself set up as part of the plan to achieve the original purpose.

Several features of this theory are worth noting in the present context. First, it suggests that prior knowledge about generic plans and goals (and the circumstances in which they are put into effect) may be prerequisites for understanding natural language passages. Secondly, the theory suggests that the process of explaining events must involve drawing inferences about intermediate plans and goals so that these can subsequently be used to structure and interpret the material that comes later in the passage. Thirdly, it suggests that both *event* and *goal* information is needed before any plans can be inferred. According to the model, one of these without the other would not provide sufficient grounds for making the inference.

The purpose of our experiments was to investigate the psychological reality of some of these ideas. Do people infer plans, etc. in the course of comprehension and do they use this information to understand later parts of the text? Do they require both event and goal information before they are able to do this effectively, or can they make some progress when just one of these is available?

3. Experiment I

The aim of the first experiment was to see if there is any indication that people draw inferences about *plans* in the course of understanding short passages. The study was based on a cued recognition technique developed by Ratcliff and McKoon (1978). Subjects were presented with a series of trials consisting of two phases: a Study phase in which they read three short passages of prose and a Test or recognition phase in which they had to indicate whether each of three test words had appeared in the preceding texts.

Half of the target words were preceded by a cue — a word which was the name of a plan entailed in one of the passages. The rationale of the experiment was that if the *plan* had been activated in any way during the Study phase, then the cue word would serve to facilitate access to all other stored information about the passage. On the basis of this hypothesis we predicted that if subjects had inferred a plan while they were reading the text, the recognition latencies in the cued condition would be shorter than those in the control (No Cue) condition.

3.1. Method

3.1.1. Materials

There were three types of study material:

(1) *A Bare Passage* which consisted of one or two sentences introducing a character and a setting and suggesting a *goal* and *plan* for achieving that goal. For example:

Prince John was ambitious for power, but his father looked as if he would go on for years. He obtained some arsenic from his friends.

(2) *Complete Passage* — This was the same as the *Bare Passage* except that there was one additional sentence which explicitly mentioned the *plan* alluded to in the first part. In the present example the new sentence was:

He thought he could get away with murder just this once.

(3) The third type of Study material consisted of an unsystematic sequence of ten content words from the *Bare Passage*. That is:

Friend,	Obtained,
Ambitious,	Father,
Arsenic,	Years,
Power,	Prince,
Go on,	Looked.

In the recognition phase the cue word was the Plan name (i.e., MURDER) and in this example the positive word was PRINCE and the negative word was ALUMINIUM.

There were 24 different sets of material of this kind. Each subject studied 8 of them in each of the three different forms and counterbalancing was used to ensure that each of the forms appeared equally often over the entire experiment.

3.1.2. Procedure

The materials were presented in capital letters on a Philips PCT 1201 Amber screen driven by an Apple IIe microcomputer. In the study phase, subjects were shown a prompt ("Now for the next story") just before each passage or wordlist. On pressing the space bar the entire passage (wordlist) was displayed. The subjects read this at their own pace — pressing the space bar when they were ready to proceed. A second and third prompt were followed by new passages (wordlists).

After each passage Ss were required to indicate whether the main character was (a) resourceful (b) stupid (c) nasty (d) brave (e) pretty ordinary. (This additional task was intended to keep the subjects paying attention.)

When the subjects had worked through the three Study items they were presented with the three test words — each preceded by the prompt "Now for the next recognition". In each case the sequence of events was as follows: the subject first pressed the space bar to initiate the test. When this happened the cue word (or the words "No cue" on the uncued trials) was displayed briefly and then followed by a blank screen for 3500 ms. The target word was then displayed until the subject pressed either the "yes" or the "no" button. At the beginning of the experiment the subjects were told that when they saw a cue word it should remind them of one of the passages or lists just studied and they were encouraged to use this information to help them in making the decision.

3.1.3. Subjects

The subjects were undergraduate students from the University of Exeter pool of volunteers, and none of them knew the precise purpose of the study. Subjects who responded particularly slowly for both trials within any of the 12 cells of the experiment (i.e., with response latencies in excess of their own mean plus 2.5 standard deviations) were eliminated from the study and replaced by new subjects. Thirty-six subjects were run altogether and 12 were rejected on the basis of this criterion.

3.2 Results and Discussion

The results were broadly in line with expectations. Table 1 shows the mean correct response time for "new" and "old" words in each of the main conditions.

	With cues		Without cues	
	1 (old)	2 (new)	3 (old)	4 (new)
Complete	1469	1399	1641	1664
Bare	1508	1463	1825	1736
Wordlist	1658	1605	1798	1619

Table 1. Results of Experiment I. Mean time (in milliseconds) taken for subjects to respond to "old" and "new" words, with and without cueing and with the three different types of experimental materials.

The results confirmed that performance was faster when subjects were able to use a cue to help them recover the critical information. ($Min F'$ $(1,45) = 5.7, p < 0.025$). In particular there was an overall cue effect for "old" words ($F1$ $(1,23) = 7.35, p < 0.025$) and a stronger one for the two passages ($F1$ $(1,23) = 8.39, p < 0.01$). However, a contrast interaction indicated that the cue effect was no greater for the two passages than it was for the word list ($F1 = 1.03$). Again as expected, this effect was less marked in the *Word List* condition where subjects are presumably least likely to infer the relevant plan (though interestingly there was some indication of a benefit even here). The cue effect was more marked in the two passage conditions, and, most important for the present purposes, there was no indication that it was any larger for the *Complete Passages* than for the *Bare Passages*. This suggests that the subjects must have inferred the plan in the latter condition.

However, these results were marred by the fact that the data were very noisy and the crucial interactions between cueing and list versus passages did not reach statistical significance. The reason for this was apparently that many of the subjects found the task extremely difficult and took excessively long to respond. (This was reflected in the high proportion of subjects excluded from the experiment on the basis of their outlier data.)

While results provided some indication that subjects may infer plans in the course of comprehension, the experimental technique did not seem to be a very promising one to pursue and so the remaining studies employed different methods of investigation.

4. Experiment II

This experiment was designed to tackle the same problem in a different way. We decided to try to avoid the effects of prolonged and deliberate processing by using a task which would only reflect influences that operate very rapidly. Specifically, we used a lexical decision task to test for the presence of an activated *plan*. The argument was that if the plan is activated during comprehension of the passage then the lexical decision time for its name should be faster than that for the corresponding decision following a control presentation (i.e., a word list). This technique for investigating the intermediate products of comprehension has already been applied successfully in our laboratory (see Sharkey and Mitchell, 1985). The earlier study was concerned with the way in which scripts are activated, deactivated and replaced by new scripts, but the rationale is essentially the same. It was argued that if an information package (i.e., a *script* or in the present case a *plan*) is still active at some point after the subject has read the text, then this should be manifested in a relatively rapid response time to words that are strongly associated with the package – in this case the plan name. Briefly then, in the present study subjects first read either a complete passage, or a word list or nothing at all and they were then presented with a letter string that was either a plan name or a nonword. It was argued that if the plan had been activated in the Complete Passage (or Story) condition then positive response latencies in this condition should be shorter than those following word lists or no prior material.

4.1. Method

4.1.1. Materials

The materials consisted of 48 passages describing a character, a setting and a predicament and strongly suggesting a goal. The passages also specified a precondition for a *plan* that would satisfy that goal. For example:

Alan had been hacking away at the brambles in the back garden for three hours with his scythe, but he seemed to have made no impression on them at all. So he went to get some petrol.

The first part indicates that the protagonist's goal is to clear the garden and the final sentence suggests that his plan is to *burn* the vegetation. In this example, therefore, the test word was BURN.

The corresponding word list was:

        petrol,   hacking,
        back,   garden,
        scythe,   impression,
        brambles,   got,
        three,   hours.

In the third condition the passage or word list was replaced by the phrase "No Passage or word list".

4.1.2. Procedure

On each trial the subjects first saw the prompt — "Now for the next passage". They then pressed the space bar to display the entire passage/word list for long enough to read it. Following a second press the screen went blank for 200 ms and was then replaced by the target word in the centre of the VDU. On a random 50% of trials this was a plan name. The subject was required to indicate as rapidly as possible whether or not the test string was an English word. The main experimental session was preceded by extensive practice with the lexical decision task in two unrelated experiments followed by fifteen trials with materials similar to those used in the main experiment itself.

4.1.3. Subjects

Eighteen subjects were recruited in the same way as in Experiment I.

4.2. Results and Discussion

The mean correct decision latencies are shown in Table 2.

	Story	Word list	No passage or Word list
Words	803	893	865
Nonwords	908	925	922

Table 2. Results of Experiment II. Mean correct response latencies (in milliseconds) to words and nonwords in the three different experimental conditions.

The results showed that the response times to words following the Story were significantly faster than those in either of the other two conditions ($Min\ F'$ $(1,40) = 9.26, p < 0.005$). This was also true for Story versus Wordlist ($Min\ F'$ $(1.40) = 6.45, p < 0.025$). Moreover, there was no difference between the last two conditions ($F1 = 1.03$).

The findings therefore provide some support for the notion that people infer plans when reading texts of this kind. However, it is possible to raise certain objections to this conclusion. In particular, it may be that neither control condition was entirely satisfactory. For example, it is conceivable that subjects respond more rapidly after reading connected prose of *any* kind and not just when some specific plan is activated. Moreover, even if the results do reflect some kind of plan activation the data do not provide any clear indication of what might be required to trigger this.

The next three experiments were designed to eliminate alternative explanations and to explore the primary effect in more detail.

5. Experiment III

In the previous experiment the text both indicated the *goal* to be achieved and provided information that was a *precondition* for a *plan* which might achieve this goal. According to Wilensky's account of comprehension outlined earlier both of these kinds of information should be essential to activate the plan. On this account it follows, therefore, that neither of them alone should produce the priming effect obtained in the second experiment. This hypothesis was tested in the present experiment by the use of four priming conditions: (1) Bare passage (as before); (2) passage with Goal only; (3) passage with Precondition information only and (4) No passage. On Wilensky's hypothesis only the first condition should produce a stable priming effect. This is because the procedure specified in the model is basically one for building a bridge between an established goal and some newly-encountered event. If either type of information is missing then the scheme offers no way of proceeding. For example, if the precondition is missing then there is no "event" to trigger the first step of the operation, and so according to the model there is no occasion to *infer* the plan. If there is no *goal* then a *plan* might be inferred at Step 2, but this cannot be the plan for any *known* goal (Step 3), so one must be *inferred* (Step 4). However, this cannot be attached to the *known* plan (Step 5) (because there is none), so the best that can happen is that the *plan* might eventually be inferred (Final Step), but even here there is no guarantee that it would correspond to the one that would be settled for in the light of a known goal, and so there is unlikely to be any activation of the *plan* in question.

In other words, according to the theory, the process of inferring the "correct" plan should depend on the availability of both *precondition* and *goal* information. As we have already mentioned this is what was tested in the current experiment.

5.1 Method

5.1.1. Materials

These were basically similar to those used in the previous experiment. For example:

(1) (Bare Passage) When the wagon train reached the gorge, the pioneers thought at first that they would be able to wade across, but after a few attempts they sent off scouts to look for tall trees to cut down.

(2) (Goal only) When the wagon train reached the gorge, the pioneers thought at first that they would be able to wade across, but they failed.

(3) (Plan precondition only) The pioneers sent off scouts to look for tall trees to cut down.

In this case the test word was BRIDGE.

Overall there were 24 different sets of test material and 24 further sets constructed in exactly the same way and used for presentation on the nonword trials.

5.1.2. Procedure

This was exactly the same as that used in the previous experiment.

5.1.3. Subjects

There were 12 subjects recruited from the same source as before.

5.2. Results and Discussion

The results are shown in Table 3. They were very disappointing and showed no significant effects at all (e.g., the difference between the "extremes" (1) and (4) was not significant ($F1 = 1.54$).

(1)	(2)	(3)	(4)
Bare Passage	Goal only	Precondition only	No Passage
942	1017	971	986

Table 3. Results of Experiment III. Mean correct response times (in milliseconds) to words in the four main experimental conditions.

The failure to obtain significant facilitation in the Bare Passage condition in this study suggests that the priming effects obtained in the previous study were probably not automatic. (If they had been, presumably they would have turned up again here, since the materials were highly similar). Changes in the design of the experiment appear to have influenced the strategies that the subjects adopted. The important changes were probably those associated with the make-up of the materials. In the previous experiment one third of all the trials (and *all* of the coherent passages) provided useful information for the following lexical decision and the indications are that subjects used this information quite consciously. Indeed, several of them reported during debriefing that they had been able to predict the test word during the experimental session. In the present study useful information was available on only a quarter of the trials and on only a third of all sensible passages. It therefore seems likely that the predictive value of a passage was markedly reduced in the current experiment, and this may have caused subjects to abandon strategies based on the predictive use of inferences.

Methodological changes in the next two experiments were designed to encourage more predictive processing by increasing the proportion of relevant passages. This was done by examining the effects of the two partial cues in separate experiments and by adding a proportion of dummy passages which were generally predictive.

6. Experiments IV and V

These two experiment were identical in their basic design. Each had three types of priming material: (1) Bare passage; (2) Goal information only (for Experiment IV) and Precondition information only (Experiment V) and (3) No passage.

6.1. Method

6.1.1. Materials

These were the same apart from minor changes and improvements. Another example:

(1) (Bare Passage) Jack had been in Dartmoor prison for ten years now and was due to be in for another ten. He was missing home so much that he had been helping his friends to dig a tunnel.

(2) (Goal only – as used in Experiment IV) Jack had been in Dartmoor prison for ten years now and was due to be in for another ten. He was missing home a great deal.

(2') (Precondition information only – as used in Experiment V) Jack had been helping his friends to dig a tunnel.

In this case the test word was ESCAPE.

In addition to the 24 experimental passages there were 9 further sets of material which made the test word highly predictable and another 15 passages which were used in connection with the nonwords trials.

6.1.2. Subjects

Eighteen subjects were obtained as before.

6.2. Results and Discussion

The results of Experiment IV (Goal only) are shown in Table 4.

For the words there was a significant Passage versus No Passage effect in an analysis treating subjects as a random effect (FI (1,17) = 14.9, $p < 0.005$). The Goal only condition was intermediate between these two, but was not significantly different either from the No Passage

	Bare passage	Goal only	No passage
Words	698	732	768
Nonwords	867	863	826

Table 4. Results of Experiment IV. Mean correct response latency (in milliseconds) to words and nonwords in the three main experimental conditions.

control ($F1$ = 2.48) or from the latency following the passage ($F1$ = 2.53). There were no significant nonword effects at all.

The results for Experiment V were comparable. They are shown in Table 5.

	Bare passage	Precondition only	No passage
Words	706	751	812
Nonwords	856	872	904

Table 5. Results of Experiment V. Mean correct response latency (in milliseconds) to words and nonwords in the three main experimental conditions.

Again there was a highly reliable overall priming effect ($F1$ (1,17) = 32.3, $p < 0.001$), but this time the partial provision of (Precondition) information was significantly different from both the control ($F1$ (1,17) = 7.8, $p < 0.025$) and from the passage ($F1$ (1,17) = 6.1, $p < 0.025$)

Taken together these results confirm that the priming effect obtained in Experiment II cannot simply be due to the use of connected prose in that study. The partial conditions in the last two experiments employed similar materials and yielded much smaller effects, suggesting that it is the *nature* of the material that is important. However, the results do not really seem to confirm Wilensky's suggestion that both *goal* and *precondition* information are needed in order for the plan to be activated during passage comprehension. In both experiments there was a tendency for the partial information to facilitate lexical decisions relative to the No passage condition. This suggests that at least in some passages the partial goal or precondition alone may have been sufficient to activate *plans*.

7. Experiment VI

The previous experiments (II & III) showed evidence of strategic effects in the use of *plan* and *goal* information, and since lexical decision could be viewed as a rather unnatural task it seems reasonable to ask whether such information is used on-line in more conventional tasks. This was tackled in the final experiment by the use of two new dependent measures – (1) the reading time for a sentence the full interpretation of which depends on the availability of *plan* information and (2) verification time for a plan statement presented without warning after the end of the text. In either case *plan* information was either supplied explicitly as part of one or two sentences early in the passage or else it was merely implied in a second version of the text from which these sentences were excluded. We argued that if *plan* information is used as the text is first being read, then the reading time for the critical sentences should depend on whether the information is directly available (in the first condition) or whether it has to be inferred from incomplete information provided (in the condition with *plan* sentences excluded).

In other words, the sentences should be read more rapidly in the first condition if *plan* information is used on-line. The prediction for the second measure (verification) follows a similar line of reasoning. If *plan* information is made available during the initial reading phase whether it is stated explicitly or inferred, then the verification task should not be affected in any way by its origins. That is, response latency should be equivalent in the two conditions.

On the other hand, if *plan* information is *not* drawn upon during the initial reading phase, then in Condition 1 it would presumably be necessary to *infer* this information in order to verify the test statement. On this account, therefore, the verification time in the implicit condition would be longer than that in the explicit condition.

A second purpose of this study was to see whether any use is made of the activated information. The experiments reported so far have shown some evidence that plan information is inferred in the course of reading. But it is already known that people draw a variety of different kinds of inferences while they process texts. For example, Bransford and his colleagues have conducted several experiments to show that readers draw inferences about instruments used in actions (Johnson, Bransford & Solomon, 1973) and about the consequences of actions (Bransford, Barclay & Franks, 1972). It could be that inferences about plans have exactly the same status as those about other kinds of information. However, according to the framework put forward by Wilensky they should play a rather more central role in subsequent processing than the others do. Specifically, according to the theory *goal* and *plan* information should be used to interpret and structure most of the information that arrives later in the text. In other words, rather than being a relatively isolated addition it is assumed to provide a framework for subsequent analysis of the material. If this is true it should be possible to demonstrate that the handling of material is influenced by inferences made earlier in the text.

7.1. Method

7.1.1. Materials

There were 32 different passages. Each started with a sentence designed to set the scene and to specify a goal. Next there was a sentence describing a plan that was adopted by the protagonist in an effort to achieve the goal. This sentence was included in the passage or excluded from it depending on the experimental condition.

Finally, there was an "Action statement" describing some aspect of the execution of the plan. This was always the precondition for the plan and the comprehension of this material therefore depends on connecting it to previous information by means of the plan – either one that is explicitly mentioned or one that is inferred. This Action statement was used for the reading time test. Following each passage there was a verification statement and the subject had to indicate whether or not it was true in the light of the preceding passage. A typical example is given below:

(Theme/goal part) The terrorists wanted a lot of money to buy weapons.

(Plan part) They decided to capture the child of a well known figure and ask for a large sum of money in return for her life.

(Action statement) They lay in wait for the daughter of the Prime Minister as she walked home from school.

Verification statements either
(1) They were planning to kidnap her ("Yes"); or (2) They wanted to give her a present ("No").

7.1.2. Procedure

Following the prompt — "Now for the next passage" — the subjects were instructed to read the passages in a self-paced reading task (with RTs recorded line by line) and then to indicate by pressing yes/no buttons whether each test statement was likely to be true given the prior information. All Action statements were divided into lines of comparable length and all Verification statements contained similar numbers of words. The experiment itself was preceded by a 15-trial practice session. For a detailed methodological discussion of the procedures employed here see Mitchell and Green (1978) and Mitchell (1984).

7.2. Results and Discussion

The results are shown in Table 6.

	Plan included	Plan not included
Reading time	1490	1597
Verification time	2001	1998

Table 6. Results of Experiment VI. The mean reading time per line (in milliseconds) for the critical action statements and the mean correct verification times for the test statements.

The results show that there was a highly reliable effect of plan inclusion on reading time ($F1$ $(1,19) = 23.5, p < 0.001$), but that the effect on verification was very much smaller and certainly not significant.

These results are consistent with the notion that plan information is inferred in the course of reading, though perhaps only when it is first needed. The evidence also suggests that an inferred plan is just as available as one that is presented explicitly.

8. General Discussion

The results seem to indicate that *plan* information is inferred during the course of understanding a passage, and that this information is then used to interpret portions of the text that occur later in the story. This information is apparently made available with rather less information than suggested in Wilensky's model, but otherwise these findings tend to confirm the importance of plan and goal information in understanding stories.

It is useful to consider this work in the light of the overall theme of breakdown in communication. The obvious implication is that for elliptical materials of the kind considered, (and indeed this would include most types of dialogue) the language understander (or generator) needs to have access to a considerable amount of information about potential plans, and there needs to be agreed procedures for activating these plans. In addition, effective communication demands that there should be standardized techniques for using these plans to make sense of new material. Without such facilities a language understander (natural or artificial) will be incapable of deriving sensible interpretations for dialogues or for connected passages like stories.

Finally, we would like to end up by considering the role of psychological work in this field. It seems possible that most of our conclusions could have been drawn without resorting to the rather long-winded methods of experimental psychology. Perhaps all that was needed was a certain amount of systematic thought about the tasks involved, in which case the answers would have been reached much more quickly than by using our experimental techniques.

However, such an exercise would only yield a *possible* solution and in any real communication or dialogue this may not correspond in any way with the one that is actually used by people. This is likely to be an important consideration because symbolic messages (i.e., all language) cannot be decoded without supplementary declarative and procedural information (such as that concerning plans and goals) which is needed to convert them into a meaningful representation. This information can be likened to a code book in a cyphering exercise. If the code book contains inaccurate or incomplete information the final message is bound to be garbled. The current work is intended to provide the basis for working systems relying on natural language communications between machines and real people (with all their foibles and limitations). With this general aim in mind, it seems sensible to find out as much as possible about the procedures and "code books" that people *actually* use rather than relying on idealised methods that could *potentially* be used in optimal circumstances. In other words we would maintain that there is an important role for psychological investigation in this kind of work.

In the present case, the experimental work provides a good deal of support for the Artificial Intelligence analysis. The possible exception to this generalization lies in the details of the procedure for inferring plans. The results suggest that humans need less information to do this than Wilensky's program does. If they also assume that others operate like this then there may be a tendency to generate material that is readily handled by people, but that is too elliptical to be interpreted by the program. In these circumstances there would almost certainly be a breakdown in communication and the most "friendly" solution to this problem would almost certainly be to modify the software so that it behaved in a more human-like fashion.

References

Bransford, J.D., Barclay, J.R., & Franks, J.J. (1972). Sentence memory: A constructive versus interpretative approach. *Cognitive Psychology, 3,* 193-209.
Johnson, M.K., Bransford, J.D., & Solomon, S.K. (1973). Memory for tacit implications of sentences. *Journal of Experimental Psychology, 98,* 203-205.
Mitchell, D.C. (1984). An evaluation of subject-paced reading tasks and other methods of investigating immediate processes in reading. In D.E. Kieras and M.A. Just (Eds.), *New methods in reading comprehension research.* Hillsdale, NJ: Erlbaum.
Mitchell, D.C., & Green, D.W. (1978). The effects of context and content on immediate processing in reading. *Quarterly Journal of Experimental Psychology, 30,* 609-636.
Ratcliff, R., & McKoon, G. (1978). Priming in item recognition: Evidence for the propositional structure of sentences. *Journal of Verbal Learning and Verbal Behavior, 17,* 403-417.
Schank, R.C., & Abelson, R.P. (1977). *Scripts, Plans, Goals and Understanding.* Hillsdale, NJ: Erlbaum.
Sharkey, N.E., & Mitchell, D.C. (1985). Word recognition in a functional context: The use of scripts in reading. *Journal of Memory and Language, 24,* 253-270.
Wilensky, R. (1978a). *Understanding goal-based stories.* PhD thesis. Yale University.
Wilensky, R. (1978b). Why John married Mary: Understanding stories involving recurring goals. *Cognitive Science, 2,* 235-266.

6

DIALOGUE IN CONTEXT

DISCOURSE SITUATION MISUNDERSTANDINGS

G. Ferrari

Department of Linguistics
University of Pisa

I. Prodanof

Instituto di Linguistica Computazionale
CNR – Pisa

In this article an intuitive model of dialogic communication is first sketched. Within this frame attention is focused on the process of 'interpretation', i.e., of relating an utterance to its contextual reality. Situation Semantics is proposed as a formalism most suitable to represent such connection with context; a few minor extensions of Situation Semantics are also proposed. Some cases of misunderstandings are analysed in terms of (partially) incorrect linking of an utterance to contextual reality, and a classification method is proposed. A distinction is also drawn between communication failures, which cause a breakdown of communication, and misunderstandings which do not cause such a breakdown.

1. Introduction

The situation in which a dialogue, and in particular a goal directed dialogue, takes place seems to affect the process of understanding the utterances which form it. Situation semantics (Barwise & Perry, 1983) provides a useful framework in which the connection between reality and the meaning of utterances, which is a stage in the process of understanding, can be studied, even if some extensions of its formalism are needed. Some cases of misunderstanding between dialogue participants can also be classified within this framework.

2. Communication, Understanding, and Interpretation

In the last ten years many studies have been carried out aimed at the computational modelling of dialogue. Some important aspects of human communication have been studied within a computational framework, such as focusing (Grosz, 1977), the inference of the speaker's intentions (Allen & Perrault, 1980; Allen, 1983), or the prevention of the hearer's misconceptions (Webber & Mays, 1983), and different techniques for their treatment have been experimented with.

All these studies have in common the underlying idea that responses to a speaker's question (utterance) are not a simple function of the meaning of the question itself, as it used to be assumed in early question-answering systems, but rather some additional processing is required by the hearer. In this way, the response shows an apparent degree of independence from the literal meaning of the question.

Generalizing this idea, we assume that any communication act consists of the following three phases:

(i)   production of a superficial message, such as an utterance, a gesture or any other act conveying some information, by an 'issuer';

(ii) understanding of the message by a 'receiver';

(iii) construction of an appropriate (re)action, which may include a response to the received message.

This is a cycle in which any phase (i) is the result of a phase (iii). It is repeated an indefinite number of times in a conversation, and at any cycle the roles of issuer and receiver are exchanged between the dialogue participants.

The term 'receiver' is used instead of the more traditional 'addressee' in order to stress the role of the hearer in a dialogue who is active in 'receiving' the message.

The starting and the ending of a dialogue are accounted for by anomalous cycles. In fact, the beginning, i.e., the production of the first superficial message (phase i) follows the construction of an 'act' which implies communication (phase iii). However, this is not a 'reaction' to a previous message, but simply an 'action', or possibly a 'reaction' to a real world situation. Conversely, the ending of a dialogue occurs when the 'receiver' of a message understands that the current utterance is the last one (phase ii) and builds a 'reaction' involving no communication (phase iii).

Previous research has shown that one of the most important activities of the understanding phase (ii) is the inference of the issuer's intentions. In particular, a mechanism has been designed which infers the plan a speaker has in mind when uttering a question and constructs a response as a function of such an inferred plan (Allen & Perrault, 1980; Allen, 1983). In the first step, the system described by Allen (1983) builds a complex action as the semantic interpretation of a given question. By means of a number of specific inference rules, this action is then recognized to be part of a more general plan which is assumed to be the complete speaker's plan. The answer is constructed as a reaction to that plan, rather than to the action directly corresponding to the utterance.

Thus, in the question

(1) "When does the train to Windsor leave?"

the following action is observed:

(2) REQUEST (A,S, INFORMREF (S,A, time 1))

with

(3) TIME (time 1) and DEPART.TIME (train 1, time 1)

(4) TRAIN (train 1) and DEST (train 1, WINDSOR)

where A and S are variables standing for the 'questioner' (A) and the 'responder' (S). REQUEST and INFORMREF are primitive actions combined together according to the content of the question. The speech act (Austin, 1962; Searle, 1969) in which the utterance is involved is also accounted for in the translation of the question. In this specific example a 'request' speech act is recognized and the following corresponding pattern is instantiated:

(5) REQUEST (speaker, hearer, action).

The basic action identified in the sentence has the pattern

(6) INFORMREF (speaker, hearer, description)

where 'description' is bound to a time specification.

The action thus associated with the question is recognized as part of the BOARD plan shown in Figure 1.

$$\text{BOARD(A,train1, TORONTO)}$$

$$\text{enable}$$

$$\text{AT(A,loc1,time1)}$$

$$\text{know}$$

$$\text{KNOWREF(A,time1)}$$

$$\text{effect}$$

$$\text{INFORMREF(S,A,time1)}$$

$$\text{want-enable}$$

$$\text{WANT(S,INFORMREF(S,A,time1))}$$

$$\text{effect}$$

$$\text{REQUEST(A,S,INFORMREF(S,A,time1))}$$

where
- EQ(train1,ENTRY-TRAIN17)
- DEPART.TIME(train1,time1)
- DEPART.LOC(train1,loc1)

Figure 1

The cooperative reaction is the constructions of the following response

(7)   "At 3.17 from gate 7"

in which not only the departure time, but also additional information about the location is provided.

The schema in (6) serves as a semantic pattern according to which the semantic interpretation process of the utterance is performed. The result is a combination of an action (INFORMREF) and some abstract objects (S,A, time 1) in a predicate-argument structure. We will assume that any similar combination of abstract designations of actions and objects organized in whatever type of structure resulting from the analysis of a given utterance is the 'semantic representation' or the 'meaning' of that utterance. In general, actions are related to verbs and objects to noun-phrases.

The inference of the speaker's plan is, instead, the product of a further reasoning activity from the utterance's 'meaning', together with the indication of the illocutionary force, and provides the relevant information from which a reaction can be constructed.

It is intuitively obvious that this is not the only reasoning activity involved in the understanding of a message. A 'receiver' probably also performs other activities such as the checking of the completeness of the information he has received from the message, the surveying of his general knowledge about the message's subject, etc. We will call all the complex activity involved in the process of understanding (including the 'meaning' of an utterance) 'reasoning' or 'inference'. The speaker's plan inference is only a small part of this process.

The changing of various parameters related to an utterance, its 'meaning' and illocutionary force remaining the same, can affect different steps of the communication loop, thus producing variants of the same dialogue.

For instance, cooperative behaviour may fail to be adopted by the 'responder'; in this case he can answer to the literal 'meaning' of the question, giving only the departure time of the train. He can even give unkind answers, as in the example

(8)   Why don't you look at the time-table?[1]

However, these answers are appropriate only in a standard (default) situation in which the 'questioner' is a traveller, the 'responder' is a railway employee, and the dialogue takes place in a railway station. The question could be addressed, instead, to another traveller. In such a case, the answer

(9)   I don't know. Let's look at the time-table. I'll help you.

would sound as cooperative as (7).

Equally, different intentions can be attributed to the speaker according to the external situation. Thus, changing the 'questioner' and the scenario, if the same question (1) is asked by a station-master of another railway employee in a situation where trains have been delayed, the inference of a BOARD plan could turn out to be inappropriate. The station-master is probably trying to activate a complex plan for rearranging gates and departure times, in order to avoid an emergency.

Finally, if (1) is asked by a traveller in a travel agency, the more likely speaker's plan is a SCHEDULE_TRIP plan.

These last two examples have shown that different speakers' plans may be inferred in accordance with different situations.

The speaker's intentions remaining constant, the construction of the appropriate reaction and of its superficial form by the hearer is also influenced by the situation. The question

(10)   Where is Cellini's Perseus?

may be uttered in Florence in Piazza Signoria or at Florence railway station, or even by some gentleman sitting in a club. In the first case, a VISIT plan can be inferred from the question, and as Cellini's Perseus is located in a corner of Piazza Signoria, a simple pointing act is required to satisfy this plan. In the second case, although the inferred plan is the same, a complex natural language communication act is initiated, possibly together with a reference to a map and even a checking of reciprocal linguistic comprehension (if the questioner is clearly a foreigner). In the third case, the simple answer

(11)   In Florence, in Piazza Signoria.

will probably satisfy the 'questioner', as no immediate VISIT plan seems reasonable.

A number of similar examples can be found, which show that the inference of the speaker's plan, as well as the construction of the reaction, depends upon various aspects of the situation in which a question is uttered.

---

[1] The examples quoted henceforth have all been directly experienced by the authors.

Generalizing this observation we will assume that in the process of understanding, on which we are now focusing, besides the 'meaning' and the representation of the speech act, and the reasoning activity that follows, involves some complex relation to reality that requires a deeper inquiry. This relation is established not only from an utterance to the real objects it refers to, but also from the utterance to the external situation in which it occurs.

Figure 2 shows a new schema of the communication process, which tries to include such a reference to reality. The 'receiver' of a message has at his disposal not only a surface version of the message, but also the reality of which the specific message is a part. He will first extract from such a message the 'meaning' and the indication of its illocutionary force. A further step of relating the message to reality is then taken. We will call this relation, represented by the arrow "interpreted on", 'interpretation' of a message. Interpretation can be viewed as a sort of instantiation, or more properly individuation, of the message's meaning. Once a correct relation between the message and the reality it is connected with is established, inference can take place in order to complete the understanding phase. Reality itself and the inference activity provide the necessary elements for (stimulate) the reaction construction. Any message, be it an initial one or a response, originates from the combination of stimuli given by reality together with some reasoning activity.

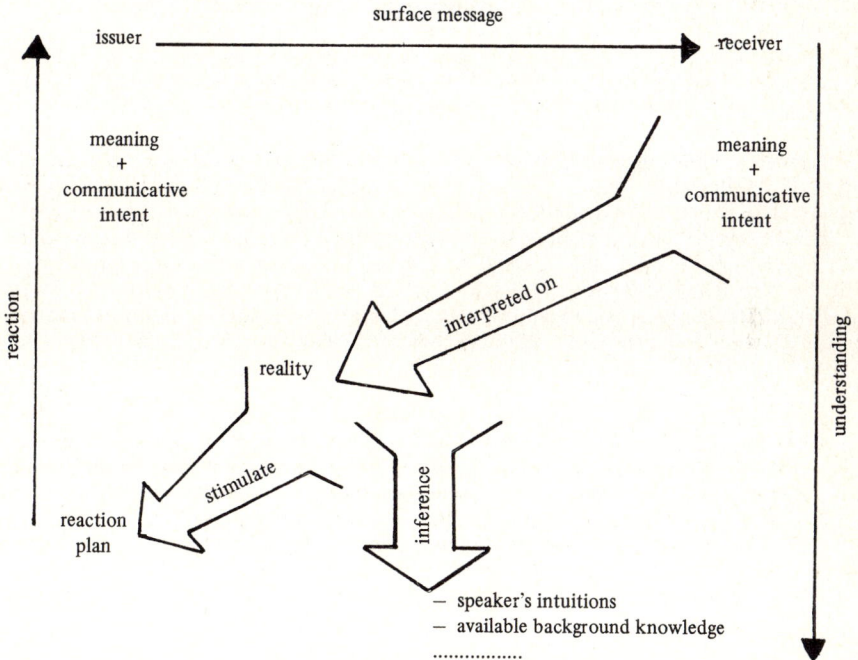

Figure 2

In human communication this connection of a message to reality is not a simple function, but has three facets, which we will call world reality, 'issuer's' and 'receiver's' view of reality. If a communication act is successful, the relation between the 'receiver's' and the 'issuer's' view of reality should coincide. Eventually, if some action intended to modify the current situation is contained in an utterance, 'issuer' and 'receiver' should agree also on the interpretation of such action, i.e., on its desired effects.

When a communication failure occurs, the overlap between 'issuer's' and 'receiver's' view of reality is only partial. The elements about which disagreement arises are the origin of the communication failure, while the overlapping elements provide a basis of mutual understanding by which a repair can be attempted.

3. Situations and Communication

We have assumed that the central activity of communication consists of understanding a message and building an appropriate reaction. Besides the reconstruction of an utterance's meaning, an interpretation step and a reasoning (inference) step play a basic role in the understanding phase.

The above examples have shown how the same utterance may be understood in different ways or may activate different reactions in accordance with the situation in which it is uttered.

In sum, an utterance, or, more generally, a message can convey different information about the needs and the beliefs of the issuer, according to a complex framework of which the message is just one element.

In this section we will focus on the nature of the connections of an utterance with reality and the mechanisms by which a link between expressions and real situations is established.

In the field of philosophy of language, many theories of communication try to account for the relation of expressions to reality. Among them, however, Situation Semantics (SS) (Barwise & Perry, 1983) provides a clear, uniform, and relatively simple formalism for representing both situations referred to by utterances and the situations in which utterances are produced. A central assumption of SS is that a single utterance may convey different information. This is called the "efficiency of language", and is described as follows: "Expressions used by different people, in different space-time locations, with different connections to the world around them, can have different interpretations, although they retain the same linguistic meaning" (Barwise & Perry, 1983, p. 5).

Also, the distinction between linguistic meaning and its use in a context, coincides with a distinction characterizing SS. This relation with reality, the context, is used in communication to pass from the "meaning" of a message to the information it carries on a specific occasion. Three general sorts of facts about utterances are recognized in a context: (i) 'discourse situation', i.e., "facts about ... who the speaker is, ... when (the utterance) occurs", (ii) 'connections', i.e., "the way the speaker (or listeners) are connected with the larger world"; 'connections' are mainly involved in reference, and (iii) 'resource situations', i.e., known facts about reality, including directly perceived facts, common knowledge about the world, or the knowledge built up by previous discourse.

The basic elements of Situation Semantics are 'situations', i.e., individual configurations of reality, both static, called 'states of affairs', and dynamic, called 'events'. The internal structure of a situation is characterized by a spatio-temporal location and a 'situation-type', i.e., a representation of how things stand in an abstract situation, regardless of where and when it occurs.

A 'situation-type' is expressed in terms of 'constituent sequences'. A 'constituent sequence' is a relation 'r' between objects 'x_i', as in "shouts at ('r'), Mr. Levine ('x_1'), Molly ('x_2')"; in a situation, a relation ('s') is established between 'constituent sequences' and the pair 0/1, or yes/no. A 'situation-type' related to a location is a 'state of affairs', while a collection of 'constituent sequences' related to different locations and yes/no pairs is a 'course of events (coe)'.

In the example

(12) in e, at 1: barks, Molly; yes
      at 1': shouts at, Mr. Levine, Molly; yes
      at 1": barks, Molly; no

the 'course of events' 'e' is described, in which Molly stops barking after Mr. Levine's shouting at her. An 'event-type' is an abstract 'coe' which contains indeterminates. Indeterminates are expressed as $\dot{x}$ symbols and may represent individuals, relations or locations. For an 'event-type' $E(\dot{a},...,\dot{r},...,\dot{l},...)$ a function 'f' assigns to indeterminates real individuals, relations, and locations; this function is called 'anchor' and has the form "$f(\dot{x}) = x$". Uniformities across individuals in events are called 'roles'; for an indeterminate $E(\dot{x},...)$, the role is expressed as $<\dot{x},E>$.

A central aspect of the relation of an utterance to the real world is captured by the 'discourse situation'. This consists in the appearing aspects of an utterance, which are formalised in terms of roles.

An abstract discourse situation is a 'state of affairs' of the form

(13) DU : = at $\dot{l}$: speaking, $\dot{a}$; yes
              addressing, $\dot{a}$, $\dot{b}$; yes
              saying, $\dot{a}$, $\dot{\alpha}$; yes

where DU stands for the situation in which discourse has an addressee. $\dot{l}$, $\dot{a}$, $\dot{\alpha}$, and $\dot{b}$ are 'indeterminates', and are also expressed in terms of 'roles', as follows:

(14) speaker = $<\dot{a}, DU>$
      addressee = $<\dot{b}, DU>$
      disc-loc = $<\dot{l}, DU>$
      expression = $<\dot{\alpha}, DU>$

Another special role is used in Situation Semantics in order to account for nominal reference. It has the form

(15) $\ddot{ref}$ = $<a_1, REF>$

where REF is a DU event-type augmented by a referring situation and ref is equivalent to

(16) referring-to, $\dot{a}$, $\dot{\alpha}$, $\dot{a}_1$; yes

If in a situation c an NP α is used to refer to some individual a_1, this is called the referent of α in c, written $c(\alpha)$. An utterance gives rise to a function from referring words α to their referents c (α); this function is called 'speaker's connections' and links an utterance to the corresponding described situation.

A complex account is given in these terms by Barwise and Perry (1983) both for different types of nominals and for verb phrases. We will simplify such a formalization by modifying (16) as follows

(17) referring-to, $\dot{a}$, $\dot{\phi}$, DS; yes

where ϕ is the set of all the referring expressions contained in an utterance, and DS is the set of referring situations. DS is, in fact, a collection of REF event-types.

It is plausible to assume that part of the process of understanding consists in the reconstruction by the hearer of speaker's ('issuer's') connections, or, more precisely, in linking the abstract objects contained in the meaning of the utterance to concrete individual objects in a situation. Although actions are not taken into account they can be treated by the same mechanism of linking abstract general actions to concrete instances of actions (see also Robinson, 1980).

In the communication cycle we will, therefore, distinguish the notion of 'speaker's (issuer's) connections' from 'hearer's (receiver's) view of the speaker's connections'.

As we have seen above, the identification of the speech act is an important element in the understanding of a given utterance. In particular, it is the most relevant hint for the inference of the speaker's expectations about the class of reactions of the hearer. For example, a REQUEST speech act is intended to trigger the construction of a reaction in the class 'response', while an ASSERT speech act will involve a reaction of assimilating the utterance as a belief of the speaker. Therefore, the construction of the hearer's reaction, and, in some cases, even the inference of the speaker's plan, are affected by the speech act.

However, the speech act, or, more precisely, the illocutionary force of an utterance, has an ambiguous status. In fact, on the one hand it qualifies the utterance as a specific communicative act in the real world, i.e., relates an utterance to a specific act, while, on the other hand, it is identified from the linguistic form of the utterance.

A possible formalization of speech acts in terms of a communicative event is only alluded to by Barwise and Perry (1983, pp. 275-286).

In a more detailed proposal by Evans (1981) speech acts are considered as transition functions from a 'situation-type' to a new 'situation-type' defined as the legitimate end state of a given speech act.

In our opinion, as the expression role

(18)   expression = $<\dot{\alpha}, DU>$

is used to present the meaning of the uttered expression, a role should be added to represent the effect an utterance is intended to have on reality, i.e., the speech act.

For these reasons we propose to add a role

(19)   spa = $<\dot{s}, DU>$

which identifies the speech act, as

(20)   DU : = ...

   acting, $\dot{a}, \dot{\alpha}, \dot{s}$; yes

The symbol α is used by Barwise & Perry (1983, p. 123) to identify constituent expressions of the sentence ϕ which is used in the utterance 'u'. As a speech act is, in general, related to 'u' as a whole, it is more correct to substitute α by ϕ, as we have informally proposed for reference. A discourse situation 'd' will, therefore, describe the reality connected to the entire utterance 'u' and its meaning ϕ.

However 'u' gives us also the set

[d_α : α is ϕ or a constituent of ϕ]

where d_α is a referring situation, (see also Barwise & Perry, 1983, p. 123, and our simplification above).

The discourse location (disc-loc) role is a spatio-temporal individuation 'l' of the discourse. Barwise and Perry (1983) seem to introduce, by this role, the flow of real external space-time locations along with any following d_α of 'd'. As we focus on the 'd' (d_ϕ) level, we may assume that any utterance 'u' in a dialogue occurs at a unitary 'l'. A dialogue is a sequence of $d_1, d_2, ..., d_n$ occurring at $l_1, l_2, ..., l_n$ where $l_i > l_{i+1}$, with $>$ = "wholly temporally precedes". However, in a generalized communication situation l_i is to be understood as only an abstract space-time unit in which the communication occurs. In reality 'issuer's' and 'receiver's' space-time locations coincide only in the case of ordinary face-to-face communication. But the participants in a communication may have different spatial locations, as in telephone communication, or different temporal locations, as in both ordinary and electronic mail. It seems reasonable, therefore, to add to d-situation the corresponding roles to represent 'speaker's' and 'addressee's' spatial and temporal locations. A non-coinciding location of the dialogue participants may affect at least the interpretation of the speaker's connections.

The process of interpretation, i.e., of connection of the meaning of an utterance to reality, consists of computing the anchor function f in order to assign the corresponding individuals, relations, and locations to the roles of a given discourse situation. It consists, in addition, of computing f(DS), i.e., the values of c (ϕ) in order to reconstruct the 'speaker's (issuer's) connections' or, more properly, to build the 'hearer's (receiver's) view of the speaker's connections'.

Let us show, with an example, how an utterance (question) can be represented in this formalism.

Sentence (11) may have, as its meaning, a schema like

(21a) INFORMREF (S, A, loc1) with
(21b) LOCATION (loc1, CELLINI'S PERSEUS).

This representation of the meaning, chosen for purely expository reasons, is assumed to be part of the following interpretation of (11):

(22)   In u : at l : speaking, a_tourist; yes
                    addressing, a_tourist, local_inhab; yes
                    saying, l_tourist, (21); yes
                    acting, a_tourist, (21), request; yes
                    referring_to, a_tourist, (21), ds; yes

where ds is the following set of situations

(23a) ds_1 := at l_1 : ....
                    saying, a_tourist, (21a); yes
                    referring_to, a_tourist, SHOW-LOC; yes

(23b) ds_2 := at l_2 : ....
                    saying, a_tourist, (21b); yes
                    referring_to, a_tourist, (21b), LOC-OF (PERSEUS); yes

where SHOW-LOC and LOC-OF(PERSEUS) are used as an action and a location in reality.

In the two cases discussed above, the location of the utterance can be either

(24)   f (l) = Piazza Isgnoria
or
(25)   f (l) = railway station

The response to these two different interpretations of the same utterance is constructed from the proposed discourse situation, together with some additional reasoning (Piazza della Signoria is the place where the mentioned monument is, and is far from the railway station), and the questioner's intended plan VISIT.

The additional reasoning we have just mentioned introduces the 'resource situation', which besides 'discourse situation' and 'speaker's connections' characterizes an utterance, and which will probably include any other aspect of reality involved in the interpretation of an utterance. These facts, however, cannot be formally included in an abstract discourse situation, as no closed set of additional roles can be recognized. On the contrary, the number and the quality of facts from reality which affect the interpretation of an utterance may dynamically vary from time to time.

We have just mentioned that, in order to answer question (11) the 'responder' must know the relative locations of Piazza Signoria and the discourse. He probably has to know other additional facts such as, what time it is, which buses stop near Piazza Signoria, at what time buses leave etc.

It is therefore obvious that they are an open-ended list, and it would be interesting to understand the strategies by which hearers select from time to time the relevant elements.

4. Discourse Situation Misunderstandings

In accordance with Barwise and Perry (1983) we have defined the process which yields the interpretation of an utterance as the anchor function $f(\dot{x})$ which binds the roles in an abstract discourse situation to elements of reality. We have added that the reference $c(\phi)$ is also involved in the same interpretation. When either the hearer or the speaker fails to correctly compute such bindings, a corresponding type of misunderstanding occurs.

We use the term misunderstanding to indicate a broader range of phenomena than communication failure. Communication failures are only those cases in which communication breaks down and cannot be resumed unless a specific repair procedure is activated. In the case of misunderstanding, on the other hand, communication may keep running and the misunderstanding be discovered some time after the occurrence of the misunderstanding point, or even remain undiscovered.

By the following examples we will try to suggest that a typology of misunderstandings can be drawn according to which role has been wrongly bound to reality.

The first role we examine is the speech act (spa) role. It is relatively simple to identify a 'request' or an 'assert' speech act on a simple linguistic basis, at least for those languages, like English, which have syntactic devices which signal them. On the contrary, it is often difficult to discriminate the speech act of an utterance in those languages which do not have the corresponding specific linguistic constructions.

In particular, Italian has two peculiarities which make the recognition of the speech act independent from the syntactic form of the utterance. The possibility of moving the subject after the verb makes the following two sentences

(26a) Giovanni e' venuto   and (26b) E' venuto Giovanni
      John has come   Has come John

almost perfectly equivalent.

In addition, nothing but intonation stresses the difference between questions, assertions, and even imperatives. Thus, the two mentioned examples may be changed into questions by simply adding a question pitch:

(27) Giovanni e' venuto? / E' venuto Giovanni?

As a consequence of this possible ambiguity, if intonation is not clear, a frequent misunderstanding arises, which is repaired by the question

(28) Stai chiedendo o lo affermi?
      Are you asking or asserting?

This is more properly a communication failure, as the conversation cannot continue unless the speech act is clarified by the speaker. This failure may be represented as a conflict between the 'issuer' view of the discourse situation, which includes

(29) acting, I, ϕ, S; yes

and a receiver discourse situation

(30) acting, I, ϕ, S?; yes

where I is any instance of an issuer, ϕ is the uttered sentence meaning, and S is any instance of a speech act.

The fact that, after the speaker's clarification, communication is in general completely restored, is a piece of evidence that both the meaning and the other roles were correctly bound to reality.

A second type of misunderstanding arises from a wrong individuation of the speaker by the hearer.

Many teachers may recall that the very first time they were lecturing the students, a scene similar to the following occurred:

The teacher, clearly embarassed, asks a student:

(31) Excuse me, where is the classroom n. 21?

The addressed student gives some unkind or aggressive answer like

(32a) What do you need to know it for?
(32b) Find it by yourself

as is traditional in many colleges to address newly matriculated students.

The teacher, even more embarassed, clarifies his position

(33) Well ... I am the teacher

The answer of the student is due to the failure of the hearer to recognize a teacher in the speaker role.

The conflict, in this case, is between the 'issuer's' (real) discourse situation

(34) ....
       speaking, a_teacher; yes

and the 'receiver' reconstruction of the 'discourse situation'

(35) ....
       speaking, a_newly_matric; yes

From the student's answer, the teacher infers the type of misunderstanding and reacts by suggesting the correct binding for the specific role indeterminate.

In some other cases, the speaker's individuation may remain uncertain, and a break down will ensue; a clarification is required in order to restore communication.

In the following dialogue fragment:

(36a) Can you show me your passport, please?
(36b) Why do you ask me this?
(36c) I am a policeman

the question (36b) has the function of providing a link for the speaker role of (36a). This remained unbound (a?) as the policeman failed to qualify himself as such (he did not wear a uniform or did not show his identity card).

A similar misunderstanding may occur about the individuation of the hearer.

In Italian railway stations the following scene is often to be observed:

A tourist in a hurry encounters somebody wearing a uniform and asks

(37) From where is the train to Naples leaving?

The person in uniform calmly answers

(38) I don't know

The tourist, disappointed, runs away, seeking someone else to question.

The tourist has in mind a situation in which it is the case that

(39) ....
    addressing, a_tourist, a_railway_employee; yes

while the 'hearer's' (real) situation is

(40) ....
    addressing, a_tourist, a_policeman; yes

This is obviously an extreme case of misunderstanding. More often the policeman is aware of the conflict between the two situations and gives some corrective answer like

(41) You should not ask me; I am a policeman. You can find a railway employee ....

In the case of face-to-face dialogue, it is difficult to imagine a misunderstanding about the discourse location.

Space and time individuations should be equally available both to the speaker and to the hearer. Some misunderstandings may arise, instead, using means of communication in which the speaker and the hearer are in different locations.

In the telephone dialogue

(42a) A: I can't find my agenda
(42b) B: Look at the shelf in front of you. I left it there, today
(42c) A: But I am at the airport; I am leaving now

the misunderstanding arises from the false assumption of A that B is in his office, while in reality he is at the airport. In other words the speaker's view of reality is that the hearer's location is his office. Utterance (42b) is intended to restore the correct link between the speaker's hearer location role and reality.

The ref role is related to a collection of referring situations DS. The anchor function, therefore, is a complex one, which performs a multiple linking from the objects mentioned in an utterance $(\alpha)_i$ and the corresponding referents $c(\alpha)_i$. Misunderstandings may arise from any mislinked word, noun or verb, in the utterance. A typology of such misunderstandings, based on a typology of how objects are individuated in reality, deserves a specific extended and detailed discussion, which lies beyond the scope of this article. A few examples will give an idea of the complexity of the problem of reference (see also Webber, 1981) and of the misunderstandings that may arise from it.

A linguistic expression may find more than one candidate referent; if, in the presence of more than one dog, the sentence

(43) What a nice dog

is uttered, it is unclear which one is referred to.

Quantification may also cause difficulties in the anchoring of an expression. If the sentence

(44) Those animals are faithful friends

is uttered in the presence of the same dogs, the ambiguity exists as to whether a universal quantifier is understood, or just those dogs are referred to.

Other types of misunderstandings, not necessarily originated by an ambiguity, can be identified.

From the above discussion, it appears that a misunderstanding can always be described as a conflict between the speaker's and the hearer's view of the discourse situation. We have analysed the simplest cases in which this is limited to one role. It is obviously possible to imagine, or to observe, more complex types of misunderstandings in which more than one role indeterminate fails to be correctly related to the corresponding individual. Another simplification consists in the fact that one of the two views involved in communication, the 'issuer's' and the 'receiver's', corresponds to reality. In fact, disagreement may arise also between speaker's, hearer's, and real discourse situation.

A table of all possible combinations of 1- to n-roles misunderstandings on two (speaker-hearer) and three (speaker-hearer-reality) dimensions provides a coherent guideline for an exhaustive classification of interpretation misunderstandings.

Another possible dimension, which appeared in some of the above examples, is the way in which disagreement occurs. In most cases a wrong individual was anchored to a given role, but in a few of them no information at all was available about the individual to be linked (x?). A deeper insight into this distinction could account for the distinction between misunderstanding and communication failure. In fact, in the case of an unanchored role, a break-down has always been observed and communication could continue only after a subdialogue had been carried on in order to substitute x? with an individual, which, by chance, can also be the wrong one. In the other case, instead, the dialogue can keep on going, although the interpretations of the utterances are misunderstood.

In this section we have sketched a method for the classification of misunderstandings originated at the interpretation level. As our basic guideline has been the notion of context as introduced and formalized by Barwise and Perry (1983), the application of this method is limited to verbal communication.

This is a narrowing of the perspective adopted in the initial sections of this article, in which a general model of communication has been attempted. However, no serious objection has been found against the extension of the same classification method to other communication ways, such as gesture, mail etc., provided that the corresponding situations are described as has been done for discourse.

5. Conclusions and Research Perspectives

In the initial sections of this paper we have drawn an intuitive model of the human communication process. We have focused on the step of understanding, which consists, in our account, of the 'semantic representation' of an utterance, its 'interpretation', and the reasoning activity related to it. We have shown that such a reasoning activity, of which the inference of the speaker's intentions (as in Allen & Perrault, 1980) is an important aspect, is heavily affected by the 'interpretation'. Finally we have focused on 'interpretation', using some of the formalized notions of SS, namely the 'discourse situation' and the 'speaker's connections', to account for it. We have also suggested that this description can be extended beyond dialogue to all forms of communication. We have concentrated on using these notions to classify various misunderstandings or communication break-downs as different types of wrong links to reality. SS has proved to be a valuable tool for describing the relations of utterances to the reality they occur in and which they describe.

However, we have provided nothing more than a static description of a discourse situation and the corresponding misunderstandings. In fact, an account of the process which links role indeterminates to the corresponding individuals and which is responsible for the wrong links, is completely missing. This process is formally represented by the function $f(x)$, i.e., the anchor, but probably corresponds to another inference process, distinct from the mentioned reasoning activity.

An important research perspective is the modelling of the complex processes required for the computing of the 'anchor'.

Returning to our general model of communication, once the central operation of providing an interpretation for a given utterance has been performed, some further reasoning activity takes place before the reaction process is started. Allen's (1983) proposed speaker's plan inference model finds its place in this phase, but, as we have already stated, it represents only one aspect of the inference activity connected with one type of purposeful dialogue. A classification of the general goals a dialogue is directed to, may provide a guideline to the enrichment of a model of such a reasoning activity. An important aspect is the identification of the mechanisms by which the participants in a dialogue realize that a misunderstanding has occurred.

An important extension of a communication model is man-machine communication. Although actual man-machine interfaces follow from tremendous simplifications of the communication process, in principle man-computer interaction can share many aspects of human communication. In particular, it seems that the realization of the communication cycle we have proposed above can be a reasonable objective of research in the field.

An important restriction, however, is imposed by the notion of reality. In fact, a computer has no knowledge of the external world, except for what may have been stored in it in form of data, programs, or higher level representations. In addition, the computer embeds its own reality in the form of different environments, such as editing, data-base query, system commands etc. In those machines in which all these environments are available in an integrated mode, the relations between them and the stored information may parallel the complexity and real world.

In any case, the SS approach we have discussed here, together with the proposed classification of misunderstandings, can be applied to man-machine communication as well.

An important consequence of the extension of a model of human communication to man-machine communication would be the possibility of transferring to the interaction with computers the most relevant results of computational dialogue modelling, without necessarily using natural language modules, whose implementation and use may appear too complex.

References

Allen, J.F., & Perrault, C.R. (1980). Analyzing intentions in utterances. *Artificial Intelligence. 15,* 143-178

Allen, J.F. (1983). Recognizing intentions from natural language utterances. In M. Brady and R.C. Berwick, (Eds.), *Computational models of discourse.* Cambridge, MA: MIT Press.

Austin, J.L. (1962). *How to do things with words.* Oxford: Oxford University Press.

Barwise, J., & Perry, J. (1983). *Situations and Attitudes.* Cambridge, MA: MIT Press.

Evans, D.A. (1981). A Situation Semantics approach to the analysis of speech acts. In *19th Annual Meeting of the ACL, Proceedings of the Conference,* Stanford.

Grosz, B. (1977). *The representation and use of focus in dialogue understanding.* Unpublished PhD thesis, University of California, Berkeley.

Robinson, A.E. (1980). *The interpretation of verb phrases in dialogs.* TN 200. SRI International.

Searle, J.R. (1969). *Speech acts.* London: Cambridge University Press.

Webber, B.L. (1981). Discourse model synthesis: Preliminaries to reference. In A. Joshi, B. Webber, and I. Sag (Eds.), *Elements of discourse understanding.* London: Cambridge University Press.

Webber, B.L. & Mays, E. (1983). Varieties of user misconceptions: Detection and correction. In *Proceedings of IJCAI '83,* Karslruhe, West Germany.

KAN: A KNOWLEDGE ACCESS NETWORK MODEL

Noel E. Sharkey
Centre for Cognitive Science
University of Essex

Amanda J.C. Sharkey[1]
MRC Cognitive Development Unit
London

A theory of communication failure in dialogue must have as its central core an account of how contextual knowledge is accessed and used in language comprehension. In the course of this chapter we shall present arguments to substantiate our claim that many instances of communication failure can be viewed as contextual misunderstandings. It is not enough however, to stress the important role played by context. We must also consider the possible processes by which context might have its effect. We shall present a version of the Knowledge Access Network model: KAN (Sharkey, forthcoming). This is a computational process model which combines theoretical insights from artificial intelligence with psychological data to provide an account of contextual processing.

"Good work," he said, and went out the door. What work? We never saw him before. There was no door.

(Richard Brautigan, 1971)

1. Some Contextual Prerequisites of Understanding

In this section we will point out some of the reasons why we think that theories of understanding should be centrally concerned with capturing the nature of context. We begin with a description of context in a very simple domain, that of words and letters, and then go on to look at context in the larger domain of language comprehension.

Let us start by discussing letters as contexts. A feature of a letter has a particular meaning within the context of a letter. For example, the central vertical feature of the letter I is the same as the horizontal line in the letter T. Thus the meaningfulness of that feature is dependent on its relation to the other features in the letter of which it is a part. So, in a sense, each feature contributes to the overall meaning of the letter. Or, to put it another way, the letter acts as a context for each feature. The same can be said for the relationship between letters and words. Words act as contexts for letters and the meaning of each letter is dependent (among other things) on the entire configuration of letters that make up the word. An example of our use of words as context can be found in Figure 1 where the 'c' in 'cool' is the same as the 'e' in 'be' and the 'a' in 'cat' is the same as the 'h' in 'the'. It is the use of the words as contexts that forces our interpretation of these letters as being different.

It is of course quite easy to see the use of context when we can label it as a 'letter' or a 'word'. But there are many other types of context which are not so easy to define. For example, an "ambitious athlete carrying on in a competition regardless of a severe injury" context. Most of

---

[1] We would like to thank the Economic and Social Research Council (United Kingdom) for supporting the continuation of this research.

TAE CAT
be cool

Figure 1. An example of the effect of word context on letter recognition.

us can probably imagine this context, but the problem is that it is not open to public inspection in the same way as a "word" context. Each of us probably has different experiences of the "brave-injured-athlete" context and so it may be considered to be unique for everyone. However, it is very likely that we all share enough of the important parts of the context to enable us to talk about it. For example, we might all expect commentaries like "she bravely soldiered on", "true championship spirit", "the hospital later announced", etc.

Our view is that social and physical contexts share many important features with letter and word contexts. That is, a particular configuration of words or sentences may evoke our mental representations of particular social or physical contexts. In the same way that a letter has meaning in terms of its position in a word, a sentence has meaning only in terms of its role in a context. We would argue that whenever an isolated sentence is read, the reader automatically computes a default context. In the remainder of this section we shall first of all give a number of examples which illustrate the significance of context of disambiguation. Then we shall point out examples of how context functions to fill in missing information which speakers/writers omit. Finally, we shall briefly describe the script approach to the study of context (Schank & Abelson, 1975).

1.1 Context and Ambiguity

Imagine a simple dialogue between two rugby-playing friends: Ciaran and James. They are sitting at the table about to eat their supper. Normally they are polite, well-behaved lads, but given the context of each other they are likely to get a bit carried away. "Pass the salt", says Ciaran. "OK" says James and immediately leaps out of his chair, throws the salt cellar roughly to Ciaran and Ciaran runs with James in hot pursuit. Ciaran eventually slams down the salt pot in the frame of the door before both boys return to the table.

This text highlights a major problem which researchers of dialogue must face. The example of passing salt seems, on the surface to be a direct speech act. But the boys' knowledge of each other changed the nature of the simple instruction. They brought a previously shared context to bear on the present one. It could only have been this shared context that enabled James to understand the intention behind Ciaran's request. In a different context however a different behaviour might have prevailed. For example, if the local vicar had come to tea and had asked James to pass the salt, he would have been very surprised if James had performed in the same way.

One view of this process would be that James had to use context to compute the intentions of the speaker in each of the situations and then respond according to those intentions. However, dialogue is not as simple as that. It relies on feedback from one party to the other. Suppose Ciaran had really intended James to pass him the salt in a way that accorded with the eating context. Now James has either misunderstood Ciaran's intentions or decided to override them with his own. In this case it was James' response which triggered Ciaran's action of running to the door frame. That is, James' response indicated to Ciaran the context James was bringing to bear in the current setting.

The potential for communication failure in this situation is obvious. Suppose James' action had not triggered the rugby context for Ciaran. The action would then have lost its metaphorical significance. If Ciaran interpreted James' response in the dining context, he might have thought that James was being unnecessarily aggressive and rude. Now if Ciaran responded according to this interpretation, a whole spiral of communication failures could ensue.

So much of language (perhaps all) is multiply ambiguous when it is viewed under the microscope of scientific enquiry. Yet in most of our everyday lives we remain unaware of the ambiguous words, phrases, and sentences with which we are continually bombarded. This is because in the normal run of things the context in which an utterance appears often permits us to interpret its meaning without us being aware of the alternatives.

Take, for example, Roger Schank's famous sentence "I saw the Grand Canyon flying to New York." Heard in isolation, we tend to think that the speaker is telling us of the situation in which s/he saw the Grand Canyon. This is because it is difficult for us to generate a context in which the Grand Canyon would be flying. But it would be quite possible, given an appropriate context, to interpret the sentence in another way. Supposing two doctors try out a new drug in their laboratory. The next day they begin to discuss some odd side effects they had experienced the previous evening. One doctor says "The front room of my house turned into a tropical rain forest." The other doctor replies, "That's nothing. I saw the Grand Canyon flying to New York." Here again we have an occasion in which successful dialogue could only occur through a shared understanding of context. If the first doctor had not experienced the side-effects he may have said something like: "Yes, I saw the Canyon flying to San Diego." To which the second doctor may have replied: "It's odd that it specifically induces Grand Canyon hallucinations", and so on. Thus failure to bring the appropriate context to bear can result in a breakdown in communication, even though it is possible for successful communication to be based on shared interpretations of quite bizarre sentences.

Contextual misconstruals of ambiguity are often used in humour. For example, take the old joke about a young woman saying to her friend "I got a bottle of whiskey for my husband". Her friend replies: "Sounds like a good deal to me". In order to understand this joke we must be able to switch between the two contexts i.e., the 'trading' context and the more likely 'giving presents' context. Recent research by Mitchell and Fagar (forthcoming) suggests that sentences such as "The groundsman chased the girl with a stick" are interpreted as 'the groundsman had the stick'. However sentences like "The groundsman saw the girl with a stick" are understood as 'the girl had the stick.' It is quite easy to change the interpretation of the first sentence by providing an appropriate context. For example, imagine that there were several girls one of whom had just done something naughty with a stick. Now we are likely to think that the girl had the stick. The second sentence is a little more difficult to alter with context because of the pragmatic constraint that "people do not see with sticks". However, if we know that the groundsman is blind we may interpret the 'seeing the girl with a stick' as a metaphor for him feeling her with it. Again, we can only interpret such sentences appropriately through the use of context. When we feel we have arrived at an understanding of sentences like these it is only because we have evoked a default context which has allowed us to make sense of them.

1.2 Knowledge-based ellipsis

Dialogues tend to be elliptical with respect to knowledge which both speakers know they share with each other. Most of us, when we tell stories, try not to bore our friends with all the boring little details which we know that they know. Consider for instance the following example.

Victoria is telling someone about her visit to the theatre. Her account is as follows: "I arrived at 6.30, and rushed up to collect my tickets so that I could get settled before the curtain went up. But the stupid girl at the box office had lost them. When she finally produced them I went into the auditorium, but the play had already started. It was so dark that I had to sit in a seat right at the back. I couldn't find my own seat until the interval". Most of us would find this story easy enough to understand, and would be able to see why Victoria sounded so annoyed by the behaviour of the girl at the box office. However, imagine that Victoria is telling the story to Ho Ti, a Chinese peasant, who has never been to a Western theatre. Although Ho Ti speaks very good text-book English, he can hardly understand Victoria's story at all. He wants to know why Victoria didn't wait to collect the tickets until after she had seen the play. He also wants to know how the girl at the box office was supposed to know that Victoria would come rushing in and expect to have some tickets in such a hurry. He asks Victoria why she keeps one of her chairs at the theatre, especially if it is always so dark. How can all the people in the audience see the play?

Victoria's story doesn't have the effect on her listener she intended. Since Ho Ti spoke English well, she assumed that he would be able to interpret her story through knowledge about the sequence of events which are typically part of a visit to the theatre. Normally, Victoria would have been right not to mention events such as her phone call earlier in the week to reserve the tickets, and the usher collecting tickets by the door of the auditorium. Similarly, most of Victoria's acquaintances would know that the ticket she bought specified a particular seat in the auditorium. If she had been telling the same story to her husband Joel, she would have been right to leave out the mundane routines she went through in the course of the evening, for she would have been able to assume that he would infer them. Thus she doesn't mention that she bought a program, nor that she had a drink in the interval. Neither does she say anything about the box of chocolates which she ate as she watched the play. She doesn't explain things such as the reason that the auditorium was dark, or why she was concerned to get the tickets before the curtain went up. She only needs to mention the departures from the normal routine for she is aware that their discrepancy makes them worth reporting.

Usually then, to keep a listener/reader interested, it is necessary to leave out all those tedious details which we all know. However, communication failure may quickly result in dialogues where both parties do not share the same cultural knowledge.

To summarize, we have presented examples that demonstrate how many instances of communication failure may result from contextual misunderstandings. We have illustrated how a contextual miscontrual may lead us to mistakenly interpret another's intentions; we gave examples of how language without context is multiply vague and ambiguous, and finally we pointed out how much of everyday conversation assumes and relies upon an adequate understanding of social context.

These examples would be difficult to account for in terms of any theory that argues that the meaning of an utterance may be computed in isolation from context. We believe that even to play the semanticist's game of single sentence interpretation, requires the generation of some default context. Our view is that in order to understand an utterance we must see its relationship to some larger contextual object. In this view, a meaningful linguistic string has one, or some combination, of the following four functions: (a) to access a context in memory; (b) to bind contextual variables: (c) to modify a context in memory; (d) to relate contexts together. We turn now to look at one theory of context which has attempted to provide a means of doing (a) and (b) within the limitations of a particular type of social context. This is Schank and Abelson's (1975) theory of scripts which provides a foundation for KAN.

1.3 Scripts as contexts

Consider the knowledge most of us share of the routines involved in visits to a restaurant, buying stamps at the post office, visiting the dentist, going to a wedding, or catching a train. Schank and Abelson (1975) proposed a theory which centred on the use of such mundane activities. In their theory, routines (or scripts) were held to be preformed stereotypical packages of knowledge that contain culturally shared information. This meant that the contexts of certain cultural events could be given a label and treated as social objects. Here was a functional theory of context application in which it was possible to specify the content.

Schank and Abelson reasoned that such scripts could be used to generate contextual hypotheses to fill in some of the information which a writer may leave out of a text. For example, when we hear a story about a restaurant, we can predict that a customer will sit at a table, look at a menu, order food from the waiting person, and so on. They are similar to that part of Clark and Marshall's (1981) common ground notion which they referred to as community membership. Scripts represent the knowledge which we can assume we share in our culture. They are just the kinds of presuppositions which cause so much trouble for foreigners.

Scripts were first implemented in SAM (Script Applier Mechanism, Cullingford, 1978). In this system scripts are accessed when appropriate contextual cues (script headers) are used to search all of the scripts in memory until a script precondition is matched. When a script is accessed it is loaded up and used as a predictive mechanism. To be more specific, the SAM system sets up expectation frames in active memory. These frames comprise a sequence of script event templates ordered according to the usual sequence of actions obtaining in a real world setting. Each event template is itself a smaller action-centered frame containing both predictions and default values for the actors and objects likely to fill in the frame slots. Contextual variables are bound in SAM when a character's action matches that of one of the role players in the script. Thus if Suzanne brings the menu she will be bound to the waitress role. Note that there is no way to modify contexts in memory, or to relate contexts together in this system.

In the next section we shall describe the KAN model which combines the script concept with a spreading activation parallel associative network account of memory. The development of this model is an attempt to capture the notion of stereotypical information without relying on "packages" of knowledge. In this way memory can be continuous while behaving 'as if' it contained script-like entities which we call knowledge structures (K-structures). Furthermore, this system differs from SAM in the way it accesses knowledge. KAN does not need to search for a matching precondition to gain entry to its knowledge. Rather it accumulates evidence for the relevance of certain clusters of information. Because of the control characteristics of KAN, as we shall see, it has the potential for relating contexts together. However at present KAN, unlike SAM, does not have a good way of binding contextual variables. This is currently undergoing active consideration in our laboratory at Essex.

2. The KAN Model

KAN (Sharkey, forthcoming) is a parallel spreading activation Knowledge Access Network model. One of its purposes is to provide a common notational form within which it is possible to incorporate findings both from script research and from work on other forms of context. KAN is based on psychological data, and has stimulated the generation of new empirically testable predictions.

KAN is essentially an interactive model (e.g., McClelland & Rumelhart, 1981). These models use a notation in which a set of representational elements such as word concepts, are represented as nodes that are linked together to form a network. Processing resources are then allocated by means of a passive spreading activation mechanism. Thus whenever a concept node receives activation from some external source it broadcasts activation to all of the neighbouring nodes to which it is linked. The extent to which a neighbouring node is activated reflects its current degree of availability in memory. Consequently, the concept node of the new word will be pre-activated and thus any decision about it will be faster.

In the remainder of the chapter, we shall describe the KAN model, look at some of the psychological data upon which it is based, and report two experiments designed to begin forming an empirical base for the extension of the model. First, however, we shall set KAN in a historical perspective.

The roots of parallel spreading activation systems may be traced back to at least two major sources, both concerned with neural systems. The first was the publication of a paper by McCulloch and Pitts (1943) which discussed the neuron as a simple binary computing element which they characterized in terms of a threshold unit. The second source is a book by Hebb (1949) in which he discussed the notion that reverbatory activity between neurons causes a permanent metabolic change (between synapses) to create cell assemblies (which were a little like schemata). The metabolic changes are essentially what have since become known as changes in synaptic strengths or weights. Thus Hebb, like the modern connectionists, held that knowledge is stored in the connections.

The ideas inherent in both the McCullough and Pitts paper and the Hebb book came to fruition in the 1960s with the development of machine learning and pattern matching by machine. Much of this work evolved around the perceptron which was originally developed by Rosenblatt and related work by Widrow in the MADALINE and ADALINE models (cf. Nilsson, 1965 for a discussion). The important point about perceptrons, for our purposes, is that a particular type of McCullough-Pitts neuron was the basic elements of these devices, i.e., the threshold logic unit (TLU). This produced either a 1 or a 0 as output. The idea is that there is an n-dimensional input vector for each TLU and an associated matrix of weights (like synaptic strengths) such that each component of the input vector is multiplied by the appropriate weight. If such a summed input exceeds a predetermined threshold, then the output of the TLU is 1, otherwise it is 0. A TLU will produce a 1 when the following expression is true.

(1) $\sum_i a_i w_{ij} > \theta$

where a_i is the activity on the ith input line into j and w_{ij} is its weight. θ is a threshold (hence Threshold Logic Unit). The relationship of TLUs to KAN will become apparent later when we discuss the Threshold Knowledge Unit.

The learning in these systems worked by adjusting the weights on the connections between feature detectors and TLUs so that a perceptron would respond only if its learned pattern was present in the input array. The interest at this time was in an automatic perceptron convergence procedure for fine tuning the perceptron weights (cf. Minsky & Papert, 1967 for a discussion of the limitations of perceptrons).

From these early beginnings three different types of connectionist architecture appear to have evolved (cf. Selman, 1985). These are the: (a) deterministic binary; (b) non-deterministic binary (i.e., the stochastic Boltzman machine); and (c) deterministic continuous. Each of these is described briefly below. It should be noted that a major difference between connectionist systems and perceptrons is that, in the former, activation flows in both directions between units whereas, in the latter, there is only a one-way flow. This makes the new connection systems more powerful.

(a) The deterministic binary system: In this system the representation of objects is stored globally across all of the network units. Since it is a binary system, a unity may be either on or off. So an object is represented as a string of binary digits. The dynamics of the system (Hopfield, 1982) are given by:

(2) $s_j(t+1) = \begin{cases} 1 \text{ if } \sum_{i \neq j} w_{ij} s_i - \theta_j > 0 \\ 0 \text{ if } \sum_{i \neq j} w_{ij} s_i - \theta_j \leq 0 \end{cases}$

where s_j is the output from the jth binary unit, w_{ij} is the strength of connection from the ith to the jth unit, θ_j is the threshold of the jth unit, and (t+1) is a time slice from the initial time t to time t + 1.

In the presence of an external input this system behaves so as to minimise its total energy, where energy E is given by:

(3) $\quad E = -\frac{1}{2} \sum_{ij} w_{ij} s_i s_j + \sum_i \theta_i s_i$

One way to think about this is that the energy of a state is a measure of the fit of the system constraints to the input, i.e., minimizing the energy of the system maximizes the "goodness of fit". To complete a partial input pattern, the system reduces its energy to a minimum and thus maximally satisfies the constraints. The algorithm for minimizing the energy is the same as for the binary TLU described above, that is, simply to have each unit adopt a true state if its input exceeds its threshold.

(b) Non-deterministic binary systems: These systems are also concerned with the global energy of the networks (e.g., Hinton & Sejnowsky, 1983; Selman, 1985; Smolensky, forthcoming). However, they introduce a non-deterministic or stochastic element into systems like (a) in order to overcome one of their major problems: getting stuck at local minima. This results from the fact that in a deterministic system only decreases in energy can occur. Thus if the system first finds a local minimum it will stay there and never find a global minimum. The way out of this situation is to allow occasional jumps to higher energy states (Hinton & Sejnowski, 1983). The dynamic behaviour is given by:

(4) $\quad s_j(t+1) \begin{cases} 1 \text{ with a probability } P_j \\ 0 \text{ with a probability } 1-P_j \end{cases}$

where P_j is a form of the Metropolis algorithm (Metropolis, Rosenbluth, Rosenbluth, Teller, and Teller, 1953) and is given by:

(5) $\quad P_j = \dfrac{1}{1 + e^{-\Delta E_j/T}}$

where ΔE_j is the energy gap between the states 1 and 0 of the kth unit, and T is the computational temperature. Note that when T = 0 system (b) behaves as system (a) and as T approaches infinity there is a random pattern of 1s and 0s.

(c) Deterministic continuous systems: These are systems in which continuous values are passed from node to node and in which a given node may have a continuously varying activation level (e.g., Feldman & Shastri, forthcoming; McClelland & Rumelhart, 1981; Anderson, 1983; Sharkey, forthcoming). Usually the activity of a unit a_i is in the range $1 > a > -1$. Thus each unit in such a system contains information about the degree of certainty about its truth given some partial input. This differs from the binary systems in which the units are always certain about their truth or falsity (although, in a way, the energy level may be considered to be a measure of certainty). Another difference which often exists between the continuous and binary systems lies in the way in which information is stored. In the latter, information about an item is stored globally across many units and thus decisions about an item are made on the basis of the state of the system. In the former, information about an item is usually stored in a single unit and in its connections with other units. Decisions are made on the basis of the activation level on a single node relative to other nodes (although this need not be the case).

The KAN model belongs to the class of continuous activation systems and as such it reaps several important advantages. First, and foremost, there is an increasing body of empirical support for continuous activation models of this type (cf. Anderson, 1983 for a review). Second, similar models have been applied to other domains such as vision (e.g., Feldman & Ballard, 1982), concept learning (Rumelhart & Zipser, 1985), speech production (Dell, 1985), and disambiguation (Cotrell & Small, 1983). A third advantage of the parallel activation models is that they use the mechanics of the brain as their underlying metaphor (as well as the computational metaphor). As Feldman and Shastri (forthcoming) point out, this can lead to very powerful constraints on possible cognitive architectures. Finally, such parallel spreading activation models have already been used to describe context effects in different domains such as letter recognition (McClelland & Rumelhart, 1981), word recognition (Neely, 1977) and fact retrieval (Anderson, 1983). Thus, by applying the formalism of interactive activation to various contextual domains, we can develop a set of content-free unifying principles. Our current research program is geared to developing the formalism in the domain of social knowledge. Converting artificial intelligence systems of text understanding into a connectionist framework enables us to devise more precise psychological processing predictions and can lead to principled refinements of those systems (although there are dangers in translating an artificial intelligence model into a process model, cf., Sharkey & Pfeifer, 1984; Sharkey & Brown, forthcoming).

We shall now spell out the mathematics and some of the general principles underlying KAN. They are similar to those used by McClelland and Rumelhart (1981). In the KAN system, the net input to a concept node is the weighted sum of the activation passed to it by its neighbours. Thus the net input n to a node j is shown by the equation

(6) $\quad n_j = \sum_i a_i \omega_{ij}$

where a_i is the activation on the ith input line to node j and ω_{ij} is the strength of connection between node i and node j.

Two further assumptions of the model are that (a) each node has a saturation point S which represents a maximum activation level which any node can attain, and (b) activation on a node decays at a rate proportional to the amount of activation on the node (this may be likened to a leaky capacitor, Sejnowsky, 1981). The effect ξ of activation input to a node i at time t may be described as

(7) $\quad \xi_{i(t)} = n_i (S - a_{i(t)})$

The activation of a node at time t + 1 (i.e., one time cycle after t) may be described by

(8) $\quad a_{i(t+1)} = a_{i(t)} - \delta(a_{i(t)} - r_i) + \xi_{i(t)}$

where r_i is the resting or base level activation at node i, and δ is the decay parameter.

Probably the best way to explain some of the general control characteristics of the system is to illustrate with a simple simulation of the behaviour of four nodes. The activation on all of the nodes on the network at time t are described by a vector $A_{(t)}$ and the activation at time n as $A_{(t+1)}$; the resting level of all the nodes is described by the vector R (resting levels are different for different nodes and are related to their frequency of access, i.e., the more frequently a node is accessed the greater will be its resting level), and the weights are described in the matrix W.

The simplified network chosen here has a one-way flow of activation between the four nodes.

$$R = \begin{bmatrix} .1 \\ .1 \\ .1 \\ .1 \end{bmatrix} \quad A_{(t)} = \begin{bmatrix} .58 \\ .1 \\ .1 \\ .1 \end{bmatrix} \quad W = \begin{bmatrix} 0 & .5 & .2 & 0 \\ 0 & 0 & 0 & .5 \\ 0 & 0 & 0 & 0 \\ 0 & 0 & 0 & 0 \end{bmatrix}$$

We can think of each column of W as being the weights on the input lines into a given node. That is, each column of W represents the weights on the input lines of the node represented by the column number from the node represented by the row number. A weight of 0 means that there is no link and thus no input activation. The network layout is illustrated in Figure 2.

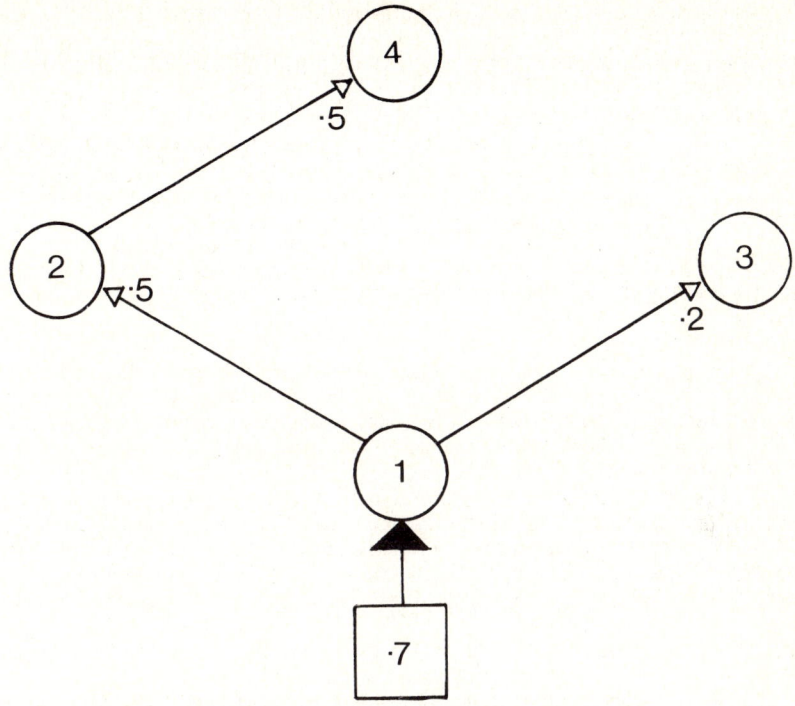

Figure 2. A simplified portion of a network showing the interaction between four representational units. The numbers <1 indicate the weights on the links.

The remaining two parameters of the model are the saturation level $S = 1$ and the decay rate $y = .2$. To demonstrate the system we shall assume that there is a constant source of activation (.7) input to node 1 for 25 time cycles and then it is allowed to decay. Otherwise activation varied continuously between a nodes resting level and 1. The activation vector at successive time cycles is computed using equations (6) to (8).

$$A_{(t+5)} = \begin{bmatrix} .58 \\ .66 \\ .6 \\ .42 \end{bmatrix} \quad A_{(t+10)} = \begin{bmatrix} .58 \\ .67 \\ .66 \\ .47 \end{bmatrix} \quad A_{(t+25)} = \begin{bmatrix} .58 \\ .67 \\ .66 \\ .47 \end{bmatrix}$$

Notice here that all of the activations reached asymptote by 10 time cycles.

This is in line with detailed research on the activation rate for associated concepts (cf. Anderson, 1983). It is important to notice also the small difference in asymtotes between nodes 2 and 3 even though there was a difference in associative strength of .3. The asymptotic activation on node 4 is considerably different than the others because the activation must filter through node 2 on route to node 4. The next set of vectors illustrate what happens when the external source activation is turned off and the activation on node 1 is permitted to decay.

$$A_{(t+50)} = \begin{bmatrix} .1 \\ .29 \\ .48 \\ .19 \end{bmatrix} \quad A_{(t+100)} = \begin{bmatrix} .1 \\ .28 \\ .47 \\ .18 \end{bmatrix}$$

The activation on these nodes is seen to be decaying back to resting level in the absence of the constant input source into the system. Note how rapidly node 1 decayed back to resting. This is because it had no other input activation.

Turning to the psychological evidence, Sharkey and Mitchell (1985) conducted four experiments which examined the possibility that humans could store their knowledge of the social world in a way which is consistent with the network account. The task they employed was the lexical decision task. This is an experimental technique that has been widely used to study the effects of context on word recognition (Becker, 1981; Becker, 1979; Becker & Killion, 1977; Fischler, 1977; Fischler & Goodman, 1978; Meyer & Schvaneveldt, 1971; Meyer, Schvaneveldt & Ruddy, 1975; Neely, 1976, 1977; Schvaneveldt & McDonald, 1981; Kleiman, 1980; Schuberth & Eimas, 1977; Schuberth, Spoehr & Lane, 1981). In this task subjects must decide as quickly as possible whether a string of letters spells an English word (e.g., RABBIT) or a non-word (e.g., ASSISTART). A response key (yes or no) is pressed as soon as the decision has been made.

The lexical decision task may be thought of as a probe that can be used to isolate and study the effects of context on the availability of a word's meaning. For example, Meyer and Schvaneveldt (1971), Becker and Killion (1977) and Becker (1979) have demonstrated that recognition is speeded up when a word is preceded by a single related word (e.g., when a target word such as DOCTOR is preceded by a context word like NURSE). In the Sharkey and Mitchell (1985) study, lexical decisions were preceded by short knowledge-based texts. These texts were constructed from a set of script norms which Sharkey and Mitchell collected. The decision words were either related or unrelated to the norms on which the texts were based. By introducing various manipulations into the texts, Sharkey and Mitchell were able to examine the conditions under which knowledge structures (k-structures) were activated and deactivated by control cues in the text.

Four major points arose out of the Sharkey and Mitchell findings. First, knowledge-based texts were shown to facilitate the recognition speed of associated words in a lexical decision task. This was tested and demonstrated in three experiments. The results of these experiments provisionally support the network model. Secondly, it was found that the prior reading of knowledge-based contexts virtually eliminated word frequency effects in a lexical decision. These findings can be accounted for in terms of the KAN model. It turns out from simulations in KAN, that differences observed in initial resting levels (i.e., word frequency differences) approach zero as activation approaches saturation (S). The more activation input to a node the less that node will show its characteristic frequency effect.

Thirdly, Sharkey and Mitchell ruled out the possibility that the facilitatory effects of knowledge-based contexts were subject to the same automatic rapid decay as single word contexts (Meyer and Schvaneveldt, 1971; Gough, Alford & Holley-Wilcox, 1981; Foss, 1982). In three experiments the priming influence of knowledge-based contexts was shown to be relatively unaffected by intervening materials up to three sentences long. However, k-structures were found to be deactivated when enough explicit cues were present in the text. In other words the pattern of activation in the network region corresponding to a k-structure is deactivated by cues from the text rather than by a process of passive decay. The fourth major point is that in one experiment the results suggested that for a short time two k-structures may be active in parallel.

Sharkey and Mitchell discussed these findings in terms of an active network region. They proposed that there is some process in which the most active superordinate node in the network can itself be selected as an internal source of activation which may be sustained over time. A number of recent studies of fact retrieval are also in line with this notion of a superordinate control node (e.g., Smith, Adams & Schorr, 1978; Reder & Anderson, 1980; Reder & Ross, 1983). The results from these studies have suggested that k-structures operate as a group of associated nodes collected under a thematic node.

It was these important insights from the psychological data that led to perhaps two of the most important and unique features of the KAN model. The first of these is the notion of a Threshold Knowledge Unit (TKU). A TKU, as its name suggests, is similar to the Threshold Logic Unit (cf. Minsky & Papert, 1969; Nilsson, 1965) and has its roots in the McCulloch-Pitts Neuron (McCulloch & Pitts, 1943). The TKU is a special unit in KAN and is essentially a boolean function which is formally true if and only if its activation level exceeds some preset threshold. Thus it shares important formal properties with deterministic binary models (we are currently investigating these shared properties).

The basic idea is that a TKU acts as a connecting station for the cluster of concepts which may be referred to as a k-structure. It is assumed that because of multiple associative links within k-structures, activation from the concept nodes summates on this central superordinate unit. When the activation on a given TKU reaches threshold θ the address of that unit is placed in a one place F-register and this unit is provided with a constant input activation F. The unit feeds this activation back into the network (in a sense it allows the system to jump out of a potential local energy minimum). This thresholded TKU will then keep the cluster of nodes associated with it in an active state until another TKU reaches threshold and displaces the first from the F-register. When a TKU is in the F-register, its subordinate cluster of associates (corresponding to a k-structure) are said to be in focus. A simplified network diagram with a TKU is illustrated in Figure 2. KAN thus provides us with a process model of knowledge focus (in the sense of Grosz, 1977). It also provides greater specificity for the knowledge access function.

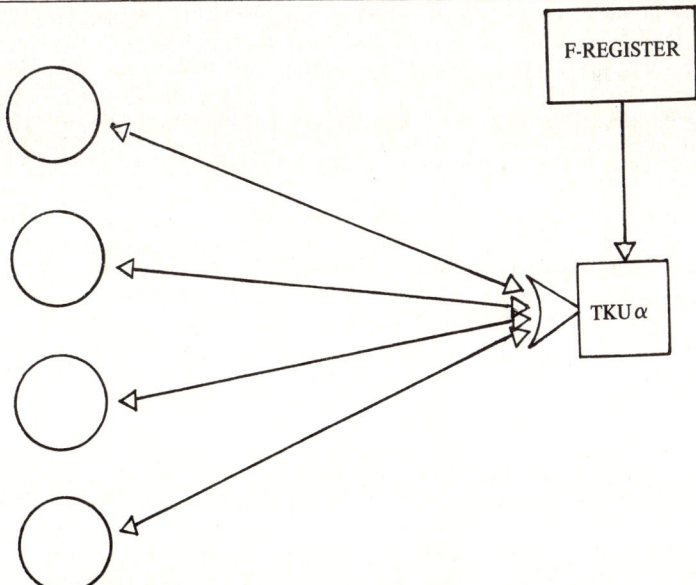

Figure 3. A simplified illustration of the connection between a set of concept nodes and their relation to a TKU and its relation to the F-register.

More formally, given two TKUs α and β

$$\text{if } a_\alpha - \delta(a_\alpha - r_\alpha) + \xi_\alpha < \theta \text{ then } n_\alpha = \sum_i a_i \omega_{i\alpha}$$

$$\text{else } n_\alpha = F_{f\alpha}$$

$$\text{until } a_\beta - \delta(a_\beta - r_\beta) + \xi_\beta > \theta \text{ then } n_\beta = F_{f\beta}$$

$$\text{and } n_\alpha = \sum_i a_i \omega_{i\alpha}$$

where F is a fixed activation value transmitted from the F-register to TKU α.

The second important insight from the Sharkey and Mitchell findings was the notion that more than one k-structure can be active in memory at the same time. Sharkey (forthcoming) characterizes this in the notion of a "hand-over period". This is where one TKU takes over the occupancy of the F-register from another. This handover may be thought of as comprising two time slices. In the first, TKU α is in the F-register, and TKU β has risen to near threshold. In the second, TKU β has reached threshold, replacing TKU α in the F-register, and TKU α has just begun to decay back to resting level. During this handover period, both network regions have empirically indistinguishable activation levels (the duration of "handover" is mainly determined by the magnitude of the decay parameter). Thus the lexical decision probe used by Sharkey and Mitchell would produce indistinguishable decision speeds for the two k-structures during handover.

In the next section of this chapter we shall describe how we have begun to extend this research in several psychological experiments on the interaction of mundane world knowledge with the process of parsing simple declarative sentences. In brief, the technique which we have developed involves embedding certain key sentences within a constrained knowledge-based text. Then the reading times for target words occurring at different points within the key sentences can be measured. The findings from this research give some indication of the way in which knowledge may interact with parsing. Some of this research is reported within a different theoretical framework in Sharkey and Sharkey (1983). These experiments represent a preliminary report of work which is still in progress. Further experimentation is currently underway.

3. Further Explorations of KAN

So far we have discussed the way in which knowledge is accessed in the KAN model. We have described psychological data from Sharkey and Mitchell's lexical decision experiments which indicate how active TKUs may influence the availability of word meanings. However, none of this research has investigated how network activation is involved in reading sentences. We now turn to two experiments which we carried out to investigate the influence of focussed knowledge on sentence processing. It should be emphasized that these experiments are not 'tests' of KAN. Rather, they explore possible extensions of KAN to parsing.

In conducting these experiments we were concerned to discriminate between three alternative hypotheses. The first hypothesis was formulated on the basis of Sharkey and Mitchell's (1985) lexical decision results. Their findings make it reasonable to suppose that in the presence of an active TKU, reading times for every TKU-related word in the sentence will be reduced. The benefits of such a word by word facilitation would be apparent in a system in which words in a sentence were individually interpreted in relation to prior context as they were read.

A second, conflicting, hypothesis is that sentence processing is handled by some independent modular system and the influence of focussed knowledge is held off until grammatical or case assignment has been completed. For example, the lexical entries of words may be held in temporary storage until the meaning of the sentence verb is recovered. When this happens a conceptual case frame could be loaded into active memory and used to assign the internal relations and word meanings within the sentence (this is how Riesbeck's (1975) English Language

Interpreter system operates). Once assignments have been completed, the k-structures may then be used to draw the inferences necessary to causally integrate the current sentence meaning with context. This end-of-sentence integration hypothesis has received some support from evidence suggesting that readers pause for longer durations at the ends of sentences than at other points during reading (Mitchell & Green, 1978; Just & Carpenter, 1980). Confirmation of this hypothesis would require that reading time facilitation be evident only in conditions where the target word was presented at the end of a sentence. Our third hypothesis is a mixture of the first two. That is, active knowledge interacts with on-line sentence processing on a word by word basis as well as assisting with case assignment and the integration of a sentence with prior context.

We investigated these alternative hypotheses by examining the priming influence of active knowledge on words at different sites within a sentence. Because of the relationship between this study and the Sharkey and Mitchell (1985) study, we constructed texts based on their norms. These knowledge-based texts served as our contextual materials (one of these texts is shown in Example (1). In both experiments reported here, we presented the texts to subjects on a computer screen. Since we were interested in the time it took subjects to read words related to active k-structures (knowledge-based-targets) compared to control-targets, we presented each sentence of the texts in two segments. This meant that the time taken to read a target (article + word) could be measured when it was either in the initial or in the final position of the sentence.

3.1 Experiment 1

In the first experiment we compared the reading times for knowledge-based-targets and control-targets both in the initial and final positions of simple declarative sentences. Considerable care was taken to find appropriate sentences in which to place the targets. One potential problem was that the final word of a sentence may be primed by other parts of the sentence in which it occurs. This has been demonstrated in several studies (Schuberth & Eimas, 1977; Kleiman, 1980; West & Stanovich, 1978; 1982; Schuberth, Spoehr & Lane, 1981). Therefore it could be difficult to tell whether the priming of final words resulted from within- or between-sentence context effects. We avoided this problem by conducting a normative study to ensure that our control-targets were more likely as sentence completions than our knowledge-based-targets. In the normative study, subjects generated single word completions to incomplete sentences which were sometimes presented in isolation, and sometimes at the end of short knowledge-based texts (see Sharkey & Sharkey, 1983 for details). The target words were selected from those generated in the normative study in such a way as to guarantee that any facilitation at the final position could be attributed solely to the active influence of a k-structure.

Twenty undergraduates took part as subjects in this experiment. The experiment materials consisted of 20 knowledge-based texts, each followed by a critical sentence containing the target words. There were four conditions altogether. In one condition the target words contained in the critical sentences were knowledge-based (and had been generated as text completions by at least 60% of the informants in the norm collections, but never generated as completions to the same sentences presented in isolation). In a second condition control-targets were included in the critical sentences (words which had been generated as isolated sentence completions in the normative study, but which had never been generated as text completions when the sentences followed a knowledge-based text). All targets were equated for word length and frequency. The constraints used in the selection of these words meant that the control-targets were more predictable from their sentence contexts than the knowledge-based-targets. Thus any reading time advantages for knowledge-based-targets over control-targets could not be attributable to within-sentence priming.

In addition, two different versions of each critical sentence were constructed so that the target word could occur in either the initial or final position. These formed the other two conditions. A typical text along with the four types of critical sentences and words is illustrated in Example (1). The targets are in italics.

Example (1):

The customer was in an irritable mood in the restaurant. But the service was very fast and efficient. His mood began to mellow. The dessert really cheered him up. He smiled and thanked the waitress as he walked out.

KNOWLEDGE-BASED, FINAL POSITION: She picked up / *the tip*.
KNOWLEDGE-BASED, INITIAL POSITION: *The tip* / pleased the waitress.
CONTROL, FINAL POSITION: She picked up / *the mess*.
CONTROL, INITIAL POSITION: *The mess* / displeased the waitress.

Other knowledge-based texts were compiled based on those used by Sharkey and Mitchell (1985). These texts were used as practice trials and as distractor texts. Finally, true/false statements were also written for each of the texts. It was felt that subjects would be more likely to read the texts for comprehension if they knew that they would be required to judge the accuracy of such statements at the end of each text.

Each sentence in the texts was presented in two separate parts. Only one part of a sentence was visible on the screen at any time. Readers progressed through the text by pressing a bar which caused the display to be immediately replaced by the next sentence part. The time taken to read each critical frame containing either the knowledge-based or control target was recorded in milliseconds.

3.1.1 Results and Discussion

Table 1 shows the mean reading times for the four main conditions. Two repeated measures ANOVAs were carried out on these data using, alternatively, subjects and materials as random effects. The primary factors in these analyses were Context and Position. These analyses yielded a significant main effect of Context for subjects ($F1$ $(1,11) = 6.12$, $p = 0.031$) and a marginal effect for materials ($F2$ $(1,19) = 3.55$, $p = 0.075$). There was also a significant effect of Position for subjects ($F1$ $(1,11) = 7.86$, $p = 0.017$) and for materials, ($F2$ $(1,19) = 8.72$, $p = 0.008$) with a marginal $MinF'$ $(1,27) = 4.13$, $p < 0.059$. The interaction of Context x Position was also significant by subjects ($F1$ $(1,11) = 5.17$, $p = 0.044$) but not by materials.

Context	Control		K-structure	
Position	Initial	Final	Initial	Final
Mean RT	686	860	655	728

Table 1. Mean Reading Times in Milliseconds for Experiment 1.

Looking first at the sentence-by-sentence hypothesis, pairwise planned comparisons for final position knowledge-based and control targets revealed significant differences both by subjects ($F1$ $(1,11) = 7.18$, $p = 0.021$) and by materials ($F2$ $(1,19) = 5.20$, $p = 0.034$) with a marginal $MinF'$ $(1,30) = 3.02$, $0.10 > p > 0.05$. The result of these comparisons, in combination with the Context by Position interaction effect, support the hypothesis that an active TKU facilitates the speed of sentence-by-sentence integration.

Furthermore, we reasoned that such end-of-sentence effects should also show up in sentences where the targets were not in the final position. We investigated this possibility by examining the reading times for final portions of the sentences which had contained initial targets. Since these end portions were of slightly different lengths (mean word length of end portions of knowledge-related sentences = 4.017, standard deviation = 1.42; mean word length of end portions of control sentences = 4.782, standard deviation = 1.303), we used a word score in the ANOVAs (word score = RT / number of words). These analyses produced a significant MinF' (1,26) = 5.077; $p < 0.05$. Thus it seems that differences between the initial knowledge-based-targets and the initial control-targets were being picked up at the ends of sentences. This lends strong support to the notion that active knowledge assists in end-of-sentence integration.

However, the picture is not so favourable for the word-by-word hypothesis. Pairwise planned comparisons revealed no statistically reliable differences between the knowledge-based and control targets in the initial position. One explanation of these findings is that the first stages of sentence processing (grammatical and case assignment) are carried out before the active k-structure is used. Nonetheless, we are reluctant to assert this explanation too forcefully. Although the differences between the initial sentence position targets were not statistically reliable, the mean response time for controls was 31 milliseconds slower than that for knowledge-based targets. We must admit to the possibility that this 31 millisecond difference might reflect a real difference in activation levels which our task was not sensitive enough to detect.

Such an explanation would provide a way of accounting for the difference between these results and the Sharkey and Mitchell lexical decision findings. We now believe that the differences between the two studies may be related to the computational complexity of the tasks (i.e., the number of processing steps). In the lexical decision task, subjects have to read the target words, and also make yes/no decisions about them. This task is computationally more complex than the task of reading a word and moving on. It is therefore likely that any differences in activation levels at concept nodes would be magnified. Thus the lexical decision task may pick up differences more reliably than our current measure. The same notion of complexity could also be used to explain why we only detected statistically reliable differences at the ends of sentences. The computational processing for sentences wrap-up (Just & Carpenter, 1980) would also be more complex than that for simply reading a word at the beginning of a sentence and moving on. At the end of a sentence a reader would also have to integrate the individual word meanings to build a conceptual structure which is appropriate for the context. If sentence wrap-up involved repeatedly accessing the active knowledge, any observable differences would be magnified on each pass.

Another possible explanation is that the control targets received some activation from the knowledge-based texts. Although they were not generated as completions to the texts in the normative study, they were still rated as reasonable completions of the texts. This difference between 'rating' and 'generating' is quite significant in terms of the current activation framework. We would postulate that in order to generate completions, subjects, after reading a text, select word candidates from among those whose concept nodes are most active. So, for knowledge-based texts, the chosen completions would be those most closely associated with an active TKU and which fitted the sentence constraints. In contrast, the rating task only demands a decision as to whether the completion is connected to the currently active k-structure. It is not possible at present to specify how such a process operates except at the most general level. The point is that although our rated controls were not as strongly associated with the k-structure as the generated targets, they were nonetheless connected. This follows from KAN in which the k-structures were continuous with memory. What it all boils down to is that we have to admit to the possibility that our control targets were being partially activated in the presence of the knowledge-based text. We conducted a second experiment, with these problems in mind, in an attempt to magnify possible initial sentence position effects.

3.2 Experiment 2

The second experiment is really an extension of the first. The only change is in our choice of control targets. We reasoned that one way around the problem of activation "leakage" would be to choose controls which were obviously not connected to the k-structures except by very remote association.

Thus control targets in the experiment were words which were unlikely completions of the texts.

These would be unlikely to receive any activation from the focused knowledge. If our reasoning is correct, in Experiment 2 we should be more likely to find a difference between knowledge-based targets and control targets in the initial portions of the sentence.

Sixteen undergraduates participated in this study. The materials they were asked to read, were the same as those used in Experiment 1, with new control targets. The controls used in Experiment 2 were unlikely completions of the knowledge-based texts. However they were reasonable completions of the critical sentences considered in isolation. Each control word was matched as closely as possible with a knowledge-based word, for word frequency and number of letters. Thus the target "the mess" in Example 1 above was replaced with "the hen". These control targets were selected by the experimenters on the basis that they did not form reasonable conclusions to the texts. The procedure employed was identical to that of Experiment 1.

3.2.1 Results and Discussion

The mean reading time for the four main conditions are presented in Table 2 below. The scores were analyzed as in the previous experiment. The main factors in the analyses were Context and Position. These analyses yielded significant main effects for Context ($MinF'$ (1,34) = 14.34, $p < 0.001$) and Position ($MinF'$ (1,34) = 6.03, $p < 0.05$). The interaction of Context x Position was also significant by $MinF'$ (1,30) = 7.70, $p < 0.01$.

Context	Control		K-structure	
Position	Initial	Final	Initial	Final
Mean RT	755	1284	688	788

Table 2. Mean Reading Times in Milliseconds for Experiment 2.

As in the previous experiment, planned comparisons were carried out on these data to examine differences between the target types in the two sentence positions. The final position contrast replicated the preceding experiment with a significant difference between the two target types ($MinF'$ (1,33) = 12.12, $p < 0.01$). However, unlike the previous experiment, the comparisons also yielded a significant difference between targets in the initial position by subjects ($F1$ (1,15) = 6.19, $p = 0.025$) and a marginal difference by materials ($F2$ (1,19) = 3.86, $p = 0.064$). Thus it appears that, in this experiment, differences are being picked up between target types in the initial sentence positions. These findings suggest that the information activated by the knowledge-based texts was facilitating reading on a word-by-word basis.

3.3 Discussion of Experiments 1 and 2

We set out to develop an empirical base for the extension of KAN to sentence processing. To this end we developed three hypotheses. These were as follows: (i) In the presence of an active TKU the reading times for every TKU-related word are reduced. Contextual facilitation would thus be evident on a word-by-word basis; (ii) Sentence processing is handled by an independent modular system, and reading time facilitation would only be evident at the ends of sentences; and (iii) A combination of the previous two hypotheses. Therefore, active knowledge serves both to facilitate reading times on a word-by-word basis, and to assist case assignment and the procedures involved in sentence wrap-up (Just & Carpenter, 1980).

In the first of our two experiments we found statistically reliable differences between targets occurring at the ends of sentences. Furthermore, differences between the end portions of sentences containing initial targets were statistically reliable. These results support the hypothesis that active knowledge units influence the integration of sentences with a prior knowledge-based context. However, the difference (31 milliseconds) between initial sentence position targets was statistically unreliable. We were concerned that these may have been 'real' differences which our task was not sensitive enough to detect. Consequently, in a second experiment, we refined our control targets to avoid possible activation 'leakage' from the active k-structures. The results suggested that active knowledge does indeed do more than influence the integration of sentences with prior context. They incline us towards the conclusion that there were differences between targets in the initial position condition of Experiment 1 but that these were of too slight a magnitude to be detected. When unlikely completions of the texts were used as controls in the second experiment, readers were found to dwell longer on initial control targets than on initial knowledge-based targets.

The two experiments reported here represent work in progress, for we are currently involved in conducting further experiments in this area. Our intention is to establish whether the effects we have found so far are really the result of contextual facilitation. There is still a possibility that the results of Experiment 2 were due to some form of inhibition stemming from the incongruity of the control targets. However, such an explanation still presupposes that the active k-structures were being used on a word-by-word basis. Incongruity is a relative concept, i.e., incongruous targets can only be detected as such with respect to congruent knowledge. A modular system, in which case assignments are handled before contextual interpretation, could not detect an initial-position incongruous word until the sentence boundary. Thus at present our preference is for an account by which word-by-word contextual facilitation is assumed. However the issue remains to be resolved. Experiments are underway to determine whether contextual facilitation is still found when the target words are moved away from the beginning of the sentence.

Further experimentation and refinement of the KAN model will increase our understanding of exactly how knowledge-based activation influences sentence processing. For the moment, the third of the three hypotheses outlined at the beginning of this section seems the most likely. Subject to further empirical evidence, we conclude that once a region of the knowledge network has been activated it serves both to help with the interpretation of individual words in a text and to assist in the integration of sentences with prior context. The combined findings from the two experiments reported here also suggest that more information is active than would be predicted from theories based on the notion of knowledge structures as discrete memory packages. If our explanation does turn out to be correct, then it seems that the control targets in Experiment 1 were picking up activation from the active TKU. These control words were never generated as completions to knowledge-based texts, and thus according to discrete theories should have been unaffected by context. In the KAN model however, memory access is not restricted to the particular knowledge-structure under current consideration. KAN treats memory as a continuum in which associated information comes under central control whilst remaining continuous with the rest of memory. It is thus possible to account for the results of our experiments within this model.

4. Conclusions

In this chapter, we have presented an account of some of the ways in which context affects language comprehension. In the first section we presented examples to illustrate our view that a linguistic string has meaning only in terms of context. We argued that the function of a language segment is to set up a context, bind contextual variables, modify a context, relate contexts together, or to perform any combination of these four. On the other hand, the function of context is both to disambiguate the language stream and to fill in information that writers/speakers omit.

We pointed out that when communication is successful, a speaker, or writer will be able to assume that the listener/reader has interpreted what s/he has said in the light of context. However, we have tried to demonstrate that a spiral of communication failure can ensue when such an assumption is unwarranted. This is because without sharing the same assumed context, the listener will infer events which the speaker does not expect to be inferred, select meanings for ambiguous words and phrases which the speaker did not intend, and set up false expectations about what the speaker is about to say. It is thus not hard to see how success and failure in communication could be the result of the operation of the same set of processes.

In the second section of this paper we presented a process model (KAN) to account for the way in which contextual effects may operate during comprehension. We believe that this model has several important advantages for research on knowledge-based context. The more important of these are outlined below.

1. Parallel access: In principle, KAN provides a means of accessing k-structures in parallel (although current simulations have been run on a serial machine). This obviates the need for slow sequential search through packets of knowledge to find preconditions for access to a packet.

2. Degrees of certainty: KAN has variable degrees of certainty about the relevance of particular contexts and inferences (a little like Feldman & Shastri's evidential semantics). KAN operates on a principle of evidence collection. The level of activation of a TKU represents the system's certainty about the relevance of the related k-structure. When a TKU reaches threshold the system 'believes' that it is true that that particular context is relevant (although it can be wrong). Then the activation on the nodes in the k-structure represent a degree of certainty about their inferential relevance.

3. Continuous memory: KAN treats memory as continuous while still behaving 'as if' there were packages of knowledge. This has two major advantages: (a) knowledge outside a k-strucutre needed to complete incoming text can be accessed without disrupting the relevance of the current k-structure; (b) it allows for more than one k-structure to be active at once so that two contexts can be related together.

4. Dynamic modification: A particular concept or action may belong to more than one k-structure. Thus an action may activate more than one structure at a given time. The relevance of a particular structure will then be determined by accrual of other evidence from the text. The advantage of this kind of wiring is that contexts may be dynamically modified in memory. However, in the current version such a modification is only temporary. KAN does not learn from repeated exposure to a similar contextual configuration. We hope that in the future we will find a way of dynamically modifying the weights on the connections between the nodes in the KAN system, thus modifying context semi-permanently.

5. Generality: KAN is described using a notational form and processing mechanisms which may be used to characterize a wide range of findings from different research, areas e.g., word recognition, semantic priming, memory for script information, knowledge-based priming, memory retrieval.

6. Sustaining of activation: The Threshold Knowledge Unit gives us the means needed to sustain activation in a k-structure while it remains relevant. It was suggested by Sharkey (forthcoming) that such a mechanism is necessary to account for disparate findings from the semantic priming literature and the script priming literature. There is also a functional argument that transitory context would be of little value. Furthermore, although we have not reported it here, it seems from some of the simulations, that sustaining acts as a passive means of word sense selection. This is being written up for report at present.

7. Deactivation: The properties of the F-register in KAN provide a simple control mechanism for switching between contexts. While this is merely a conceptual device, it overcomes some of the problems of having inhibitory links between TKUs (as in the Sharkey & Mitchell, 1985, subnode competition model). The main problem was that once a subnode seized control its competitors needed a lot of activation for a coup to succeed (especially when we consider sustaining). With the introduction of the F-register, TKUs accumulate activation independently. We are currently examining ways in which TKUs could be responsive to a global system parameter without loosing the power of the F-register.

8. Empirically-based: The KAN model has been structured according to evidence collected in a number of psychology experiments. As such the model is open to exploration and refinement as new evidence is collected. Such is our view of the study of cognition, that a model is like a platform which allows us to delve into unknown areas (cf. Sharkey & Pfeifer, 1984; Sharkey & Brown, forthcoming, for arguments as to why artificial models need an empirical basis.)

9. Generates predictions: Because KAN has an empirical base and because it has been constructed from processing principles which other psychological models share, we can make and test new psychological predictions.

10. Parallel k-structures: The 'handover' phase of KAN provides a mechanism that accounts for the way in which two k-structures can have empirically indistinguishable activation levels for a short period of time.

11. Formalism: KAN can provide a content free model of context dependent understanding which has a mathematically well-defined processing mechanism. This enables us to compare the fundamental structure of KAN with other computational systems. It also allows us to expand KAN as new computational discoveries are made in connectionism.

In the final section of this chapter we presented what was essentially work in progress. We detailed two experiments in which we explored possible extensions of KAN to on-line sentence processing. We found strong support for the hypothesis that active k-structures facilitate sentence wrap-up. The combined results of the two experiments also suggested that k-structures exert an influence on a word-by-word basis. However, the issue as to whether the word-by-word influence is facilitatory or inhibitory remains in doubt. We tend to favour the former, both for functional reasons and because of minor indications of priming in the first experiment. However, the resolution of the issue awaits further empirical research. Experiments are underway to determine whether contextual facilitation is still found when the target words are moved away from the beginnings and ends of sentences. Our aim is to investigate the effect of a context on words which are more centrally placed e.g., at the end of an adverbial phrase.

Work on KAN is continuing in our laboratory at Essex University both on the computational and on the experimental front. More work needs to be done on the semantics of the system which is a hybrid of conceptual dependency (Schank, 1975) and evidential semantics (Feldman & Shastri, forthcoming). We are also working to combine the KAN model with the Goal Integration Network model (cf. Sharkey & Bower, 1984; Sharkey & Bower, forthcoming). There are plans to empirically explore such links in a series of experiments to be conducted at Essex and Stanford. In this way the current model will be extended to cover the processes involved in comprehending text in the light of the goals of the speaker/writer or characters in the text. Furthermore we are developing the parser as we collect more evidence on the interaction of KAN with sentence processing.

References

Anderson, J.R. (1983). *The architecture of cognition.* Cambridge, MA: Harvard University Press.
Becker, C.A. (1979). Semantic context and word frequency effects in visual word recognition. *Journal of Experimental Psychology: Human Perception and Performance, 5,* 252-259.
Becker, C.A. (1981). Semantic context effects in visual word recognition: An analysis of semantic strategies. *Memory and Cognition, 8,* 493-512.
Becker, C.A., & Killion, T.M. (1977). Interaction of visual and cognitive effects in word recognition. *Journal of Experimental Psychology: Human Perception and Performance, 3,* 389-401.
Brautigan, R. (1971). *Loading mercury with a pitchfork.* New York: Simon and Schuster.
Clark, H.H., & Marshall, C.R. (1981). Definite reference and mutual knowledge. In A.K. Joshi, I. Sag, and B. Webber (Eds.), *Elements of discourse understanding.* Cambridge University Press, Cambridge.
Cottrell, G.W., & Small, S. (1983). A connectionist scheme for modelling word sense disambiguation. *Cognition and Brain Theory, 6,* 89-120.
Cullingford, R.E. (1978). *Script application: Computer understanding of newspaper stories.* Technical Report 116, Yale University, Department of Computer Science.

Dell, G.S. (1985). Positive feedback in hierarchical connectionist models: Applications to language production. *Cognitive Science, 9,* 3-23.

Feldman, J.A., & Ballard, D. (1982). Connectionist models and their properties. *Cognitive Science, 6,* 205-254.

Feldman, J.A., & Shastri, L. (forthcoming). Semantic and neural networks. In N.E. Sharkey (Ed.), *Advances in cognitive science.* Chichester: Ellis Horwood.

Fischler, I. (1977). Associative facilitation without expectancy in a lexical decision task. *Journal of Experimental Psychology: Human Perception and Performance, 3,* 18-26.

Fischler, I., & Goodman, G.O. (1978). Latency of associative activation in memory. *Journal of Experimental Psychology: Human Perception and Performance, 4,* 455-470.

Foss, D.J. (1982). A discourse on semantic priming. *Cognitive Psychology, 14,* 590-607.

Gough, P.B., Alford, J.A., & Holley-Wilcox, P. (1981). Words and contexts. In O.J.L. Tzeng and H. Singer (Eds.), *Perception of print: Reading research in experimental psychology.* Hillsdale, NJ: Erlbaum.

Grosz, B. (1977). *The representation and use of focus in dialogue understanding.* Technical note 15, SRI International Artificial Intelligence Center.

Hebb, D.O. (1949). *Organization of behavior.* New York: John Wiley.

Hinton, G.E., & Sejnowski, T.J. (1983). Analyzing cooperative computation. *Proceedings of the Fifth Annual Conference of the Cognitive Science Society,* Rochester, NY, May.

Hopfield, J.J. (1982). Neutral networks and physical systems with emergent collective computational abilities. In *Proceedings of the national Academy of Sciences, 79,* 2554-2558.

Just, M.A., & Carpenter, P.A. (1978). Inference during Reading: Reflections from Eye Fixations. In J.W. Senders, D.F. Fisher, and R.A. Monty (Eds.), *Eye Movements and the Higher Psychological Functions.* Hillsdale, NJ: Erlbaum.

Just, M.A., & Carpenter, P.A. (1980). A theory of reading: From eye fixations to comprehension. *Psychological Review, 84,* 320-354.

Kleiman, G.M. (1980). Sentence from contexts and lexical decisions: Sentence-acceptibility and word relatedness effects. *Memory & Cognition, 8,* 336-344.

McClelland, J.L., & Rumelhart, D.E. (1981). An interactive model of context effects in letter perception: Part 1. An account of basic findings. *Psychological Review, 88,* 375-407.

McCulloch, W.S., & Pitts, W.H. (1943). A logical calculus of ideas immanent in nervous activity. *Bulletin of Mathematical Biophysics, 5,* 115-133.

Metropolis, N., Rosenbluth, A.W., Rosenbluth, M.N., Teller, A.H., Teller, E. (1953). Equation of state calculation by fast computing machines. *Journal of Chemical Physics, 21,* 1087.

Meyer, D.E., & Schvaneveldt, R.W. (1971). Facilitation in recognizing pairs of words: Evidence of a dependence between retrieval operations. *Journal of Experimental Psychology, 10,* 227-234.

Meyer, D.E., Schvaneveldt, R.W., & Ruddy, M.G. (1975). Loci of contextual effects on word recognition. In P.M.A. Rabbitt & S. Dornic (Eds.), *Attention and performance V.* New York: Academic Press.

Minsky, M., & Papert, S. (1969). *Perceptrons.* Cambridge, MA: MIT Press.

Mitchell, D.C. & Fagar, D. (forthcoming). Psycholinguistic work on passing with Lexical Functional Grammar. In N.E. Sharkey (Ed.), *Advances in cognitive science.* Chichester: Ellis Horwood.

Mitchell, D.C., & Green, D.W. (1978). The effects of context and content on immediate processing in reading. *Quarterly Journal of Experimental Psychology, 30,* 609-636.

Neely, J.H. (1976). Semantic priming and retrieval from lexical memory: Evidence for facilitatory and inhibitory processes. *Memory and Cognition, 4,* 648-654.

Neely, J.H. (1977). Semantic priming and retrieval from lexical memory: Roles of inhibitionless spreading activation and limited-capacity attention. *Journal of Experimental Psychology: General, 106,* 226-254.

Nilsson, N.J. (1965). *Learning machines.* New York: McGraw Hill.

Reder, L.M., & Anderson, J.R. (1980). A partial resolution of the paradox of interference: The role of integrating knowledge. *Cognitive Psychology, 12,* 447-472.

Reder, L.M., & Ross, B.H. (1983). Integrated knowledge in different tasks: Positive and negative fan effects. *Journal of Experimental Psychology: Human Learning, Memory and Cognition, 9,* 55-72.

Riesbeck, C. (1975). Conceptual analysis. In R.C. Schank (Ed.), *Conceptual information processing.* North Holland: Amsterdam.

Rumelhart, D.E., & Zipser, D. (1985). Competitive learning. *Cognitive Science, 9,* 75-112.

Schank, R.C. (1975). *Conceptual information processing.* North Holland, Amsterdam.

Schank, R.C., & Abelson, R.P. (1975). Scripts, plans and knowledge. *Proceedings of the Fourth International Conference on Artificial Intelligence.*

Schvaneveldt, R.W., & McDonald, J.E. (1981). Semantic context and the encoding of words: Evidence for two modes of stimulus analysis. *Journal of Experimental Psychology: Human Perception and Performance, 7,* 673-687.

Schuberth, R.E., & Eimas, P.D. (1977). Effects of context on the classification of words and non words. *Journal of Experimental Psychology: Human Perception and Performance, 2,* 243-256.

Schuberth, R.E., Spoehr, K.T., & Lane, P.M. (1981). Effects of stimulus and contextual information on the lexical decision process. *Memory and Cognition, 9,* 68-77.

Sejnowski, T.J. (1981). Skeleton filters in the brain. In G.E. Hinton & J.A. Anderson (Eds.), *Parallel models of associative memory.* Hillsdale, NJ: Erlbaum.

Selman, B. (1985). *Rule-based processing in a connectionist system for language understanding.* Technical Report CSRI-168, Computer Systems Research Institute, University of Toronto.

Sharkey, N.E. (forthcoming). A model of knowledge-based expectations in text comprehension. In J.A. Galambos, J.B. Black, and R.P. Abelson (Eds.), *Knowledge Structures.* Hillsdale, NJ: Lawrence Erlbaum.

Sharkey, N.E., & Brown, G. (forthcoming). Why artificial intelligence needs an empirical foundation. In M. Yazdani (Ed.), *Artificial intelligence: Principles and applications.* London: Chapman-Hall.

Sharkey, N.E., & Bower, G.H. (1984). The integration of goals and actions in text understanding. In *Proceedings of the Sixth Annual Cognitive Science Society.*

Sharkey, N.E., & Bower, G.H. (forthcoming). The goal integration network. In P.E. Morris (Ed.), *Cognitive modelling.* Chichester: John Wiley.

Sharkey, N.E., & Mitchell, D.C. (1985). Word recognition in a functional context: The use of scripts in reading. *Journal of Memory and Language, 24,* 253-270.

Sharkey, N.E., & Pfeifer, R. (1984). Uncomfortable bedfellows: Cognitive psychology and artificial intelligence. In M. Yazdani and A. Narayanan (Eds.), *Artificial intelligence: Human effects.* Chichester: Ellis Horwood.

Sharkey, N.E., & Sharkey, A.J.C. (1983). *Levels of expectation in sentence understanding.* Cognitive Science Technical Report No. 21, Yale University.

Smith, E.E., Adams, N., & Schorr, D. (1978). Fact retrieval and the paradox of interference. *Cognitive Psychology, 10,* 438-464.

Smolensky, P. (forthcoming). Formal modeling of subsymbolic processes: An introduction to Harmony Theory. In N.E. Sharkey (Ed.), *Advances in cognitive science.* Chichester: Ellis Horwood.

West, R.F., & Stanovich, K.E. (1978). Automatic contextual facilitation in readers of three ages. *Child Development, 49,* 717-727.

West, R.F., & Stanovich, K.E. (1982). Source of inhibition in experiments on the effect of sentence context on word recognition. *Journal of Experimental Psychology: Learning, Memory, and Cognition, 8,* 385-399.

COMMUNICATION FAILURE AT THE PERSON-MACHINE INTERFACE: THE HUMAN FACTORS ASPECTS

Margaret M. Gardiner

ITT Europe

Bruce Christie

Department of Office Technology
and Administration
City of London Polytechnic

The human factors aspects of communication failure can often be traced to specific problems with the choice or method of implementation of particular types of dialogue and of input and output devices. The form of dialogue between a user and an electronic system can be broken down into "species" of interaction. Communication failure can then be seen to result in part from a mismatch between the particular "species" of interaction chosen for a given task, and the user's preferred form of communication when performing similar tasks in non-electronic environments. The first part of the chapter considers these different "species" of interaction. The second part of the chapter then focusses more specifically on the task of information retrieval from large databases. It considers three approaches which illustrate in different ways how advances in technology are making possible more natural forms of communication between human and machine.

1. Introduction

This chapter focusses on the technology available to support communication between human and machine. It addresses those aspects of communication failure which can be traced to specific problems with the choice or method of implementation of particular types of dialogue and of input and output devices. It is this focus on the technology of the interaction between the human and the machine that gives the chapter its Human Factors perspective.

The "person-machine dialogue design" is the structure within which the machine and its user interact. It is designed by the product designer and constrains the sequences of actions, commands and system feedback that can take place when users attempt to communicate with the product. Such dialogue designs used to involve largely text input and output, in accordance with a formal language that the user had to spend some time learning. Nowadays, with advances in the technology available, more interesting and more natural possibilities exist. These may involve a variety of different types of input and output formats, both "verbal" and "non-verbal". Hence it is possible to communicate with a system using a purely graphical "language", as well as successively closer approximations to "natural" language. It is also possible to point (literally or with suitable mechanical devices) to items on the system's visual display to give a command or effect a selection, or to speak and have one's speech understood by the system. These possibilities are discussed further, later in this chapter.

The interaction between a machine and its user is a dialogue in the true sense of the word, since the user normally has some need which (s)he wants to communicate to the machine, the machine possesses the means to satisfy the need, and the interaction normally involves a two-way process. Like conventional person-person dialogue, person-machine dialogue may also involve communication failure.

2. Species of Interaction

The dialogue is supported physically by the input and output devices used by the system.

Developments in technology are making new forms of dialogue possible.

The input and output devices supporting a dialogue are normally linked together in functional sequences. These combinations form "species of interaction", analogous to biological species. These species have been and are continuing to evolve rapidly under the changing forces of technology push and market pull. In the competition for market share, the design of the user-interface is becoming as important as the provision of the underlying functionality, and only those species of interaction that are the fittest for human use will survive. (The authors would like to acknowledge the contribution of James Harvey — Manager of ITT Europe's Artificial Intelligence Technology Centre — to the development of the concept of "species of interaction".).

2.1 Keying — Seeing

Keying input through a keyboard and seeing the output presented on a visual display is still the dominant species of user-system interaction. This is true whether the users are highly experienced or not, and whether the task is simple or complex.

Implications for Communication Failure. This species has the advantage that it is very familiar to many office workers — especially secretarial and clerical staff but increasingly professionals and managers as well. The main disadvantage is that it does not make full use of the wide range of communication "channels" (e.g., speech, gesture) that form a natural part of human communication in other contexts. The limitation is further exacerbated by the type of cathode ray tube display technology normally used which precludes very high resolution graphics and so severely limits visual communication from the machine to the human.

2.2 Selecting—Seeing

The most common method of interaction within this class uses a keyboard for input, a visual display unit for output, and a menu from which the user can select the appropriate option.

An early example of this type of interaction was Prestel (an example of videotex). This was aimed at the inexperienced user and was very successful in terms of being easy to learn. Experienced users, however, quickly found the method irritatingly slow and long-winded. More recent developments have produced more sophisticated systems which reduce these problems to some extent.

Prestel and other videotex systems rely on a numeric keypad by which the user indicates the number of the menu items (s)he wishes to select. More recently, a different approach to selection has been developed, using a cursor controlled by a hand-held "mouse" or other pointing device which transmits x-y coordinates to the system. Moving the cursor allows the user to indicate the desired display options. The cursor remains in position until moved by the user with the selection device or moved by the system in response to some user action.

Alongside the development of the selection devices, there have been developments in display technology (including the bit-mapped display) to produce increasingly high resolution. This has led to the development of graphics displays and complementary interaction methods to enable the manipulation of graphics images.

Implications for Communication Failure. Providing the user with a means of selecting from a set of predefined options reduces the load on the user's memory, at least in some respects. Specifically, it removes the need for the user to remember the precise "command strings" (option labels) in detail. This significantly reduces failures of communication due to simple syntactic errors. On the other hand, the user still has to intuit — or remember from previous occasions — what the different options mean (what they lead to). This is by no means foolproof in practice, as is evident from the significant amount of backtracking and failure to find "targets" which has been observed when users are asked to search for "targets" (e.g., specified pages in a videotex hierarchy) which are known to exist (e.g., the Bush and Williams experiments, cited by Christie, 1981). The use of menus, by itself, is therefore only a partial solution to the problem of communication failure.

2.3 The Rise of the WIMP

The combination of the mouse selection device and graphics display has produced a new kind of product in the office. This is the WIMP interface (Windows, Icons, Mouse and Pull-Down/Pop-Up menus). The GEM package, from Digital Research, which runs on a number of machines (e.g., IBM XT/AT, Atari 520ST) was perhaps the first such product to become widely incorporated into a variety of different office machines.

Windows. Windows can be defined as partitions of variable size and position in a high resolution display. They enable multiple views of one or more process operations to be maintained in an interactive system; in this way the user can activate a number of tasks and switch contexts between the windows representing the tasks. Windows can be used to display text (for example, multiple documents), line drawings, and graphics images.

Icons. Icons are graphical representations of real-world artefacts encoded in a symbolic form. They can be used to represent a wide range of objects such as pages of text, books, documents, and files.

Mouse. A mouse is used to point to icons which the user wishes to "activate" or "open". Typically, an icon is "highlighted" (e.g., changes colour or changes to reverse video) as soon as the cursor moves onto it. The user can then either move the cursor somewhere else or press one of the mouse buttons to "activate" or "open" the icon concerned.

Menus. Although many of the functions are displayed as icons, commands are used to describe the actions which can be performed with the functions. These commands are displayed as groups in a "menu bar" at the top of the screen with only those relevant to the user's current context being on display at any given time. A command is selected using the mouse, causing a menu list to be displayed of all commands relevant to that group; each of these can then be selected using the mouse.

Implications for Communication Failure. The WIMP type of user interface reduces the probability of communication failure between human and machine (compared with more conventional interfaces) by bringing together many of the advantages of the techniques described above. The environment created is ideal for inexperienced users since they are provided with familiar images and are guided through each stage of the interaction by an appropriate set of menu options presented as a list. The use of icons representing familiar objects and incorporated into a spatial representation of the more conventional environment (e.g., office) facilitates the use of existing psychological schemata. Add to this an integrated set of "tools" which allows the user to perform tasks avoiding computer jargon, and an interface is provided which begins to meet the needs of the inexperienced user.

2.4 Touching – Seeing

Touching what we see is one of our most basic forms of communication; it provides a natural and direct means of indicating selection and choice and is the means by which we operate many of the mechanisms available to us through switches, knobs and other forms of control. Technologies have been developed to detect touch and these are combined with display devices to provide a mode of interaction which gives the user direct control. Alongside these developments has been the addition of colour to displays generally and to high-resolution displays in particular; colour is seen to be particularly important because of the added real-world fidelity given to the displayed images which may be selected by touch.

Touching is indicated to the system by some form of touch-screen device. Two categories of these can be distinguished: those that consist of a screen of clear material overlaying the display, and those that consist of sensors surrounding the face of the display. They all operate by transmitting x-y coordinates whenever a touch is detected. The touch may be *via* a stylus but is more usually the user's finger.

An enhanced bit-mapped display is the basic element around which the display terminals in this group are constructed; the enhancement is provided by additional bits used in the display process to generate colour.

Colour. Colour extends the amount of information that can be extracted from a graphics image. It is useful for applications where the user must distinguish rapidly among several categories of data and especially where items are dispersed on the screen or where meaning must be extracted from complex relationships.

Videodisc and optical disc images. Videodiscs and optical discs as part of an interaction method can be used to present information in novel ways exploiting their nature as interactive high-quality image devices.

Implications for Communication Failure. Touching forms a normal part of human communication in many contexts, and is relatively easy to incorporate into communication with a machine. By providing redundant and/or unambiguous information to the system, it can do much to reduce the probability of communication failure resulting from a misinterpretation by the machine of a command given by the user.

2.5 Speaking – Hearing

Speaking is the most natural form of human communication using the medium of natural language.

Two main technologies are used to provide speech recognition, using either Acoustic Representation or a Knowledge-Source Driven Representation.

Acoustic Representation. This approach applies general signal analysis techniques to the speech signal in an attempt to isolate a unique pattern; this is compared with a stored reference set of patterns representing the words in the vocabulary for recognition. In this way, the speech utterances can be identified when a match is found. A particular weakness in terms of the probability of communication failure is that this approach does not attempt to check the syntax or semantics of the spoken utterances or relate them to the context of the task.

Knowledge-Source Driven Representation. This approach takes into account the dynamic use of language, the environment of the task, and the context of the utterance in order to interpret the spoken utterance. Hence many sources of knowledge about the spoken utterance are taken into account, each level contributing to recognition and making up for deficiencies in information from the other levels. The probability of communication failure at this level is therefore much reduced.

The development target for all of these technologies is aimed at supporting continuous spoken utterances from any speaker using an unlimited vocabulary, minimising the chances of communication failure due to limitations on any of these factors. However, a number of stages can be identified in the progress towards this target aimed at providing usable, although limited, speech recognition. These stages are: Isolated Word Recognition (IWR); Continuous Speech Recognition (CSR); and, for each of these, either Speaker Dependent or Speaker Independent form.

Three technologies are used to generate speech from a computer system and these are: Digital Compression; Phoneme Synthesis; and Pre-recorded Speech.

Digital Compression. With this technology, a set of spoken words is digitally encoded, compressed and stored. This technique models the human vocal tract by simulating the parameters used in producing speech.

Phoneme Synthesis. With this technology, speech is synthesised by concatenating phonemes, the basic unit of speech sound.

Pre-recorded Speech. With this technology, spoken sentences or phrases can be recorded in full and stored for later "generation".

Implications for Communication Failure. The status of development of the technology of speech recognition suggests use of speech recognition when the following criteria apply:

When the user's hands and eyes are dedicated to the task leading to so-called "head-up, hand-free" applications.

When the user has high information overload and speech can be used to provide an alternative channel of communication.

When intermediate steps in the task can be eliminated, such as avoiding the need to stop and key-in data, thus reducing task time.

When the users are required to move around the task environment.

When the users are reluctant to use conventional input methods (such as senior managers' reluctance to use keyboards).

When handicapped users are involved who are unable to use another input medium.

The criteria are useful in directing the use of the technology for a particular task.

As far as speech generation is concerned, the technology is continually improving but the following general points apply. Digital compression tends to result in only moderate intelligibility, lacking in intonation. Phoneme synthesis is the most flexible technique but tends to sound unnatural and is often of low intelligibility since it lacks prosodic information. Pre-recorded speech is the most natural but is the most limited in flexibility as all the words and sentences have to be spoken in advance in order to make the recordings.

2.6 Looking – Seeing

Looking is used by us to locate items of interest and take in information about our environment, whether that information is moving images, other people, or text. Looking is therefore normally considered an input medium to the user. It can also be considered as an input to the system in so far as information located with the user's eyes can be used as a control action by the system. (It is also true that humans take account of direction looking, for example when attempting to "catch a waiter's eye" or in conversation when noticing that the other's attention has been distracted.) Looking provides an alternative modality in system communication to support other modes.

A technology has been developed known as the "pupil-centre corneal-reflection method" which is able to detect the user's direction of gaze when looking. It uses an infra-red light source to detect the movement of the eye between the pupil centre and the cornea; the signals are picked up by a television camera and computer-analysed.

Where people look provides an index of what they are interested in and where they are going visually to gain necessary information. It can be used in conjunction with a display of objects to indicate selection. Feedback is provided to the user either by a cursor which follows the user's gaze or by a change in the display field.

Where and how the user looks causes the system to zoom in on some window of interest; at first pass the system cuts to a full-screen view; a further option is to have a "special effects" type of zoom where the frame of the small dynamic window expands to full-screen size. Two competing principles determine the stimulus for zooming in. One is based upon zooming automatically after the user has looked at an image for a defined time; the other requires a deliberate action using some other input means such as a joystick, pointing, or speech.

There are two ways in which the user "leaves" a window which is currently zoomed in: one, temporarily, for some reason irrelevant to the display (for example, someone has entered the room, or the user takes a telephone call); two, when the user leaves the window for the display at large or for some other window. The system takes action related to where the user's eyes go; that is, if the user is distracted the system maintains its current hold; if another view is sought, the system changes. The user is kept aware of the rest of the action when zoomed in, by a video "double exposure mix", controllable by the user; that is, a background full view is maintained, controllable by joystick.

Implications for Communication Failure. The human can exert very fine control over eye movement, and in principle can use this to send very precise locational information to the system. In some cases, this could form a natural component in person-machine communication — as it does in normal person-person dyadic interaction, for example (where, as mentioned above, direction of gaze can "cue" appropriate behaviour from the other person). However, it could also be potentially disrupting if not implemented appropriately. Much perceptual processing of the environment seems to be geared towards maintaining a subjectively stable perceptual field despite changes occurring at the retina. Violating this principle by linking eye movements (and the associated changes at the retina) too directly or inappropriately to changes in the environment (the display) could be expected to disrupt communication in some cases.

2.7 Gesturing

Gesturing, either pointing or gesticulating, with the hands is another form of human communication; it is generally used in conjunction with speaking to indicate the desired item from a group, such as in "I'll have that one". As such it minimises the amount of voice communication required and increases the efficiency of communication.

A device for detecting users' gestures has been developed as a prototype (Bolt, 1980). It is based on measurements made of displacements in a magnetic field. One device remains static in the gesturing environment, another is strapped to the user's wrist; the displacement between these two devices in the magnetic field generated is used to determine where the user is pointing. Feedback is provided to the user by the cursor, which moves according to where the user is pointing.

Gesturing is another form of selecting and touching, although the technologies are not as well developed as for these other forms of communication; its experimental use to date has been to supplement speech recognition, where the deficiencies in the technology can be compensated by the additional information provided by gesturing. For example, if a user points to something and says "that", the speech recognition device only has to recognise a single word. In the absence of gesturing, the recognition device would have to interpret a lengthy description (Bolt, 1980).

Implications for Communication Failure. Broad pointing gestures in principle can be detected fairly accurately by a computer-based system and could provide useful supplementary information in addition to speech or typed inputs. Such broad movements would most appropriately be used in an environment where there is an emphasis on large-scale spatial displays. In the more usual situation of an office product with a relatively small display area, there might be a case for developing analogues of day-to-day gestures. In the case of pointing, this has already led to the development of dialogues based on pointing by means of a mouse.

2.8 Multimedia Interaction

User input and output would benefit from devices supporting the whole range of human communication involving keying, selecting, touching, speaking, looking and gesturing for input — and seeing and hearing for output. The essence is to enable the user to use any of the devices without constraint, according to the user's preferences and skill and the demands of the task. Where different devices are used to achieve the same action, the effects should be consistent across the range of devices; this so-called "modeless" interaction is an essential prerequisite for effective multimedia uses.

A spatial data management system. SDMS (Spatial Data Management System) was developed at the Massachussetts Institute of Technology (MIT) as a demonstration system some years ago and is an example of multimedia interaction. It was established in the MIT Machine Architecture Group's experimental area known as the Media Room. It uses a large screen occupying a complete wall to present a "Dataland" to the user. This consists of a variety of full colour images and sources of data arranged spatially to suit the user. The user can explore Dataland by using the joystick on the right-hand arm of the user's chair; movement in any direction causes a "helicopter-like flight" over the surface of Dataland in the direction of the joystick control. Whilst the large screen reflects movement in Dataland, the right-hand monitor presents a "world view" of Dataland by always showing its complete status as a single plan view. A "you are here" marker indicates the relationship of the objects on display on the large screen to the world view.

As an alternative to using the joystick the user can touch any of the objects on the world view display (that is, on the right-hand monitor). This causes the large screen to change to display the object selected in full detail. The joystick located on the left arm of the chair enables the user to zoom in and out of the large screen display. This causes the item to increase or decrease in size and detail correspondingly, including colour, sound, and movement.

Objects in Dataland may be of the type that allow user interaction with the data; such data types contain a so-called "key map" which is presented to the user when such objects become fully visible on zooming in. The key map may be a table of contents in the case of a book or document, it may be a time counter dial associated with a television set enabling a user to move to a different time portion of a program. Other data types include "processes" represented as icons of familiar objects such as a calculator or telephone.

The left-hand monitor, incorporating a touch screen, reflects the key map and process allowing the user to manipulate them appropriately. For example, the buttons of the calculator can be touched for computation and those of the telephone for communication. Similarly, the contents list of the book can be touched to reveal a particular chapter or page. A data tablet, upholstered to serve as a lap pad, is used to make notes and annotate objects on display; these are then associated with the appropriate objects and retrieved every time the object is referenced. The notes can also be made as a spoken commentary through a telephone attached to the data tablet; these are again stored and played back appropriately.

Sound is used in Dataland to provide an added cue to object recognition; as the user approaches an object by traversing the information space of Dataland any sound emitted by the object increases in volume. Hence a user can be guided by sound as well as by space.

Implications for Communication Failure. Evaluation of the use of SDMS has been limited. It has been reported to show a quick learning time, particularly with respect to navigation using the joystick; the importance of the auxiliary monitor providing the world view to complement the large screen is stressed (Bolt, 1979). Equally effective was the use of icons within the spatial framework; Bolt reports that users respond by referring to items by their spatial location – for example, "the letter to the right and north of the green square". The third important component is the use of sound to provide an added medium of communication.

Many of the ideas developed in the SDMS Dataland are beginning to be found in commercial systems such as the Xerox Star and, more recently, the Apple Lisa and Macintosh, and products such as GEM.

Multimedia interaction in principle matches to the greatest extent the richness of normal person-person communication. However, just as in normal person-person communication, it requires the system to integrate a large amount of information from a variety of different sources, and to do so very quickly in real time. This is a problem given the current state of development of technology, and a problem which is exacerbated by the requirement to load the user down with various sensing devices so that the system can pick up the necessary information. Developments in either or both of these areas would be a useful step forward in facilitating person-machine communication and reducing communication failure.

2.9 Natural Language

Natural language dialogue is still a highly experimental area, whose requirements transcend conventional user-system dialogues and abut the realm of artificial intelligence. User interpretation of natural language is sensitive to context and prone to ambiguities (e.g., Longuet-Higgins & Isard, 1970). It may also not be an optimal idea to attempt to make communication between a user and an electronic system too natural as, for example, the credibility of the information provided by the system may be reduced in the eyes of the user if the system is seen to "talk the same language". Inexperienced system users may also credit the system with more intelligence than it really has, and may find it frustrating and off-putting when the system "refuses" to give them the information they expect from it, on the basis of such inappropriate expectations (e.g., Martin, 1973).

Constrained natural language dialogue may offer a better solution in the long term, but this may also be seen as essentially just a sophisticated command language, with its attendant pros and cons.

3. Matching the System to the User and the Task

The above represent the main species of interaction that are available to the designer of person-machine dialogues. Although they have been described separately for the sake of clarity, they are not mutually exclusive, but can exist together in the same system. This is important as a means of enabling the system in some degree to meet the needs of different users and the demands of different tasks. Hybrid dialogues and parallel dialogues represent two key ways in which this can be achieved.

3.1 Hybrid Dialogues

During the period of interaction with a system, a user normally undertakes a variety of transactions (e.g., enter data, retrieve information, etc.), each of which may present the user with very different sets of requirements and may be best served by a very specific type of dialogue. In this case, it does not make sense to constrain the dialogue to one single style only; it is preferable to use the style which is most appropriate to each transaction, or set of similar transactions (see e.g., Eason & Damodaran, 1981, for further discussion). This mix of dialogue styles constitutes a hybrid dialogue.

The current trend towards integrated computer systems with distributed resources has increased the complexity of person-machine interactions and will allow a greater diversity of tasks to be performed at any given workstation. Users will also be more familiar with certain system facilities than with others, and so responsibility for initiative in the interactions will have to be devolved to different degrees between the user and the system, depending on the user's familiarity with the particular facility (s)he is working on. Also, different tasks will require different dialogue styles to facilitate interaction in accordance with their particular salient features. Hybrid dialogues come into their own in these circumstances, and have the additional advantage that they make it easier for the user to identify the system context (s)he is in, based on the available dialogue style. A significant disadvantage in the use of hybrid dialogues is that they make it very difficult to maintain system consistency and simplicity, two variables known to influence user acceptability of the system and the ease with which users can achieve error-free, flexible transactions.

3.2 Parallel Dialogues

As users gain experience with a system, their requirements change. They will want to exercise ever greater control over the interaction, use the system in a more flexible manner, and take advantage of some of the more sophisticated facilities available in the system. This entails a change in initiative from the system to the user. To cater for this requirement, two, or several dialogue styles are necessary, available in parallel.

Users can then progress effortlessly from one skill level to another, always having at their disposal a style of dialogue which maximises the efficiency of their transactions.

A classic example of parallel dialogues is menu-selection in conjunction with command-language input: the "shorthand"-type of interaction (see also Hall, 1978). This type of parallel dialogue is used very effectively in systems like the Apple Lisa and the Macintosh, where they also allow the user to match the style of input to the particular application package being used. Hence for graphics, where the user tends to use the mouse as an input device, it is convenient to pull down a menu and select options in this way; in word-processing mode, where most inputs are made through the keyboard, the command language shorthand can be more convenient. The price the user pays for this facility is loss of consistency between different parts of a given system, so that it is difficult to abstract a single unified model of the system's mode of operation. On the other hand, an advantage of this type of dialogue is that infrequent use of the system need not entail complete loss of accuracy when the user goes back to use the system after a long pause: the menu-selection mode can be used to refresh the user's memory of specific commands and sequences, as well as their command language "shorthands", with little or no loss of speed and efficiency.

3.3 Conclusions

Advances in technology mean that a variety of different "species of interaction" will be possible between the human and the machine in the electronic office environments of the next five to ten years. Careful choice of an appropriate species will help the designer to provide a Human Factors input-output environment that will support dialogues minimising the extent and frequency of communication breakdown.

The various "species of interaction" that are becoming possible differ in the attentional, memory, and other cognitive demands they make of the user. The various species differ in how appropriate they are in different circumstances, relating to the nature of the task and the nature of the user. And wherever a particular species is appropriate in principle, an indefinitely large number of particular dialogues can be constructed — most of which will be sub-optimal in terms of how well they "mesh in" with the way the human is designed.

The onus is on the designers of electronic products to try to optimise their design as far as possible, so that the design fits in well rather than poorly with the way the human is built. In terms of the flow of communication between the user and the electronic office system, this means designing the dialogue to be compatible with the cognitive make-up of the human, taking account of the kind of task the user needs to do. In the following sections we consider some of the key approaches that have been developed specifically with database query in mind.

4. Dialogues for Data Query

Reisner (1981) and Shneiderman (1980) provide useful reviews of approaches to dialogues aimed at retrieval of information contained in large databases. We will not attempt to cover all possible approaches here but will instead focus on three approaches which are sufficiently different to give a broad idea of the range of variation possible, and which illustrate some of the key problems involved in this area. The three approaches selected are:

SEQUEL, a formal query language developed by Chamberlin and colleagues e.g., Chamberlin, (1976) of IBM and extensively research by, amongst others, Reisner (e.g., 1977);

CUPID, a graphical query language developed by McDonald and Stonebraker (1975); and

RABBIT, a "retrieval-by-reformulation" language developed by Williams (1984).

4.1 SEQUEL: A Keyword Approach

SEQUEL adopts the keyword approach to query language. As an example of this, the query "Find all employees who work for Mike Smith and who make less than $20,000" is formulated in SEQUEL terminology as:

```
SELECT        NAME
FROM          EMP
WHERE         MGR='SMITH'
AND           SAL 20000
```

Experiments have been carried out to test whether formal query languages offer any advantages over less formal ways of interacting with a computer, such as Query by-example. Reisner (1977) reported that a frequent cause of communication failure when using formal query languages, such as SEQUEL, was a tendency for subjects to make mistakes by attempting to derive the syntax of the queries from normal English syntax.

Reisner explained these failures in terms of a model of what users do when attempting to convert a particular query into formal notation. The model is very similar to some of the schema models of long-term memory. Its basic assumption is that users generate mental "templates" of the structure of the question they wish to ask of the system, and then attempt to determine the specific parameters that need to be inserted into the template. Failure occurs because natural English questions are not easy to translate directly into the format required by the query language.

According to this model, the likelihood of communication failure can be reduced if users are given the right template structure they should use — as happens, for example, with Query-by-example types of dialogue (see Thomas & Gould, 1975).

However, for both Query-by-example and other types of database query languages, a recurring problem users have concerns the description to the system of quantification and set membership parameters that are essential to the correct formulation of a request (e.g., Thomas, 1976; Gould & Ascher, 1975; Reisner, 1977; Thomas & Gould, 1975). Few users are familiar with the formal logic structure of many query languages. Furthermore, research on human thinking and problem-solving suggests that in many situations people do not, in any case, use formal logic — they think in more probabilistic and fuzzy ways (e.g., Braine, 1978).

At a more general level, Barnard et al. (1981) and Bott (1979) suggest that many of the results obtained with experiments testing users' learning and use of command and formal query languages may be understood in terms of "reconstructive memory". This means that — while experienced users may have at their disposal, and be able to retrieve, the command or syntax directly — they tend instead to make inferences about possible options based on what they do remember and their expectations about what the command terms or syntax should be. Thus, if the dialogue allows the user to abstract simple rules for generating commands, it will also lead to the correct types of inference being made. This type of dialogue will also reduce the load placed on the user's working memory, releasing "processing power" for other necessary operations, like planning the next stages in the interaction, considering system feedback, etc.

By and large, languages such as SEQUEL are favoured by programmers, to whom the keyword approach comes more or less naturally. Proponents of the more abstract types of query languages such as Query-by-example claim that their more abstract approach is more directly compatible with the way non-programmer users think, and that keywords may "clutter" the normal flow of reasoning (e.g., Zloff, 1975; 1978). In addition, the contextual cues involved in the Query-by-example approach may facilitate the abstraction of a clearer "model" of the system, or of a given application, increasing the ease with which users will be able to use the system to full advantage.

4.2 CUPID: A Graphics Approach

CUPID adopts a very different approach to database interrogation. It is based on interactive graphics, and relies on users' spatial sense, and their ability to abstract complex information on the basis of the spatial relationships among the data presented.

CUPID'S interactive graphics aid the user in producing a "template" resembling a flow-chart, which contains the information necessary to formulate the query. Different-shaped boxes, arrows and spatial position are then used to represent the relevant dependencies. Users interact with CUPID *via* a lightpen, which allows them both to move and generate the diagrams that represent a query. Figure 1 gives an example of CUPID notation for the query "Print the names of ski resorts in New York state which have at least one expert trail" (from Shneiderman, 1980).

Surprisingly, CUPID has not been extensively researched and the key psychological question of whether keyword-oriented dialogues or other text-based query languages offer any advantages or disadvantages over positional or abstract notation has so far received exceptionally little interest from the Human Factors and academic communities.

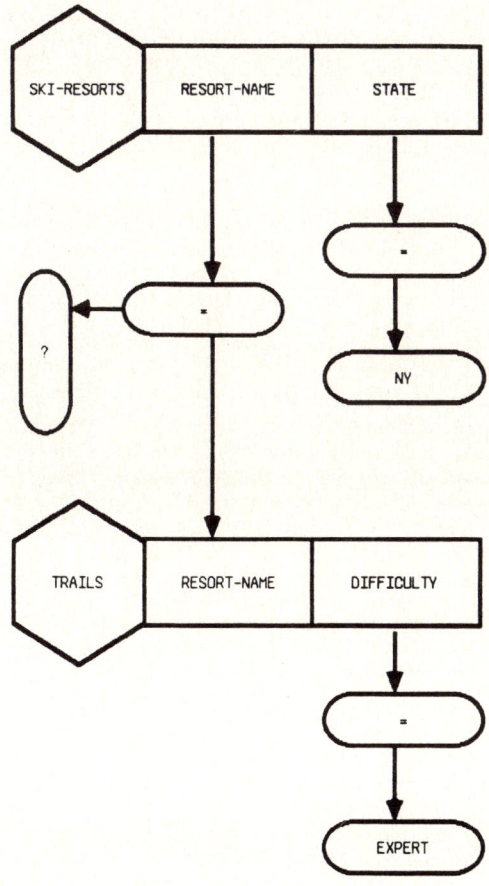

Figure 1. An example of the CUPID format for a simple query.

At one level, however, CUPID is disappointing in the use it makes of a potentially very powerful medium. Graphical representation of complex information is one way in which, in principle, the dialogue between the human and the machine can be facilitated: humans have a highly developed visual and spatial processing system. In addition, as we mentioned above, a recurrent stumbling block in user-system dialogue for information retrieval is the representation of quantification and set membership parameters (e.g., Thomas, 1976; Gould & Ascher, 1975). Yet there is very little in CUPID's notation that utilises the power of the graphics medium to help the user avoid these two problem areas.

Imaginative use of graphics — even at the low level of using Venn-Type diagrams to indicate set membership — could be a valuable new approach to this type of dialogue. Disappointingly, very little research or application work appears to have been done in this area.

4.3 RABBIT: Interactive Reformulation

RABBIT is a database retrieval interface developed at the Xerox Palo Alto Research Center. It is based on the notion of "retrieval by reformulation" — an approach that allows the user to refine interactively partial descriptions of the target item or items (s)he wants to retrieve. This is done by allowing the user to critique, using special menus, successive examples and counter-examples of items that the system considers fit the description the user has given it of the items being searched for (Williams, 1984). In this way, failure is recognised to be an integral part of the process of finding a wanted item — failure on any particular iteration is utilised to refine the communication from the user to the system, normally reducing the probability or size of failure on the next iteration.

RABBIT uses the same kind of windowing techniques used in the Apple Lisa and Macintosh. Examples of possible target items that the system considers fit the query are presented to the user in "active form" windows. A pop-up menu displays the critiquing options the user may utilise to refine the query. Separate windows then contain both an example of an item retrieved that satisfies the query as initially formulated, and additional information as to the number of matches that the query has produced in the database.

Figure 2 gives an example of what a RABBIT interface looks like when a personal search is being conducted. Bold-type entries are the categories that define the information required (e.g., person, employee, etc.). Italicised, indented items are entries, or fields on which the user has already commented, providing valuable context information relating to the consequences of previous actions. Several levels of comment and description are allowed. The top window in the example represents the query: "Find a person who happens to be an employee, aged over 21, working in some corporation in either California or Delaware". James Tancon, the person described in the lower window, satisfies these conditions, and more, as do the 14 matches in the window next to it. The pop-up menu in the middle allows the user to narrow down the search field and refine the description initially given by, for example, prohibiting or requiring specific parameters.

Implications for Communication Failure. The concept of retrieval by reformulation is based on a particular approach to the psychology of memory (e.g., Williams and Hollan, 1981; Williams, 1981; Norman & Bobrow, 1979). It is particularly suitable for tasks where what the user wants to do is browse for information or explore a database. There are other query languages which allow browsing (e.g., ZOG — see McCracken and Akscyn, 1984), but these tend to involve an alternative approach, requiring the user to hop from node to node in a large network of connections that structure the items in a database. RABBIT is particularly interesting because it allows users to find what they want by describing it, thus providing a good electronic simile of the way people normally try to retrieve information from their own memories.

It is a comparatively new approach and as such has not yet been fully researched. Williams (1984) does provide a series of example "stories" which address different areas where research might be needed, or where there seem to be key differences between the approach adopted in RABBIT and approaches to other query languages which may account for the power of the interface.

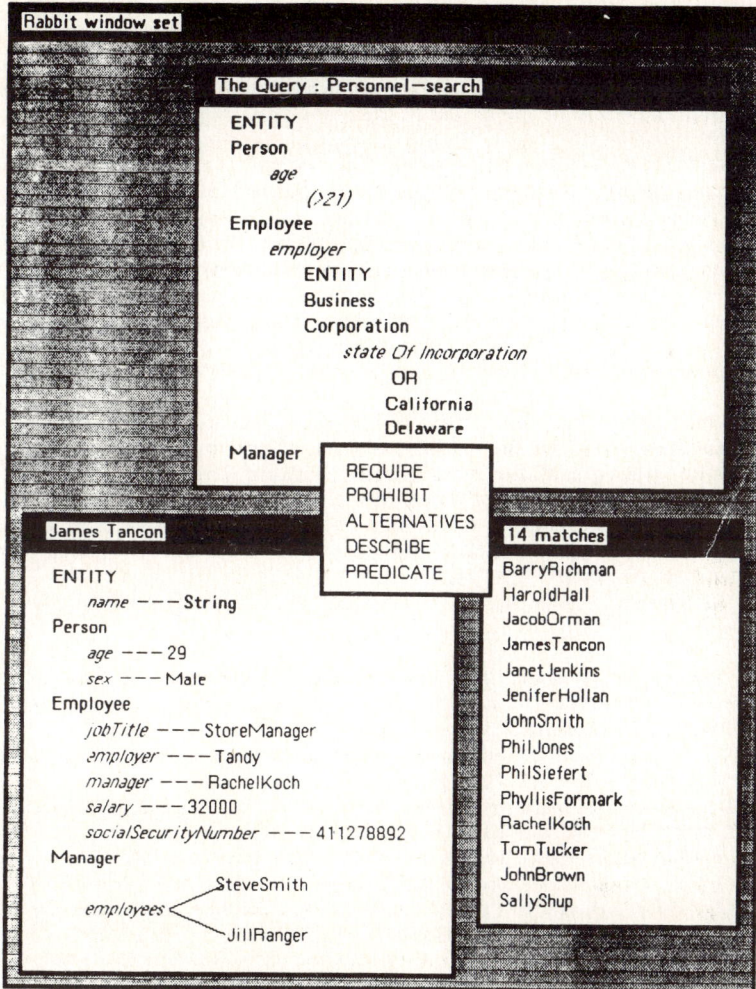

Figure 2. A RABBIT Display.

Developed as a text-based interface, it could be extended to incorporate graphics as well. The trend in interface design in this area does seem to be in this direction. FileVision, for example, (a software package developed for the Macintosh) combines the "form-filling" approach with a graphical, spatial description of data which users can access directly by pointing with a mouse. It provides a powerful tool for information retrieval, which could be made even more powerful if it were to incorporate at least some of the features of an interface like RABBIT.

5. Conclusions

Different approaches to dialogue design for database query are possible, emphasising: keyword search; spatial representation of relationships using graphics; and progressive reduction of communication failure through iterative reformulation of the query in the light of example items retrieved by the system.

What limited research has been done on these approaches suggests that providing a "template" onto which users can map their requirements may reduce communication failure and improve learning times (see also Fitter, 1978). In the absence of such a template, users tend to overgeneralise from a knowledge of the syntax of the query they formulate in their native language.

Graphics seem likely to reduce communication failure if used appropriately to provide a more condensed way of describing a query, tapping into the non-verbal kind of information processing that seems to characterise much human problem-solving. However, current interface designs using graphical techniques do not generally use them as effectively as they could. This is even more true when considered in relation to the much improved high-resolution, colour displays that are now becoming available even on low cost machines. So use of graphical techniques may prove to be an important step forward in reducing communication failure between human and machine, but a much more imaginative approach to their use is needed. Early hints of what might be to come can be seen in systems such as RABBIT and FileVision.

From a Human Factors viewpoint, effective human-machine dialogue depends upon compatibility between the human and the machine in terms of information processing, both at the general, semantic level, and at the more specific, syntactic level. Until recently, both the technology available and the state of development of Human Factors restricted the range of possible approaches to the kind of syntax that could be used. Advances in the technology available and in Human Factors now point the way to new forms of syntax that are better suited to human cognitive processing.

References

Barnard, P.J., Hammond, N.V., Morton, J., & Long, J.B. (1981). Consistency and compatibility in human-computer dialogue. *International Journal of Man-Machine Studies, 15,* 87-134.

Bolt, R.A. (1979). Spatial Data Management Systems. *Library of Congress,* No. 78-78256.

Bolt, R.A. (1980). Put that there: Voice and gesture at the graphics interface. *Computer Graphics* (ACMSigraph) ACM 0-8971-021-4/40/007-262, 262-270.

Bott, R.A. (1979). *A Study of Computer Learning: Theory and methodolgies.* Center for Human Information Processing Technical Report No. 82, University of California, San Diego.

Braine, M.D.S. (1978). On the relation between the natural logic of reasoning and standard logic. *Psychological Review, 85,* 1-21.

Chamberlin, D.D. (1976). SEQUEL 2: A unified approach to data definition, manipulation and control. *IBM Journal of Research and Development, 20,* 560-574.

Christie, B. (1979). *Face to File Communication: A Psychological Approach to Information Systems.* Chichester/New York: John Wiley & Sons Ltd.

Eason, K.D., & Damodaran, L. (1981). The needs of the commercial user. In M.J. Coombs and J.L. Alty (Eds.), *Computing Skills and the User Interface.* New York: Academic Press.

Fitter, M. (1978). Information systems and individual behaviour. In J. Banbury and R.K. Stamper (Eds.), *Management Information Systems.* New York: Academic Press.

Gould, J.D., & Ascher, R.N. (1975). Use of an IQF-like query language by non-programmers. *IBM Research Report,* RC 5279.

Hall, P.A.V. (1978). Man-computer dialogues for many levels of confidence. *SCICON Report.* London: SCICON Consultancy International Limited.

Longuet-Higgins, C., & Isard, S.D. (1970). *The Monkey's Paw.* Conference on Man-Computer Interaction organised by the I.E.E., September 1970.

McCracken, D.L., & Akscyn, R.M. (1984). Experience with the ZOG human-computer interface system. *International Journal of Man-Machine Studies, 21,* 293-310.

Martin, J. (1973). *Design of Man-Computer Dialogues.* Englewood Cliffs, NJ: Prentice-Hall.

McDonald, N., & Stonebraker, M. (1975). CUPID: The friendly query language. *Proceedings of the ACM Pacific Conference,* San Francisco.

Norman, D.A., & Bobrow, D.G. (1979). Descriptions: An intermediate stage in memory retrieval. *Cognitive Psychology, 11,* 107-123.

Reisner, P. (1981). Human factors of data-base query languages: A survey and assessment. *Computing Surveys, 13,* 13-31.

Reisner, P. (1977). Use of psychological experimentation as an aid to development of a query language. *IEEE*

Transactions on Software Engineering, SE-3, 218-229.
Shneiderman, B. (1980). *Software Psychology: Human factors in computing and information systems.* Cambridge, MA: Winthrop.
Thomas, J.C. (1976). Quantifiers and question-asking. *IBM Research Report.* RC 5866.
Thomas, J.C., & Gould, J.D. (1975). A psychological study of Query-by-Example. *Proceedings of the National Computer Conference,* Anheim, 439-445.
Williams, M.D. (1981). Instantiation: A data base interface for the novice user. *Xerox Palo Alto Research Center Working Paper.*
Williams, M.D. (1984). What makes RABBIT run? *International Journal of Man-Machine Studies, 21,* 333-352.
Williams, M.D., & Hollan, J.D. (1981). The process of retrieval from very long term memory. *Cognitive Science, 5,* 87-119.
Zloff, M.M. (1975). Query-by-Example. *Proceedings of the National Computer Conference.* Montvale, NJ: AFIPS Press.
Zloff, M.M. (1978). Design aspects of the Query-by-Example database language. In B. Shneiderman (Ed.), *Databases: Improving usability and responsiveness.* New York: Academic Press.

NONVERBAL BEHAVIOUR IN DIALOGUE

N.P. Sheehy
Department of Psychology
University of Leeds

A conceptual analysis of the ontogenesis of the relationship between verbal and nonverbal behaviour is used to reveal similarities and differences between these behavioural streams. It is argued that gestures should be considered a special category of nonverbal behaviour and that these are essentially 'verbal' acts. Problems associated with modelling human-computer interaction are considered. It is argued that no single model adequately captures the most salient aspects of both user and computer conduct. This limits design initiatives aimed at exploiting the potential of multi-media user interfaces. Theoretical and computational difficulties relating to the use of gestural input to computers are considered and implementations which attempt to achieve partial solutions are described. It is argued that formal descriptions of both verbal and nonverbal behaviour cannot be achieved but successful tactical implementations may be accomplished through coordinated multi-disciplinary efforts.

1. Introduction

That aspect of human conduct described as 'communication' is concerned with the creation, negotiation and management of shared meanings about the world. Here communication is not used as a synonym for language. Language is a part of communication and the capacity to share meanings through gestures antedates the emergence of language. The emergence of the capacity to communicate is intimately related with the capacity to coordinate and organize actions in the form of a single social activity with shared significance. This view is derived from Mead's (1934) and Vygotsky's (1966) analyses of language and cognition. It is also associated with Austin's (1962) and Searle's (1970) demonstrations that utterances are social events with ontological significance, rather than verbal representation of private cognitions (Winograd, 1980).

'Nonverbal behaviour' refers to biologically determined expressions of emotions and learned signs. The distinction between emotional expression and learned signs is operationalized through a dichotomy between voluntary and involuntary nonverbal behaviour. Spontaneous, involuntary nonverbal behaviour is associated with emotional expressions and is considered part of the emotion. Symbolic gesturing is associated with arbitrarily defined sets of intentional and voluntary actions. Buck's (1982) view is that human communication involves the simultaneous expression of both kinds of nonverbal behaviour in association with the stream of verbal behaviour, and his is representative of contemporary views. There is some evidence that each kind of nonverbal behaviour is functionally independent by the first 18 months of life and that differentiation continues throughout childhood (Lewis & Rosenbaum, 1977).

Ekman and Friesen (1969) have described nonverbal behaviour according to the manner in which meaning is encoded. Extrinsically coded acts signify meanings which are not apparent from the acts themselves. Some extrinsic acts are arbitrarily encoded, such as the thumbs-up sign signalling 'OK'. Others are iconically coded, as when a television producer signals 'end' with a throat-cutting movement with the index finger or when someone draws a referent object in mid-air. Intrinsically coded acts convey meanings which are revealed in the acts themselves. For example, one could substitute a fist-to-hand slap in the sentence 'She gave him a punch'.

Ekman and Friesen's (1969) classification of nonverbal behaviour is the most widely accepted. Emblems, their first category, are characterized by being easily substitutable by spoken words and are essentially iconic in nature. Illustrators, their second category, are directly related to speech and illustrate what is being spoken. Efron (1941) has identified six types of illustrator: batons (rhythmically synchronous movements), deictic movements, spatial movements, ideographic movements (that trace a line of thought), kinetographic movements (depicting bodily movements) and pictographs. Ekman and Friesen's third cateogry, regulators, maintain and regulate dialogue and include head nods and small postural changes. Affect displays involve facial expressions of emotion principally. Adaptors are probably the most complex of the nonverbal behaviours. They are not intentionally communicative but are often seen as important aspects of our personality. Examples of adaptors are face-touching, head scratching etc. Adaptors appear to accompany social-cognitive activity and may reflect surface features of covert mental processes.

2. Development of Language and Gesture

Essential to the emergence of language is the ability to create and use gestures (Trevarthen & Hubley, 1978). Gestures have meanings and form which are related directly to the social actions from which they are derived. Thus, gestures are actions that imply actions. They are not simply convenient 'nonverbal' behaviours which facilitate communications. Language develops in a rich social-relational context and interaction format plays a central role in language acquisition (cf. Edwards, 1978). Moreover social situations do not lose their formative significance following childhood. Social formats and socio-cultural conventions continue to play a crucial role in structuring adult communications (cf. Harre, 1985).

It is widely accepted that a capacity to communicate is essential to the development of language but little is known of the details of the relationship. Similarly, the role of linguistic input in language acquisition is poorly understood. Partly this is because there are few opportunities available to study the effects of variation in linguistic input on language acquisition. However, investigations of language acquisition in deaf children reveal some of the consequences of degraded linguistic input for the acquisition of language and gesture. Goldin-Meadow and Mylander (1984) have provided some evidence that deaf children develop gestural communications systems similar in structure to the communications systems of children learning language in more enriched linguistic environments. Since it is argued here that gestural communication is essential for the development of natural language this finding is not surprising. It appears that deaf children develop three kinds of lexical signing: deictic signing, characterization signing and marker signing (Goldin-Meadow, 1979, 1982; Goldin-Meadow & Feldman, 1975). Goldin-Meadow and Mylander (1984) have also suggested that gestural communication in deaf children has a number of robust features. The first of these is discreteness and concatenation. Deaf children display a tendency to communicate using discrete gestures which are concatenated to form larger gestural strings. This use of sign sentencing to convey meaningful relations parallels the frequently observed practice in normal-hearing children of concatenating discrete lexical units to form sentence strings. In both normal-hearing and deaf children this phenomenon occurs even when there exists the possibility of incorporating the discrete units into single form.

Second, deaf children's gestural systems can be highly structured. Deaf children sign in predictable ways. They use surface devices usually observed in natural language, such as ordering of elements and strategies for the production and deletion of elements.

Third, the communications systems of both deaf and normal-hearing children accommodate recursive propositions. In both natural language communications systems and deaf children's gesture systems it is possible for the child to generate a proposition and derive new strings that contain the propositional unit.

These three properties of the deaf child's gesture system reflect an ability to invent resilient features of human communications systems, even in impoverished stimulus environments. In addition to these resilient properties there are fragile aspects of natural language systems. Fragile properties require linguistic models for input, and include, for example, the development of auxiliary and movement rules (Curtiss, 1977; Newport, Gleitman & Gleitman, 1977; Goldin-Meadow & Mylander, 1984).

One reason for examining highly developed gestural systems, such as those invented by deaf children, is that they identify the relative importance of linguistic input for the acquisition of good communications rules. The evidence has not gone unchallenged (cf. Bates & Volterra, 1984), but it is sufficiently strong to support the conclusion that at the developmental level a distinction between 'verbal' and 'nonverbal' behaviour is difficult to sustain. Children do not acquire a language system and, independently, a correlated system for signalling nonverbally. The two systems are related intimately. The capacity for suppressing or supplementing one way of communicating with another, for example for suppressing overt gestures, does not reveal anything of the relationship between the two.

There is a second reason for examining deaf children's gesture systems and this has to do with the resilient features of those systems. The three resilient properties of those systems (discreteness, structure and recursion) and the lexical signs they support (deictic signing, characterizing and marker signing) suggest how communication between people and computers might be developed. Specifically, good person-computer interaction should attempt to incorporate these aspects of gesture systems.

Despite the developmental evidence indicating the closeness of the connections between the verbal and nonverbal streams it has often been thought that nonverbal behaviour is largely irrelevant to an understanding of adult dialogue. This view has often been supported by three kinds of argument. First, it may be argued that any piece of nonverbal behaviour can be expressed verbally but the obverse is not the case. However, the gestural systems of deaf children reveal this not to be the case: Second, it may be argued and demonstrated that it is possible to suppress all nonverbal communication but still produce complex, meaningful communications. However, deaf children are capable of communicating in rich ways without recourse to speech utterances. Third, it may be argued that the developmental evidence indicates that nonverbal behaviour becomes less important once a language has been acquired. This developmental change is interpreted as indicating that nonverbal behaviours are simply expressions of early precursors of later complex linguistic forms. The opposite view, taken here, is that dialogue involves parallel indexing of cognitions in several channels. An analysis of dialogue requires examination of the matrix of channels used, the degrees to which they are articulated and the ways in which they combine and interact.

McNeill (1985) has articulated a particularly strong view of the relationship between nonverbal and verbal behaviour. McNeill argues that gestures and speech share a computational stage: they develop internally as psychological performances. Although two behavioural streams are produced they are experienced as a single stream of action. McNeill's analysis is based partly on a distinction between two kinds of thinking: imagistic and syntactic. These are coordinated in referential and dialogue-oriented gestures. He describes 'imagistic' thinking and global and synthetic and 'syntactic' thinking as linear and segmented. Dialogue is thought to comprise a global-synthetic representation with a linear-segmented string. Thus, one finds that thought has grammatical features and grammar has cognitive features. This is a reflection of the fact that thinking and speaking are continuous systems. Thus, symbolic gesturing is viewed as a product of common psychological processes that also produce symbolic verbal utterances. Gestures and speech provide parallel views of a unitary underlying psychological process. This is a controversial point of view which has been espoused by some other psychologists (Argyle, 1975; Kendon, 1983) who have been concerned to show that nonverbal behaviour is not merely complementary to and supportive of speech utterances. In effect McNeill argues that gestures are verbal and this contradicts a more prominent view which conceptualizes the nonverbal stream as an independent signing system governed by its own laws. It is also contrary to those linguistic analyses which consider the analysis of language structures exclusively in terms of speech sounds and grammar.

In support of the argument for viewing gestures and speech as parts of a unitary psychological process McNeill musters five arguments. First, gestures occur only during speech. Second, gestures and speech have parallel pragmatic and semantic functions. Third, gestures synchronize with parallel linguistic units. Fourth, gestures are similar to speech in that parallel decrements occur following neurological damage associated with Broca's and Wernicke's aphasia. Fifth, there is developmental evidence to show that gestures develop in parallel with speech.

The weight of empirical evidence is towards a conclusion that gestures and speech are part of a common psychological process and that the analysis of dialogue should begin from this perspective. One implication of working from this perspective is that person-computer dialogue can be seen to attenuate a wide range of human communicative skills. For instance, the restricted seating arrangement associated with terminal use means that users are forced to work within a constrained number of communications channels for long periods of time. There are few opportunities to engage in gestural-type interaction, although touch sensitive screens have broadened the number of communications channels available. It may be that increasing the number of input media available to users facilitates not only more accurate and efficient interaction but also more versatile problem-solving by users. In the next section the relationship between deictic gesturing — the kind most frequently accommodated in user interfaces — and speech is considered in greater detail.

3. Deixis: Pointing and Voicing

It is well known that speech and gesture are synchronized in subtle ways and usually we are not aware of this until asynchronous communication occurs (Condon and Ogston, 1971; Kendon, 1980; McNeill, 1979). However, little is known about the details of the processes involved in achieving synchrony between gesture and speech in adult communications. Investigations of deictic gestures provide some indications. McNeill (1981) describes "iconic" gestures as object-focussed in that they are concrete depictions of meanings expressed vocally. Deictic gestures are a special class of gestures in so far as they are often compulsory. 'Here' and 'there' usually require the speaker to make a pointing gesture in order to identify the referent.

Levelt, Richardson and Heij (1985) have distinguished between two hypotheses about the relationship between speech and deictic gesturing. The first is the modular hypothesis, which suggests that a relationship between speech and gesture is established at the central planning stage, but that execution is modular or independent. Interactive adaptation between systems is not possible because the systems operate in a ballistic fashion: once begun a sequence of actions cannot be adjusted to achieve better synchrony. The second hypothesis is that the speech and gestural systems are synchronized through mutual adjustment (by ongoing feedback) during the execution of gestural sequences and speech utterances. Through a series of experiments Levelt et al., (1985) found that "speech and deictic gesture are interactive in the planning phase, but well nigh ballistic in the execution phase" (p. 162). Moreover, it appears that synchronization is achieved by adapting speech to gesture; gesture is affected marginally by speech.

These findings are probably generalizable only to tasks which are similar to those studied in Levelt's et al. experimental settings. For example, they cannot be extended to McNeill's 'iconic gestures' where one part of a gesture may relate to a particular utterance and another part of the gesture may relate to a different utterance. However, their experimental task and apparatus approximated the kind of situation frequently encountered by users of touch-sensitive screens. There, users are often required to make a choice among a number of alternatives on display and this is what Levelt et al. (using LEDs) required of their subjects. There is also an important difference between the two situations. Computer users usually do not use speech input and it is not clear how deictic gesturing and a psychomotor skill, such as typing, may be related. It seems highly probable that there is a sharp conflict between a preference to perform two actions (point and type) concurrently and the physical constraints against doing this. Although systems which permit users to exploit possibilities for different kinds of communication are widely available these accommodate naturalistic interaction at a superficial level. Later in this chapter a command and control workstation is described. It attempts to overcome some of these difficulties by facilitating more flexible interaction within and between input media.

4. Modelling Nonverbal Behaviour

In order to incorporate aspects of nonverbal communication in human-computer systems it is necessary to simplify, by modelling, the functions and processes of nonverbal behaviour in dialogue. This raises a number of difficulties which have to do with the nature of human conversation and the nature of nonverbal behaviour. Nonverbal behaviour is different from speech in that speech is digital but nonverbal behaviour is normally not. Spoken language is digital in the sense that it is composed of discrete word units. Although some forms of nonverbal behaviour are readily identifiable as discrete gestures (e.g., deictic gesturing), nonverbal acts are less well defined. They are produced and perceived as a continuous, uninterrupted stream of activity and they represent a fundamentally human quality in interpersonal dialogue. Consequently, attempting to isolate and model some important features of interpersonal dialogue for inclusion in human-computer systems raises substantial difficulties.

In considering ways of modelling human communication it is important to distinguish representational from functional models. The charge that models are readily confusable with reality is apposite when modelling human-computer interaction. Faulty interpretation is easy when one takes an analogy too literally. Shannon and Weaver's (1949) mathematical model of communication is probably the most frequently adopted model for human-computer interaction. The model lends itself to this application because Shannon and Weaver's formulation is a response to some of the design needs of electronic engineers. It is useful to conceptualize human-computer interaction as comprising information sources, transmitters (encoders), signal processors, receivers (decoders) and message distinations. The mathematical model is a good representational model of the computer's part and a good metaphorical model for the human's role. The model serves functions (representational or metaphorical) which are defined by the perspectives taken and this is a source of potential difficulty. It is easy to over-extend the representational features to incorporate the role of the person and to over-extend the metaphorical features to incorporate the role of the computer. Such over-extension invalidates the application. People are more than information processors and computers are not human-like.

The notion of communication as a linear process conflicts with the reality of human dialogue. The dialectic, reciprocal aspects of communication are better represented by nonlinear models. Thus, Schramm's (1963) modification of Shannon and Weaver's model accommodates simultaneous message sending and receiving. Dance and Larson (1972) proposed a helical spiral model of human communication. It is not as sophisticated as the Schramm or Shannon and Weaver formulations, but it emphasizes elements of (1) continuity, (2) unrepeatability and (3) accumulation which are essential to human dialogue. Another model, due to Westley and Maclean (1957), emphasizes the fact that dialogue occurs in the context of social formats and every communicative act is socially symbolic. A more complex version of this idea is contained in Becker's (1970) mosaic model. In that model communicative acts are imagined to link message elements from different social situations. Communication is modelled as a cube, composed of message element building blocks. The most important feature of the model is the way it demonstrates how messages with identical meanings can be constructed from almost infinite combinations of different elements. Also it shows how messages with identical elements may have different meanings. There are other models, such as the transactional model of Ruesch and Bateson (1951) which emphasize the relational and irreversible aspects of human communication arising from the negotiation and attribution of meanings to acts.

The nonlinear models of communication are better approximations of human dialogue than the information processing model. However, they are not particularly suitable for analyzing the role of the computer and they have not been formally described. Perhaps their most important contribution is that they raise an issue of definition. In many respects the information processing model is not a model of dialogue. It is a model of influence based on physical reactivity. Within the model it is possible to describe, perfectly adequately, relationships of influence among elements without using the concept of dialogue. More general objections to treating human-computer interactions as dialogue have been raised by Smith (1980). Human-computer interaction is not communication in the full sense. Machine operators are rarely thought to 'communicate' with the machine. The influence relationship between machine and user may be camouflaged and a presentation which mimics aspects of a communicative exchange may be possible. However, like theatrical presentations, they rely for their success on acceptance of an agreed set of conventions and willingness to suspend disbelief. In human-computer interaction it is possible to encourage users to suspend disbelief about the fact that they are 'communicating' with the machine. To a large extent designing for user friendliness is concerned with convincing users that the interaction is truly conversational. It is partly on this basis that Pinsky (1983) justifies treating human-computer interaction as a special 'mechanical' form of dialogue.

Conversational programmes, such as Eliza (Gaines & Shaw, 1984) although simple in structure, initially 'fooled' a large percentage of human interactants into believing that they were conversing with the computer. However, natural language processors are brittle in the sense that, when the interaction becomes problematic or complex, it loses its conversational and communicative texture and the presentation becomes cosmetic or thin. Human dialogue is conducted at a routine, socially immersed, non-problematic level of awareness. Attention is focussed on the topic and not the dialogue *per se*. Interaction between computers and their users is intrinsically problematic because the 'dialogue' lacks credibility. The fact that users are prepared to tolerate fragile systems and even 'unfriendly' environments reflects principally on their motivation and imagination.

5. Channel Inconsistency and Dialogue Failure

Studies of the role of nonverbal behaviour in dialogue failure have concentrated on the concept of channel inconsistency. Early work (e.g., Ekman and Friesen, 1969) had a strong clinical emphasis and was concerned with the detection of cues to deception. Subsequently the focus shifted to an investigation of broader issues although concern with deceptive communication remains a salient topic in the literature. Consequently researchers have been concerned to identify which communication channels are most important in both truthful and deceitful communication. Mehrabian (1972) has suggested that the cumulative evidence points to a relative weighting approximated in the equation:

$$\text{Total} = 0.07 \text{ verbal} + 0.38 \text{ vocal} + 0.55 \text{ facial}$$

This is a somewhat crude linear model but it captures an important feature of the importance of nonverbal behaviour in human dialogue. Whereas the verbal component is used principally to communicate matters of interest the nonverbal component is concerned with the broader interpersonal meaning associated with the verbal exchange. In judging deception, for example, it is often not what one says but how one says it that is important.

A large number of studies have investigated the relative contributions of information derived from the face (e.g., Mehrabian & Ferris, 1967), speech content (Krauss, Apple, Morency, Wenzel & Winton, 1981) and vocal quality (Zuckerman, Amidon, Bishop & Pomerantz, 1982). Combinations of speech, face and body posture have also been investigated (Ekman, Friesen, O'Sullivan & Scherer, 1980; O'Sullivan, Ekman, Friesen & Scherer, 1985). The evidence suggests the relative importance of nonverbal and verbal behaviour is a complex function of the attributes being rated, the information channels available and the emotional state and truthfulness of the person being rated. It is not simply that one kind of variable is invariably superior to another. For example, non-verbal behaviour appears to play a more important role in judges' attributions when the person being observed is speaking frankly, than when that person is being deceptive. O'Sullivan et al. (1985) have summarized the interrelationship thus: "What you say is an important source of information about you when you're lying about bad feelings, but not when you're telling the truth about good feelings" (p.60). In contrast, when deception does not entail a high level of affect judges find it difficult to discriminate truthful from deceptive answers at above chance level (Ekman, Friesen & Scherer, 1976).

Ekman and Friesen (1969, 1974, 1982) hold that control of facial expression is essential in effective deception. When people attempt to be deceptive they invest considerable effort in managing facial activity and often this is effective. In contrast, when a person speaks truthfully the face is a particularly important source of information. However, concentration of effort on facial management may lead to unintended leakage in other channels. Deception may be revealed in poor management of limb movement and posture. However, the evidence is equivocal. It seems that, in some cases, observers may ignore 'leakage' cues because the level of inconsistency between speech content, facial expression and body movements is too difficult to manage simultaneously. In general it seems that when there are inconsistencies between the verbal and nonverbal channels the verbal message is likely to be outweighed only by the entire nonverbal message, rather than a single channel (Mehrabian & Wiener, 1967).

6. Channel Availability

When interactants have a restricted range of channels available to them a number of consequences follow. Morley and Stephenson (1969, 1970, 1977) found that eliminating visual contact between interactants produces more task-oriented dialogue but it also decreases conversational spontaneity and decreases the probability of achieving a negotiated solution acceptable to both parties (Rutter & Robinson, 1981). Surprisingly, it is not the availability of visual contact *per se* that is important but the opportunities such contact affords interactants to exploit a larger sample of social cues (Rutter, Stephenson & Dewey, 1981). However, there is an optimal distance for eye-to-eye communication. Close physical proximity increases opportunities for using gaze as a communication channel but it is also associated with higher levels of stress (Lewis & Fry, 1977). Nonverbal channels interact in complex, non-combinative ways. For example, Mehrabian (1972) combined positive and negative verbal messages with positive and negative vocalizations. Inconsistent negative messages (positive verbal, negative vocal) produced a perceived negative effect overall. Inconsistent positive messages (negative verbal, positive vocal) produced an overall positive effect.

Failure, viewed from a nonverbal perspective, is associated with the management of a personal presentation. When failures occur people are highly motivated to make effective repairs because the failure is often seen as a personal reflection on their communicative competence. Viewed in this way Goffman (1971) has described a repair process which he has termed 'remedial interchanges'. These involve the remediation of failures through, for example, account giving, apologies and requests to violate a rule before doing so. In these circumstances the failure is used to illuminate a point in the conversation. Lead-ins such as 'I know it sounds odd but ...' and 'You're not going to believe this but ...' reveal how interactants may use failures as a communication device by ensuring that the failure is not construed by the listener as a poor self-presentation. This has important implications for the comparative analysis of failure in interpersonal and human-computer interaction. Linguistic analyses are unlikely to reveal the different meanings experienced within the two contexts. At an interpersonal level dialogue failure is experienced as a failure in self-presentation and interactants are motivated to retain a positive image and to avoid seeming incompetent. In human-computer interaction failures are less likely to have this meaning. Users will usually be less concerned with presenting a positive, competent performance before the computer and they are likely to experience failures and repairs differently. A qualification here is that users who find a particular exchange with a computer convincing may temporarily respond to the computer as if it were another person (Scheibe & Erwin, 1979). This is an indication of good dialogue design but its success is also its weakness because users are likely to hold unreasonable expectations of the computer's capabilities. Sophisticated user interfaces are often characterized by their breadth of coverage of human factors and cognitive ergonomic variables but they tend to lack sufficient depth to sustain high level problem-solving and interaction with the user. The empirical evidence considered in this section suggests that accommodating forms of nonverbal communication in user-computer interfaces will tend to increase message redundancy, increase users' sense of involvment in the interaction (and thereby help the user to sustain higher levels of motivation) and encourage more spontaneous and creative dialogue.

The SRI command and control workstation (CCWS) project is an example of a computer-based multi-media information system which attempts to accommodate a range of user-input while protecting the user and the computer from high rates of interactional failure (cf. Poggio, Aceves, Craighill, et al., 1985). The goal of the system is to facilitate access to databases through natural language queries and graphic data representations. The design of the CCWS is based on four premises. First, the system should allow users to access, process and exchange information in several media, singly and in combination. Second, the system should support real-time conferencing. Third, the system should use autonomous workstations so that interaction failures can be isolated and repaired locally. Fourth, the system should provide intelligent assistance across a wide range of communications tasks. At the level of the user interface these premises are incorporated in four goals. (1) Comprehensiveness: The user should have access to a comprehensive range of facilities, but shielded from excessive exposure. (2) Consistency: Tools should perform similarly in different media (e.g., 'delete' characters and mouse buttons). (3) Naturalness: In order to facilitate effective dialogue it should be possible to use pointing, synthetic language and natural language interchangeably as appropriate. (4) Individualization: Interaction should be customized to individual demands and to changing demands. The success of the project hinges on the flexibility and adaptability of the user-computer system. The emphasis is on interchangeability and flexibility so that users choose solutions appropriate to task demands. In this respect the CCWS mimics an important feature of interpersonal dialogue: the availability of a variety of communication modes which can be used to predict and avoid failure or diagnose and repair failure. The provision of an adaptive ('customized') interface allows both the computer and the user to learn from failures and use this information for future interactions. A fundamental issue in the design of multi-media interfaces is that they should accommodate concurrent processing in several channels. This issue has both hardware and software implications. In considering the relationship between deixis in speech and gesture it was pointed out that each system appears to operate in a ballistic fashion, once plans have been co-ordinated centrally. Currently systems such as the CCWS are restricted by hardware limitations on parallel processing and limitations on the user to work with a number of discrete input media (e.g., users find it difficult to use both a mouse and keyboard simultaneously). The potential of gestural input lies in the opportunity it may afford users to use several input media simultaneously and in cumulative interaction.

A consideration of the potential for extending person-computer interaction to incorporate gestural components must tackle a number of fundamental considerations. The first relates to the general nature of gesture. Gesture is a continuous and highly complex phenomenon which cannot be formally defined and is not computationally tractable. Thus, any attempt to accommodate gestural input will be only partially successful. The second relates to the nature of gestural input in particular. A fundamental feature of nonverbal behaviour, including gesture, is that it is encoded and decoded unconsciously. Awareness of activity at the level of nonverbal behaviour usually disrupts a smooth conversational flow. For example, talking about eye-contact patterns will almost certainly disrupt both the patterning of gaze and the verbal flow. A paradox of gestural input is that to work effectively users should not be aware that they are using gestures (they should occur 'naturally') nor that the computer is responding to their gesticulation. However, such an activity (gesticulating to a computer) seems 'unnatural' and sub-conscious awareness that the computer is 'watching' the user, however unobtrusively that may be, would almost certainly disrupt interactions which might otherwise have gone smoothly. It could be argued, therefore, that analysis of human dialogue and understanding of human-computer interaction have little to do with one another (cf. Dreyfus, 1979). However, Winograd (1980) has described some ingredients for a "calculus of language acts" which would accommodate parallel and transportable research in both 'phenomenic domains'. Whether it is possible to identify the ingredients for a calculus of gestural acts is not clear but efforts, albeit uncoordinated, are being made in this direction.

7. Gestural Input: Examples

Gestural input is associated with the decoding of nonverbal acts encoded by a user. There are two main design considerations: the nature of the gestures to be input and the form of the input. One of the simplest systems available is micro-computer based and is intended for the non-vocal, severely physically handicapped person (Gravill, Griffiths, Potter & Yates, 1985). The system measures the electrical potential between the cornea and retina of the eye by means of two surface electrodes placed horizontally adjacent to the lateral aspect of each eye. Two switch functions are produced, one when the eyes move to the left and the other when they move to the right. From the point of view of the handicapped user the effort required to input alphanumeric characters is acceptable. However, even if the system could be modified and extended to make it cost effective for a non-handicapped person the surface electrodes would probably be sufficiently intrusive to discourage users from exploiting its potential.

An alternative approach is to use the infra-red limbus reflection method. This uses an infra-red light source to detect direction of gaze and therefore overcomes the problems associated with surface electrodes. The system is based on the hypothesis that what users look at is an indicator of what they are interested in. With such a system it is necessary to build a complex range of software filters to discriminate between saccadic eye-movements, random search and concentrated gazing. Even so there is no guarantee that what users are looking at necessarily reflects what they are seeing. More generally, the retinal image bears an indirect and often poor relationship to the user's perceptual image or subjective perceptual field. Nevertheless the conception of the system and its implementation indicates that the potential for utilizing gestural input has yet to be realized. However, probably the largest worry for systems using infra-red limbus reflection concerns the as yet unknown consequences of long-term exposure of the eyes to focussed infra-red light.

Eye-gaze is only one kind of nonverbal behaviour. Gestural input using more gross body movements has also been experimented with. Probably the best known example is the 'Put that there' system which uses gesture as part of a command language (Bolt, 1980). In this system a sensor on the user's wrists acts as a pointing device and the spoken command 'put that there' relies on parallel processing of speech and gestural input. More complex systems have also been developed (Purcell, 1985) although the impetus for these arises not from a wish to develop systems capable of decoding human movement in real time but rather from a need to improve cost-efficiency in gesture encoding. Graphical animation systems which stimulate human movement (Herbison-Evans, 1982; Calvert, Chapman and Patla, 1982; Korein and Badler, 1982) and facial gestures (Parke, 1982) require vast investments of time and effort in order to produce high fidelity simulations. Purcell (1985) has described a body tracking system which, in conjunction with a conventional camera, is used to track the motion of light emitting diodes distributed across the surface of a specially prepared garment. The method, called 'scripting-by-enactment', requires the animator to enact the movement s/he wishes to simulate and the system generates an image of that movement for inclusion in the animated sequence. The system is still at an experimental stage and is too intrusive and costly to be used in a majority of commercial environments. However, a scaled version of the system might be useful in flexible manufacturing systems which require complex interaction between intelligent robots and human co-workers. More generally, the system is valuable in providing a database of human movements and gestures. Such a database would provide a useful test-bed for the hypothesis that there are neurally-based synergistic components of action used in the coordination of some intentional actions (cf. Lee, 1984). This hypothesis, if supported, would provide a firm basis from which to investigate further the possibilities for computer detection and classification of intentional action sequences.

The most experimental and costly systems for the detection and decoding of human movement use as input television quality images. They can be considered highly experimental because optimal parallel architectures for this form of image processing have yet to be described (cf. Yalamanchill & Aggarwal, 1985). Such systems are costly because the computational demands associated with their use are very high (cf. Cheng & Huang, 1984; Radig, 1984; Nagao, 1984). However, a number of experimental studies have indicated that unobtrusive image sequence analysis of real world human motion is a feasible direction for research and development (cf. Badler & Smolier, 1979; O'Rourke & Badler, 1980; Webb and Aggarwal, 1981; Akita, 1984; Tsukiyama & Shirai, 1985). Aleksander (1983) has described a system (WISARD) developed at Brunel University which is capable of near-real-time detection of changes in facial expressions and has been used successfully to bring about branching in a programme.

A majority of first-generation image-processing systems comprise four main parts: an input transducer, a frame store, a feature extractor and a classifier. The input transducer, usually a good quality television camera, is normally capable of producing 512 x 512 picture joints or pixels every 1/25th of a second and each pixel is given an 8-bit code. Thus, for each frame 256K of memory is required. High frame acquisition rates place very heavy memory loads on the system. Fortunately it is usually possible to sample frames at a slower rate and temporarily place the digitized image in a frame store, thereby reducing the memory requirement. The feature extractor is a programme which detects and extracts pre-defined geometrical features from an image held in the frame store. Extraction of simple features can usually be accomplished within the frame store (e.g., detection of angles and edges). The classifier section performs more higher-level analyses. Classification is costly because each pass on the data vector (the digitized image) entails a calculation on 256K data points. Currently high-level classification is extremely slow but prospects for close to real-time classification at moderate levels of complexity are good (cf. Buxton & Wiejak, 1985). However, such systems require mainframe support. High-level classification is accompanied by substantial database requirements. Knowledge-based systems for the analysis and classification of images require very large amounts of domain specific knowledge (Neumann, 1985) because images underconstrain the scenes they represent (cf. Tsotsos, 1984). However, the development of sophisticated image processing systems will have to move in the direction of knowledge-based architecture if they are to operate at acceptable speeds. But as Sloman (1983) has pointed out, we need to know a good deal more about what should go into these databases — more about what needs to be represented and how it can be represented usefully.

8. Gestural Input: Prospects and Conclusions

The potential for unobtrusive gestural input is tied to a number of issues. The most important of these relates to the storage and access of image-gesture databases. Kawai and Tamura (1985) tackled this in a different context, namely a PC based sign language generation system. Their system receives keyboard and speech input and converts this to sign language or finger spelling which is displayed on a CRT in real time. The system stores 40 sign language patterns and 50 finger spellings on two floppy disks. Accessing and displaying the appropriate pattern is fast because each sign pattern can be described in terms of four sub-patterns and this information is used to structure the database and to generate the image. However the system requires further development if the size of the database is to be expanded significantly. Specifically, the development of appropriate database architectures coupled with better retrieval techniques (e.g., Lee, 1980) must be a priority.

Architectural and processing advances may be accomplished by considering the potential for the development of 'grammars' of action and gesture. 'Grammars' of action, based on pushdown automata, have been formulated (Skvoretz & Fararo, 1980; Skvoretz, 1984), but further research is required to determine the limitations as well as the advantages associated with knowledge-based representation derived from such formulations. The future is encouraging. For example, Arlinghaus (1985) has achieved a semi-formal description of the structure of nonverbal communication through eye-contact using graph-theoretic tools and theorems.

The emphasis in this chapter has been on gesture decoding. However the encoding-decoding distinction introduces arbitrary and unhelpful discontinuities in understanding processing features of nonverbal behaviour in dialogue. For example, in considering the role of nonverbal behaviour in human computer interaction it was shown that advances in image decoding have been directly linked to the needs of animators to encode human movement in real-time. The encoding and decoding of gestures at the user interface present different kinds of problems, however. One needs to distinguish between functional and structural features of gestures. An examination of the role of nonverbal behaviour in dialogue may reveal important functions of that system which could usefully be incorporated into person-computer interaction. However, the implementation of functional equivalence should not be tied to a need to produce morphologically similar devices. For instance, pointing can be achieved in a number of ways: it is more important to understand the function of deixis rather than to concentrate on sophisticated simulation. Nonverbal behaviour is concerned with regulating and timing dialogue and with providing an interpersonal 'commentary' on the vocal interaction. Taking account of these functions implies a need to provide concurrent feedback on user-computer interaction in order to facilitate better interaction. Whether the feedback looks like nonverbal behaviour is irrelevant.

Although the problems associated with accommodating gestural input at the user-machine interface have been identified it has been assumed that these are surmountable. However, solutions will require a strategic, multi-disciplinary response. Working in isolation, disciplines will arrive at isolated and partial solutions and claims of success are likely to be exaggerated; problems which are eliminated in one domain may surface in other domains. The difficulties are not only computational. A more thorough understanding of the inter-relationships between language, thought and gesture is required in order to plan for efficient and effective interaction with human users; the development of multi-media systems hinges on this. The need for such understanding identifies a related need, namely that for fundamental research which is informed by the nature of the problem, rather than theoretical curiosity.

References

Akita, K. (1984). Image sequence analysis of real world human motion. *Pattern Recognition, 17,* 73-83.
Aleksander, I. (Ed.) (1983). *Artificial vision for robots.* London: Kogan Page.
Argyle, M. (1975). *Bodily communication.* New York: International University Press.
Arlinghaus, S.L. (1985). Eye-contact graphs. *Behavioral Science, 30,* 108-117.
Austin, J.L. (1962). *How to do things with words.* Cambridge MA: Harvard University Press.
Badler, N.I., & Smolier, S.W. (1979). Digital representation of human movement. *Association for Computing and Machine Computation Survey, 11,* 19-38.
Bates, E., & Volterra, V. (1984). On the invention of language: An alternative view. *Monographs of the Society for Research in Child Development, 49,* 130-142.
Becker, M.H. (1970). Sociometric location and innovativeness: Reformulation and extension of the diffusion model. *American Journal of Sociometry, 35,* 267-282.
Bolt, R.A. (1980). Put that there: Voice and gesture at the graphics interface. *Computer Graphics* ACM 0-8971-021-4/40/007-262, 262-270.
Buck, R. (1982). Spontaneous and symbolic nonverbal behavior and the ontogeny of communication. In R.S. Feldman (Ed.), *Development of nonverbal behavior in children.* New York: Springer-Verlag.
Buxton, H., & Wiejak, J. (1985). Towards computer vision. In P.J. Offen (Ed.), *VLSI image processing.* London: Collins.
Calvert, T.W., Chapman, J., & Patla, A. (1982). Aspects of the kinematic simulation of human movement. *IEEE Computer Graphics and Applications, 2,* 41-50.
Cheng, J.K., & Huang, T.S. (1984). Image registration by matching relational structures. *Pattern Recognition, 1,* 149-159.
Condon, W.S., & Ogston, W.D. (1971). Speech and body motion synchrony of the speaker-hearer. In D.L. Horton and J.J. Jenkins (Eds.), *Perception of language.* Columbus, Ohio: Merrill.
Curtiss, S. (1977). *Genie: A psychoanalytic study of a modern-day "wild-child".* New York: Academic Press.
Dance, F.E., & Larson, C.E. (1972). *Speech communication: Concepts and behavior.* New York: Holt, Rinehart, & Winston.
Dreyfus, H. (1979). *What computers can't do: A critique of artificial reason.* San Francisco: Freeman.
Edwards, D. (1978). Social relations and early language. In A. Locke (Ed.), *Action, gesture and symbol: The emergence of language.* New York: Academic Press.
Ekman, P. & Friesen, W.V. (1969). Nonverbal leakage and clues to deception. *Psychiatry, 32,* 88-105.
Ekman, P. & Friesen, W.V. (1974). Detecting deception from the body and face. *Journal of Personality and Social Psychology, 29,* 288-298.
Ekman, P. & Friesen, W.V. (1982). Felt, false and miserable smiles. *Journal of Nonverbal Behaviour, 6,* 238-252.
Ekman, P., Friesen, W.V., & Scherer, K. (1976). Body movement and voice pitch in deceptive interaction. *Semiotica, 16,* 23-27.
Ekman, P., Friesen, W.V., O'Sullivan, M., & Scherer, K. (1980). Relative importance of face, body and speech in judgments of personality and affect. *Journal of Personality and Social Psychology, 38,* 270-277.
Gaines, B.R. & Shaw, M.L.G. (1984). *The art of computer conversation.* London: Prentice Hall.
Goffman, E. (1971). Remedial interchanges. In E. Goffman (Ed.), *Relations in public microstudies of the public order.* New York: Harper and Row.
Goldin-Meadow, S. (1979). Structure in a manual communication system developed without a conventional language model: Language without a helping hand. In H. Whitaker & H.A. Whitaker (Eds.), *Studies in neurolinguistics* (volume 4). New York: Academic Press.
Goldin-Meadow, S. (1982). The resilience of recursion: A study of a communication system developed without a conventional language model. In L.R. Gleitman & E. Wanner (Eds.), *Language acquisition: The state of the art.* New York: Cambridge University Press.
Goldin-Meadow, S. & Feldman, H. (1975). The creation of a communication system: A study of deaf children of

hearing parents. *Sign Language Studies, 8*, 225-234.

Goldin-Meadow, S. & Mylander, C. (1984). Gestural communication in deaf children: The effects and non-effects of parental input on early language development. *Monographs of the Society for Research in Child Development, 49*, 1-121.

Gravill, N., Griffiths, P.A., Potter, R., & Yates, A. (1985). Eye control of microcomputers. *Computer Bulletin, 1*, 15-16.

Harre, R. (1985). Situational rhetoric and self-presentation. In J.P. Forgas (Ed.), *Language and social situation*. New York: Springer.

Herbison-Evans, D. (1982). Real-time animation of human figure drawings with hidden lines omitted. *IEEE Computer Graphics and Applications, 2*, 27-33.

Kawai, H. & Tamura, S. (1985). Deaf-and-mute sign language generation system. *Pattern Recognition, 18*, 199-205.

Kendon, A. (1980). Gesticulation and speech: Two aspects of the process of utterance. In M.R. Key (Ed.), *Nonverbal communication and language*. The Hague, Mouton.

Kendon, A. (1983). Gesture and speech: How they interact. In J.M. Wieman & R.P. Harrison (Eds.), *Nonverbal interaction*. California: Sage.

Korein, J.U. & Badler, N.I. (1982). Techniques for generating the goal-directed motion of articulated structures. *IEEE Computer Graphics and Applications, 2*, 71-82.

Krauss, R.M., Apple, W., Morency, N., Wenzel, C., & Winton, W. (1981). Verbal, vocal and visible factors in judgments of another's affect. *Journal of Personality and Social Psychology, 40*, 312-320.

Lee, E.T. (1980). Similarity retrieval techniques. In G. Goos & J. Hartmanis (Eds.), *Lecture notes in computer science: No. 80 pictorial information systems*. New York: Springer-Verlag.

Lee, W.A. (1984). Neuromotor synergies as a basis for coordinated intentional action. *Journal of Motor Behavior, 16*, 135-170.

Levelt, W.J.M., Richardson, G., & Heij, W.L. (1985). Pointing and voicing in deictic expressions. *Journal of Memory and Language, 24*, 133-164.

Lewis, S.A. & Fry, W.R. (1977). Effects of visual access and orientation on the discovery of integrative bargaining alternatives. *Behavior and Human Performance, 20*, 75-92.

Lewis, M. & Rosenbaum, L.A. (1977). *Interaction, conversation, and the development of language*. New York: Wiley.

McNeil, D. (1979). *The conceptual basis of language*. Hillsdale, NJ: Erlbaum.

McNeill, D. (1981). Action, thought and language. *Cognition, 10*, 201-208.

McNeill, D. (1985). So you think gestures are nonverbal? *Psychological Review, 92*, 350-371.

Mead, M. (1934). *Mind, self and society*. Chicago: Chicago University Press.

Mehrabian, A. (1972). *Nonverbal communication*. Chicago: Aldine-Atherton.

Mehrabian, A. & Ferris, S.R. (1967). Inference of attitudes from nonverbal communication in two channels. *Journal of Consulting Psychology, 31*, 248-252.

Mehrabian, A. & Wiener, M. (1967). Decoding of inconsistent communications. *Journal of Personality and Social Psychology, 6*, 109-114.

Morley, I.E. & Stephenson, G.M. (1969). Interpersonal and interparty exchange: A laboratory simulation of an industrial negotiation at plant level. *British Journal of Psychology, 60*, 543-545.

Morley, I.E. & Stephenson, G.M. (1970). Formality in experimental negotiations: A validation study. *British Journal of Psychology, 61*, 383-342.

Morley, I.E. & Stephenson, G.M. (1977). *The social psychology of bargaining*. London: Allen & Unwin.

Nagao, M. (1984). Control strategies in pattern recognition. *Pattern Recognition, 1*, 45-56.

Neuman, B. (1985). Vision systems: State-of-the-art and prospects. In T. Bernold & G. Albers (Eds.), *Artificial intelligence: Towards practical application*. Amsterdam: North-Holland.

Newport, E.L., Gleitman, H., & Gleitman, L. (1977). Mother I'd rather do it myself: Some effects and non-effects of maternal speech style. In C.E. Snow & C.A. Ferguson (Eds.), *Talking to children*. New York: Cambridge University Press.

O'Rourke, J. & Badler, N.I. (1980). Model-based image analysis of human motion using constraint propagation. *IEEE Transactions on Pattern Analysis and Machine Intelligence, PAMI-2*, 522-536.

O'Sullivan, M., Ekman, P., Friesen, W., & Scherer, K. (1985). What you say and how you say it: The contribution of speech content and voice quality to judgement of others. *Journal of Personality and Social Psychology, 48*, 54-62.

Parke, F.I. (1982). Parameterized models for facial animation. *IEEE Computer Graphics and Animation, 2*, 61-68.

Pinsky, L. (1983). What kind of "dialogue" is it when working with a computer? In T.R.G. Green, S.J. Payne,

& G.C. van der Veer (Eds.), *Psychology of computer use.* London: Academic Press.

Poggio, A., Aceves, J.J.C.L., Craighill, E.J., Moran, D., Aguilar, L., Worthington, D, & Hight, J. (1985). CCWS: A computer-based multimedia information system. *IEEE Computer, November,* 92-103.

Purcell, P. (1985). Gestural input to interactive systems. *Computer Bulletin, 1,* 3-4 & 7.

Radig, B. (1984). Image sequence analysis using relational structures. *Pattern Recognitoin, 1,* 161-167.

Ruesch, J. & Bateson, G. (1951). *Communication: The social matrix of psychiatry.* New York: Mouton.

Rutter, D.R. & Robinson, B. (1981). An experimental analysis of teaching by telephone: Theoretical and practical implications for social psychology. In G.M. Stephenson & J.H. Davis (Eds.), *Progress in applied social psychology,* (volume 1). Chichester: Wiley.

Rutter, D.R., Stephenson, G.M., & Dewey, M.E. (1981). Visual communication and the content and style of conversation. *British Journal of Social Psychology, 20,* 41-52.

Scheibe, K.E. & Erwin, M. (1979). The computer as altar. *Journal of Social Psychology, 108,* 103-109.

Schramm, W. (1963). *The science of human communication.* New York: Basic.

Searle, J.R. (1970). *Speech acts.* Cambridge MA: Cambridge University Press.

Shannon, C.E. & Weaver, W. (1949). *The mathematical model of communication.* Urbana, IL: University of Illinois Press.

Skvoretz, J. (1984). Language and grammars of action and interaction: Some further results. *Behavioral Science, 29,* 81-97.

Skvoretz, J. & Fararo, T.J. (1980). Language and grammars of action and interaction: A contribution to the formal theory of action. *Behavioral Science, 25,* 9-22.

Sloman, A. (1983). Image interpretation: The way ahead? In O.J. Braddick & A.C. Sleigh (Eds.), *Physical and biological processing of images.* New York: Springer-Verlag.

Smith, H.T. (1980). Human computer communication. In H.T. Smith & T.R.G. Green (Eds.), *Human interaction with computers.* London: Academic Press.

Trevarthen, C. & Hubley, P. (1978). Secondary intersubjectivity: Confidence, confiding and acts of meaning in the first year. In A. Locke (Ed.), *Action, gesture and symbol: The emergence of language.* London: Academic Press.

Tsotsos, J.K. (1984). Knowledge and the visual process: Content, form and use. *Pattern Recognition, 17,* 13-27.

Tsukiyama, T. & Shirai, Y. (1985). Detection of the movements of persons from a sparse sequence of TV images. *Pattern Recognition, 18,* 207-213.

Vygotsky, L.S. (1966). Development of the higher mental functions. In *Psychological research in the USSR.* Moscow: Progress.

Webb, J.A. & Aggarwal, J.K. (1981). Structure from motion of rigid and joined objects. *Proceedings of the 7th IJCAI,* 686-691.

Westley, B.H. & Maclean, M.S. (1957). A conceptual model for communication research. *Journalism Quarterly, 34,* 31-38.

Winograd, T. (1980). What does it mean to understand language? *Cognitive Science, 4,* 209-211.

Yalamanchill, S. & Aggarwal, J.K. (1985). Analysis of a model for parallel processing. *Pattern Recognition, 18,* 1-16.

Zuckerman, M., Amidon, M.D., Bishop, S.E., & Pomerantz, S.D. (1982). Face and tone of voice in the communication of deception. *Journal of Personality and Social Psychology, 32,* 347-357.

7

A SOCIO-LINGUISTIC PERSPECTIVE

NEGOTIATION AND BREAKDOWN IN
SPEECH EVENT CONSTRUCTION

John Wilson

Department of Communication*
University of Ulster

Despite a growing multi-disciplinary interest in the study of conversation there is not as yet any clear definition of exactly what a 'conversation' is. The aim of this paper is to present a model which allows us to delimit conversation as an individual speech event. We argue that conversation is but one kind of speech event among others found in everyday talk, and that consequently participants involved in everyday talk must have ways of recognising when they are or are not involved in conversation. Principles for the negotiation of conversational talk are analysed in relation to actually recorded everyday speech, with consideration given to degrees of success or failure in the negotiation process.

1. Introduction

Despite a growing interest in the study of conversation little or no attempt has been made to specify the exact nature of *conversation* as a speech event. Prevalent has been the intuitive assumption that one need only record samples of everyday interaction in order to gain access to conversational materials. This is, of course, to some extent true; however, the reality is that any situation, formal or informal, is 'event variable'. By *event variable* I simply mean that in any interactive situation a variety of different speech events may be taking place, interacting with and overlapping each other. Hymes (1972) made this point some time ago. He argued that conversation (or any other event type) is not wholly constrained by context but rather by rules for the performance of that event. A conversation could arise during a church sermon, in a courtroom, or in any other *formal* context. Equally, it is therefore theoretically feasible that formal speech event types could arise in *informal* contexts; one could have interviews, interrogations, perhaps even sermons and cross examinations. A *speech event*, as the term is used here, relates to rules of speaking not to the context itself. Consequently, to say that conversation is that type of speech event which arises in informal contexts tells us very little about conversation, since other speech event types may also arise in such contexts.

What is required is a model of conversation which specifies the rules for the performance of conversation, and elsewhere I have argued that such a model may be predicated upon a basic principle that conversation is that speech event where an *equality of speaker rights* is maintained (see Wilson & Gunn, 1983; Wilson, 1985a, 1985b; see also below). In this paper I want to draw on this basic heuristic principle in order to distinguish conversation from some other speech event

---

* Currently on sabbatical leave at Department of Speech Communication, Southern Illinois University, (at Carbondale).

types which may also be found in informal contexts. Since we define conversation here in relation to an equality of distribution in speaker rights, it follows that any speech event which does not accord with this principle is not conversation.

In distinguishing conversation from other speech events, the aim is to indicate how conversation differs from those other speech events found in informal contexts. Clearly this is an important exercise, as any study of everyday interaction must be capable of distinguishing what is conversational from what is not. A failure to do this will make it unclear whether any findings which emerge apply to conversation or to some other speech event.

The main aim of this paper will be to answer the question of how participants involved in everyday interaction actually know, at any given moment in the on-going talk, which speech event they are involved in. Using our model of conversational behaviour we will analyse how participants move their talk from one speech event type to another; a process which will be referred to as *out-moding*. We are not only interested, however, in the successful process of out-moding but also in the fact of a breakdown or misinterpretation of out-moding procedures occurring. By focusing on breakdown we hope to further elucidate the nature of out-moding itself, that is by attempting to understand what is taken for granted when the process is successful (Gumperz & Tannen, 1979; Milroy, 1984).

Before we proceed any further, however, it is necessary for us to get a clearer picture of what exactly is meant by a model of conversation predicated upon a principle of equality in speaking rights.

2. Conversation: Delimitation According to Speaker Rights

Work within interpretive sociolinguistics has so far distinguished (qualitatively: see Gumperz, 1982) several different kinds of speech event: interviews (research or employment), committee meetings, lectures, seminars, courtroom encounters and so on. One feature which is utilised in the description of all of these various speech event types is the relative distribution of 'speaking rights'. In the interview it is the interviewer who asks the questions, in the classroom the teacher controls the topic, and in the committee meeting conventionally understood rules for the order and selection of speaking turns are adhered to.

Now one thing most analysts of everyday talk are agreed on is that conversation is located at the end of a continuum of informal interaction types. Consequently, it makes sense to assume that within conversation formal procedures for the control of ongoing talk would be at a minimum, if not, indeed, non-existent. Therefore, this leads us to the conclusion that within conversation speaker rights must be distributed equally.

It is important to stress here that we are not suggesting that participants involved in conversation have, or contribute, an equal number of speaking turns; rather we are arguing that any individual has an equal right to initiate talk, interrupt, respond, question or refuse to do any of these. We are not claiming, of course, that there are no controls available or employed within conversation, or that individuals in other speech events cannot initiate, respond etc. There is however, a difference in the relative distribution of the right to carry out such actions. For example, in the classroom pupils may initiate talk, or even interrupt the teacher's talk, but the pupil's right to do this is controlled within the speech event. Pupils cannot simply interrupt whenever they want to, their contribution must be seen as relevant to the ongoing talk: this relevance is assessed not by the class as a whole but by one individual, the teacher. In the classroom speaker rights are highly constrained, they are distributed only, or mainly, through one participant. In conversation on the other hand, no one individual is given such rights, these rights are accessed on an equal basis.

Speaker rights may sound, at first, remarkably similar to *speaker turns*, speaker turns being defined in terms of the turn taking model developed by Sacks, Schegloff, and Jefferson (1974, 1978). The basic rules of this model may be summarised as follows (see appendix for notational conventions):

(1) a. Current speaker selects next
    b. Next speaker self selects
    c. Current speaker self selects and continues

Despite a superficial similarity between speaker turns and speaker rights, speaker rights differ from speaker turns in several respects. Fundamentally, speaker turns are controlled by the principles of the turn-taking system; speaker rights, on the other hand, are controlled within (and at the same time contribute to) the structure of the speech event itself. We can explain this difference more clearly by considering the ramifications of silence (a lack of talk) within speech events.

The distribution of silence within ongoing interaction has been discussed mainly by *conversational analysts,* and linked by them to choices within the turn-taking system (Daden, 1975; Levinson, 1983; Schegloff, Sacks & Jefferson, 1977). It is argued, for example, that if rule (1a) of the turn taking system is chosen (current speaker selects next: see above) then any silence following the selection of this rule will be directly attributed to the next speaker so selected. If a speaker selects a next, by using a question for example, and no answer is forthcoming then the resultant silence is the responsibility of that individual who did not respond. The distribution of silence within the turn-taking system has communicative consequences. A refusal to answer may indicate a lack of knowledge, a lack of approval for the question, an attitude adopted towards the questioner and so on (see Levinson, 1983, ch. 3; Daden, 1975). The distribution of silence may also, however, be understood outside the turn taking system and related specifically to the distribution of speaker rights within the construction of speech events.

If we look at the speech event of interviewing for example, there is general agreement that the pattern of talk involves the interviewer questioning and the interviewee answering. It was noted above that within the turn-taking system the interviewer's questions would expect following talk (questions acting to select the next speaker); but it might also be argued that within the speech event of interviewing the interviewee's answers also expect following talk, the completion of an answer being a signal for the next question. At the level of speech event construction the answer acts to initiate the next turn. Consequently, within the interview silence may be said to be distributed in two senses. Firstly, in terms of the system of turn talking, where a question acts to select a next speaker, and secondly, in terms of the speech event itself, where an answer signals a next question.

One objection to this interpretation is that since an answer selects the next speaker it is simply a special case of rule (1a) of the turn-taking system. If this is so it is a special case because it occurs within a particular kind of speech event. An answer will not select a next speaker in all speech events, therefore it is the structural expectations of the speech event which create the special case. It makes sense, then, to treat the distribution of speaker rights in such cases as a manifestation of the speech event and not the turn-taking system.

The argument is that the distribution of talk (and therefore of attributable silence) resides both within the turn-taking system and within the speech event (that is according to the conventional distribution of rights to speak within particular structural contexts).

Our argument has been that speaker rights differ from speaker turns, and that the distribution of speaker rights varies with the kind of speech event underway. Consequently, speaker rights (and their relative distribution) act as a heuristic in deciding which type of speech event is taking place. It is for this reason that we make use of the notion of speaker rights in describing the individual nature of conversation as a speech event, a speech event in which there is an equality in the distribution of speaker rights.

This, then, is the broad outline of the model of conversation we will draw on in describing the distribution of speech events within informal talk. Let us return, therefore, to our main task of locating and describing the ways in which talk is moved between speech events, the process we have referred to as out-moding.

3. Out-Moding Talk

An out-mode will be defined as any utterance which differs in structure (or content) from the ongoing speech event established by previous or surrounding talk. The function of the out-mode is to change the status or definition of the developing speech event. Two main types of out-mode will be distinguished here: *institutionalised out-modes* and *conversationally tried out-modes*. Institutionalised out-modes establish or re-establish a particular power or status relationship between both speaker and hearer. This power or status differential is institutionally recognised; for example, teacher/pupil or officer/soldier. Conversationally tried out-modes introduce speech events which, although differing from conversation (as it is defined in this paper), supplement the flow of conversation as a process of ongoing talk. We are thinking here of such speech events as jokes, narratives, banter (a specific kind of 'teasing' (see below)) and so on.

4. Institutionalised out-modes

In any asymmetrically biased power relationship the participant who holds the position of greatest power may, if that participant so wishes, relax any differential which exists in an attempt to create a conversational context. Now how far one can actually go in removing the institutionalised differences established in certain social contexts is a matter of debate. But what we are interested in are the kinds of structural procedures employed where an individual does wish to remove established institutionalised differences. After all, a teacher may chat with pupils, an officer with soldiers, parents with their children, and so on.

At any point in such a negotiated conversation (negotiated in the sense that both speaker/hearer must agree that they are in conversation) any latent power differential may be reestablished, thus returning the control of talk (to whatever degree) to one individual. Any utterance employed to talk out of the mode of conversation in this way is an *instutionalised out-mode*. Consider the following:

(2) T:FC
    a. D. Jesus that was close
    b. C. Who is that? He's good
    c. R. That's that player what ya ma call him?
    d. D. Some Russian sounding name
    e. R. Prash-prat Pratchkova or something
    f. M. RUSSELL look at the time GET to YOUR BED

This extract is taken from a family conversation. The context is as follows: it is Saturday evening, the family have just finished supper and they are watching a soccer match on television. The talk previous to turn (f) has been generated by the match and reveals an overall conversational pattern (only some of the previous talk is included in our example). Turn (f) represents what we are calling an institutionalised out-mode. It is made up of two command clauses with an initial emphatic stress on *Russell*. The use of command clauses of this type represents a problematic situation in conversational talk. If there is an equality of speaker rights one individual is unlikely to have the right to command others. However, if we treat (f) as an out-mode functioning to establish or re-establish parental power, that is, to justify the right to command Russell to go to bed, then we may account for (f) outside the framework of conversation. Within conversation Russell would have the right to either question or reject command forms such as (f), but where parental power is established and accepted this option is removed. This is not to say that children do not reject parental instructions, of course they do! However, where both parent and child recognise parental power then the rejection of parental commands is unlikely. When commands are rejected it is generally because (under certain circumstances) the child refuses to accept parental power.

(3) T:FC
    D. Where were you last night till twelve o'clock?
    K. Out with X
    D. What doing?

K. Nothing
M. Tell your father what you were doing!
K. It has nothing to do with youse
M. Don't be cheeky

The failure to achieve agreement on parental power, and therefore the parents right to command, may lead to *communicative breakdown,* or to a renegotiation of the parent/child status. The nature of such breakdowns and renegotiations will be discussed below; but for the present we should note that an utterance such as (f) acts to establish a power differential and so moves talk out of the mode of conversation.

A similar example may be given from the classroom context:

(4) T:SC
a. C. Where did you get your hair done Miss
b. T. Down Anne Street ( )
c. J. Are they any good?
d. T. *I* think so ... where do you get yours done
e. J. My sister does it
f. T. Oh (.) I think its nice
g. J. Thanks
h. K. See P's got scalped again
i. J. Aye she looks like a hedgehog (laughs)
j. T. I think it looks alright
k. C. Ya must be joking its awful
l. T. Well ahahah I have seen better (laughs) ...
m. T. Anyway RIGHT back to your SEATS (1.5) OK LISTEN EVERYBODY roll time.

As with our family context an argument can be made that previous to turn (m) talk was in a conversational mode. (m), like (2f), employs a command structure and places an emphatic stress on *right* : (m) functions to establish the teacher's institutionalised status. The interaction transcribed in (4) takes place at the beginning of a morning lesson, we join the talk at a point where all the pupils have not yet arrived in the classroom. While the teacher is waiting she is prepared to relax the status differential between herself and the pupils in order to become involved in conversation. But once all the pupils have arrived the teacher reasserts her position of power by employing (m) as an out-mode. Those pupils who have been involved in converstaion with the teacher now recognise that this speech event is ended, and they return to their seats, ready to take on the role of pupil in a somewhat different kind of speech event. This recognition is dependent on an understanding of the role of the out-mode employed at (m). (m) cannot be accounted for within the framework of conversation since it does not respect an equality of speaker rights: it can be accounted for, however, within the framework of the classroom where the teacher has the power to direct and demand certain behaviours from the pupils.

Of some twenty-five institutionalised out-modes analysed in both home and school contexts, twenty-one were in the imperative mood, with the other four being defined as indirect or embedded commands (see Sinclair & Coulthard, 1975; Labov & Fanshel, 1977). For example:

(5) a. T. Would you like to get started (writing)
b. M. Would you run down to the shops for me

These forms are defined as indirect commands because the option of a negative response is not really open to the hearer, as example (6) shows:

(6) T:SC
T. Would you like to put that on the board Tommy?
To. No not really sir
T. COME ON move it get yourself up to the board

Equally one could imagine the consequences of a refusal to (5b) within a home context:

> (7) M. Would you run down to the shop
>   C. I don't want to
>   M. Well I'm not asking you I'm telling you to get going

Associated with this overwhelming use of command structures as institutionalised out-modes is a lowering of pitch with an accompanying increase in loudness. Further, eighteen of the out-modes analysed contained (as a pre-move) some form of attention getting device such as *right, well, OK*. Since, in most cases, talk is already underway, these attention getting devices serve to pre-mark a change of direction within the talk, not of the ongoing topic but of the speech event.

To summarise then, institutionalised out-modes are defined as being in the imperative mood, prosodically marked by a lower pitch movement, associated with an accompanying increase in loudness, and optionally pre-marked by an attention getting device.

5. Conversationally tied out-modes

Conversationally tied out-modes introduce conversationally related speech events. They are conversationally related in that they do not disrupt participant status, indeed they act in an opposite manner and are potentially integrative within the ongoing talk. For example, participants may (if they so wish) utilise these events to activate further topic frames within the construction of conversational talk. However, these events operate under distinct and separate performance rules of their own, and, consequently, they are considered outside the framework of conversation as it has been defined in this paper.

Many of the events we are calling conversationally tied have, at a common sense level, been treated as being indicative of conversation itself. For example, telling a joke is an activity often associated with conversational activity. But again this is a result of equating context with event. Jokes are associated with informality, informal contexts are associated with conversational talk, therefore jokes are conversational phenomena. However, as we have already argued, conversations as specific speech events may arise in any context: equally, therefore, jokes may also arise in any context. Those events which we discuss in this section do display similar structural qualities to conversation, but they function in quite different ways from conversational talk and must be considered separately from the notion of conversation as a speech event.

In considering conversationally tied out-modes we will look at how jokes and banter are introduced into ongoing talk, and at how this introduction indicates to the participants that they are involved in (or about to become involved in) a speech event outside conversation. These two events have been chosen because their informal character makes them obvious candidates for definition as conversational forms. We will argue, however, that these events function independently in their own right. It is not intended that a detailed analysis of the internal structure of these events should be given, as such an analysis is available elsewhere (Sacks, 1971, 1974; Milroy & Milroy, 1977; Milroy, 1980). The aim is to describe how these events are out-moded, that is, to show how participants recognise they are outside the frame of conversation and how they, accordingly, change their interpretation of certain utterances to fit the frame of the out-moded event.

5.1 Jokes

It is difficult to arrive at a definition of what a joke actually is, since any utterance has the potential to be funny. But the fact that someone does think an utterance is funny does not automatically make it a joke. Of course the situation may be represented the other way round; an utterance intended by (S) to be funny may not be thought of as such by (H); does this mean, then, that it is no longer a joke? Or rather only a joke by definition for (S) but not for (H)? In order to overcome this problem a joke is defined here as any utterance (or combination of utterances) whose internal structure or content may be shown to have been organised for the amusement of others. Admittedly, this definition is somewhat general; however, according to such a definition (H) may not find a particular structure funny, but it will remain a joke where it can be shown that it has been organised for that purpose. Equally, if (H) does find a particular structure funny, and it can be shown that the internal structure or content of the utterance was not intended as such, then it is not a joke. The question is, of course, how does one show that any structure has been organised to function as a joke? Well, as a first stage we can show how it differs from surrounding talk, that is, how (S) indicates that he is in the mode of joke telling.

In order to consider how this is achieved we will analyse two particular joke styles, what I will call the *Q-joke* and the *narrative joke*. These two styles have been selected because they occur most frequently in the data. The Q-joke is defined according to its question/answer format with the answer acting as the *punchline*. The narrative joke involves the telling of a story, making use of specific narrative structures, with the pay off or punchline being a structural consequent of the story, i.e., being dependent on the spatial and temporal organisation of the narrative (see Labov & Waletsky, 1966).

5.2 Q-jokes

The Q-joke is dependent on the principle of the question/answer format, although the production of the joke does not require that these principles be adhered to completely. For example, (S) need not wait for (H) to answer (indeed the joke is based on the premise that (H) cannot answer) and may supply the answer (punchline) himself.

(8) T:FC
    a. P. Why do Irishmen not eat toast?
           Because they lost the recipe.

    T:SS
    b. G. How does an Irishman talk when his brain is taken out?
           With an English accent.

Where (H) does respond to the question it is usually to admit that he does not know the answer, and recognises a punchline is on the way.

(9) T:SS
    a. G. How does an Irishman clean his underpants?
        A. I don't know
        G. Hangs them up and beats the shit-shite out of them

    T:SC
    b. T. What is black and brown and looks well on a catholic?
        D. Go on tell me
        T. An Alsatian

For the Q-joke to stand any chance of *coming-off* it is necessary that (H) recognise that he is not being asked a genuine question. If (H) were to give such an interpretation he might actually attempt to answer the question or query why it has been asked. Such misinterpretations do occur and we will discuss some of these later. However, the very fact that an attempt to answer a Q-joke represents a misinterpretation indicates that these jokes are speech events separate from conversation. Within conversation as a speech event questions are said to presuppose an unknown variable which requires a value (see Wilson, 1982), and it is expected that (H) has the ability to supply a value for this variable; that is, in asking the question (S) believes that (H) has access to information which will allow him to fill in the missing value and thus answer the question. These principles cannot be applied, however, to (8) and (9); indeed the affect of these questions is dependent on the assumption that (H) does not know the answer. In those situations where (H) does know the answer he will know the punchline and therefore the joke, and to this extent the joke will fail:

(10) T:IC
    M.   What do you throw a drowning catholic?
    J.   His wife and family, heard it.
    M.   It's good though ain't it?

How, then, does (S) manage to out-mode talk in order to make sure that participants recognise that no answer is expected? Intonationally and structurally the question element in a Q-joke is basically the same as for normal conversational question forms (see Bolinger, 1978). Where the Q-joke differs is at the propositional level. Within the surface structure of Q-jokes there are clausal elements which are either accepted as *joke generic,* for example, Englishman, Irishman, Scotsman and so on, or which suggest unusual incidents or implausible events:

(11) a.   What did the three foot dwarf say to the prostitute?
     b.   Did you hear about the man with no legs?
     c.   How does a Martian eat his dinner?

Of course it is possible that a Q-joke may be about relatively normal events or even specific individuals:

(12) T:IC
    a.   Have you seen Stevie Wonder's new car? (a blind pop singer)
    b.   No.
    c.   Neither has he (laughter).

However, in examples such as (12), (S) intends the question to be treated as normal; the joke value here comes in the follow up response to the answer, and with this the interaction is retrospectively out-moded. This method of Q-joking is less frequent in that it depends on retrospective out-moding, that is, on an initial treatment of the question as normal. Consequently, it runs a higher risk of being misinterpreted. In the following example the joke depends upon (H) making use of a frame of knowledge which involves a specific individual, and focusing in particular on that individual's reputation as someone who talks a lot about masturbation:

(13) T:SC
    a.   Did you hear about X?
    b.   No.
    c.   He fell down the stairs and broke his pyjamas (laughs).
    b.   What do you mean? Was he hurt?
    c.   No ya silly cunt its a joke.
    b.   I don't get it.
    c.   Forget it the-n.
        (2.0)
    b.   OH AYE a aye I see what ya mean.

Even in these examples, however, although (H) treats the question as a genuine one, (S) is aware that there is no possibility that (H) could have an answer (unless again he knew the joke).

The out-moding of Q-jokes is achieved, then, by asking questions about implausible events or activities, about conventionally recognised joke generic groups (these may even be represented by what seem to be concrete noun forms but which have no specific referent: Rastus (blacks), Murphy (Irishman), or by leading (H) to a retrospective making of what was assumed to be a genuine question.

5.3 Narrative jokes

Narrative style jokes involve the telling of a story. The internal structure of the joke is similar to narratives in general (see Labov & Waletsky, 1966) but again participants are somehow aware that what they are listening to is to be interpreted differently.

Narrative style jokes are frequently marked by an opening question for example:

(14) a. Did you hear the one about
b. Have you heard this one
c. Do you know the one about

These questions are very similar to what I have referred to elsewhere (Wilson, 1985a) as *pre-topic checks*. The function of a pre-topic check is to assess levels of shared knowledge before talk proceeds on a specific topic. The pre-topic check is necessary in that if (H) did not have access to a relevant topic frame then the initiation of ongoing talk on this frame would be difficult, i.e., one cannot talk about a topic without some knowledge of that topic. In the same way, if an individual wishes to tell a joke he assumes that those who will hear the joke have not been told it before. In order to verify that assumption (S) may employ a pre-question of the kind found in (14). But this pre-question differs in import from the pre-topic check. A positive response to a pre-topic check ensures continuation of the topic; however, a positive response to the pre-joke check will generally ensure that the joke does not continue. Here again this difference in outcome (as with the ability to answer the question in the Q-joke) indicates that we are dealing with an independent speech event. Below is an example of a narrative joke:

(15) T:SC
G. ... there was this wee man an he could speak but he was deaf and one day he lost his nanny goat and went to the vicar and says, 'I've lost my nanny goat' (0.5) and the vicar says, 'Well we'll announce it in church on Sunday'. The wee man says, 'How will I know when you are announcing it',and the vicar says, 'I'll wipe my brow'. (1.0). So when Sunday came it was dead hot and the vicar stood up in church and he says, 'By the way next Sunday we have a lady missionary coming from Africa' and he accidently wiped his brow you see cause he was dead, dead hot and em (1.0) the wee man that lost his nanny goat stood up and siad, 'You'll easy know her cause one tit's longer than the other and her belly is scraped with barbed wire'. (laughter).

At the level of out-moding, Q-jokes and narrative jokes have much in common. Narrative jokes, like Q-jokes, often involve implausible events or activities, and they too include conventional joke generic form. However, perhaps the clearest indication of the out-moding of narrative jokes is the pre-joke maker; this may be a question as in (14), but it may also be in an imperative form:

(16) a. Wait till I tell ya this wee joke
b. Here listen to this one
c. Hey here's a great joke for you

These forms act to explicitly out-mode talk. Not only do they actually contain the word joke (and where they do not, participants may recognise a forthcoming joke form, the conventional form, *one* often used as a substitute for joke or funny story), but the fact they are in the imperative mood indicates a possible movement of talk.

The question to be asked is what guides the selection of either an imperative or question based introduction to narrative jokes, since they both mark out-moding in the same way? The obvious answer is, of course, that in one situation (S) is not sure whether (H) has heard the joke before, therefore he asks a question; in the other he is convinced that (H) has not heard the joke, and further that (H) would be very much interested in hearing the joke, therefore (S) employs an imperative form.

This is, however, only part of the answer. Marking a joke as out-moded does not guarantee that those other participants present will agree to the out-moding. Since participants are in conversation, and therefore in a situation of equality in speaker rights, it is not obligatory that they listen to jokes. And in some situations participants may explicitly refuse the out-moding:

(17) T:SS
E.   No now that's a dirty one

T:IC
A.   I'd rather not hear another one of your jokes if you don't mind

In this sense, joke based pre-questions act not only to assess whether (H) already knows the joke, but also to (indirectly) assess whether he wants to hear it or not.

As we have seen, then, the narrative joke is marked in a similar way to the Q-joke. This is perhaps not surprising: although they are different styles of joke telling, they are, after all, the same speech event.

5.4 Banter

Banter, as the term is employed here, refers to the use of an individual utterance or series of utterances from one or more than one speaker which aim to create some amusement by downgrading (with an explicit statement, or indirectly through innuendo) the *face* of some other participant (Goffman, 1970; Brown & Levinson, 1978). The downgrading of face is, of course, particularly problematic for conversation. Banter, however, depends on participants recognising they are out of *conversation*, i.e., that the utterance is not to be taken *seriously* but should be understood as being *non-serious* and lighthearted. In many ways banter is analogous to speech events such as *sounding* (Labov, 1972b, 1973; Kochman, 1975, 1981) or tantalism (Edwards, 1979). It is perhaps not as structured or as articulated as these other events (and it certainly has not been studied in the same detail), but an individual involved in bantering someone else is displaying his wit, a particular feature of sounding (Goffman, 1974). We will discuss the relation between banter and sounding further under *miscommunication in out-moding* (see below). For the moment we may note that banter, like sounding, revolves around the non-serious downgrading of a participant's *face* (or the *face* of someone close to the participant, for example, girlfriend or member of his/her family).

The non-serious nature of banter is displayed mainly through prosodic cues, although as Milroy and Milroy (1977) note, banter also usually involves shorter speaking turns. The prosodic features of banter are discussed by Milroy and Milroy (1977, see also Milroy, 1980). They argue that banter is marked by a rapid tempo, with clear features of loudness and pitch. These features act to indicate that the utterance is out-moded, i.e., not to be interpreted within the frame of conversation, but within the framework of banter. Here are some examples:

(18) T:SS
a.   T.   I was gonna punch him
     D.   You couldn't punch a bus ticket
     R.   Aye the last time you lifted your arm you fell over (laughter)

T:SS
b.   E.   Look at the legs pipe cleaners
     G.   What are you talking about liquorice legs (laughs)

T:SC
c.   G.   Are you playing on Saturday?
     I.   I don't know
     A.   Sure you haven't been asked
     I   I have
     J.   Aye they never asked her (laughter)
     I.   What the fuck would you know
     J.   You're dropped.

Banter may arise in any context, although it most frequently occurs in informal adolescent contexts. In the Milroys' data (Milroy & Milroy, 1977) banter also occurred most frequently with younger informants. However, we should not assume older participants do not employ banter. Banter seems to occur most readily in those situations where there is no underlying potential power differential. This may be because the nature of banter would be more likely to activate any potential power differences which existed. This is a consequence of banter's face threatening quality. But this is not to say, of course, that banter does not occur in such contexts. Consider the following from a school classroom:

(19) T:SC
- T. Hey Sam I hear you scored yesterday
- S. Aye Sir a cracker
- B. It bounced off yer head
- S. What would you know you can't even see the ball
- T. (   ) It's big enough anyway (laughs)
- S. Funny aren't we don't forget I've seen you play (1.0) I've seen better one legged footballing ducks
- B. Aye they call him eat the ball
- T. You're just jealous

Banter, then, is out-moded through prosodic marking. Such marking indicates to (H) the non-serious nature of the utterance, i.e., that although face-threatening the utterance is not an attack on the speaker, but a display of (S)'s wit within the framework of banter.

6. Out-Moding and Miscommunication

In this section miscommunication in the production of out-modes is discussed. Miscommunication is discussed here because, as Gumperz and Tannen (1979) point out, 'by studying what has gone wrong when communication breaks down, we seek to understand a process that goes unnoticed when successful' (p. 308). In other words, by studying miscommunication in the production of out-modes we may clarify and verify our earlier analysis as well as producing an account of certain communication difficulties which arise in conversation interaction.

Following Milroy (Mimeo: P11 published as Milroy, 1984) a distinction will be drawn in the analysis of miscommunication between *misunderstandings* and *communicative breakdown*. Misunderstandings "involve a simple disparity between the speaker's and the hearer's semantic analysis of a given utterance ... communicative breakdown on the other hand occurs when participants perceive something has gone wrong". This is a useful distinction, and it is one which will be followed here. We are interested in those situations where something has gone wrong in the production of those kinds of out-modes previously discussed.

6.1 Institutionalised out-modes and communicative breakdown

Institutionalised out-modes arise in situations where there exists a potential asymmetrical power relationship which can be realised at any moment within the ongoing talk. In order for an institutionalised out-mode to work it is necessary that participants accept and recognise the validity of this potential power or status difference. In utilising an institutionalised out-mode, breakdown may occur at several levels. Consider the following examples:

(20) T:SS
- (a) 1. P. Give me that bat
- 2. R. Yes masa (with affected accent)
- 3. P. Ahaha alright would you pass me the bat?

T:SS
- (b) 1. J. Run down there and get us a drink
- 2. D. Fuck off who do you think you are wee girl?
- 3. J. OK don't get excited I only asked

In these examples breakdown arises because (S) employs a form which implies institutionalised out-moding, but the situation is such that no potential asymmetrical power relationship exists to justify such a form. In both (20a) and (20b) the participants have relatively equal status roles (both examples are from informal adolescent contexts); in such contexts we would not expect to find such commands. Both J and P employ forms which suggest they have a higher status than D and R, but no such potential status difference exists. J's and R's utterances are therefore treated as unjustified institutional out-modes.

Let us consider each example in more detail. In (20a) R reacts to P's utterance with a sarcastic comment. The comment is sarcastic in that it highlights the unjustifiable claims of P's turn. R's utterance implies an extreme form of status differential, master/slave. Of course no such status difference of this kind exists; indeed, no status difference of any kind exists between P and R, and this is the point. P recognises the force of R's sarcastic comment and is in agreement with its claim. As a result P reactivates utterance (20a:1) as an indirect request — the kind of form one would have expected in a symmetrical status relationship. The very contrast between the command and the indirect request indicates that something was wrong with utterance (20a:1). Although it is not necessary for both participants to agree that there has been a breakdown and to repair it, in this case we have an explicit example where they do.

(20b) is similar to (20a), it also involved the misuse of a command form in a situation where power and status are symmetrically distributed. What is interesting about (20b) is the way in which the participants treat the breakdown. The command (utterance 1) is clearly face threatening, and D responds to it in a forthright manner. D explicitly rejects J's right to employ such a form and then questions J's criteria for the use of this form in the first place. This question is actually a rhetorical form, i.e., one where the answer is already known — the answer in this case being that no matter who J may think herself to be, she has no power over D. J's response to all this is particularly fascinating. A surface interpretation of (20b:3) implies that J is surprised by D's reaction, in that J believes that she only asked a question. Clearly, J did not ask a question. Why then should she claim that she did? One answer is, of course, that it defends her against the attack made by D, in that if she only asked a question such an attack is unjustified. What J seems to mean by (20b:3) is that she intended (20b:1) to be treated as an indirect request, something like *would you go and get me a drink?* i.e., the kind of form we would have expected in this situation. This may indeed have been J's intention, and the command may have been a mistake, but D gives a perfectly valid interpretation of (20b:1) in the light of this particular informal context. It is this confusion over intention and actual realisation that leads to the breakdown.

Both (20a) and (20b) arose from the inappropriate use of commands in a symmetrical power based context. The employment of a command form implies dominance of (S) over (H). In both (20a) and (20b) (S) did not hold such a position of dominance. In both cases (S) stepped out of the frame of conversation without appropriate justification. The intention of the speaker in both cases was probably not to imply dominance, but nevertheless the forms they used allowed for such an interpretation.

It should be noted that the fate of every command form utilised in situations of symmetrical power relationships need not be the same as those in (20a) and (20b). The hearer has the right to ignore the command as a slip of the tongue. That is, he may comply with the command without believing that he is in a position of deference with regard to the speaker. The difference in reaction found in (20a) and (20b) is a reflection of the kinds of choice open to hearers. In (20a) (H) recognises that the command is potentially face threatening but he does not believe it is intentionally so, and so his reaction is to utilise lighthearted sarcasm. In (20b), on the other hand, (H) is offended by the command, presumably because he believes it to be intentional. Consequently, the reaction is much more forceful than that found in (20a).

Let us move on now to consider those breakdowns which occur where an asymmetrical power relationship does exist. The following extract takes place in a school classroom:

(21) T:SC
    a.   N.   Did you hear W's in jail?

b.	T.	What?
c.	A.	Aye caught muggin an old lady
d.	T.	You're joking
e.	N.	No seriously he's a thug just like B there (laughs)
f.	B.	⌊Watch it hog features
g.	A.	So we'll not be seeing him for a while
h.	T.	Put your feet off the desk B
i.	B.	Why? The boots are clean
j.	T.	Look because I'm telling ya son alright
k.	B.	Fucks sake (whispered)
l.	T.	Move it son
m.	B.	Fuck me
n.	T.	What did you say?
o.	B.	Nothing

Earlier, it was argued that institutionalised out-modes were frequently marked by pre-forms such as *right, well, now* or *O.K.*, that they were lower in pitch (relative to surrounding talk) with an increase in loudness, and that they were frequently in the imperative mood. In (21) the teacher intends (h) as an out-mode. His intention was probably not to stop the conversation altogether (as with earlier examples), but just for that moment he wanted to re-establish his status in order to achieve a particular goal. (h), however, while in an imperative form, is not marked by any of the other features associated with institutionalised out-moding. Apart from its imperative form, (h) is similar in tone and loudness to surrounding talk. It is perhaps for this reason that B does not treat (h) as an institutionalised out-mode. This is reflected by (i), where B, believing himself still to be within the framework of conversation, questions the command itself. This reply is treated by the teacher as an indication of disobedience and he reacts to it by asserting his right to issue such commands. This time his utterance is lower in pitch with a relative increase in loudness. B's response (k) is muttered, almost whispered, and this reflects, perhaps, a confused exasperation. B's reaction at (k) is understandable. The breakdown in communication resulted from T's failure to employ specific institutionalised out-mode markers. Consequently, B cannot understand why T should be angry with him for simply asking a question. On the teacher's part, of course, he does not recognise that he failed to make clear his intention at (h), and therefore sees B's response (i) as threatening his right to perform institutionalised out-modes. And so confusion on both sides leads to breakdown.

A number of objections might be made to this analysis. Firstly, if (h) is not marked as an out-mode can we be sure it is intended as one? Secondly, how do we know that (i) is not simply disobedience? The answer to the first question is that if T had not intended (h) as an out-mode there is no justification for his outburst at (j). Further, if it is not an out-mode then we would have to have some way of accounting for the occurrence of this command form within what was, up to the occurrence of (h), a conversational frame. Of course, we might argue that (h) is treated as an out-of-frame command by B, and that is why he questions its use. But B is not so much questioning the *command* qua command form, as the reason for the command, and this is a slightly different point. As for the second objection, if (i) were intended to convey disobedience then we might expect its prosodic structure to give some indication of this. We might expect an increase in loudness for example, or the use of a particular stress on *why*. (i) is, however, very similar to the surrounding talk. The stress in this utterance falls on *boots,* and the voice quality is indicative of jocularity rather than a refusal to obey.

Let us consider now a similar example from the family context:

(22) T:FC

a.	L.	Did ya go to the club last night?
b.	K.	Aye it wasn't much good
c.	M.	Who was there?
d.	K.	Ah ... the usual crowd
e.	M.	You're lucky to have something like that

f. K. I know
g. L. ⌊I know dead lucky
h. M. Look at that wee darling
i. K. Wee darling she can be a cheeky wee brat
j. ? I'm not
k. M. L would ya make us a cup of tea?
l. L. Ach I couldn't be bothered I'm tired
m. M. Go on
n. L. Ach no
o. M. LOOK make me a cup of tea I'm not asking I'm telling ya
p. L. Aye alright will, it's always me

I want to argue here that (k) is intended to convey institutionalised out-moding. The utterance itself carries no out-moding markings and is in the form of an indirect request, which is suggestive of a symmetrical as opposed to asymmetrical relationship. The intention here is implicit rather than explicit, it is based on the institutionalised power difference between parent and child. This argument is based on the extract as a whole rather than the individual utterance itself, in particular the following utterance (m) which displays an increase in loudness, and utterance (o) which is an explicit out-moding of the request made at (k). What I want to suggest is that the mother intends (k) as an indirect command rather than a request. However, this intention is not marked in any way in the surface structure of the utterance, therefore, not surprisingly, L treats (k) as a request. L refuses to fulfil this request and adds a justification or mitigation of this refusal. The fact that M did not intend (k) to be treated as a request is indicated specifically at (o), where M explicitly out-modes the talk by specifying that she was not asking for the action to be carried out but commanding it.

This analysis is based on a retrospective analysis of what has already occurred. Conversational analysts criticise this method because only analysts look forward, conversationalists construct talk as an ongoing process. But surely part of this ongoing process is the assessment of particular interpretations given to your utterances, and where you feel that a specific interpretation is incorrect (whether this is a speaker or hearer's fault), to reactivate your utterance in the light of this misinterpretation; that is, in the light of what has already occurred. Now an analysis of the kind given above does have the advantage that it has access to what actually occurred; but in looking at (o) in relation to (k) we are not looking forward we are looking back with the conversationalist. If we treated each utterance in this extract on a moment by moment basis we would analyse (k) as an indirect request, and structurally that is what it is. In arguing that M intended it as an indirect command we are accepting what M herself was actually telling us at (o). This is not an analyst's assessment but the actual informant's *own assessment*.

This example is a particularly tricky one, although we have already come across a similar example in (20b). Earlier, we noted that J claimed she had only asked a question when in fact she had employed a command. In (22) we have the opposite. Here M is claiming that she intended a command when in fact she made an indirect request. In both cases the difference between what the speaker believes to have been the case and what actually was the case results in some form of breakdown.

Now an objection may be made here that I am manipulating the interpretation to fit the argument. Could we not simply account for (o) by assuming that (k) is a genuine indirect request and that it is the refusal at (1) which activates M's use of an explicit command. In this way we could account for the data without suggesting that (1) was intended as an out-mode. But this may be countered by asking why the refusal at (1) should activate (o)? Speakers may, if they so wish, refuse to fulfil indirect requests. L not only does this but also mitigates her refusal, which is something acceptable and associated with such refusals. What is not associated with this pattern is a following command. There is no reason why this refusal should be treated as unsatisfactory unless M had intended (k) to be understood as an out-mode, an utterance which asserted her status. In this case a refusal threatens this status, and one would expect, in this situation at least, a reaffirmation of that status, if not a reprimand for refusal to comply with the command. In (22), (o) is an affirmation of status.

As a final example of breakdown within institutionalised out-moding I will consider briefly those contexts where (H) does not recognise (S)'s right to utilise his/her power. Let us take an example from the family context:

(23) T:FC
    M. Who brought you home last night?
    C. Nobody
    M. It must have been somebody
        (2.0)
    M. Come on tell me
    C. It's got nothing to do with you, it was just a friend alright
    M. You're getting very cheeky

Within the range of parental concern or control it is of course very difficult to draw the line between where the range of the parental domain begins and where it ends. As most parents are aware the line is usually defined more sharply as the child gets older. This extract is between a mother and her 16-year-old daughter. They have a very good and mature mother/daughter relationship. However, in this extract C asserts her right to withhold access to certain information which she believes her mother has no power to command her to give up.

Breakdown at this level of out-moding is very difficult to define in that what is breakdown for one family may not be for another. It is interesting to note, however, that this analysis suggests that breakdown will occur when there is disagreement over the range of parental power in relation to the individual rights of the emerging adult. This is not particularly new. But one interesting suggestion which might follow here is that within those areas where parental power is no longer accepted to the same extent, parents should adopt equality structures (work within a conversational frame) rather than continuing to employ out-modes.

6.2 Conversationally tied out-modes and communicative breakdown

As with institutionalised out-modes, conversationally tied out-modes succeed only in so far as (S) and (H) agree that the talk has been moved from one event type to another. There exist specific techniques for marking such moves, but on occasion these may (for whatever reason) be misread by either the speaker or the hearer; in such situations we have the potential for communicative breakdown. In this section we shall consider how such breakdowns occur within the production of narrative jokes, Q-jokes, and banter.

I will begin the discussion by considering the narrative joke. I start here because, interestingly, I can find no examples of breakdown in the out-moding of such jokes. There are some examples of breakdown in the presentation or interpretation of the actual narrative itself, but this is breakdown of a different kind in that the joke mode is already established.

There is a good reason why narrative jokes should be successful in achieving out-moding. The use of pre-joke markers, whether in imperative or interrogative form, explicitly make it clear that a joke is about to follow:

(24)   a. Did I tell ya the joke about ...
           b. Wait till I tell ya this one ...
           c. Here wait till I tell ya this wee joke ...

With such explicit marking it is not surprising that examples of breakdown in out-moding are hard to find.

The Q-joke is also relatively successful in achieving out-moding, although here one does find some examples of miscommunication. We have already noted one example of this previously. Here is another:

(25) FN:
    a. R. Did ya hear about the Irish army, they invaded Falkirk
    b. P. What?
    c. R. Falkirk
    d. P. You mean, OH aye (laughs)
    e. R. Ya silly cunt

The misinterpretation arises here because P treats (a) as a pre-topic marker rather than a Q-joke. Consequently, P attempts to locate a frame for the concept of Falkirk. Now the joke functions on the similarity of the names *Falkirk* and *Falkland*, but it is just this similarity which may have confused P. The difficulty may be highlighted by drawing on analogy with syntax. In syntax participants apply particular interpretive strategies to sentences (Aitchinson, 1976). These strategies are based on particular assumptions, for example, '... that the first noun will go with the first verb in a NP-VP actor action sequence as part of the main clause' (p. 203). As a result, when presented with a sentence like (26), which is perfectly grammatical, we find it difficult to comprehend.

(26) The pig pushed in front of the piglets ate all the food

Utilising a similar argument for conversation we could suggest that participants also utilise particular strategies, i.e., they apply the *rules of conversation* to any pre-question, as a first step, and unless the pre-question has a specific joke generic marking, or refers to highly implausible events (e.g., *there was this wee man with no legs*), then it may be treated as a pre-topic marker. Consequently, any Q-joke which refers to real individuals or events runs a high risk of misinterpretation.

Certainly those few examples I have been able to find may be accounted for in this way. In (25), because *Falkirk* sounds very similar to *Falkland*, P may have assumed some point was being made on this fairly serious topic. Despite the fact that *Irish Army* is used, this interpretation holds. I would argue that *Irish Army* is not necessarily joke generic. We cannot assume in conversation that everytime something Irish is mentioned a joke is in progress.

Here is a further example which may be analysed and accounted for in the same way.

(27) FN:
    An. Did ya hear that they want to put one of my wee lads in a home. He says he'll be buggered if he's going to Kincora (laughs).
    T. That's terrible
    An. Kincora
    J. Kincora ya dummy
    T. Fuck that's right, (laughs) I'm sitting here thinking (laughs)
(Kincora is the name of a boy's home in Belfast where there was a sexual scandal. A number of supervisors were convicted of sexually assaulting the boys.)

As with (25), (27)'s initial pre-marker does not contain any indication of implausibility or any joke form, thus T is free to give a topic interpretation. This he does, failing to draw the necessary link between *Kincora* and *buggery*.

Breakdown is more likely to occur for Q-jokes where they are not specifically marked. This is not surprising — research has shown that invited inferences (based on world knowledge) require more processing than necessary inferences (inferences dependent only on the language) (Chaffin, 1979; Singer, 1979), therefore a joke marked by a specific word or clause will be simpler to process than one which requires implications to be drawn from the relation between concepts in the real world (i.e., Kincora/buggery, Falkirk/Falkland).

Banter has been defined as the use of an explicitly face-threatening form whose prosodic marking indicates that it is intended non-seriously. Clearly, if an utterance intended as banter is interpreted seriously (within the conversational frame) then a communicative breakdown is possible. Since the forms explicitly threaten face they will be responded to defensively or agressively where they are taken seriously. Breakdown may occur for two reasons: firstly, because of an inadequate prosodic marking of the form intended as banter, and secondly, banter may fail because (H) is not willing to accept (despite correct marking) the claims of the banter.

(28) T:SS
a. T. Are you playing tomorrow?
b. H. Aye
c. D. Sure you can't play football (laughs)
d. H. I'm fucking better than you (agitated)
e.   What the fuck makes you think you're good
f. D. I didn't say I was
g. H.   [Well fuck up then
h. D. Hey I was only fucking joking ya not take a joke
j. H. It didn't sound like a joke to me

In (28) (c) was intended as banter; it does not, however, contain any of the markings suggested by Milroy and Milroy (1977). It is delivered in a moderate tempo with a loudness and pitch relatively similar to earlier (conversational) talk. As with a number of our other examples, we argue that (c) is meant as banter because of the justification given later at (h). This justification suggests that (c) was not conveyed seriously, but was meant non-seriously. D failed, however, to mark (c) appropriately and so H interprets it as *face*-threatening. It is interesting to note, here, the comment made at (i). H explicitly indicates that if (c) was meant as banter it should have *sounded non-serious;* this is informant reinforcement of the claim that banter has a particular sound structure. Consider now (29):

(29) T:IC
P. Fuck me look at yer man (old man with no arms sitting in a wheelchair)
J. Aye he'd be handy to have around the house (laughs)
D. Armless ain't he (laughs)
L. That's not funny
D. What?
L. That
J. It's only a fucking joke
L. Well what's funny about that

In (29) J and D are having fun at the expense of an old man (who is out of hearing distance) and L objects to this. There is nothing structurally wrong with the forms employed by D and J, therefore it must be the case that L believes only certain kinds of banter are acceptable. This is an extreme example perhaps (and some may argue we are dealing with jokes rather than banter), but it is a fact that not all banter delivered correctly and therefore intended as non-serious will be accepted. It is difficult to assess whether this is a reflection of the makeup of different individuals or whether there is a linguistic pattern involved.

Work on sounding offers one possible solution (Labov, 1973). Labov has argued that *ritual insults* are defined as such because they are false, that is, their claims are not verifiable in the real world. Where these claims become verifiable, that is, where an insult is based on an actual attribute of one's parents say, then the claim becomes an *actual* as opposed to a *ritual insult*. Labov makes this claim on evidence which shows that where actual insults occur the recipient moves his talk out of the mode of sounding and reacts defensively or aggressively. Kochman (1975, 1981) argues that Labov has misinterpreted the evidence however, and suggests that actual actual insults may play a central role in the game of sounding. He points out that if (S) can get (H) to react defensively to an *actual* insult then (S) will have won the game.

As we have already noted, banter is, in some senses, similar to sounding in that it displays a speaker's *wit*. It is also interesting to note, in the light of Kochman's claim, that where a participant reacts aggressively or defensively to banter this reaction is often negatively assessed by others:

(30) FN:
St. You're putting on the weight
J. Aye
A. Too much old beer
St. A right wee fatty got
A. Mr. Piggy (laughs)
St. No a wee fat runt (laughs)
J. What are you talking about baldy you're just an ignorant cunt
G. You've *cracked* (the term *cracked,* as it is used in this example, actually implies J is taking things too seriously)
J. Who?
A. *Can't take it*
St. Hey J's cracked (in this situation 'to crack' is a sign of over-sensitivity and is negatively assessed)

But can we account, however, for that banter which is rejected in relation to its *truth value*. Such an approach has an obvious appeal, it suggests that banter must not only be prosodically marked but must be propositionally marked as untrue or as unlikely to be true. Despite the appeal of this distinction the evidence available does not support this conclusion. Banter is frequently based on true claims or claims which, while not quite true, are not necessarily false:

(31) T:IC
A. Are ya going to the ( ) tonight?
R. I don't know I'm busy
T. Oh aye that's right we don't ask you we ask your Ma (laughs)
R. Funny cunt

In (31) both A and T have access to certain background information, that is, that R has been helping his mother paint the house. T's claim (which I am treating as banter) is not explicitly false, in that the issue of helping his mother paint the house effects R's choice. The claim is also, however, not explicitly true in that R is, to some extent, responsible for his own decisions.

The issue of why some banter is not accepted by certain participants is at present unresolved. It is unclear whether there is any structural pattern involved or whether it is simply the result of a variety of extraneous factors such as the mood (H) happens to be in, participants present, the relation between (S) and (H) and so on.

7 Summary

In this paper we have argued that conversation is only one of several speech event types to be found within informal interaction. Consequently, it is important in the analysis of everyday talk to be clear on which speech event research refers to; we cannot assume we have conversational materials simply because we have recordings of everyday talk.

It follows, of course, that we should also be capable of distinguishing between conversation and those other events found in informal talk, and this has been a central issue in our discussion. We have outlined and explored the kinds of linguistic methods employed in moving talk from one event type to another, what we call out-moding. Also, we have provided secondary evidence for our analysis by showing that certain breakdowns which occur in ongoing talk are a consequence of processing errors with the out-moding procedure.

The findings presented here suggest that there is much to be gained from defining conversation as a specific speech event, as opposed to treating conversation as something equivalent to everyday talk.

Appendix – Notational Conventions

The data for this paper come from naturally occurring Belfast adolescent conversations. The term 'natural' is used here to refer to data which have not been elicited by the use of experimental methods (see Wilson, 1985b; chap. 2).

Transcription

[	–	marks overlap
(0.0)	–	pauses or gaps in seconds (2.0) and tenths of seconds (0.1)
CAPS	–	relatively high amplitude
?	–	(a) indicates question
		(b) if placed in front of an utterance it indicates that the identity of the speaker is unclear
( )	–	(a) uncertain passages of transcript
		(b) also used to supply necessary contextual information
_____	–	draws attention to some phenomenon of direct interest to the discussion
(.)	–	micropause
(( ))	–	researcher's interpretation of data where recording quality is unclear

Data :Recorded

The following abbreviated forms will be used to indicate context of recording:

(1) T:SS — tape from youth summer scheme
(2) T:FC — family context
(3) T:IC — informal context (general)
(4) T:SC — school context
(5) — constructed examples are unmarked

References

Aitchinson, J. (1976). *The articulate mammal: An introduction to psycholinguistics*. London: Hutchinson.
Bolinger, D. (1978). Yes/No questions are not alternative questions. In H. Hiz (Ed.), *Questions*. Dordrecht: Reidel.
Brown, P., & Levinson, S.C. (1978). Universals in language usage. Politeness phenomena. In E.N. Goddy (Ed.), *Questions and politeness: Strategies in social interaction*. Cambridge: Cambridge University Press.
Chaffin, R. (1979). Knowledge of language and knowledge about the world. A reaction time study of invited and necessary inferences. *Cognitive Science, 3,* 311-328.
Daden, I. (1975). Conversational analysis and its relevance for teaching teaching English as a second language. Unpublished MA thesis, University of California, Irvine.
Edwards, W.F. (1979). Speech acts in Guyana: Communicating ritual and personal insults. *Journal of Black Studies, 10,* 20-39.
Goffman, E. (1970). *The presentation of self in everyday life*. Harmondsworth: Penguin.
Goffman, E. (1974). *Frame analysis*. New York: Harper Colophon.
Gumperz, J.J., & Hymes, D. (1972). *Directions in sociolinguistics*. New York: Holt, Rinehart & Winston.
Gumperz, J.J., & Tannen, D. (1979). Individual and social differences in language uses. In C. Fillmore, D. Kempler, & S.Y. Wang (Eds.), *Individual differences in language ability and language behaviour*. New York: Academic Press.
Gumperz, J.J. (1982). *Discourse Strategies*. Cambridge: Cambridge University Press.
Hymes, D. (1972). Models of the interaction of language and social life. In J.J. Gumperz, & D. Hymes, (Eds.), *Directions in sociolinguistics*. New York: Holt, Rinehart & Winston.
Kochman, T. (1975). Grammar and discourse in vernacular black English. *Foundations of Language, 13,* 95-118.
Kochman, T. (1981). *Black and white styles in conflict*. Chicago: Chicago University Press.
Labov, W., & Waletsky, J. (1966). Narrative analysis: Oral versions of personal experience. In J. Helm (Ed.), *Essays on the verbal and visual arts*. Seattle: University of Washington Press.
Labov, W. (1973). The linguistic consequences of being a lame. *Language in Society, 2,* 81-113.
Labov, W., & Fanshel, D. (1977). *Therapeutic Discourse: Psychotherapy as conversation*. New York: Academic Press.
Levinson, S.C. (1983). *Pragmatics*. Cambridge: Cambridge University Press.

Milroy, J., & Milroy, L. (1977). Speech and context in an urban setting. *Belfast Working Papers in Language and Linguistics, 2*(1), 1-30.

Milroy, L. (1980). *Language and social networks.* Oxford: Basil Blackwell.

Milroy, L. (1984). Comprehension and context: Successful communication and communicative breakdown. In P. Trudgill (Ed.), *Applications of sociolinguistics.* London: Academic Press.

Sacks, H. (1971-72). *Lecture notes.* Mimeograph, Department of Sociology University of California, Irvine.

Sacks, H. (1974). An analysis of the course of a jokes telling in conversation. In R. Bauman, & J. Sherzer (Eds.), *The ethnography of speaking.* Cambridge: Cambridge University Press.

Sacks, H., Schegloff, E.A., & Jefferson, G. (1974). A simplest systematics for the organisation of turn taking for conversation. *Language, 50*(4), 696-735.

Sacks, H., Schegloff, E.A., & Jefferson, G. (1978). A simplest systematics for the organisation of turn taking for conversation. In J. Schenkein (Ed.), *Studies in the organisation of conventional interaction.* New York: Academic Press.

Schegloff, E.A., Sacks, H., & Jefferson, G. (1977). The preference for self correction in the organisation of repair in conversation. *Language, 53*(2), 361-382.

Sinclair, J. McH., & Coulthard, M. (1975). *Towards an analysis of discourse.* London: Oxford University Press.

Singer, M. (1979). Processes of inference during sentence encoding. *Memory & Cognition, 7*(3), 192-200.

Wilson, J. (1982). Come on now answer the question! An analysis of constraints on answers. *Belfast Working Papers in Language and Linguistics, 5*(1), 70-101.

Wilson, J., & Gunn, B. (1983). Defining topic in conversation: The intonational ordering of entailments. Paper presented at Symposium on Discourse Analysis, Hatfield Polytechnic.

Wilson, J. (1985a). On the boundaries of conversation: Delimiting conversation as a speech event. Paper presented at International Symposium on Discourse Processing, St. Patrick's College, Dublin.

Wilson, J. (1985b). Conversation matters: Towards a definition of everyday conversation. Unpublished doctoral dissertation, Department of English, The Queen's University, Belfast.

COMPUTATIONAL SOCIOLINGUISTICS AND COMMUNICATION FAILURE: ON THE RESOLUTION OF INCOMPLETENESS IN AUTOMATIC DISCOURSE PARSING

Brian Torode

Department of Sociology,
Trinity College, Dublin

Any method of discourse analysis must practise its own solution to the problem of communication failure. An influential recent study (Labov & Fanshel, 1977) resolves incompleteness in naturally occurring therapy talk by an interpretive procedure which treats both complete and incomplete spoken utterances as the *documents* of underlying logical propositions. The procedure relies on detailed ethnographic knowledge of the social context in which the talk occurs. As such, it suffers from a number of drawbacks. In this paper a different solution to the problem is demonstrated. Incomplete syntactic units are repaired by matching them with adjacent complete sentences. Discourse analysis then proceeds on the basis of the completed sentences. The implication of this approach is that sociological interpretation should follow linguistic discourse analysis, rather than precede it.

1. Ideological and Interactionist Discourse Analysis

There are two sharply divided traditions within recent sociological discourse analysis. One attempts to give an *objective* description of quasi-grammatical structures which constrain the articulation of ideologies in speech or writing. In England, its leading practitioner is Bernstein (1973). A paradigm for this approach is *Morphology of the folk tale*, by the Russian formalist writer Propp (1968).

The other approach attempts to provide an 'intersubjective' interpretation of discourse *from the participants' own point of view*. Its practitioners today include the 'ethnomethodologists' Schegloff (1972) and Jefferson (1980) and the 'interactionist' Goffman (1971). A model for recent work in this tradition is the study *Therapeutic discourse* by Labov and Fanshel (1977).

The two sides take up opposed positions on the role of social context in sociolinguistic analysis. For the 'intersubjectivists', all language is language-in-social-context which is properly the object of sociological rather than linguistic study. Labov and Fanshel go so far as to claim that, in orderly conversation, "there are no connections between utterances. ... obligatory sequencing is not found between utterances, but between the actions which are being performed" (p. 70).

By contrast the "objectivists" argue that the distinction between "context-free' and "context-bound" language is fundamental to the ideological claims made in ordinary language and they use linguistic criteria to divide one from the other. Bernstein's (1973) distinction between "elaborated" and "restricted" codes in the speech of school boys is the most famous attempt to do this. (Torode, 1984, distinguishes "extra-ordinary" from "ordinary" language on this basis.)

Relations between the two approaches have been strained. A significant controversy concerning the role of language skills in educational attainment occurred in New York in the 1960s. Bernstein (1973) argued that the working class are deprived of *linguistic means of production* just as, for Marx, they had been deprived of *material means of production*. American educationalists claimed this legitimated their view that "Non-standard" (Black) English was inferior to standard English in its capacity for rational thought.

Labov (1972) countered this claim by employing a method of propositional analysis to show that Black English was capable of subtle and complex logical statements. Silverman and Torode (1980) argued that in an important sense, Bernstein's position was more persuasive than Labov's.

We noted a paradox in Labov's position, namely that in order to argue for the superiority (subtle propositional complexity) of an instance of Non-standard English over an instance of Standard English, he had to imply the superiority of his own language to that of either of the speakers. For it is only through Labov's writing that the explicitly propositional content of the speech is revealed.

Labov is a fluent modern exponent of *linguistic universalism,* the view that all languages are equivalent in their capacities, hence morally equal to one another. This view grants a privileged position to the language of the linguist who reveals the underlying universal features hidden by particular linguistic expressions. This is disguised by the claim to celebrate linguistic particularism. Labov's method interprets particular linguistic expressions as manifestations of universal underlying propositions.

The analytical technique used in "The logic of non-standard English" (Labov, 1972) was merely illustrative. But in 1977, Labov and Fanshel published *Therapeutic discourse,* an exemplary demonstration of the method, now called 'comprehensive discourse analysis'. Sociologically, Labov and Fanshel base their method on the dualistic approach of Goffman (1971), who interpret the observable particularities of face-to-face personal interaction, such as politeness rituals, as surface manifestations of underlying egoistic drives: the Hobbesian war of all against all. From Goffman, Labov and Fanshel drew their theory that real underlying conflict is *mitigated* by the veneer of consensus which structures the appearance of everyday life.

Goffman and Labov and Fanshel examine surface features of social life in detail, but assume that these details lack ultimate significance, because that significance lies in an unobservable and asocial underlying reality which is presupposed by the analyst's interpretation. (Torode, 1976, proposes that Goffman's surface/deep dichotomy be reconstructed as a way of analysing division within the observable surface of language.)

2. 'Comprehensive' versus 'Surface' Discourse Analysis

The crucial stage in Labov and Fanshel's methodology, for present purposes, involves the following relation between observation and interpretation:

observed: interpreted:

TEXT
PARALINGUISTIC CUES   EXPANSION INTO TEXT

For Labov and Fanshel, text comprises *propositions*. Propositions either appear explicitly on the surface of the talk or are imputed to underly that surface, where they are signalled by paralinguistic cues.

In practice, surface propositions are found overwhelmingly in the therapist's talk whereas it is in the patient's talk that underlying propositions are inferred. This implies a privileged relation between Labov and Fanshel and the therapist, in two senses: (1) Labov and Fanshel's 'expansion' technique is primarily a means of *translating* the inarticulate (i.e., implicitly propositional) talk of the interviewee into the articulate (i.e., explicitly propositional) talk of the therapist; (2) The work of Labov and Fanshel in *analysing* the discourse, complements the work of the therapist in *eliciting* the discourse. (This is similar to Labov's relationship with his interviewer, John Lewis, in Labov, 1972.) For both therapist and discourse analyst, the point of the therapeutic interview is to bring underlying propositions to the surface. But the fact that there are two distinct means to the same end also implies a certain conflict between therapists' and analysts' conceptions of the therapeutic discourse.

Labov and Fanshel (1977), like Labov (1972), are *moral universalists*. For them, any speech is as articulate as any other. Non-standard (Black) English is as articulate as (White) standard English, because both can be translated into Labov's propositional terms. (As noted above, this really implies that the speech of the universal linguist is superior to any of the speech he analyses.)

For Labov and Fanshel, patient speech which relies heavily on paralinguistic cues is as articulate as the most professional therapeutic talk (the distinction between these two types of speech mirrors that between the "restricted" and "elaborated" codes of Bernstein, 1973). But the therapist is committed to *changing* the patient's speech. She is a *moral particularist:* she must believe that her own distinct discourse is a superior one. This is especially the case because by formulating explicitly matters which other discourses fail to address, her discourse can help to solve problems which those other discourses perpetuate. (In Bernstein's terms, she offers the patient forms of linguistic capital with which to transform her linguistic world.)

Labov and Fanshel's 'propositional' interpretation of Rhoda's speech is primarily based on the discourse of social-psychological role theory. For instance, Rhoda's opening remark is treated as follows:

TEXT 1.1[a]   EXPANSION
[unaltered except "..." used for   *[unaltered]*
"..." or longer sequences]

I don't ... know whether   I am not sure
I – I think   but I claim that
I did the right thing,   (1) I did what you say is right
jistalittle ...   or (?1) what may actually be right
situation came up ...   when (4) I asked my mother to
                                              help me by coming home after she
                                              had been away from home longer than
                                              she usually is, creating some small
                                              problems for me, and

an' I tried to uhm ...   I tried
well try to ... use   to use the principle
what I've learned here,   that I've learned from you here
                                              (S) that I should express my needs
                                              and emotions to relevant others, and
see if it worked   see (S?) if this principle worked.

In their expansion, (n) indicates numbered 'local propositions', while (S) indicates a general proposition, namely that "One should express one's needs and emotions to relevant others".

In Labov and Fanshel's interpretation, the patient's talk builds logical connections between propositions. The expansions involving local proposition (4) and general proposition (S) have been learned by the discourse analysts from Rhoda's subsequent talk, yet are attributed by them to the opening of the therapy session.

Thus their expanded text analysis is *synchronic:* the development of an argument by Rhoda (or by the therapist) is eliminated by their extrapolation of every fragment to the complete argument.

Jefferson's (1980) approach to "trouble talk", by contrast, would note "jistalittle" as a specifically downgraded statement of trouble, characteristic of what she calls a "troubles premonitory sequence". "Just a little" minimizes the matter. Jefferson persuasively argues that narrative troubles talk always alternates between "troubles" and business as usual", and that initially, in order to build trust between teller and recipient, "trouble" must be formulated within the context of business as usual (cf. the account of trust given by Garfinkel, 1967). Only subsequently can the teller trust the recipient with a larger scale version of the trouble.

Jefferson analyses troubles talk as a form of extended story-telling: she does not, of course, claim to account for the practise of therapy. But neither do Labov and Fanshel. Their analysis is directed to penetrating beneath the surface veneer of discourse to discover underlying propositions about family role relations. But their propositional account of these role relations is static. The situation of mother, school-age daughter, and aunt at home is better understood, but not changed, by their analysis.

On Labov and Fanshel's account, it is reasonable to infer that change comes when Rhoda grows up and leaves home, i.e., by means of extra-discursive events impinging from the outside, rather than from the therapist's intervention in the discourse situation. But it is also reasonable to ask that analysis should account for therapy as a practise devoted to changing the world through discourse. I suggest that the change which the patient wants to achieve through therapy can be discovered from an analysis of her discourse.

Rather than presuppose knowledge of the social context in which Rhoda and her therapist find themselves, I develop a method for formal analysis of the talk in its own terms. A pioneering attempt at such analysis is Propp's account of the Russian folk tale (Propp, 1968), to which I now turn.

3. A Formal Model of Discourse

Writing in 1920, in the post-revolutionary USSR, Vladimir Propp sought to systematise the study of the Russian folk-tale (Propp, 1968). Influenced by a historical and nationalist conception of linguistics which had swept Europe in the nineteenth century, the ethnographers of late Tsarist Russia had traversed the countryside collecting volumes of fairy stories. Propp's analytical project in part expressed ennui at the sheer quantity of material gathering dust on Moscow library shelves.

Propp claimed his method of analysis would eschew the merely external classification of tales by topic (e.g., dragon tales, animal tales, landlord tales, etc.) because such categories could never be exclusive and exhaustive (one could always find a tale about a dragon *and* a landlord). Instead he claimed to provide an internal examination of the formal structure of the tale. It is convenient to illustrate his method with the story of 'Snow White' which, though not a Russian tale, is well-known to a Western audience.

According to Propp, the rich variety of *characters* found in the fairy tale — which remains the preoccupation of those who externally classify tales by their content — disguises an essential poverty of form. Every tale, he proposed, involved some selection from only 31 *functions,* which must always occur in the same order.

Certain functions were, he claimed, essential: the tale must begin with A = harm or injury by the villain to a member of the family, (e.g., 'Snow White' begins when the jealous Queen has her abandoned in the forest) and must end with W = wedding (Snow White marries the prince).

Others are optional. Many of these occur in pairs, e.g., *interdiction* (the prohibition against Snow White allowing anyone into the dwarfs' cottage) need not always occur, but if it does then it is always followed by *violation* (Snow White admits the old woman who gives her a poisoned apply). Interdiction without violation would be pointless in the narrative economy of the tale.

Propp's use of functions is axiomatic rather than empirical. In many stories, including 'Snow White', functions appear to occur in other than his required order. For instance the poisoned apple incident is a *second* instance of function A, considerably separated by other functions from the first occurrence.

Propp's solution is to posit a complex structure of more than one plot. So 'Snow White' involves two simple plots. One is centred on the 'family' of Snow White as princess in the home of her evil stepmother. The other is centred on the 'family' of Snow White as quasi-mother in the home of the dwarfs. The heroine of the first tale is the (female) 'victim', Snow White. The hero of the second is the (male) 'seeker', the Prince who rescues her after the dwarfs have embalmed her in a glass case. These are the two main variants of Propp's fairy tale structure.

On this reading, Propp's axioms are consistently upheld, and the structure of the story is shown to be more complex than might at first appear. Although the method is purely formal, involving no interpretation of the 'meaning' of the story, a natural interpretation can be made once the morphological analysis is complete.

It seems that, in general, the fairy tale is 'about' family structure. Beginning with some threat to the family (harm or injury to a member) and ending with a restoration and celebration of the family (the wedding) it reaffirms this institution central to peasant society. The story of Snow White specifically enacts complementarities and conflicts between the family of origin and the family of destination in the life of a girl in patrilineal society.

Propp's method offers a paradigm for formal discourse analysis, if we allow that a 'tale' is one instance of discourse, involving the following features:

(1) distinction between lexical items (Propp's *characters*) and syntactic rules for their combination (Propp's *functions*) in the description of a discourse;

(2) axiomatic specification of syntactic rules, notably (i) compulsory presence of certain rules; (ii) conditional occurrence of others in combination if they occur at all; (iii) obligatory sequencing;

(3) the combination of discourses constituting a specific act of speech or writing is to be determined by the strict application of the axioms, not by external classification;

(4) the formal analysis does not involve any interpretation of the discourse. Interpretation should follow the formal analysis, upon which it should be based.

Discourse analysis of naturally occurring speech, inspired by Propp's formalism, can begin by treating pronouns as the characters in stories which conversationalists tell one another (cf. Torode, 1984, pp. 86ff). In the fragment with which Rhoda's interview begins (quoted above), the pronoun 'I' occurs nine times in the 49 words of the fragment. Following the external classificatory approach to fairy tales, this 'I' talk could be designated as a narrative unit.

But internal examination of the speech shows that these nine occurrences are not all of the one type. The 'I' not only represents characters *in* stories: it also represents the teller of the story, within the complex whole of the speech. (Harvey Sacks has insisted on the relation between teller and character as a critical feature of ongoing narrative; Sacks, 1970). This complex whole can be likened to the 'Snow White' plot, and two simpler stories can be distinguished within it.

One story is that of the *acting* 'I', in such a clause as "I did the right thing". The other story is that of the *reflecting* 'I', in clauses such as "I don't know whether —" or "I think —". Having distinguished these, we observe that empirically they occur together in the talk, as in "I think I did the right thing". The 'I' discourse as a whole is thus internally subdivided into I_reflective and I_active clauses, which are bound together as 'prefix' and 'suffix' in the sentences of the talk.

In these terms, the first fragment of the talk can be regarded as consistently structured. The talk as presented by Labov and Fanshel is fragmented by (i) *repetition* (I—I) (what I — what I); (ii) *hesitation* phenomena (the therapist's 'mhm''); and (iii) unresolved *discontinuities* (I don't know whether I=; I tried to=). Labov and Fanshel also mark pauses, by variable numbers of dots, but this will be ignored here.

Formal analysis shows the fragmentation to be merely apparent. Axiomatically, the discourse is coherent. In particular, the three kinds of fragmentation are simply markers of the gap between reflective and active clauses, as defined above:

reflective
I don't know whether I=
I think

<what> I
<what> I've learned here

active
I did the right thing

I tried to=
try to use <what?

The repetition (I–I) and the discontinuity (whether=) mark the fact that two alternate 'reflective' prefixes ("I don't know whether" and "I think") can attach themselves to the same 'active' suffix ("I did the right thing").

The repetition (what I – what I) and the discontinuity (to=) mark that two alternate 'active' prefixes ("I tried to" and "I try to") can attach themselves to the same 'reflective' suffix (<what> I learned here).

The construction of Rhoda's sentence thus involves the selection of components to fill each of two functions (here called 'prefix' and 'suffix'), from a range of available alternatives. This precisely resembles the way Propp's fairy tales are constructed by selecting characters to fill each of up to 39 functions, from a range of available alternatives.

Interpretation can now proceed on the basis of this formal discourse analysis. Only at this stage is it appropriate to bring in contextual knowledge. This is therapy talk. 'Here' is the therapy session, both now and in the past, contrasted with an implied 'there' in the family setting.

Thus the division of the 'I' discourse field into two stories, which appears as a division within the individual utterances of Rhoda's speech, precisely replicates the division of the patient's life into two spheres: that of her family, where she is active (doing things), and that of the therapy session, where she is reflective (thinking, knowing, learning things).

Propp's formalist approach to narrative structure offers a paradigm for automatic discourse analysis. As a preliminary to the demonstration of such an analysis, the next two sections will introduce Prolog grammar rules, and their use in phrase structure parsing.

4. Syntactic Parsing in Prolog

The principles of syntactic parsing, i.e., the recognition and grammatical analysis of well-formed sentences, by means of Prolog grammar rules, are well-established (Clocksin & Mellish, 1981). For instance, the sentences [i, think] and [it, worked], represented here as Prolog *lists* of alphabetic constants (i.e., words) can be parsed by means of the following rules:

```
sentence — → pronoun, verb.
pronoun — → [i].
pronoun — → [it].
verb — → [worked].
verb — → [think].
```

The listing of more than one clause for any one predicate (here 'pronoun' and 'verb') offers a choice to Prolog, which will try to match the top clause first. The subsequent clauses are available on failure of the first.

The query:
? — phrase (sentence, X).
will then result in the reply

                X = [i, worked]
If user now types; <return>, Prolog will find other solutions
                X = [i, think];
                X = [it, worked];
                X = [it, think].
The last of these is unacceptable: to avoid it, the rules must be elaborated to make Person a parameter both of pronoun and of verb, and to require agreement between them in sentence (as in rule 2.112 in the Appendix below).

Prolog translates the 'grammar rules', expressed in terms of the symbol '— →' (meaning "is rewritten as") into the following 'clauses' (see Appendix for guide to Prolog notation):

    sentence (A,B): − pronoun (A,C), verb (C,B).
    pronoun ([it|A], A).
    pronoun ([i|A], A).
    verb ([worked|A], A).
    verb ([think|A], A).

Here ':—' means "implies", and [X|Y] means the list whose head, comprising a single item, is X, and whole tail, comprising a list of none, one, or many items is Y. The 'phrase' query is translated into:

    ?− sentence (_33, [ ])

where '_33' identifies variable number 33, and [ ] identifies the empty set, called "null". The query is answered by the following steps:

    Call: sentence (_33, [ ]).
    Call: pronoun (_33,_34).
    Exit: pronoun ([i:_34],_34).   (So _33 = [it|_34])
    Call: verb (_34, [ ]).
    Exit: verb ([worked], [ ]).   (So _34 = [worked])
    Exit: sentence ([i, worked], [ ]).

Further solutions are obtained by backtracking:

    Redo: sentence ([it, worked], [ ]).
    Redo: verb (_34, [ ]).
    Exit: verb ([think], [ ]).   (So now _34 = [think])
    Exit: sentence ([i, think], [ ]).

And so on.

Prolog thus adds two parameters to each predicate defined by a grammar rule. Both parameters are lists. It is convenient to call them 'CarriedForward' and 'BroughtForward', or (Cfd, Bfd) for short. Each call to a predicate on the right hand side of the 'sentence' rule 'brings forward' a list of unparsed words from the next predicate, adds to its head none, one, or several words, and 'carries forward' the new list to the preceding predicate. The parsing is complete when the final [ ] is brought forward.

In discourse parsing, it is often necessary to 'retrieve' information held in the two hidden parameters. Sometimes this involves writing the clauses in the explicit form. (Rules 12.21, 12.22, 3.11, and 4.2 below are explicit grammar rules.) But often this is not necessary. The predicate append/3 (i.e., having three parameters) can do the work.

Following Clocksin and Mellish (1981) 'append' is defined as follows:

    /*5.21*/ append ([ ],L,L).

/*5.22*/ append ([A|B],C,[A|D]):− append (B,C,D).

The first rule states that any list appended to null is the same list. The second states that to append list C to list [A|B], first append C to tail B, with result D, then replace head A onto D. This is an excellent example of a recursively defined rule, which implies a definite solution procedure. As Clocksin and Mellish point out, in writing the second rule to define 'append', one has to assume the procedure 'append' already does the job which it is being defined to do.

Using 'append' as defined here, the following will be true:

append ([1,2,3],[4,5,6] [1,2,3,4,5,6]).

The predicate will discover this solution from any determinate starting point, thus:

?− append ([1,2,3],[4,5,6],X).   X = [1,2,3,4,5,6].
?− append ([1,2,3],X,[1,2,3,4,5,6]).   X = [4,5,6].
?− append (X,[4,5,6],[1,2,3,4,5,6]).   X = [1,2,3].

The third of these will be most useful here.

Consider the problem of searching for a sentence embedded in an incoherent string of words. The following procedure, additional to those given above, will match any string X:

gap (_,_).
parse − → gap, sentence, gap.
sentence − → [].
    ?− phrase (parse,X).

The 'gap' predicate, defined in terms of two uninstantiated variables, will match anything. Prolog will initially set the two variables of the first 'gap' equal to one another, i.e., gap (X,X), and try to match the whole of X as a sentence plus anything (the second 'gap'). If this fails, then on backtracking the first gap will match the first work of X, and Prolog will try to match the rest of X to a sentence plus anything.

However, this parse is uninformative. An informative parse results from the following expanded definition of 'gap' which is proposed by Dahl and Ambramson (1984):

gap (A,B,C):− append (A,C,B).
Then,
sentence (Sentence, A,B):− append (Sentence, B.A).
parse (Prefix, Sentence, Suffix) − →
    gap (prefix), sentence (Sentence), gap (Suffix).

Now the query
?− phrase (parse (Prefix, S, Suffix), [nonsense, i, think, rubbish]). will result in the response:

Prefix = [nonsense]
S = [i, think]
Suffix = [rubbish]

In each clause, append (Balance, Bfd, Cfd) has been used to 'subtract' Bfd from Cfd and reveal the Balance added.

5. Phrase Structure Grammar

A phrase structure parser defines 'sentence' not directly in terms of terminal *word-types* (pronoun, verb, etc.), but indirectly in terms of *phrase-types* (noun phrase, verb phrase, prepositional phrase) which are themselves defined in terms of other phrase-types, and word-types. A phrase structure grammar thus requires 'TOP', 'INTERMEDIATE', and 'BOTTOM' level rules to define sentences, phrases, and words respectively.

Generalised Phrase Structure Grammar (Gazdar & Pullum, 1982) additionally defines 'META' rules which operate on any level. Thus a meta-rule is needed to enable the conjunction 'and' to link identical word-types ("the black and white dog") phrase-types ("the black dog and the white cat"), and sentences ("the black dog bit and the white cat scratched"). For simpler presentation, meta-rules will be avoided here, except where they provide the most economical grammar.

The discourse parser incorporates three main elaborations of the familiar phrase structure syntactic parser, as follows:

(1) Rules above the top sentence level analyse relations between sentences. They especially involve:

    (1.1) the listing of adjacent sentences in discourse;
    (1.2) attempts to resolve incomplete sentences by matching them with previous or subsequent complete sentences;
    (1.3) lexical comparisons to discriminate non-overlapping discourse fields.

(2) Lexical root rules below the bottom word level are required to identify distinct forms of the same noun (e.g., "cat", "cat's", and "cats") or verb (e.g., "try", "tries", "tried").

Lexical parsing applies grammar rules to the lists of characters making up words. Prolog makes these lists available by means of the built-in predicate name/2. Prolog also regards any material input within double quotations as the list of characters making up that material.

In the discourse parser roots are defined by dictionary entries, e.g., root (situation). Regular complex words are built by adding a suffix to a root, using the following basic rule:

    root (SuffixList, A,B):—
        root (Root),
        name (Root, RootList),
        append (RootList, SuffixList, RootPlusSuffixList),
        name (RootPlusSuffix, RootPlusSuffixList),
        append ([RootPlusSuffix],B,A).

The root-plus-suffix rule could be called in the course of parsing [it, works], using a definition of verb such as:

    verb (singular,3) — → root ("s") .

In attempting to match this clause, the dictionary is consulted to find a Root, say 'think'. This is converted to RootList, [t,h,i,n,k], by the built-in Prolog predicate 'name' (actually a list of ASCII codes is involved). The suffix specified in double quotes, here "s", appears as the SuffixList [s] which is appended to RootList to give RootPlusSuffixList, namely [t,h,i,n,k,s]. This is now converted to RootPlusSuffix, i.e., 'thinks'. Finally 'append' is used to see whether the list [RootPlusSuffix], i.e., [thinks], matches the difference between Cfd and Bfd in the list being parsed. Here, this will fail. But backtracking will search all possible root/1 dictionary entries, and eventually [works] will be matched.

The predicate root/3 defined above is used in grammar rules to define regular verb and noun forms, e.g.:

    noun(plural) — → root ("s").
    verb (singular,3) — → root ("s").

This procedure avoids the need to distinguish verbs from nouns. The parser will recognise strings such as [the, situation, situations, the, situation] and [the, think, thinks, the, think] as sentences.

(3) Special significance is attached to those complex sentences which contain sentences as constituents of themselves. Three instances of this phenomenon arise in Rhoda's first utterance. They involve rules for reflectives, infinitives, and relatives.

(3.1) *reflective constructions*
The word string "i,think,it,worked" would not be correctly parsed as [[i,think] [it worked]], a list of two sentences. Instead, [it,worked] should be recognised as 'reflected upon' by [i,think]. This is achieved by identifying the transitivity of those verbs which participate in the Verb + Sentence construction as 'reflective'.

>   verb phrase — → verb (reflective), sentence.

With this rule, [i,think,it,worked] is correctly recognised as a single sentence.

(3.2) *infinitive constructions*
The rules employed here require no 'infinitive' construction as such. Instead an infinitival phrase such as "to, use, what, i, learned, here" is treated as follows:

>   prepositional-phrase — →
>   preposition (to) + sentence (imperative).

In conjunction with other rules, this will permit the parser to recognise [i,tried,to,use,it] as a sentence.

The rationale for this procedure is first that every kind of imperative phrase is equivalent to "to" + imperative in this way, and second that this equivalence is directly utilised in everyday discourse, as in the following imaginary instance:

>   A: Do it.
>   B: I tried to.

In such an exchange, A 'throws' the imperative as a conversational missile. However B's "to" catches it like a ball in a game. The infinitive form neutralises the imperative: by this means, a grammatical device turns (potential) conflict into (achieved) consensus. (This type of interchange is discussed more generally in section 12 below.)

(3.3) *relative clauses*
Gazdar (1981) proposes a phrase structure treatment of relative clauses which is important to the present demonstration. This involves the 'slash' category, meaning roughly subtraction of one component (say noun phrase) from the verb phrase or sentence which would properly contain it. It is illustrated by the following rules:

>   sentence — → noun _phrase, verb_phrase.
>   verb_phrase — → verb, noun_phrase.
>   noun_phrase — → relative_pronoun, sentence_slash_noun_phrase.
>
>   sentence_slash_noun_phrase — → verb_phrase.
>   sentence_slash_noun_phrase — → noun_phrase, verb_phrase_slash_noun_phrase.
>
>   verb_phrase_slash_noun_phrase — → verb.

A wordlist such as [i,use,what,i,learned] is recognised as a single sentence by this means.

6. Demands of a Discourse Parser

The rules outlined above will recognise a string of words as a succession of sentences, including the three types of complex sentences indicated, and print a list of the sentences so distinguished.

However this is not sufficient for present purposes. The sentence list merely divides the original string, giving no indication of structure. In particular, words are listed as presented, with no indication of their lexical roots, and sentences are listed as presented, with no indication of their complex structure.

The familiar syntactic parse tree, achieved by adding a Tree parameter to each clause in the grammar rules (Clocksin & Mellish, 1981) serves many valuable purposes, including the display of results and as input to matching procedures. However, it does not directly serve the purposes of discourse analysis.

Instead, the procedure employed here is to synthesise a discourse parsing, step-by-step, by precisely replicating those steps which the grammar rules use in analysing the original word string (cf. the analytical/synthetic method utilised in Torode, 1984).

As a first step, the two hidden parameters in each grammar rule from 'sentence' down to 'root' are repeated by two explicit (Cfd,Bfd) parameters. The effect of this will be simply to replicate the original word string. Thus, the query:

?− phrase (discourse(Cfd,Bfd), [i,think,it,worked]).

will produce the response:

Cfd = [i,think,it,worked]
Bfd = [ ]

Next, the two discourse parsing parameters are manipulated in three cases: (1) lexical rules; (2) complex sentence rules; (3) discourse rules.

6.1 Lexical Rules

Four instances must be distinguished:

(i) conjunctions, determiners, and certain other word-types, are treated as having no roots, (rules 3.22, 3.25, below).

The result of this manipulation is that the phrases [a,situation] is represented in the discourse parsing as simply [situation]. The determiner is thus treated as an insignificant element of the discourse.

(ii) pronouns, prepositions, and certain other word-types, are treated as having roots identical to themselves, (rules 3.23, 3.24, below).

The result of this is that [i,try,to] is represented as [i,try,to] in the discourse parsing. (Below, a more elaborate treatment of the pronoun 'i' will be proposed.)

(iii) regular nouns and verbs have their roots defined by amended root predicates. The root/6 rule, which should be compared with the rule for root/3 given above, is as follows:

```
/*3.11 ROOT + SUFFIX RULE */
/*3.11*/ root(Transitivity,SuffixList,Cfd,Bfd,A,B):-
    root(Transitivity,Cfd,Bfd,[Root],[]),
    name(Root,RootList),
    append(RootList,SuffixList,RootPlusSuffixList),
    name(RootPlusSuffix,RootPlusSuffixList),
    append([RootPlusSuffix],B,A).
```

This rule calls 3.260, root/5, which calls root/2 dictionary entries. The result of this is that the Sentence [the, situations, worked] is represented by the SentenceRoots [situation,work]. Thus

Number and Person disappear from the Roots. However, they are integral to the syntactic analysis, and must be passed to the discourse parser. This is done by creating the *list* [Number, Person, Transitivity, Roots] and passing this as a single parameter called Features to the 'discourse' predicate. (Cf. rules 12 and 13 below. In a fuller version, Tense would be treated in the same way as Number and Person, but here it is ignored.)

(iv) irregular nominal and verbal forms are handled by special dictionary entries alongside the rules for regular cases. This is illustrated by the following fragment of the tense rules, not all of which are needed in the present instance:

```
/* irregular forms */
past_participle(come,come):— !.
past_participle(try,tried):— !.
past(come,came):— !.

/* regular forms */
past — → past_participle.
past_participle — → root ("ed").
```

The cut, !, is needed to prevent backtracking generating 'regular' forms of the irregular past-participles and past tenses.

6.2 Complex Sentence Rules

The three cases of complex sentences which include other sentences are *reflective, infinitive,* and *relative* clauses. In each case, the discourse parser should mark the included sentence as a unit. This can best be done by placing it as a list within the parsing for the complex sentence as a whole.

To illustrate this manipulation, compare two treatments of reflective verb phrases:

```
verb_phrase(reflective,Cfd,Bfd) — →
    verb(reflective,Cfd,X),
    verb_phrase_branch(reflective,X,Bfd).

verb_phrase_branch(reflective,Cfd,Bfd) — →
    sentence(Cfd,Bfd).
```

The use of the '_branch' here and elsewhere is a programming convenience which avoids Prolog having to repeat the 'verb' step on backtracking. The two discourse parsing parameters simply replicate the grammar rules, so that "think,it,worked" is parsed as [think,it,work].

Now change the '_branch' rule to the following:

```
verb_phrase_branch(reflective,[Sentence|Bfd],Bfd) — →
    sentence(Sentence,[]).
```

The result is that Sentence is passed as a single item to the first discourse parsing parameter of verb_phrase_branch. The parsing of "think,it,worked" will now be [think,[it,work]]".

This manipulation is employed in rules 2.213 (relative), 2.2212 (reflective) and 2.2326 (infinitive) below, with the result that:

[i,think,it,worked] is represented as [i,think,[it,work]]
[i,try,to,use,it] is represented as [i,try,to,[use,it]]
[i,use,what,i,learn] is represented as [i,use,[i,learn]]

A different treatment of the relative clause will be proposed below.

7. The Resolution of Incompleteness in Discourse Parsing

Number, Person, Case, Transitivity, Tense, Gender and the other syntactic features can be handled as additional parameters of the predicates defined by grammar rules. The two rightmost parameters will always be the hidden (Cfd,Bfd) pair, and by convention the two adjacent parameters will be reserved for the discourse parsing.

In a systematic parser, all sentence, phrase-type, and word-type features should be listed in a single parameter. Indeed, for greater generality, PhraseType should be simply another feature in the list, taking values such as 'sentence', 'noun_phrase', 'verb', etc. All phrase-types would then be defined by clauses of a single parameter called, say, pt. This has merit because many rules for agreement and combination or phrase-types involve operations on lists of features, not single features. A canonical order of features within the list is implied by the structure of syntactic rules themselves (Gazdar & Pullum, 1982).

In the present prototype parser, these points will be ignored. Tense will not be discussed at all, though it is patently relevant to the analysis of Rhoda's remark. Adjectives, adverbs, and most conjunctions will be omitted, since their consideration does not significantly alter the arguments.

The discourse, syntactic, and lexical grammar rules provided below should in principle be able to parse Rhoda's first utterance, which Labov and Fanshel (1977, p. 363) render as follows:

I don't ... know, whether ... I-I think I did-the right thing, jistalittle ... situation came up .... an' I tried to umh .... well, try to ....use what I-what I've learned here, see if it worked.

But here, the task will be simplified to the analysis of the following utterance:

[i, know, i, i, think, i, did, it, a, situation, came, up, i, tried, to, try, to, use, what, i, learned, see, if, it, worked]

To parse this correctly will require the following features:

Feature	Phrase Types to which applicable	Values allowed to Feature
Number	sentence,np,yvp noun,pronoun,verb	imperative,singular plural
Person	np,noun,pronoun,verb	1,2,3
Case	noun,np,pronoun	nominative,accusative
Transitivity	sentence,verb,vp	[intransitive], [transitive], [reflective], [incomplete], and *lists* thereof

These features, and the rules which enforce their agreement, are straightforward, with the following exceptions:

(i) imperative is classed as a Number. This is a shorthand which would have to be abandoned if the need arose to distinguish singular from plural imperatives.

(ii) [incomplete] is classed as a Transitivity. This shorthand, together with (iii) below, greatly simplifies incompleteness resolution procedures.

(iii) Transitivity is always a *list*. The list contains only one item unless the head is reflective.

This rule also applies to the tail. Thus the list comprises a transitivity preceded by a list of no, one or any number of 'reflective's.

The result of applying the rules defined so far is:

```
| ?- test.

Sentence:                 [i,know,i]
Number,Transitivity:      singular, [reflective,incomplete]
Roots:                    [i,know,[i]]

Sentence:                 [i,think,i,did,it]
Number,Transitivity:      singular, [reflective,transitive]
Roots:                    [i,think,[i,do,it]]

Sentence:                 [a,situation,came,up]
Number,Transitivity:      singular, [intransitive]
Roots:                    [situation,come,up]

Sentence:                 [i,tried,to,try,to,use,what,i,learned]
Number,Transitivity:      singular, [transitive]
Roots:                    [i,try,to,[try,to,[use,[i,learn]]]]

Sentence:                 [see,if,it,worked]
Number,Transitivity:      imperative, [reflective,intransitive]
Roots:                    [see,[it,work]]

yes
```

This is informative, but also frustrating. The display clearly suggests connections which are not made by the parser. In particular, it appears that:

(i) the first two sentences share discourse structure, so the "i, know, i" in sentence one is an alternative reflective prefix to "i, think, i" in sentence two

(ii) the last sentence is incorrectly parsed as an imperative. Rather, it shares discourse structure and in particular the prefix "i, tried, to, try, to" with the previous sentence.

The discourse parser can be modified to reveal these implied links by developing predicates for forward and backward matching of sentences. The forward case is the model for both. The parser should not accept incomplete sentences, i.e., sentences whose transitivity is [incomplete] or a list ending with 'incomplete'. Instead, such a sentence, and its features, should be placed on one side.

Accordingly, two new parameters are created for 'discourse' namely IncompleteFeatures and IncompleteSentence. Now when 'discourse' has called 'sentence', an option must be exercised depending on the transitivity of Sentence. Branching is achieved by creating the predicate discourse branch.

When transitivity is a list which when reversed matches [incomplete|_], rule 13.3 'discourse_ branch' calls 'discourse' with SentenceFeatures as IncompleteFeatures and Sentence as IncompleteSentence. Otherwise rule 13.4 'discourse_branch' calls 'discourse' in the usual way.

When rule 12.21 'discourse' is called with IncompleteFeatures = [], it behaves in the usual way. But when this parameter is not null, rule 12.22 'discourse' first parses the next sentence in the discourse, then attempts to match the incomplete items with it.

The predicate 4.41 'forward_match' attempts to match the previous incomplete item with some part of the sentence following it. The predicate 5.5 'gap' (due to Dahl & Abramson, 1984) is used to achieve this.

Thus 'forward_match' will first attempt to parse the beginning of the next sentence as having the same number, person, and transitivity as its incomplete predecessor. If this fails, backtracking will try substrings starting with the second word of the sentence, and so on. If a match is made, then the last incomplete sentence can be substituted for the matching substring of the new sentence, by appending Prefix to its beginning, and Suffix to its end. This 'implied' sentence is now parsed to ascertain new values for Number, Person, and Transitivity.

In the present case, [i,know,i] implies [i,know,i,did]. This matching vindicates the treatment of 'reflection' and of 'reflective' transitivity by the discourse parser. The procedure employed here matches material parsed as [reflective, transitive] with the parsing [reflective, incomplete] to resolve 'incomplete' with 'transitive'.

In effect, this is because, by rule 2.113, 'incomplete' matches solo noun-phrase (here "i"). The technique can be refined by more elaborate specification of incompleteness, as is done in the case of 'incomplete_solo_verb_phrase' below.

Had the reflective structure of both the first two sentences not been acknowledged by the discourse parser, then the preferred resolution of incompleteness (perhaps by positing repetition of "i") would be "i,know,i,think,i,did,it". The problem for discourse parsing is that this sentence is syntactically well-formed, but is implausible in everyday speech. (It might be acceptable in specialist discourse, e.g., a university philosophical discussion about "thinking".) A criterion will be proposed below for rejecting such a parsing, but it is not required here since the present procedure directly discovers the correct solution.

In the forward matching procedure, Number, Person, and Transitivity are replicated, but of course Roots is not. The predicate 'match' passes back the newly CompletedFeatures and CompletedSentence to 'discourse' which prefixed them to the stack of complete sentences, then calls discourse_branch with the current sentence in the usual way. The 'forward_match' routine successfully resolves the incompleteness of the first sentence:

```
Sentence:              [i,know,i,did,it]
Number,Transitivity:   singular, [reflective,transitive]
Roots:                 [i,know,[i,do,it]]
```

It does not alter the remainder of the parsing.

Procedure 4.2 (backward_match) is similar to the above, but more complex because it attempts to match from the end of the preceding sentence first. Since 'gap' takes the initial value gap ([]), it is necessary to reverse the preceding sentence, apply 'gap(ReverseSuffix), gap (ReverseSentence), gap (ReversePrefix)', to discover the candidate ReverseSentence, then reverse this and parse it.

8. Under- and Over-Elaborate Resolutions of Incompleteness

The backward-match procedure finds no immediate application in Rhoda's first utterance because, though intuitively they are inadequately parsed, [try,to,use ...] and [see, if, [it, works]] are analysed as *imperatives,* not as *incomplete.* The proposal therefore arises that imperatives should be treated as incomplete sentences by the parser. This is implemented in rule 13.1.

In order to implement backward matching, the previous sentence must be carried forward even when it is complete. To distinguish this case from the carrying forward of an incomplete sentence, set IncompleteFeatures = [].

Clause 13.2 contains the 'standard' backwards resolution procedure for incomplete sentences, i.e., those with incomplete Transitivity as defined earlier. Clause 13.1 applies the same procedure to imperative sentences, i.e., those whose Number is 'imperative'.

Thus, the final sentence of Rhoda's utterance is parsed by 12.21 as 'imperative'. At this point, PreviousSentence will be [i,tried,to,try,to,use,what,i,learned]. Branch 1 of 'discourse_branch', i.e., rule 13.1, is then called. It calls rule 4.2 'backward_match' which attempts to find an imperative in the tail of PreviousSentence. The first attempt, [learned] does not begin with an imperative, neither do [i,learned] nor [what,i,learned]. But [use,what,i,learned] is an imperative. So 'backward_match' succeeds, generating [i,tried,to,try,to,see,if,it,worked] as the implied completion of the imperative.

The implication of this procedure, if it correctly captures the process which ordinary pariticipants in discourse employ, is that the imperative suffers from a kind of 'incompleteness', and can only be assigned as a last resort, when no nearby sentence offers the prospect of 'completion'.

The parser now produces the following result:

| ?- test.

Sentence: [i,know,i,did,it]
Number,Transitivity: singular, [reflective,transitive]
Roots: [i,know,[i,do,it]]

Sentence: [i,think,i,did,it]
Number,Transitivity: singular, [reflective,transitive]
Roots: [i,think,[i,do,it]]

Sentence: [a,situation,came,up]
Number,Transitivity: singular, [intransitive]
Roots: [situation,come,up]

Sentence: [i,tried,to,try,to,use,what,i,learned]
Number,Transitivity: singular, [transitive]
Roots: [i,try,to,[try,to,[use,[i,learn]]]]

Sentence: [i,tried,to,try,to,see,if,it,worked]
Number,Transitivity: singular, [transitive]
Roots: [i,try,to,[try,to,[see,[it,work]]]]

yes

Arguably this has too successfully resolved incoherence! The 'tried to try to' expressions conceal a problem.

The double embedding of "i,tried,to,try,to,use,it" involves a similar difficulty to "i,know,i,think,i,do", as noted above. It is syntactically acceptable, but discursively implausible. But whereas all doubly embedded reflectives are suspicious (except perhaps those in which "you know" provides the outer layer, cf. Torode, 1985), multiply embedded infinitives are commonplace.

It seems it is the lexical repetition of the root 'try' within the sentence which casts doubt upon the double embedding in this case. If so, then a simple test can be applied to each candidate sentence parsing. Repetition of any root should cause the parser to backtrack and seek another solution.

The test which is to be applied to discourse structure resembles that which Gazdar and Pullum (1982, p. 21) apply to syntactic structure. They argue that "exhaustive constant partial ordering" (ecpo) should characterise the strings generated by a syntax if that syntax is to be adequately formalised by means of phrase structure rules.

They divide their rules into two types, namely Immediate Dominance (ID) of one level over another, and Linear Precedence (LP) within each level. An infringement of either offends the rules for the predicate 'ecpo'.

A full implementation of this test involves seeking discourse structures across a significant range of talk or text, by merging structures revealed in each individual utterance and then applying the test to the merged structure. The simple test applied here is derived from such a method. It merely searches for repetition of items within the SentenceRoots structure.

The following sentences would fail such a test:

> The cat sat on the cat.
> Jack jumped and Jill jumped.
> He hit him accidentally.

Each of these is syntactically acceptable. But in plain English each is problematic, for a simple reason. The lexical repetition of the same word suggests repeated reference to the same entity in the world; but the syntactic differentiation of the position of the two words suggests reference to distinct entities in the world.

The defect is remedied by the insertion of markers to resolve the ambiguity: "**One** cat sat on the **other** cat"; "Jack and Jill **both** jumped", or if not then "Jack jumped and Jill **also** jumped"; "he hit him**self** accidentally". In the absence of such markers, and if an alternative complete parsing can be constructed, the defective sentence can safely be rejected.

This objective can be achieved by applying the predicate 'ecpo' to the discourse structure of each sentence. This is done by inserting it in clauses 12.21, 12.22, 4.11, and 4.2

The definition of ecpo in rules 4.4 and 4.5 involves the predicate 5.9 'member' (Clocksin & Mellish, 1981), but is otherwise unexceptional. It searches the structure for repetitions and succeeds if it finds none. If it does find repeated words, it marks them with a preceding '***', prints out the marked structure with an error message, and fails.

Unfortunately, this test will cause massive failure in the parsing, due to the appearance of the personal pronoun 'i' at every level of discourse structure, in repeated violation of the ID rule.

Personal pronouns, and in particular 'I', are evidently an exception to the ecpo rule. This should not occasion surprise: philosophers from Descartes to Husserl have made play with paradoxical 'I' statements. In particular, the fact that 'I' represents the speaking subject, and hence in a sense discourse itself, involves it in a dual role when it appears as an object represented in the discourse.

The solution adopted here is a formalisation of the distinction between the 'active' and 'reflective' 'I's made earlier (section 3 above). It requires that each nominative personal pronoun be marked with the transitivity of the sentence whose subject position it occupies. This is achieved by first inserting a personal pronoun marker in the relevant 'root' dictionary entries (cf. rule 3.2401), then applying the predicate 4.31 'pronoun_marker' to the discourse parsing of each noun-phrase occurring within a sentence (rules 2.112 and 2.122) to mark the root following the marker with the transitivity of the sentence.

The discourse parsing of the sentence "i,think,i,did,it" now becomes [i_reflective, think, [i_transitive, did, it]], thereby representing the 'divided self' as a visible feature of discourse structure.

Now only the sentence [i_transitive, tried, to, try, to, use, what, i_reflective, learned] will fail the ecpo test, due to its repetition of the root 'try'. (The test permits repetition of the preposition "to".) This failure forces backtracking in clause 12.21. The alternative parsing [i,transitive, tried, to] results. The next sentence is then parsed as [try, to, use, what, i_reflective, learned], but this is an imperative. According, rule 13.1 initiates backward matching, which constructs [i_transitive, try, to, use, (what) [i_transitive, learn]] as an implied sentence passing the ecpo test.

Next, when [see, if, it, worked] is encountered, 4.2 backward_match matches it directly to the satisfactory solution of the previous sentence. The complete parsing is now as follows:

```
| ?- test.

*** indicates repeated item(s) infringing ECPO. Parse rejected.
    [i_transitive,try,to,***,try,to,use,i_reflective,learn]

Sentence:              [i,know,i,did,it]
Number,Transitivity:   singular,[reflective,transitive]
Roots:                 [i_reflective,know,[i_transitive,do,it]]

Sentence:              [i,think,i,did,it]
Number,Transitivity:   singular,[reflective,transitive]
Roots:                 [i_reflective,think,[i_transitive,do,it]]

Sentence:              [a,situation,came,up]
Number,Transitivity:   singular,[intransitive]
Roots:                 [situation,come,up]

Sentence:              [i,tried,to]
Number,Transitivity:   singular,[transitive]
Roots:                 [i_transitive,try,to]

Sentence:              [i,try,to,use,what,i,learned]
Number,Transitivity:   singular,[transitive]
Roots:                 [i_transitive,try,to,[use,[i_reflective,learn]]]

Sentence:              [i,try,to,see,if,it,worked]
Number,Transitivity:   singular,[transitive]
Roots:                 [i_transitive,try,to,[see,[it,work]]]

yes
```

Incompleteness has been automatically eliminated, and further analysis can be undertaken on the six complete sentences here presented.

9. A Normal Form for Discourse

The first axiom of Propp's (1968) formalist method is that the sequence of functions is always identical. As indicated above, a methodological preoccupation of Generalised Phrase Structure Grammar (Gazdar & Pullum, 1982) is the establishment of a canonical order for features. Ethnomethodological conversational analysis has insisted on sequencing rules. In the formal discourse analysis proposed here, a canonical sequence of transitivities in complex sentences is assumed. It can be represented thus:

[reflective [transitive]]

This axiomatic ordering is apparently violated in two distinct ways in the last two sentences of Rhoda's first utterance, as automatically completed by the parser.

The situation may be represented as follows:

[i,try,to,use,what,i,learned]

[transitive[transitive[reflective]]]
[i_transitive,try,to,[use,[i_reflective,learn]]]

[i,try,to,see,if,it,worked]
[transitive[reflective[transitive]]]
[i_transitive,try,to,[see,[it,work]]]

Two grammatical constructions lead to these inversions of canonical order. In this section the relative clause construction of the form "I use what I learn here" will be re-examined. In the following section, the quasi-auxiliary construction "try to" will be considered.

The precedence order [reflective[transitive]] is explicitly contradicted by parsing the relative clause in the fifth sentence as [i_transitive,try,to,[use,what,[i_reflective,learned]]]. This suggests that such a parsing is incorrect. The 'what' clause cannot be reflected upon by the 'use'. However the 'learn' clause could reflect upon the 'use' if the sentence were rewritten to read:

[i_transitive,tried,to,[use,what]]
[i_reflective,learned,here,to,[use,what]]

where 'what' is understood to be an unknown, but the same unknown in each case. Such a rewriting can be defended.

In Gazdar's (1981) treatment, which has been followed up to now, the word "what" serves in the main clause as a relative pronoun, hence as a noun-phrase, and it also points forward to a 'trace' pronoun in the subordinate clause. This approach serves to resolve the unknown 'what'. In effect, it regards the subordinate clause as an answer to the interrogative "what" of the main clause. ("What did I use?" "What I learned here").

Pêcheux (1981, chapters 4-5) has proposed an alternative treatment of subordinate clauses. 'What' statements can make meaningful and coherent contributions to discourse even when the unknown identity remains unresolved. Pêcheux discusses at length a sentence which troubled Frege (1952), namely "He who discovered the elliptic form of the planetary orbits died in misery". Gazdar's approach could retrieve the interrogative within this: "Who died in misery?". "He who discovered the form of the orbits".

Pêcheux's interest is a different one. Though 'he' can here be resolved to Johannes Kepler, it is not necessary to so in order to grasp the claim to divine retribution suggested by the sentence. The sentence structure suggests that the *reason* why 'he' died in misery was *because* 'he' discovered the form of the orbits, even though this suggestion is not an explicit one.

Pêcheux treats the subordinate clause as offering the answer to a "why" question posed by the main clause. Valuable as this treatment may be in hermeneutic interpretation of talk or text, it might seem impossible to incorporate into an automatic method, whereas Gazdar's approach seems evidently grounded in the form (subordinate clauses are incomplete sentences, which the main clause can complete) and lexicon ('what' serves both as relative and interrogative) to hand.

But there is a grammatical form which expresses answers to "why" questions in a very general way, namely the preposition "to". The content of the "why" question posed by the subordinate clause can also be retrieved automatically: it is simply the infinitive form of the verb in the main clause.

Pêcheux's insight can therefore be formalised in a way which has very general applicability. For instance, Frege's sentence would be rewritten as:

[He discovered the form of the planetary orbits (only) to [die in misery]]

Likewise the famous advertising phrase "What we want is Watneys" would be rewritten as:

[We want (it) to [be Watney's]]

An automatic procedure similar to that employed in marking pronouns would be required to rewrite the relative clause ("slash" sentence in Gazdar's terminology) as the main clause of a new sentence, with the old main clause transformed into an infinitival construction.

The implication of such a rewriting rule would be that the relative clause, like the imperative, is treated as a kind of incomplete sentence in need of repair. (As much is already implied by the "slash" designation). It violates the canonical sequence in which transitivities can be combined in complex sentences, according to the formalism which the discourse parser imposes.

To put this another way, the relative clause performs an inversion of the normal form of discourse. This is confirmed by the tense structure of Rhoda's remark, which has not been considered in the automatic analysis performed here. "What I learned here" is a retrospective insertion in mid-sentence. The main clause, "I try to use − " is present tense. The relative is inserted after this but, both by its tense and by its inversion of transitivities, it positions itself before the main clause.

This treatment of relative clauses demonstrates that the application of an axiomatic approach can reveal complex narrative structure in the everyday utterance just as Propp's use of the same approach revealed complex narrative structure in the fairy tale.

10. Diachronic Discrimination of Discourse Trains

To understand the treatment of the second infringement of the canonical transitivity sequence, it is necessary to examine aspects of the discourse excluded from the simplified analysis presented above. Accordingly, some adjectives and adverbs are restored to Rhoda's remarks, and her second utterance, namely "Now, I don't know if I did the right thing", (Labov & Fanshel, 1977, p. 363), is included. It must be stressed that, with the exception of the interpretation of discourse positions as "here" and "there", all that follows is within the scope of automatic analysis.

The discourse parsing incorporating the rewritten relative clause, can be arranged as follows:

Here   There

[i_reflective,not,know, [i_transitive,do,right,thing]]
[i_reflective,think,   [i_transitive,do,right,thing]]

                        [situation,come,up]

                        [i_transitive,try,to]
                        [i_transitive,try,to,[use,what]]
[i_reflective,learn (to)   [use,what]]
                        [i_transitive,try,to,[see,[it,work]]

[i_reflective,not,know, [i_transitive,do,right,thing]]

As this example shows, once the discourse has been rewritten in normal form, that form is open to sociological interpretation. Specifically, the discourse consistently distinguishes two subject positions, which can be labelled "Here" ('inhabited' by the character i_reflective), and "There" ('inhabited' by the character i_transitive). This interpretation provides a context for the way in which the verb "try" will be treated, even though that treatment can be an automatic one.

A first possibility would be to eliminate the infringement of canonical sequence by deeming that "try" was not after all a transitive verb, but an auxiliary. It is natural to regard auxiliaries as having no roots, so that they are excluded from the discourse parsing. (When acting alone, as in Rhoda's first two sentences, "do" is of course not an auxiliary but a transitive verb in its own right.)

The consequence of this would be to dismantle the coherence which the parser has elaborately constructed around the imperative "see if it worked". It would mean that although this imperative borrowed an "I" from the preceding sentence, this became actualised as an i_reflective (in [i_reflective,see(if)[it,work]]) whereas the borrowed "I" itself was an i_transitive (in [i_transitive,use]). The infinitival structure of [i,try,to,[do]] would be collapsed into [i,do], as must presumably be necessary with future tense "going" constructions.

The superior course is to acknowledge the transitivity of "try". Interpretively, this means that [try to[use]] and [try to[see]] are both articulated by i_transitive, from the There position. From this position, transitivity (use) and reflectivity (see) are articulated. As can be determined from the two non-canonical transitivity sequences [transitive, [transitive]] (in [try,to,[use]]) and [transitive,[reflective,[transitive]]] (in [try,to,[see,(if),[it,work]]]), Rhoda here articulates canonical discourse ([use], and [see,(if),[it,work]]) from a shifted position.

Thus in principle, automatic analysis can discriminate the repertoire of subject positions from which discourse is articulated, as these subject positions are constituted in the discourse itself. But analysis of ongoing discourse events (such as the articulation of the "There" position in two sentences, to be subsequently abandoned again), requires a diachronic analysis. Such analysis can proceed by an extension of the method already employed.

The main tool of synchronic discourse analysis has been a search for mutually exclusive subsets of lexical root items, structured in a consistent ordering within the sentence. Such structures were revealed, after adopting special measures to exclude determiners and some other items, to permit pronouns to take multiple positions in the structure, and to 'normalise' the deviant word orderings of relative clauses.

A diachronic analysis can also be based on a search for mutually exclusive subsets of lexical root items. Hobbes (1968) regarded discourse as a "train" of thoughts or words which continued until "interrupted". Adjacent sentences may be said to constitute a train when their roots share one or more common members. (In Torode, 1984, such a unit is called a stanza.) As in the case of the synchronic analysis, a special treatment will be required for personal pronouns.

Two relationships between adjacent trains are of special interest in discourse analysis. Interruption (Silverman & Torode, 1980, chapter 1 and *passim*) arises when one train gives way to another. In principle it implies conflict between rival ways of defining the situation. Insertion (Schegloff, 1972) arises when one train is inserted into and therefore contextualised by another train. In principle it implies consensus, i.e., a mutual relationship between two definitions of situation. However such consensus will ordinarily involve relations of power and dependency between the rival trains.

The alternation between interruption and insertion is not a timeless one, but is constructed in an ongoing manner in talk itself. Interruptive moves prospectively open new discourse, whereas insertion is constituted retrospectively by closing moves. The "catching" of an imperative by an infinitive, discussed in section 5 above, is an example of interruption being turned into insertion. Such ongoing construction can be demonstrated in a diachronic analysis of Rhoda's first utterance.

Start: Train 1 (overlap: do the right thing)
Subject Position: HERE

[Here   [There]]
[i_reflective,not,know   [i_transitive,do,right,thing]]
[i_reflective,think   [i_transitive,do,right,thing]]

Open Interruption: Train 2
Subject Position: THERE

[There]
[situation,come,up]

Open Interruption: Train 3 (overlap: use,try)
Subject Position: THERE

[There[There]]
[i_transitive,try,to,[use,trace]]

[Here   [There]]
[i_reflective,learn,   (to)[use,trace]]

[There [Here [There]]]
[i_transitive,try,to,[see,if,[trace,works]]]

Close Insertion: Train 4 (overlap: know,thing)
Subject Position HERE

[Here   [There]]
[now, i_reflective do not know if   [i_transitive,do,right,thing]]

The synchronic structuring of [Here[There]] within completed and normalised sentences has already been analysed. Now the [HERE [THERE] HERE] diachronic structuring of subject positions is also visible. Rhoda articulates the point of view of HERE in Train 1. In Train 2 she interrupts the talk, shifting to an "objective" THERE position. Train 3 marks another interruption: from within the THERE position, a "subjective" reflection on the world is articulated. Train 4 returns to HERE, catching the THERE talk as an "insertion sequence" (Schegloff, 1972).

In the course of Train 3, the discourse has undergone two shifts, and has become located far from the Here and Now therapy session. At this point the therapist intervenes with "Mhm" (Labov & Fanshel, 1977, p. 363). Rhoda immediately reverts to her first subject position. Retrospectively, an insertion sequence has been constituted: Rhoda's remarks in Trains 2 and 3 did not initiate a lasting departure from the therapeutic setting, but merely documented a reality external to that setting which concerns the therapeutic discourse proper, namely what I "know" here and now in therapy itself, (cf. Garfinkel, 1967 on "documentary interpretation").

Each train, automatically discriminated, establishes a distinct spatial and temporal location. The whole is an insertion sequence, beginning and ending in the "here" and "now" (both words appear explicitly) of the therapy session. The whole is also a narrative, a story with a definite beginning (train 2) and end (train 4). The core of the narrative is train 3. Train 3 constitutes a repetition, there in the home, of the dichotomy between here and there which is constituted in therapy. The verb "try" establishes this repetition as a failure.

Insertion thus constitutes closure of conversational meanderings by a return to the original starting point, by answering the question, completing the sequence, documenting the interpretation, ending the story. Interruption by contrast is open, fissile, playfully leading talk and text off and away. The play between them can never be finally resolved, though perhaps both the ethnomethodologists and Labov & Fanshel over-confidently assume that a closed consensus based structure will finally be achieved.

Discourse analysis need make no such assumption: the methods demonstrated here can in principle record both closing and opening, structuring and destructuring, and so throw new light on therapeutic discourse and the situation of the patient which that discourse ongoingly constitutes.

Rhoda's dilemma is not that she has done the wrong thing, or even the possibility that she has done the wrong thing, but that when immersed in the family setting she is unable to determine whether she has done the right or the wrong thing. Her task as therapy patient is not to change her behaviour so she does the right thing, but to articulate the discourse with which to discriminate between what is right and what is wrong, to see what works and what does not work. There is potential conflict of interest here, as in all therapy, since if she develops this discourse then she will no longer need therapy.

Her progress in mastering this discourse, and also the ways in which her therapist facilitates or negates such mastery, can be monitored through the course of the therapy session by means of formal discourse analysis, as demonstrated by this specimen analysis of her first utterance.

APPENDIX

A Prototype Discourse Parser Written in Prolog

Guide to Prolog Notation
For a complete guide to Prolog cf. Clocksin & Mellish (1981).

The symbol "=" used below is merely explanatory: It is not part of Prolog.

General
/*...*/   = comment
:— public   = declaration to compiler

Constants and Names of variables, functions
alpha   = lower case initial: constant, predicate, or function
Alpha   = capital initial: variable, numeric, alpha or list
_   = underline: uninstantiated variable, may take any value
"xyz"   = double quotes: the list of ASCII codes for [x,y,z]
'Ayz'   = single quotes: the alphabetic constant Ayz

Rules
alpha/n   = the predicate alpha (_,_,_,...) with n parameters
alpha :—   = normal rule defines Prolog predicate alpha
alpha —→   = grammar rule defines Prolog predicate alpha (A,B)
{...}   = encloses normal clause within Prolog grammar rule

Lists
[x,Y,z]   = list comprising constant x, variable Y, constant z
[X|Y]   = list comprising head variable X, tail list Y.
[ ]   = "null", the empty list.

Built-in predicates/0
!   = cut. Satisfied once only. Fails on backtracking.
fail   = fail.
nl   = line feed ('new line')

Built-in predicates/1
var (X)   = succeeds if X is an uninstantiated variable
write (X)   = print the variable X (may be constant or list)

Built-in predicates/2
phrase (alpha, X) = succeeds if alpha (X,[ ]) is true, intended for use when alpha is a grammar rule such that alpha —→ X.

name(Name,NameList) = when Name has an alphabetic or numeric constant as its value, then NameList is corresponding list of ASCII codes

```
/* 0. INPUT/OUTPUT */
/*0.01*/ :- public test/0,parse/1,discourse/6.

/*0.1*/ test:- parse([i,know,i,i,think,i,did,it,a,situation,came,up,i,tried,to,try,to,use,
what,i,learned,see,if,it,worked]).

/*0.2*/ parse(X):-
   phrase(discourse([],[],[],Y),X),
   pp(Y).

/*0.31*/ pp([]):- !.
/*0.32*/ pp([[Sentence,[No,Transitivity,Roots]]|T]):-
   nl, write('Sentence:'),tab(14),write(Sentence),
   nl,write('Number,Transitivity:'),tab(3),write(No),write(','),write(Transitivity),
   nl,write('Roots:'),tab(17),write(Roots),
   nl,
   pp(T),!.

/* 1. DISCOURSE PARSING RULES */

/*11*/ discourse(_,_,Res,RevRes)-->
   [],
   {reverse(Res,RevRes)}.

/*12.21*/ discourse([],PreviousSentence,T,Res,A,B):-
   sentence(No,Person,Transitivity,SentenceRoots,[],A,C),
   ecpo(SentenceRoots),
   append(Sentence,C,A),
   discourse_branch(PreviousSentence,Sentence,No,Person,Transitivity,SentenceRoots,T,
Res,C ,B).

/*12.22*/ discourse(IncompleteFeatures,IncompleteSentence,T,Res,A,B):-
   sentence(No,Person,Transitivity,SentenceRoots,[],A,C),
   ecpo(SentenceRoots),
   forward_match(IncompleteFeatures,IncompleteSentence,CompletedFeatures,Completed
Sentence,A,C),
   append(Sentence,C,A),
   discourse_branch(CompletedSentence,Sentence,No,Person,Transitivity,SentenceRoots,
[[CompletedSentence,CompletedFeatures]|T],Res,C,B).



/* BRANCH 1: imperative sentence treated as incomplete (cf case 2) */
/*13.1*/ discourse_branch(PreviousSentence,LastSentence,imperative,Person,Transitivity,
SentenceRoots,T,Res)-->
   {backward_match([_,_,_,_],LastSentence,CompletedFeatures,CompletedSentence,Previous
Sentence,[])},
   discourse([],CompletedSentence,[[CompletedSentence,CompletedFeatures]|T],Res).

/* BRANCH 2:incomplete sentence: try to backward_match with previous sentence */
/*13.2*/ discourse_branch(PreviousSentence,LastSentence,No,Person,Incomplete,Sentence
Roots,T,Res)-->
   {incomplete(Incomplete)},

   {backward_match([No,Person,Incomplete,SentenceRoots],LastSentence,CompletedFeatures,
CompletedSentence,PreviousSentence,[])},
   discourse([],CompletedSentence,[[CompletedSentence,CompletedFeatures]|T],Res).
```

/* BRANCH 3: incomplete sentence: try to forward_match with subsequent sentence */
/*13.3*/ discourse_branch(PreviousSentence,LastSentence,No,Person,Incomplete,Sentence
Roots,T,Res)-->
   {incomplete(Incomplete)},
   discourse([No,Person,Incomplete,SentenceRoots],LastSentence,T,Res).

/* BRANCH 4: complete sentence: proceed normally */
/*13.4*/ discourse_branch(PreviousSentence,LastSentence,No,Person,Transitivity,Senten
ceRoots,T,Res)-->
   discourse([],LastSentence,[[LastSentence,[No,Transitivity,SentenceRoots]]|T],Res).

/* 2. SYNTAX RULES */

/*2.1 'TOP LEVEL' (sentence) RULES */

/*2.11 SENTENCES PROPER */
/*2.111*/ sentence(imperative,2,Transitivity,Cfd,Bfd)-->
   vp(imperative,2,Transitivity,Cfd,Bfd).
/*2.112*/ sentence(Number,Person,Transitivity,CfdMarked,Bfd)-->
   np(Number,Person,nominative,Cfd,X),
   vp(Number,Person,Transitivity,X,Bfd),
   {pronoun_marker(Transitivity,CfdMarked,Cfd,X)}.
/*2.113*/ sentence(Number,_,[incomplete],Cfd,Bfd)-->
   np(Number,_,_,Cfd,Bfd).
/*2.114*/ sentence(Number,_,[incomplete_solo_vp],Cfd,Bfd)-->
   vp(Number,_,_,Cfd,Bfd).
/*2.115*/ sentence(incomplete,_,[incomplete],[Word|Bfd],Bfd)-->
   [Word].

/* 2.12 SENTENCE SLASH CATEGORIES */
/*2.121*/ s_slash_np(Number,Person,Transitivity,Cfd,Bfd)-->
   vp(Number,Person,Transitivity,Cfd,Bfd).
/*2.122*/ s_slash_np(Number,Person,Transitivity,CfdMarked,Bfd)-->
   np(Number,Person,nominative,Cfd,X),
   vp_slash_np(Number,Person,Transitivity,X,Bfd),
   {pronoun_marker(Transitivity,CfdMarked,Cfd,X)}.

/*2.2 'INTERMEDIATE LEVEL' (phrase structure,Cfd,Bfd) RULES */

/*2.21 NOUN PHRASES */
/*2.211*/ np(Number,Person,Case,Cfd,Bfd)-->
   pronoun(Number,Person,Case,Cfd,Bfd).
/*2.213*/ np(Number,Person,_,Cfd,Bfd)-->
   det,
   noun(Number,Person,Cfd,Bfd).

/* RELATIVE clause: whole Sentence treated as unit */
/*2.213*/ np(Number,3,_,[Sentence|Bfd],Bfd)-->
   relative_pronoun(Number,3,_),
   s_slash_np(Number,_,_,Sentence,[]),!.

/* 2.22 VERB PHRASES */
/* 2.221 VERB PHRASES PROPER */
/*2.2211*/ vp(Number,Person,FinalTransitivity,Cfd,Bfd)-->

```
        verb(Number,Person,FirstTransitivity,Cfd,X),
        vp_branch(FirstTransitivity,FinalTransitivity,X,Bfd).


/* REFLECTIVE clause: whole Sentence treated as unit */
/*2.2212*/ vp_branch(reflective,[reflective|NextTran],[Sentence|Bfd],Bfd)-->
    reflective_conjunction,
    sentence(_,_,NextTran,Sentence,[]),!.

/*2.2213*/ vp_branch(Transitivity,[Transitivity],Cfd,Bfd)-->
    np(_,_,accusative,Cfd,X),
    prep(_,X,Bfd).

/*2.2214*/ vp_branch(Transitivity,[Transitivity],Cfd,Bfd)-->
    prep(_,Cfd,Bfd).

/* 2.222 VERB PHRASE SLASH CATEGORIES */
/*2.2221*/ vp_slash_np(Number,Person,[Transitivity],Cfd,Bfd)-->
    verb(Number,Person,Transitivity,Cfd,X),
    prep(_,X,Bfd).

/* 2.3 PREPOSITIONAL PHRASES (including infinitives) */
/*2.313*/ prep(Preposition,Cfd,Bfd)-->
    preposition(Preposition,Cfd,X),
    prep_branch(Preposition,X,Y),
    prep(NextPreposition,Y,Bfd).
/*2.314*/ prep([],Bfd,Bfd)-->[].


/*2.323*/ prep_branch(_,Cfd,Bfd)-->np(_,_,accusative,Cfd,Bfd).
/*2.324*/ prep_branch(to,Cfd,Bfd)-->infinitive(Cfd,Bfd).
/*2.325*/ prep_branch(_,Bfd,Bfd)-->[].

/* INFINITIVE clause: whole Sentence treated as unit */
/*2.326*/ infinitive([Sentence|Bfd],Bfd)-->
    sentence(imperative,2,_,Sentence,[]),!.

/*2.327*/ infinitive(Bfd,Bfd)-->[].

/* 3. LEXICAL RULES */

/* 3.1 LEXICAL PARSING RULES */

/*3.11 ROOT + SUFFIX RULE */
/*3.11*/ root(Transitivity,SuffixList,Cfd,Bfd,A,B):-
    root(Transitivity,Cfd,Bfd,[Root],[]),
    name(Root,RootList),
    append(RootList,SuffixList,RootPlusSuffixList),
    name(RootPlusSuffix,RootPlusSuffixList),
    append([RootPlusSuffix],B,A).


/* 3.12 NOUN PARSING */
/*3.121*/ noun(singular,3,Cfd,Bfd)-->root(_,Cfd,Bfd).
/*3.122*/ noun(plural,3,Cfd,Bfd)-->root(_,"s",Cfd,Bfd).
/*3.123*/ noun(singular,3,Cfd,Bfd)-->gerund(_,Cfd,Bfd).

/* 3.13 VERB PARSING */
```

```
/*3.131*/ verb(singular,3,Transitivity,Cfd,Bfd)-->root(Transitivity,"s",Cfd,Bfd).
/*3.132*/ verb(Number,Person,Transitivity,Cfd,Bfd)-->
          {not_third_person_singular(Number,Person)},
          root(Transitivity,Cfd,Bfd).
/*3.133*/ verb(No,_,Transitivity,Cfd,Bfd)-->
          {not_same(No,imperative)},
           past(Transitivity,Cfd,Bfd).

/*3.135*/ gerund(Transitivity,Cfd,Bfd)-->root(Transitivity,"ing",Cfd,Bfd).
/*3.136*/ past(Transitivity,Cfd,Bfd)-->root(Transitivity,"ed",Cfd,Bfd).


/* 3.2 DICTIONARY */


/*3.220 DETERMINERS */
/*3.2201*/ det-->[a].
/*3.2202*/ det-->[the].

/*3.230 PREPOSITIONS */
/*3.2300*/ preposition(Root,[Root|Bfd],Bfd)-->
           [Root],{pr(Root)}.

/*3.2301*/ pr(in).
/*3.2302*/ pr(on).
/*3.2303*/ pr(to).
/*3.2304*/ pr(up).


/*3.240 PRONOUNS */
/*3.2401*/ pronoun(singular,1,nominative,[xyz_personal_pronoun,i|Bfd],Bfd)-->[i].
/*3.2402*/ pronoun(singular,3,_,[it|Bfd],Bfd)-->[it].
/*3.2403*/ relative_pronoun(singular,3,_)-->[what].

/*3.250 CONJUNCTIONS */
/*3.25001*/ reflective_conjunction-->[if].
/*3.25002*/ reflective_conjunction-->[].


/*3.260 NOMINAL AND VERBAL ROOTS */

/*3.260*/ root(Transitivity,[Root|Bfd],Bfd)-->
   [Root], {root(Transitivity,Root)}.

/*3.2601*/ root(transitive,situation).
/*3.2603*/ root(transitive,try).
/*3.2604*/ root(transitive,use).

/*3.2605*/ root(intransitive,come).
/*3.2606*/ root(intransitive,work).

/*3.2607*/ root(reflective,know).
/*3.2608*/ root(reflective,learn).
/*3.2609*/ root(reflective,see).
/*3.2610*/ root(reflective,think).

/*3.2611*/ past(intransitive,[come|Bfd],Bfd)-->[came].
/*3.2612*/ past(transitive,[do|Bfd],Bfd)-->[did].
/*3.2613*/ past(transitive,[try|Bfd],Bfd)-->[tried].
```

/* 4. SPECIAL UTILITIES */

/*4.11*/ forward_match([NI,PI,TI,RI],IncompleteSentence,[NC,TC,RC],CompletedSentence)-->
   gap(Prefix),
   sentence(NI,PI,TI,RNew,[]),
   {ecpo(RNew)},
   gap(Suffix),
   {append(Prefix,IncompleteSentence,MidSentence)},
   {append(MidSentence,Suffix,CompletedSentence)},
   {sentence(NC,PC,TC,RC,[],CompletedSentence,[])}.
/*4.12*/ forward_match([NI,PI,TI,RI],IncompleteSentence,[NI,TI,RI],IncompleteSentence).

/*4.2*/ backward_match([NI,PI,TI,RI],IncompleteSentence,[NC,TC,RC],CompletedSentence,LastS
LastSentence,[]):-
   reverse(LastSentence,RevLastSentence),

   gap(RevSuffix,RevLastSentence,A),
   gap(RevSentence,A,B),
   gap(RevPrefix,B,[]),

   reverse(RevSentence,Sentence),
   sentence(NI,PI,TI,_,[],Sentence,[]),
   not_incomplete(TI),

   reverse(RevPrefix,Prefix),
   reverse(RevSuffix,Suffix),
   append(Prefix,IncompleteSentence,MidSentence),
   append(MidSentence,Suffix,CompletedSentence),
   sentence(NC,PC,TC,RC,[],CompletedSentence,[]),
   ecpo(RC).

/*4.31*/ pronoun_marker([Transitivity|_],CfdMarked,Cfd,Bfd):-

   member(xyz_personal_pronoun,Cfd),
   append(NounPhrase,Bfd,Cfd),

   gap(RootsPrefix,NounPhrase,E),
   gap([xyz_personal_pronoun],E,F),
   gap([Pronoun],F,G),
   gap(RootsSuffix,G,[]),

   name(Pronoun,PronounList),
   name(Transitivity,TransitivityList),
   append(PronounList,"_",PronUList),
   append(PronUList,TransitivityList,PronounMarkedList),
   name(PronounMarked,PronounMarkedList),

   append(RootsPrefix,[PronounMarked],MidRoots),
   append(MidRoots,RootsSuffix,MarkedNounPhrase),

   append(MarkedNounPhrase,Bfd,CfdMarked),!.
/*4.32*/ pronoun_marker(_,Cfd,Cfd,Bfd).

/*4.41*/ ecpo(Structure):-
   ecpo(Structure,Members),

```
          !,
          ecpo_test(Members).

/*4.42*/  ecpo(Structure,Members):-
          ecpo(Structure,[],Members),!.

/*4.43*/  ecpo([],Result,Result).
/*4.44*/  ecpo([H|T])-->
          ecpo(H),
          ecpo(T).
/*4.45*/  ecpo(Word,Members,[Word,'***'|Members]):-
          not_same(Word,to),
          member(Word,Members).
/*4.46*/  ecpo(Word,Members,[Word|Members]).

/*4.51*/  ecpo_test(L):-
          member('***',L),
          reverse(L,RevL),
          nl,write('*** indicates repeated item(s) infringing ECPO. Parse rejected,'),nl,tab(5),write(RevL),nl,
          !,fail.
/*4.52*/  ecpo_test(_).


/* 5 GENERAL UTILITIES */

/*5.11*/  not_third_person_singular(singular,3):- !,fail.
/*5.12*/  not_third_person_singular(_,_):- !.

/*5.21*/  append([],L,L).
/*5.22*/  append([A|B],C,[A|D]):- append(B,C,D).

/*5.31*/  not_same(X,_):- var(X).
/*5.32*/  not_same(_,X):- var(X).
/*5.33*/  not_same(X,X):- !,fail.
/*5.43*/  not_same(_,_).

/*5.5*/   gap(Residue,A,B):- append(Residue,B,A).

/*5.61*/  incomplete(incomplete).
/*5.62*/  incomplete(IncompleteList):-
          reverse(IncompleteList,[incomplete|_]).

/*5.71*/  not_incomplete(X):-
          incomplete(X),
          !,fail.
/*5.72*/  not_incomplete(_).


/*5.81*/  reverse([],[]):- !.
/*5.82*/  reverse([A],[A]):- !.
/*5.83*/  reverse([A|B],C):-
          reverse(B,D),
          append(D,[A],C),!.


/*5.91*/  member(X,[X|_]).
/*5.92*/  member(X,[_|T]):- member(X,T).
```

Acknowledgements

This paper aims to make explicit the methodology implied by "Automatic Discrimination of Discourse Fields", paper presented to the International Symposium on Discourse Processing, 15-16 July 1985, held at St Patrick's College, Dublin. I am grateful to Kathy Davis, Nigel Gilbert, Tony Hak, Paul ten Have, David Singleton, Ruth Torode, and Steve Yearley for their critical comments on the earlier paper. Thanks to Mike Brady for assistance in all aspects of Prolog programming.

References

Bernstein, B. (1973). *Class, codes, and control.* London: Palladin.
Clocksin, W. & Mellish, C. (1981). *Programming in Prolog.* New York: Springer-Verlag.
Dahl, V. & Abramson, H. (1984). *On gapping grammars* (Technical Report TR-84-5). Burnaby: Simon Fraser University Computing Science Department.
Frege, G. (1952). On sense and reference. In P.T. Geach & M. Black (Eds.), *Translations from the philosophical writings of Gottlob Frege.* Oxford: Blackwell. (Original work published 1892).
Garfinkel, H. (1967). *Studies in ethnomethodology.* Englewood Cliffs: Prentice-Hall.
Gazdar, G. (1981). On syntactic categories. *Philosophical transactions of the Royal Society, B295,* 267-283.
Gazdar, G., & Pullum, G. (1982). *Generalised phrase structure grammar: A theoretical synopsis.* Bloomington: Indiana University Linguistics Club.
Goffman, E. (1971). *The presentation of self in everyday life.* Harmondsworth: Penguin.
Hobbes, T. (1968). *Leviathan.* Harmondsworth: Penguin. (Original work published 1651).
Jefferson, G. (1980). On trouble-premonitory response to inquiry. *Sociological Inquiry, 50,* 153-185.
Labov, W. (1972). The logic of non-standard English. In P.P. Giglioli (Ed.), *Language and social context.* Harmondsworth: Penguin.
Labov, W. & Fanshel, D. (1977). *Therapeutic discourse.* New York: Academic.
Pécheux, M. (1981). *Language, semantics, ideology.* London: Macmillan.
Propp, V. (1968). *Morphology of the Folk Tale.* Austin: Texas University Press. (Original work published 1920).
Sacks, H. (1970). *Spring 1970 lectures 1 to 6.* Unpublished transcribed lectures. Irvine: University of California.
Schegloff, E. (1972). Notes on a conversational practise: Formulating place. In D. Sudnow (Ed.), *Studies in social interaction.* New York: Free Press.
Silverman, D., & Torode, B. (1980). *The material word: Some theories of language and its limits.* London: Routledge.
Torode, B. (1976). Teachers' talk and classroom discipline. In M. Stubbs & S. Delamont (Eds.), *Explorations in classroom observation.* London: Wiley.
Torode, B. (1984). *The extra-ordinary in ordinary language* (Konteksten series No. 5). Rotterdam: Erasmus University.
Torode, B. (1985). Articulating educational discourses. In T. Hak, J. Haafkens, and G. Nijhoff (Eds.), *Working papers in discourse and conversational analysis* (Konteksten series No. 6). Rotterdam: Erasmus University.

NAME INDEX

Abelson, R.P. 205, *212*, 237, 238, *241*, 257, *268*, 288, 290, 291, *306*, *307*
Abramson, H. 368, 375, *390*
Aceves, J.J.C.L. 331, *337*
Adams, N. 297, *307*
Aggarwal, J.K. 333, *337*
Aguilar, L. 331, *337*
Aho, A.V. 113, *118*
Airenti, G. 237, *241*
Aitchinson, J. 356, *359*
Akita, K. 333, *335*
Akscyn, R.M. 320, *322*
Albers, G. *336*
Aleksander, I. 333, *335*
Alford, J.A. 296, *306*
Allen, J.F. 123, 124, 127, *146*, *147*, 216, *219*, 222, *230*, 246, 247, 249, *256*, 271, 272, 284, *285*
Alshawi, H. 68, *76*
Alry, J.L. *322*
Amidon, M.D. 330, *337*
Anderson, A. 162, 169, 173, 174, *182*, *183*
Anderson, J.R. 293, 294, 295, 297, *305*, *306*
Anick, P.G. 102, *119*
Appelt, D.E. 124, 127, *146*
Apple, W. 330, *336*
Argyle, M. 327, *335*
Arlinghaus, S.L. 334, *335*
Ascher, R.N. 318, 320, *322*
Austin, G.A. 224, *230*
Austin, J.L. 234, *241*, 272, *285*, 325, *335*

Badler, N.I. 333, *335*, *336*
Ballard, D. 293, *306*
Banbury, J. *322*
Bara, B.G. 237, *241*
Barclay, J.R. 266, *268*
Barnard, P.J. 318, *322*
Barron, F. *230*
Barwise, J. 52, *58*, 271, 276, 277, 278, 279, 280, 283, *285*
Bates, E. 41, *46*, 327, *335*
Bates, M. 102, *118*, *120*, 123, *147*
Bateson, G. 329, *337*
Bauman, R. *360*
Beaugrande, R. de 60, 63, *76*
Becker, C.A. 296, *305*
Becker, M.H. 329, *335*
Bernold, T. *336*
Bernstein, B. 361, 362, 363, *390*
Berwick, R.C. 160, *219*, *285*
Berwik, R.C. *256*

Bever, T.G. 83, 90, *97*
Beyth-Marom, R. 222, *230*
Birnbaum, L. 101, 112, *120*, 202, 204, *212*
Bishop, S.E. 330, *337*
Black, J. 104, 109, 115, *120*, 207, *212*, *307*
Black, M. *390*
Bobrow, D.G. 35, *46*, 154, *159*, 320, *322*
Bobrow, R.J. 109, 113, 114, *118*, *120*, 123, *147*
Boggs, W.M. 102, *119*
Boguraev, B.K. 45, *46*, 68, *76*
Bolinger, D. 348, *359*
Bolt, R.A. 314, 315, *322*, 333, *335*
Bott, R.A. 318, *322*
Bouma, H. *97*
Bouwhuis, D. *97*
Bower, G.H. 207, *212*, 305, *307*
Brachman, R.J. 123, 137, *146*, *147*
Braddick, O.J. *337*
Brady, M. 160, 219, *256*, *285*
Braine, M.D.S. 318, *322*
Bransford, J.D. 266, *268*
Brautigan, R. 287, *305*
Briscoe, E.J. 68, *76*
Brown, G. 41, *46*, 102, *120*, 294, 305, *307*
Brown, J.S. 102, 113, *118*, 136, *146*, 251, *256*
Brown, P. 350, *359*
Bruce, B.C. 5, 7, *24*, 81, 82, 95, 96, *97*, 109, *118*, 128, *147*, 224, 228, *231*
Bruce, G. 36, 45, *46*
Bruner, J.S. 224, *230*
Burton, D. 229, *230*
Burton, R.R. 102, 113, *118*, 251, *256*
Buxton, H. 334, *335*
Buck, R. 325, *335*
Bullowa, M. *47*

Calvert, T.W. 333, *335*
Camaioni, L. 41, *46*
Carberry, S. 22, 23, *24*, 117, 118, 157, *159*, 192, 197, *199*, 217, *219*, *230*, 246, 247, *256*
Carbonell, J.G. 7, *24*, 59, 66, 67, 77, 101, 102, 117, *118*, *119*, 194, *199*
Carlson, L. 50, 51, 52, *58*
Carpenter, P.A. 299, 301, 302, *306*
Cater, A.W.S. 68, 74, *77*
Celce–Murcia, M. 82, *97*

Name Index

Cercone, N. 76, 77
Chaffin, R. 356, *359*
Chamberlin, D.D. 317, *322*
Chang, C.L. 102, *119*, 190, *199*
Chapman, J. 333, *335*
Charniak, E. 102, *119*
Cheng, J.K. 333, *335*
Chenoweth, N.A. 82, *97*
Christie, B. 229, *230*, 310, *322*
Chun, A.E. 82, *97*
Cicourel, A.V. 44, *46*
Clark, H.H. 291, *305*
Clocksin, W. 366, 367, 368, 371, 377, 383, *390*
Cohen, L.J. 62, 77, 227, *230*
Cohen, P.R. 6, *24*, 123, 124, 128, *146*, *147*, 234, *241*, 246, *256*
Colby, K.M. 26, *32*
Cole, P. *46, 199*
Collins, A. 251, *256*
Colombetti, M. 237, *241*
Condon, W.S. 41, *46*, 328, *335*
Cook, C. 102, *120*
Coombs, M.J. *322*
Cooper, W.E. *97*
Corsaro, W.A. 26, *32*
Cottrell, G.W. 293, *305*
Coulthard, M. 229, *230, 231*, 345, *360*
Craighill, E.J. 331, *337*
Cullingford, R.E. 291, *305*
Curtiss, S. 327, *335*

Daden, I. 343, *359*
Dahl, V. 368, 375, *390*
Damodaran, L. 316, *322*
Dance, F.E. 329, *335*
Davis, J.H. *337*
Day, R.R. 82, *97*
De Jong, G.F. 210, *212*
Delamont, S. *390*
Dell, G.S. 293, *306*
Dement, W. *230*
Dennett, D.C. 226, *230*, 252, *256*
Dewey, M.E. 331, *337*
Dickson, W.P. 37, *46, 47*
Dornic, S. *306*
Dowty, D. 56, *58*
Dreyfus, H. 332, *335*
Dummett, M. 221, *230*
Dyer, M.G. 207, *212*

Eason, K.D. 316, *322*
Eastman, C.M. 7, *24*, 100, 102, 107, *119*
Edwards, D. 326, *335*
Edwards, W. 222, 229, *230*, 350, *359*
Efstathion, J. 222, *231*
Egan, O. 10, 14, *24*, 227, *230*
Eimas, P.D. 296, 299, *307*

Einhorn, H.J. 222, *230*
Ekman, P. 325, 326, 330, *335, 336*
Erwin, M. 331, *337*
Estes, W.K. 3, *24*
Evans, G. *230*
Evans, D.A. 222, *231*, 278, *285*

Faerch, C. 224, 227, *230*
Fagar, D. 289, 304, *306*
Fanshel, D. 26, *32*, 345, *259*, 361, 362, 363, 364, 365, 373, 380, 382, *390*
Fararo, T.J. 334, *337*
Feldman, H. 326, *335*
Feldman, J.A. 293, 305, *306*
Ferguson, C.A. *336*
Ferrari, G. 50, *58*, 229, *230*
Ferris, S.R. 330, *336*
Fertig, S. 124, *147*
Fikes, R. 154, *160*, 246, 247, *256*
Fillmore, C.J. *32, 46, 119, 359*
Finemann, L. 100, *119*
Fischler, I 296, *306*
Fischoff, B. 222, *230*
Fisher, D.F. *306*
Fitch, F.B. 229, *230*
Fitter, M. *322*
Flavell, J.H. 37, *46*
Fodor, J.A. 83, 90, *97*, 162, *183*
Forgas, J.P. *336*
Forrest, M. 14, *24*
Forster, K. 83, *97*
Foss, D.J. 296, *306*
Franks, J.J 266, *268*
Fraser, B. 60, *77*
Frege, G. 379, *390*
French, P. *46*
Friesen, W.V. 325, 326, 330, *335*, 336
Fromkin, V.A. 81, *97*, 102, *119*
Fry, W.R. 325, 331, *336*

Gahagan, J. 235, *241*
Gaines, B.R. 330, *335*
Galambos, J.A. *307*
Gallaher, T.M. 42, *46*
Gallaire, H. *256*
Gardiner, M. 10, 14, *24*, 229, *230*
Garfinkel, H. 42, *46*, 364, *390*
Garrett, M.F. 83, 90, *97*
Garrod, S. 162, 169, *182, 183*
Garvey, C. 43, *46*
Gawron, J.M. 100, *119*
Gazdar, G. 369, 370, 373, 376, 378, 379, *390*
Geach, P.T. *390*
Gelman, R. 38, *46*
Gentner, D. 134, *147*
Giglioli, P.P. *390*

Name Index

Gilbert, G.N.   *46*
Gleitman, H.   327, *336*
Gleitman, L.R.   327, *335, 336*
Goddy, E.N.   *359*
Goffman, E.   235, *241,* 331, *335,* 350, *359,* 361, 362, *390*
Goldin, S.E.   251, *256*
Goldin-Meadow, S.   326, 327, *336*
Golinkoff, R.M.   42, *46*
Goodman, B.   9, 19, 20, *24,* 123, 128, 139, 145, *147,* 222, 228, 229, *230*
Goodman, G.O.   296, *306*
Goodnow, J.J.   224, *230*
Goos, G.   *336*
Gough, P.B.   296, *306*
Gould, J.D.   318, 320, *322, 323*
Granger, R.H.   117, *119*
Gravill, N.   332, *336*
Green, D.W.   267, *268,* 299, *306*
Green, T.R.G.   *336, 337*
Grice, H.P.   44, *46,* 188, 191, *199*
Griffiths, P.A.   332, *336*
Grimshaw, A.D.   26, *32*
Grosjean, F.   91, 92, 97
Grosz, B.   6, 8, 19, *24,* 55, *58,* 124, 127, 133, 136, 138, 141, *147,* 154, 157, *159, 160,* 197, *199,* 218, *219,* 222, *231,* 271, *285,* 297, *306.*
Gumperz, J.J.   26, *32,* 35, *46,* 342, 351, *359*
Gunn, B.   341, *360*

Haafkens, J.   *390*
Haas, A.   123, 139, *147*
Hak, T.   *390*
Hall, P.A.V.   317, *322*
Hammond, K.   223, *231*
Hammond, N.V.   318, *322*
Harper, J.   229, *230*
Harre, R.   326, *336*
Harris, J.   6, *24,* 83, 85, 86, 94, 95, *97*
Harris, L.R.   99, 100, 103, *119,* 187, *199*
Harrison, R.P.   *336*
Hartmanis, J.   *336*
Hayes, P.J.   7, *24,* 45, *46,* 101, 113, *119*
Heath, C.   *46*
Hebb, D.O.   292, *306*
Heidorn, G.E.   100, 102, 104, 109, 117, *119*
Heij, W.C.   328, *336*
Helm, J.   *359*
Henderson, A.   154, *160*
Hendrix, G.G.   100, 104, 112, 117, *119*
Herbison-Evans, D.   333, *336*
Herdan, G.   221, *231*
Hight, J.   331, *337*
Hinton, G.E.   293, *306, 307*
Hintikka, J.   *58*
Hirschberg, J.   216, *219*

Hiz, H.   *359*
Hobbes, T.   381, *390*
Hobbs, J.R.   59, 64, 65, 76, 77, 222, *231*
Hollan, J.D.   320, *323*
Holley-Wilcox, P.   296, *306*
Hookway, C.   *256*
Hopfield, J.J.   292, *306*
Horrigan, M.K.   254, *256*
Horton, D.L.   *335*
Huang, T.S.   333, *335*
Hubley, P.   326, *337*
Humphreys-Jones, C.E.   25, 28, 30, *32*
Hymes, D.   341, *359*

Indurkhya, B.   59, 64, 65, 66, 77
Ingria, R.   123, *147*
Isard, S.D.   316, *322*
Israel, D.J.   123, 124, *147,* 246, *256*

James, M.   100, 104, 113, *120*
Jefferson, G.   45, *46,* 342, 343, *360,* 361, 363, 364, *390*
Jenkins, J.J.   *335*
Jensen, K.   100, 102, 109, 117, *119*
Jones, K.P.   77
Johnson, C.   43, *46*
Johnson, M.   59, 60, 62, 65, 66, 77, 111, *119,* 266, *268*
Johnson-Laird, P.N.   161, *183*
Joshi, A.   *24, 58, 147,* 157, *160, 183,* 188, *199,* 213, 214, 218, *219, 231,* 253, *256, 285, 305*
Just, M.A.   *268,* 299, 301, 302, *306*

Kaplan, J.   214, 218, *219*
Kaplan, R.M.   35, *46,* 102, *120*
Kaplan, S.J.   102, 113, *119,* 149, *160,* 190, *199*
Kasper, G.   224, 227, *230*
Katz, J.J.   162, *183*
Kawai, H.   334, *336*
Kay, M.   35, *46*
Kempler, D.   *32, 46, 119, 359*
Kendon, A.   327, 328, *336*
Key, M.R.   *336*
Kieras, D.E.   *268*
Killion, T.M.   296, *305*
King, J.   100, *119*
King, M.   *119*
Kleiman, G.M.   296, 299, *306*
Kleinmuntz, B.   222, 224, *230*
Kleinmuntz, D.N.   222, *230*
Klovstad, J.   102, *120*
Kochman, T.   350, 357, *359*
Kolodner, J.L.   207, *212*
Konolige, K.   234, *241*
Korein, J.U.   333, *336*
Krauss, R.M.   330, *336*

Kroch, A.A.	108, *119*	McDonald, N.	317, *322*
Kwasny, S.C.	104, 107, 117, *119*	McDowell, J.	*230*
		McGregor, G.	26, *32*
Labov, W.	26, *32*, 345, 347, 348, 350, 357, *359*, 361, 362, 363, 364, 365, 373, 380, 382, *390*	McKeown, K.R.	124, *147*, 154, 157, *160*, 197, *199*
		McKoon, G.	259, *268*
Lakoff, G.	59, 60, 62, 65, 66, *77*, 111, *119*	McLean, D.S.	7, *24*, 100, 102, 107, *119*
Lamping, J.	100, *119*	McNeill, D.	327, 328, *336*
Lane, P.M.	296, 299, *307*	McTear, M.F.	37, 43, 45, *46*
Langford, D.	42, *46*	Mead, M.	325, *336*
Larson, C.E.	329, *335*	Mehrabian, A.	300, 331, *336*
Lebowitz, M.	101, 112, *120*, 207, *212*	Mellish, C.	366, 367, 368, 371, 377, 383, *390*
Lee, E.T.	334, *336*		
Lee, W.A.	333, *336*	Mercer, R.	214, *219*
Lehnert, W.G.	*24*, *46*, 52, *58*, *77*, 97, *147*, 231, 256	Metropolis, N.	293, *306*
		Meyer, D.E.	296, *306*
Leobner, E.	100, *119*	Miller, G.A.	59, 60, 61, 63, *77*
Levelt, W.J.M.	328, *336*	Miller, L.A.	100, 102, *119*
Levi, J.N.	76, *77*	Milroy, J.	346, 350, 357, *360*
Levinson, S.C.	59, 61, 63, *77*, 229, *231*, 343, 350, *359*	Milroy, L.	26, *33*, 342, 346, 350, 351, 357, *360*
Lewis, D.K.	162, 163, 164, 169, *183*	Minker, J.	*256*
Lewis, M.	*46*, 325, 331, *336*	Minsky, M.	292, 297, *306*
Lindman, H.	222, 229, *230*	Mischel, T.	*46*
Linell, P.	164, *183*	Mitchell, D.C.	261, 267, *268*, 289, 296, 297, 298, 299, 300, 301, 304, *306*, *307*
Lipkis, T.	138, *147*		
Litman, D.J.	127, *147*, 246, 247, 249, *256*		
Locke, A.	335, *337*	Montgomery, M.	*230*
Long, J.B.	318, *322*	Monty, R.A.	*306*
Long, M.H.	82, *97*	Moran, D.	331, *337*
Longuet-Higgins, C.	316, *322*	Morency, N.	330, *336*
Luppescu, S.	82, *97*	Morgan, J.L.	*46*, *199*
Lyons, J.	27, *33*, 59, 60, 63, *77*	Morley, I.E.	331, *336*, *337*
Lytinen, S.	207, *212*	Morris, P.E.	*307*
		Morton, J.	318, *322*
Maclean, M.S.	329, *337*	Moser, M.	123, 139, *147*
MacLure, M.	42, *46*	Mouradian, G.	113, *119*
Makhoul, J.	102, *120*	Myers, T.	*33*
Malone, T.	154, *160*	Mylander, C.	326, 327, *336*
Mamdani, A.	222, *231*		
Mark, W.	142, *147*, 246, *256*	Nagao, M.	333, *336*
Markman, E.M.	39, *46*	Narayanan, A.	*307*
Marshall, C.R.	291, *305*	Nash-Webber, B.	102, *120*
Marslen-Wilson, W.D	82, 83, 84, 87, 90, 91, 92, 94, 96, *97*	Neely, J.H.	294, 296, *306*
		Neuman, B.	334, *336*
Martin, J.	316, *322*	Newport, E.L.	327, *336*
Matthews, K.	154, *160*	Nijhoff, G.	*390*
Mauldin, M.L.	102, *119*	Nilsson, N.J.	246, 247, *256*, 292, 297, *306*
Mays, E.	18, 19, *24*, 149, *160*, 190, *199*, 200, 214, *219*, 271, *285*	Norman, D.A.	35, *46*, 320, *322*
McAllester, D.	123, *147*	Ochs, E.	*46*
McClelland, J.L.	291, 293, 294, *306*	Offen, P.J.	*335*
McCoy, K.F.	21, 22, *24*, 149, 157, *160*, 188, *199*, 214, *219*	Ogston, W.D.	328, *335*
		Olds, J.	*230*
McCracken, D.L.	320, *322*	Olds, M.	*230*
McCulloch, W.S.	292, 297, *306*	O'Rourke, J.	333, *336*
McDonald, D.	251, *256*	Ortony, A.	59, *77*
McDonald, J.E.	296, *306*	O'Sullivan, M.	330, *335*, *336*

Paivio, A.	59, 60, 62, 63, 77	Robinson, J.J.	102, *119*
Pang, D.	222, *231*	Rommetveit, R.	162, 164, *183*
Papert, S.	292, 297, *306*	Rosenbaum, L.A.	325, *336*
Parke, F.I.	333, *336*	Rosenberg, R.	214, *219*
Patla, A.	333, *335*	Rosenblum, L.	*46*
Paulson, E.A.	100, *119*	Rosenbluth, A.W.	293, *306*
Pause, P.E.	221, *231*	Rosenbluth, M.N.	293, *306*
Payne, S.J.	*336*	Ross, B.H.	297, *306*
Pêcheux, M.	379, *390*	Ross, J.R.	101, 116, *119*
Pereira, F.	109, *120*	Ruddy, M.G.	296, *306*
Pereira, L.M.	109, *120*	Ruesch, J.	329, *337*
Perrault, C.F.	123, *147*, 222, *230*, 234, *241*, 246, 247, *256*, 271, 272, 284, *285*	Rumelhart, D.E.	59, 60, 63, 77, 291, 293, 294, *306*
		Rutter, D.R.	331, *337*
Perry, J.	52, *58*, 271, 276, 277, 278, 279, 280, 283, *285*	Saarinen, E.	221, *231*
Peters, S.	56, *58*	Sacerdoti, E.D.	246, *256*
Pfeifer, R.	294, 305, *307*	Sacerdoti, E.E.	100, 104, 112, 117, *119*
Phillips, L.	229, *230*	Sacks, H.	45, *46*, 342, 343, 346, *360*, 365, *390*
Piaget, J.	37, *46*		
Pinsky, L.	330, *336*	Sacks, S.	59, 77
Pitts, W.H.	292, 297, *306*	Sag, I.	*58*, 100, *119*, *147*, *160*, *183*, *231*, *285*, *305*
Poggio, A.	331, *337*		
Polanyi, L.	6, *24*	Sagalowicz, D.	100, 104, 112, 117, *119*
Pollack, M.	216, 217, *219*, 246, 252, 255, *256*	Sager, N.	104, *119*
		Sander, L.W.	41, *46*
Pomerantz, S.D.	330, *337*	Sanford, A.	169, *182*
Potter, R.	332, *336*	Savage, L.J.	222, *230*
Prince, E.	218, *219*	Scha, R.J.H.	6, *24*
Prodonoff, I.	*58*	Schank, R.	35, 45, *46*, 59, 68, 69, 70, 72, 73, 74, 76, 77, 101, 112, *120*, 202, 204, 205, 206, 207, 209, *212*, 237, 238, *241*, 257, *268*, 288, 289, 290, 291, 305, *306*
Propp, V.	361, 364, 365, 366, 378, *390*		
Pullman, G.K.	100, *119*		
Pullum, G.	369, 373, 376, 378, *390*		
Purcell, P.	333, *336*		
		Schegloff, E.A.	45, *46*, 342, 343, *360*, 361, 381, 382, *390*
Quine, W.V.O.	226, 227, *231*		
		Scheibe, K.E.	331, *337*
Rabbitt, P.M.A.	*306*	Schelling, T.C.	162, *183*
Radig, B.	333, *337*	Schenkein, J.	*360*
Ramshaw, L.A.	117, *119*	Scherer, K.	330, *335*, *336*
Ratcliff, R.	259, *268*	Schieffelin, B.	*46*
Reddy, D.	45, *46*	Schmoize, J.	123, *147*
Reddy, M.J.	67, 77	Schorr, D.	297, *307*
Reder, L.M.	297, *306*	Schramm, W.	329, *337*
Reichman, R.	6, *24*, 127, 133, *147*	Schubert, L.	76, 77
Reilly, R.	3, 10, 14, *24*, 36, 229, *230*	Schuberth, R.E.	296, 299, *307*
Reisner, P.	317, 318, *322*	Schvaneveldt, R.W.	296, *306*
Reiter, R.	216, *219*, 250, *256*	Schwartz, R.	102, *120*
Richardson, G.	328, *336*	Searle, J.R.	63, 77, 234, *241*, 272, *285*, 325, *337*
Riesbeck, C.	35, 45, *46*, 204, *212*, 298, *306*		
		Segre, A.M.	210, *212*
Riley, P.	225, 226, 229, *231*	Sejnowski, T.J.	293, 294, *306*
Ringle, M.H.	5, 7, *24*, 36, 45, *46*, 77, 81, 82, 95, *96*, *97*, 128, *147*, 224, 228, *231*, *256*	Selfridge, M.	204, *212*
		Selman, B.	292, 293, *307*
		Senders, J.W.	*306*
Robinson, A.E.	278, *285*	Shannon, C.E.	329, *337*
Robinson, B.	331, *337*	Shantz, C.U.	37, *47*
Robinson, E.J.	38, 39, *46*		

Name Index

Sharkey, A.J.C.   298, 299, *307*
Sharkey, N.E.   261, *268*, 287, 291, 293, 294, 296, 297, 298, 299, 300, 301, 304, 305, *306, 307*
Shastri, L.   293, 305, *306*
Shatz, M.   38, *46*
Shaw, M.L.G.   330, *335*
Sheehy, N.   14, *24*, 229, *230*
Sherzer, J.   *360*
Shirai, Y.   333, *337*
Shneiderman, B.   317, 319, *323*
Sidner, C.L.   8, *24*, 123, 124, 127, 139, *147*, 157, *160*, 197, *200*, 218, *219*, 246, 247, *256*
Silverman, D.   362, 381, *390*
Simon, J.C.   *97*
Sinclair, J.McH.   229, *231*, 345, *360*
Singer, H.   *306*
Singer, M.   356, *360*
Skvoretz, J.   334, *337*
Sleigh, A.C.   *337*
Slocum, J.   100, 104, 112, 117, *119*
Sloman, A.   334, *337*
Small, S.   293, *305*
Smith, E.E.   297, *307*
Smith, H.T.   329, *337*
Smith, N.   *199, 219, 256*
Smolensky, P.   293, *307*
Smolier, S.W.   333, *335*
Snow, C.   41, *47, 336*
Solomon, S.K.   266, *268*
Sondheimer, N.K.   6, *24*, 104, 107, 109, 113, 114, 116, 117, *120,* 187, 188, 189, *200,* 228, *231*
Sonnenschein, S.   38, *47*
Sowa, J.F.   76, *77*, 190, *200*
Spoehr, K.T.   296, 299, *307*
Stallard, D.   139, *147*
Stamper, R.K.   *322*
Stanovich, K.E.   299, *307*
Staros, C.J.   117, *119*
Starr, K.   124, *147*
Stephenson, G.M.   331, *336, 337*
Stevens, A.   251, *256*
Stokes, W.T.   42, *47*
Stonebraker, M.   317, *322*
Stubbs, M.   *390*
Sudnow, D.   *390*
Suppes, P.   221, 223, *231*

Tamura, S.   334, *336*
Tannen, D.   26, *32*, 35, *46*, 342, 351, *359*
Tarone, E.   82, *97*
Taylor, G.B.   117, *119*
Teller, A.H.   293, *306*
Teller, E.   293, *306*
Thomas, J.C.   318, 320, *323*
Thompson, B.H.   6, 7, *24,* 99, 102, *120,* 193, *200*
Thompson, H.   35, *46*
Toolan, M   229, *231*
Torode, B.   361, 362, 365, 371, 376, 381, *390*
Tou, F.   154, *160*
Trawick, D.J.   118, *120*
Trevarthen, C.   41, *47*, 326, *337*
Trudgill, P.   *33*
Tsotsos, J.K.   334, *337*
Tsukiyama, T.   333, *337*
Turner, T.J.   207, *212*
Tversky, A.   157, 158, *160*
Tyler, L.K.   82, 83, 84, 87, 90, 91, 92, *97*
Tzeny, O.J.L.   *306*

Ullmann, J.   113, *118*

Vaina, L.   *58*
van der Veer, G.C.   *337*
Van Lehn, K.   136, *146*
Vilain, M.   123, 139, *147*
Voge, W.M.   100, 104, 113, *120*
Volterra, V.   41, *46*, 327, *335*
Vygotsky, L.S.   325, *337*

Waldinger, R.   247, *256*
Wales, R.J.   *97*
Waletsky, J.   347, 348, *359*
Walker, D.   *24,* 102, *120*
Walker, E.C.T.   *97*
Wall, R.   56, *58*
Wanner, E.   *335*
Ward, G.   218, *219*
Waltz, D.L.   59, *77,* 101, *120*
Wang, W.S-Y.   *32, 46, 119, 359*
Warren, D.H.D.   109, *120*
Wasow, T.   100, *119*
Weaver, W.   329, *337*
Webb, J.A.   333, *337*
Webber, B.L.   18, 19, *24,* 58, 109, *118,* 123, 141, *147, 160, 183,* 190, *200,* 213, 214, 216, *219, 231, 256,* 271, 283, *285, 305*
Weiner, E.   157, *160*
Weinstein, S.   157, *160,* 218, *219*
Weischedel, R.M.   6, *24,* 100, 104, 109, 113, 114, 115, 116, 117, *120,* 187, 188, 189, *200,* 213, *219,* 228, *231*
Welsh, A.   83, 91, 94, *97*
Wenzel, C.   330, *336*
Wessels, J.   91, *97*
West, R.F.   299, *307*
Westley, B.H.   329, *337*
Weizenbaum, J.   42, *47*
Whitaker, H.A.   *335*
White, R.   *32*
Whitehurst, G.   38, *47*
Widdowson, H.   26, *33*

Wiejak, J.	334, *335*
Wieman, J.M.	*336*
Wiener, M.	331, *336*
Wilensky, R.	258, 262, 265, 267, *268*
Wilkins, D.E.	247, *256*
Wilks, Y.A.	59, 64, 65, 70, *77*, 100, 102, *120*
Williams, M.	154, *160*, 317, 320, *323*
Wilson, J.	341, 347, 349, 359, *360*
Winograd, T.	26, *33*, 35, *46*, 104, *120*, 154, *160*, 233, *241*, 325, 332, *337*
Winton, W.	330, *336*
Wish, M.	154, *160*
Wolf, J.	102, *120*
Woods, W.A.	102, 104, 114, 116, *120*, 123, *147*, 162, 164, *183*
Woolf, B.	251, *256*
Worthington, D.	331, *337*
Yalamanchill, S.	333, *337*
Yates, A.	332, *336*
Yazdani, M.	*307*
Yoshii, R.	117, *119*
Yule, G.	41, *46*
Zaefferer, D.	221, *231*
Zipser, D.	293, *306*
Zloff, M.M.	318, *323*
Zuckerman, M.	330, *337*
Zue, V.	102, *120*

SUBJECT INDEX

act	69, 72, 73, 76	conversation	
general	76	structure	41
primitive	59, 68, 70, 72–75	telephone	233
ADALINE	292	cooperation	165, 174, 225–227, 230
analogy	60, 61, 63, 66, 76, 134, 150, 151	cooperative	
		agent	213
inference	64	behaviour	214
reasoning	65, 66	dialogue	189, 222, 226
anchor function	279, 280, 282, 284	expert	215–217, 219
augmented transi-		interaction	44, 213
tion network	104, 107, 108, 112–116, 118, 229	question-	
		answering	49, 230, 251, 254
parser	108, 109, 111	response	273, 274
banter	350, 351, 355–358	creative explanation	211
beliefs	188, 214, 234, 235, 237, 240, 249, 255	CUPID	317–320
		Cyrus	207
domain	214	databases	14, 16–18, 21, 45, 99, 100, 102, 107, 113, 149, 150, 173, 190, 191, 284, 317, 318, 320, 321, 331, 333, 334
erroneous	188, 189, 251, 255		
Black English	361–363		
Boltzman machine	292		
BORIS	207	Dataland	315
CASPAR	101	decision making,	
CCWS	331, 332	rational norms	222
cluster analysis	169, 174	decision theory	223, 224
cohort theory	92, 94	demon	204–206
communication		description	
acquisition	327	imprecise	129
natural	309, 312	location	167–170, 174, 182
non-verbal	37, 309	over specific	134
person-machine	327	repair	125
referential	37, 45	types	177, 178
conceptual		development	
case	69, 70, 298	gesture	326
dependency	68, 71, 72, 74–76, 204, 305	language	326
primitives	59, 68, 73	dialogue	
representations	76	coherence	26
structure	301	context	190, 192, 199, 257
context	49, 53, 63, 68, 76, 82–89, 91, 92, 94, 111, 158, 159, 162, 188, 194, 218, 276, 287–289, 299, 301, 303, 304, 316, 341, 361	cooperation	123
		decision theory	221, 222, 229
		focus	6, 8, 10
		formal theory	49, 52
dependent	157	game	221
effects	83, 150, 153, 294, 296, 303	goals	225, 228
external	9, 11, 16, 20	goal-directed	222, 271, 284
facilitation	305	information-	
internal	5, 10, 11, 16, 22	seeking	189, 191
model	191–193, 199	instructional	19
switching	311	intentions	55
task-oriented	215, 312	interactional	
theory	291	nature	45
social	3, 37, 53, 288, 290	natural	188, 191, 194, 199, 222, 223, 246, 252, 255
word	287, 288		

natural language 164, 316
negotiation 188
person-machine 25, 32, 46, 81, 316, 320
plan-based 222, 223, 229
process 23
rules 49, 51
situation 271
structure 3, 10, 13, 23, 55, 57, 127, 223
task-oriented 11, 124, 127, 136, 331
discourse
  analysis 362, 365, 366, 371, 381, 382
  automatic analysis 366, 380, 381
  context 85, 86, 88–92
  dependencies 86
  domain 155
  grammar 52
  location 279, 282
  model 5, 8, 9, 20, 21, 96
  parsing 367, 369, 370, 372–380
  partial model 203
  rules 371, 373
  semantic 111
  situation 52, 53, 55, 58, 191, 276–281, 283, 284
  sociological
    analysis 361
  structure 93, 376, 377
  therapeutic 382
DYPAR 102
ELI 204, 298
ELIZA 330
ellipsis 7, 100, 113, 175, 180, 246, 257, 267, 268, 289
EPISTLE 102, 117
errors
  constraint
    violation 101, 114–116, 118
  omitted
    constituents 100
  punctuation 7, 99, 100
  quantification 17, 18
  second language 225, 227
  selection
    restrictions 60
  speech shadowing 84
  spelling/
    typographical 99, 101, 116, 118
  types of 103
  ungrammaticality 7, 100, 102
  vocabulary 100, 101, 115, 118
  word order 101
ESPRIT 229
expectations 39, 203, 206, 214–216, 230, 291
expert system 21, 102, 149, 213, 219, 245
failure
  avoidance 161, 180, 219, 332
  comprehension 94

conversation 81, 93, 96
cost 223, 224, 226
definition 36
degrees of 222
description 136
detection 15, 20, 39, 45, 58, 228, 229
developmental
  perspective 36
dialogue 330
distal 223, 224
expectation 203, 204, 207–211
extensional 118
human-computer
  interaction 331
illocutionary 225
inference 36
input 5, 43, 44, 82, 83, 96, 224
intensional 188, 190
lexical 6, 36, 81, 92–96, 112, 228
model 5, 36, 43, 44, 96, 224, 257, 318
negotiation 36, 42, 345
non-verbal
  communication 42
parse 109
peer interaction 43
perceptual 5, 36, 82, 93, 96, 228
pragmatic 53, 54, 61, 225
presupposition 50, 53, 54, 113, 218
proximal 223, 224, 228
reference 9, 18, 20, 32, 38, 39, 43, 44, 136, 138, 141, 144, 146, 161, 171, 180, 222
repair 14, 18, 23, 36, 37, 42, 43, 45, 53, 54, 58, 81, 93–97, 113, 117, 136, 211, 332
semantic 53, 54, 109, 111, 225
syntactic 5, 7, 36, 53, 54, 82, 96, 225, 228
false
  assumptions 18
  conclusions 213–215, 217, 218
FileVision 321, 322
FWIM 136
focusing 127, 128, 138, 141, 157, 197, 216, 218, 271, 297, 302
folk tale 364
formal descriptions 51, 365
  analysis 169
  Bayesian 222
  discourse 364, 378, 383
  inferences 214
  quasi- 50–52, 55, 57, 58
  games 237
gaze direction 313, 314, 333
gestures 41, 228, 283, 314, 325, 327, 333, 334
  decoding 334

deixis	332, 334	inheritance	154
dialogue-oriented	327	INTELLECT	99
encoding	334	interaction	
function	328	informal	342, 358
referential	327, 328	modeless	315
symbolic	325	multi-media	314, 315, 335
goal	36, 58, 67, 124, 127, 135, 174, 188, 189, 191, 198, 202, 205, 223, 228, 234, 235, 238, 240, 245, 249, 253, 255, 257, 258, 262, 263, 265, 266, 268	person-machine	213, 257, 309, 310, 330, 332
		species of	310, 316, 317
		interface	
		multi-media	332
		natural-language	99, 110, 112, 159
conversation	150, 353	person-machine	6, 35, 284
directed	123	INTERLISP	112
inference	192	input/output	
positions	164, 166, 171, 178, 179	devices	309, 310, 314
protagonist's	261	IPP	207, 208
structure	5, 10, 11, 16, 20–23	jokes	346, 347, 350, 357
task	192	narrative	347–350, 355
user's	216	question/answer	347–350, 355, 356
GPSG	369	KAN	290–292, 294, 296–298, 301–305
hermeneutics	379		
human factors	309, 317, 319, 322	KL-ONE	114, 137–139, 141, 142, 146
ill-formed		knowledge	44, 66, 202, 203, 222, 235, 238, 249, 251, 252, 257, 290, 301
absolute	99, 101, 111, 113, 116		
conceptualization	75		
grammatical	228	about people	233, 237, 239, 241
input	6, 53, 99, 100, 102, 104, 112, 115–118, 123, 221	active	302
		base	142, 146, 150, 157, 158, 213, 215–218, 334
plan	253		
pragmatic	23, 101, 157, 187–191, 199	contextual	366
relative	99, 101, 111	cultural	290
semantic	187, 189	discourse	139
syntactic	53, 117, 187, 189	domain	65, 66, 101, 139, 191, 250, 334
illocutionary			
act	234	expert's	217
force	54, 58, 274, 275, 278	encoded	204
image processing	333	hierarchical	139
induction	221, 223, 226	implicit	201
inference	39, 63, 69–74, 76, 142, 205, 217, 257, 264, 272, 299, 303, 304	lexical	68, 83
		linguistic	82, 139, 140, 145, 202
		mutual	163, 171, 288, 289
false	188, 213, 214, 216, 219	perceptual	139, 140
pragmatic	103, 118	pragmatic	54, 139
process	60, 69, 74	representation	138, 146, 189
rules	247, 272	semantic	64, 76
uncertain	222	shared	134
information		social	294, 296, 364
contextual	25, 92, 93, 152	sources	83, 136, 137, 139, 140, 312
dependencies	85, 90, 91	state	36
inferred	257, 258	stereotype	236, 291
non-verbal	4, 322	structure	203, 205, 206, 208, 211, 236, 240, 291, 296–299, 303–305
old/new			
distinction	85, 87–89	syntactic	64, 76, 322
perceptual	145	taxonomic	137
pragmatic	117	world	63, 64, 66, 76, 137, 202, 205, 233, 276, 284, 298
processing	322		
semantic	85	LADDER	112, 113
syntactic	85	language acquisition	326, 327

lexical
  parsing 369
  root 369, 371, 381
  rule 371
learning
  explanation-based 210
  inductive 208
  program 209
LISP 104, 107
logic 49
LUNAR 102
MADALINE 292
maze game 164–166, 172–174, 180, 182
meaning
  dynamic 162, 164, 180, 182, 206, 207
  Gricean theory 191
  encoded 325
  linguistic 276
  negotiated 325
  static 162, 164, 180, 182
  structure 223
  theories 161, 162, 164, 180
memory
  context 304
  explanation pattern 208–212
  load 310, 317, 318
  long-term 318
  organisation 203, 204
  parallel associative network 291
  script 291, 304
  semantic 62
  structure 204, 207–209
meta-rule 100, 102–104, 107, 109, 111–117, 187
metaphor 59–63, 65–67, 75, 76, 157, 289, 293
  comprehension 62, 63, 67, 74, 76
  spatial/visual 64, 76
MINITAB 14, 15
miscommunication
  circumvention 146
  focus 130, 132, 133
  out-mode 351, 355
  recognition 123
  reference 124, 127, 128, 134, 135, 283
  repair 35, 123, 128, 131, 135
misconceptions
  attribute 150, 151
  can be 18, 22
  correction 213
  is 18
  object 150
  prevents 213
  property 21, 149–152, 159
  property value 21, 149, 152
misspells 6, 99, 112

misunderstanding 175, 180, 182
  anticipation 35
  causing communication breakdown 17, 26
  complexity 26
  components 28, 31
  contextual 133, 290
  detection 39
  definition 27
  intonation 280
  structure 26, 28, 31
  penalties 182
model
  computational 305
  connectionist 292, 294
  conversation 343
  information-processing 3, 187, 329
  language 114
  mental 169, 208
  metaphorical 329
  neural 292
  non-verbal communication 329
  of individuals 222, 239, 241
  user 37, 45, 117, 150–154, 159
Montague grammar 51, 56, 58
MOPs 207, 208, 211
MOPtrans 207, 209
multidisciplinary approach 335
natural language interface, robust 23, 123, 124, 146, 189
negotiation
  conventions 169–172, 174, 177–180, 182
  conversation 331
  speaker rights 344
  reference 136, 164
NOAH 246
non-verbal behaviour 11, 182, 325–327, 330, 332–334
object
  description 21, 38
  perspective 22, 150, 153–158
  reference 43, 124
  similarity 157
out-mode 343–346, 348, 349, 352–354, 358
perception 292
perlocutionary
  effect 234
Personae 233, 236, 238–241
plan 23, 123, 198, 202, 205, 228, 230, 234–236, 245, 246, 255, 257, 266, 268
  active 261, 262, 265, 267
  inference 55, 189, 192–196, 199, 246–253, 255, 259, 261–263, 266–

	268, 272–274	semantic	6, 9, 10
invalid	246, 251, 254, 255	verbs	65
precondition	262, 263, 265	response	
recognition	124	appropriate	49, 57, 245, 246, 253–255
task-related	195, 196, 199	latency	87
underlying	199, 254, 255	ROMPER	149
valid	254	RUS	113, 115
pragmatic	62	SAM	291
overshoot	22, 23, 101, 188–191, 193, 196, 199	scripts	205–207, 209, 211, 223, 237, 261, 290, 291, 296
infelicity	60	SDMS	315
predictions	39	semantics	49, 62, 63, 85
presuppositions	50, 52, 214	anomalous	60, 61
contextual	53, 54, 291	model-theoretic	56
linguistic	53, 54	primitives	64
pragmatic	53, 54, 55	sense extension	59, 66, 67, 75, 76
PROLOG	109, 139, 366, 369	sentence rules	371, 372
parsing	366–368, 370	SEQUEL	317, 318
proposition	27–29, 32, 195, 199, 247, 250, 362	SHRDLU	233
		situation semantics	50, 52, 55–58, 271, 276, 277, 284
analysis	362–364		
attitudes	51, 54, 57, 111, 226	sociolinguistics	37, 40, 41, 43, 44, 342
erroneous	27, 28, 190, 191, 193–196, 198	SOPHIE	102
		speaker rights	341–345
psychological		speech	
analysis	26, 267	acts	41, 222, 223, 226, 227, 229, 230, 272, 278, 280, 288
data	291, 297, 298		
reality	258	comprehension	6, 82, 83, 91–93, 95, 96, 97, 102
question answering	214, 215, 246, 251, 271		
RABBIT	317, 320–322	deixis	332
realization, states of	29, 32	errors	81, 82
reasoning	214, 217	event	341–343, 346, 347, 350, 358
default	215, 216	everyday	342, 358, 375
faulty	150	generation	313
reference search	138	input	328, 334
referring expressions,		natural	365
precision	172, 182	perception	82
REL	193	processing	82, 83, 96
relaxation		production	81, 96
arc test	107, 109, 111, 115	shadowing	83, 85, 87, 91, 94
constraint	104, 116, 117	spreading activation	291–295, 304
description	136–141, 145, 229	states	73, 76
mechanism	141, 146	primitive	68
reference	222	stereotype	235, 236
reminding	208, 210, 211	STRIPS	246
representation	74	taxonomy	155
act	73, 75	communication	
domain	65	failure	5, 7, 11, 16, 18, 20, 23, 128, 146
internal	3, 5, 8, 10, 13, 23		
language	187, 188	knowledge base	135, 138, 139, 142, 144
lexical	5	therapy talk	362, 363
meaning	6, 68–70, 76, 204, 279	threshold knowledge unit	292, 297, 300–303
mental	288		
plan	247	threshold logic unit	292, 293
semantic	64, 159, 194, 198, 273, 284	transitional	
speech-act	275	probabilities	201, 202
state	59, 75, 76	transition schemata	51
syntactic/		turn taking	36, 41, 113, 225, 226, 342, 343

understanding
  plan-based   208
  predictive   201
  program   209
  text   201–203
WIMP interface   311
WISARD   333
written language
  bias   164
word recognition   82–85, 91–94, 304, 312
  latency   85, 296
word monitoring   84, 85, 88, 90, 91
  latency   87, 88
world model   22, 23, 59, 102, 187–192, 194, 195, 234, 252

ZOG   320

J
RF